Biodiversity and Conservation of the Yucatán Peninsula

Gerald Alexander Islebe • Sophie Calmé •
Jorge L. León-Cortés • Birgit Schmook
Editors

Biodiversity and Conservation of the Yucatán Peninsula

Springer

Editors
Gerald Alexander Islebe
El Colegio de la Frontera Sur ECOSUR
Chetumal, Quintana Roo
Mexico

Sophie Calmé
Université de Sherbrooke
Sherbrooke, Québec
Canada

Jorge L. León-Cortés
El Colegio de la Frontera Sur ECOSUR
San Cristóbal de las Casas, Chiapas
Mexico

Birgit Schmook
El Colegio de la Frontera Sur ECOSUR
Chetumal, Quintana Roo
Mexico

ISBN 978-3-319-06528-1 ISBN 978-3-319-06529-8 (eBook)
DOI 10.1007/978-3-319-06529-8

Library of Congress Control Number: 2015951833

Springer Cham Heidelberg New York Dordrecht London

Printed on acid-free paper

Springer International Publishing AG Switzerland is part of Springer Science+Business Media (www.springer.com)

Preface

More than 4000 years of human environment interaction have left an undeniable imprint on the natural resources of the Yucatán Peninsula, an area of remarkable biodiversity and of astonishing natural beauty. Home to the millennium Mayan culture biodiversity, it has strongly been linked to human activities but at the same time to careful natural management. Different forest types and ecosystems cover the Yucatán Peninsula and provide habitat and protection for emblematic vertebrate species, like the jaguar, tapir, howler, and spider monkeys, all of which represent essential components of the region's natural view. This important but fragile condition of the peninsula urges us to effectively conserve and protect flora and fauna but also to bear in mind the needs and management practices of local populations. The Yucatán Peninsula represents a perfect example of how conservation is connected to human activities, though this circumstance strongly indicates that both forest and coastal ecosystems represent sources of valuable economic income and well-being for the rural population. Although not all productive activities can be considered sustainable or ecologically sound, the enormous and incredibly detailed local knowledge is an example of how use and conservation are simultaneously taking place.

The Yucatán Peninsula with an extension of 145,000 km^2 is home to about 2500 species of plants, >3000 insect species, six species of ungulates, five felids, three primate species, 555 bird species, 24 species of amphibians, and 118 species of reptiles. These figures provide a clear idea on the importance of biodiversity of the Yucatán Peninsular to the world but at the same time challenge the needs for documenting levels of richness of other "highly diversified" components of biodiversity, i.e., invertebrates, fungi, and bacteria. At a wider context, natural disturbances like hurricanes, some of them of major magnitude, and fires have an impact on biodiversity. The northern part of the Yucatán Peninsula tends to be the most affected hurricane area. Forests have shown a remarkable resilience to these large-scale natural disturbances, but the added effects of both human-induced land use changes and natural disturbances have caused considerable changes at a variety of spatial scales, from local to regional scales. On top of the mentioned threats, global climate change adds another challenge to the task of land use changes and biodiversity conservation. Drastic changes along the eastern Yucatán coast can be observed, with tourism as the main driver of land use change. At present, 39 protected areas can be found in the Yucatán Peninsula, a 21.6 % of the total area, which helps conservation efforts and support carbon dioxide sequestration, among many other ecosystem benefits.

On the academic side, the Yucatán Peninsula has attracted scholars since long times. Botanical, zoological explorations started in the nineteenth century, with the legacy of the ancient Maya culture as the main inspiration for many archaeologists, seeking to understand the rise and fall of this Mesoamerican high culture. The increasing demand on a "fresh view" on the regional biodiversity and conservation convinced us to offer a book on a wide stream of important biological conservation topics. Our aim was, in close collaboration with 46 authors from twenty different institutions, to offer a comprehensive overview on the knowledge on biodiversity and conservation of the Yucatán Peninsula. Few studies have integrated state-of-the-art accounts on ecology, biodiversity, and conservation of the Yucatán Peninsula, and in some others future tasks and research needs have been identified. We hope that this book will be useful not only to ecologists, conservation biologists, and decision makers but also to postgraduate students, lecturers, and teachers, encouraging further research, supporting conservation of this and other areas in the tropics and subtropics.

Chetumal, Mexico Gerald Alexander Islebe
Québec, Canada Sophie Calmé
Chiapas, Mexico Jorge L. León-Cortés
Chetumal, Mexico Birgit Schmook

Acknowledgments

The editors are grateful to all authors for their kind cooperation and their willingness to share their knowledge. The Consejo Nacional de Ciencia y Tecnología (CONACYT) and the Commission for the Knowledge and Use of Biodiversity (CONABIO) are acknowledged for funding many of the projects which led to results published in this book. We thank El Colegio de la Frontera Sur for providing us facilities to keep our research ongoing. At Springer we are grateful to Roberta Gazzarolle and Susan Westendorf for their support and patience. Susannah McCandless and William Parsons are acknowledged for English editing. The editors want to thank Alejandro Aragón-Moreno, Nancy Cabanillas-Terán, Juan Carlos Joo-Chang, Manuel Girón-Intzin, Manuel Ramos, and Mauro Sanvicente for their help and support.

Contents

List of Contributors

Efraím Acosta-Lugo Pronatura Península de Yucatán, A.C., Mérida, Yucatán, Mexico

Marisol E. Almaraz-Almaraz Departamento de Conservación de la Biodiversidad, El Colegio de la Frontera Sur Unidad San Cristóbal de las Casas, San Cristóbal de las Casas, Chiapas, Mexico

José Luis Andrade Centro de Investigación Científica de Yucatán, Unidad de Recursos Naturales, Mérida, Yucatán, Mexico

María Andrade-Hernández Pronatura Península de Yucatán, A.C., Mérida, Yucatán, Mexico

José Arellano-Morín Secretaría de Desarrollo Urbano y Medio Ambiente del Gobierno del Estado de Yucatán, Mérida, Yucatán, Mexico

Martha Bonilla-Moheno Instituto de Ecología A.C., Red de Ambiente y Sustentabilidad, Veracruz, Xalapa, Mexico

Ubaldo Caballero Departamento de Conservación de la Biodiversidad, El Colegio de la Frontera Sur Unidad San Cristóbal de las Casas, San Cristóbal de las Casas, Chiapas, Mexico

Sophie Calmé El Colegio de la Frontera Sur, Chetumal, Quintana Roo, Mexico

Université de Sherbrooke, Sherbrooke, Québec, Canada

Guillermo Castillo-Vela Departamento de Conservación de la Biodiversidad, El Colegio de la Frontera Sur Unidad Campeche, Campeche, Mexico

Rogelio Cedeño-Vázquez Departmento Ecología y Sistemática Acuática, El Colegio de la Frontera Sur Unidad Chetumal, Chetumal, Quintana Roo, Mexico

Pierre Charruau Centro del Cambio Global y la Sustentabilidad en el Sureste A. C., Villahermosa, Tabasco, Mexico

Cuauhtémoc Chavéz Departamento de Ciencias Ambientales Unidad Lerma, Universidad Autónoma Metropolitana, Edo de México, Mexico

Juan Manuel Dupuy Centro de Investigación Científica de Yucatán, Unidad de Recursos Naturales, Mérida, Yucatán, Mexico

Rafael Durán-García Centro de Investigación Científica de Yucatán, Unidad de Recursos Naturales, Mérida, Yucatán, Mexico

Griselda Escalona El Colegio de la Frontera Sur, Campeche, Campeche, Mexico

Luisa I. Falcón-Álvarez Instituto Nacional de Ecología, Universidad Nacional Autónoma de México, México DF, Mexico

William Folan Universidad Autónoma de Campeche, Centro de Investigaciones Históricas y Sociales, Campeche, Mexico

Gerardo García-Contreras Pronatura Península de Yucatán, A.C., Mérida, Yucatán, Mexico

Eduardo García-Frapolli Centro de Investigaciones en Ecosistemas, Universidad Nacional Autónoma de México, Morelia, Michoacán, Mexico

Claudia González-Salvatierra Instituto Potosino de Investigación Científica y Tecnológica, San Luis Potosí, Mexico

Héctor A. Hernández-Arana Departamento de Sistemática y Ecología Acuática, El Colegio de la Frontera Sur Unidad-Chetumal, Chetumal, Quintana Roo, Mexico

Gerald Alexander Islebe Departamento Conservación de la Biodiversidad, El Colegio de la Frontera Sur Unidad Chetumal, Chetumal, Quintana Roo, Mexico

Paula Jackson Ecology, Evolution and Organismal Biology, Kennesaw State University, Kennesaw, GA, USA

Jerry Kaster School of Freshwater Science, University of Wisconsin, Milwaukee, WI, USA

Gunther Koehler Senckenberg Forschungsinstitut und Naturmuseum Frankfurt, Frankfurt, Germany

Jorge L. León-Cortés Departamento de Conservación de la Biodiversidad, El Colegio de la Frontera Sur Unidad San Cristóbal de las Casas, San Cristóbal de las Casas, Chiapas, Mexico

Eurídice Leyequién Centro de Investigación Científica de Yucatán, Unidad de Recursos Naturales, Mérida, Yucatán, Mexico

Haydée López-Adame Departamento de Sistemática y Ecología Acuática, El Colegio de la Frontera Sur Unidad-Chetumal, Chetumal, Quintana Roo, Mexico

Barbara MacKinnon-H Amigos de Sian Ka'an A.C., Mérida, Yucatán, Mexico

Martha Elena Mendéz-Gonzaléz Centro de Investigación Científica de Yucatán, Mérida, Yucatán, Mexico

Swany Morteo Montiel Red de Ecología Funcional, Instituto de Ecología A.C., Xalapa, Veracruz, Mexico

Georgina O'Farrill University of Toronto, Toronto, ON, Canada

Luciana Porter-Bolland Instituto de Ecología A.C., Red de Ecología Funcional, Xalapa, Veracruz, Mexico

Pablo Jesús Ramírez-Barajas Departamento Conservación de la Biodiversidad, El Colegio de la Frontera Sur Unidad Chetumal, Chetumal, Quintana Roo, Mexico

Casandra Reyes-García Centro de Investigación Científica de Yucatán, Unidad de Recursos Naturales, Mérida, Yucatán, Mexico

Rafael Reyna-Hurtado Departamento Conservación de la Biodiversidad, El Colegio de la Frontera Sur Unidad Campeche, Campeche, Mexico

David W. Roubik Smithsonian Tropical Research Institute, Balboa, Republic of Panama

Smithsonian Tropical Research Institute, Unit 9100, DPO, AA, USA

Miguel A. Ruiz-Zárate Departamento de Sistemática y Ecología Acuática, El Colegio de la Frontera Sur Unidad-Chetumal, Chetumal, Quintana Roo, Mexico

Odilón Sánchez Sánchez Universidad Veracruzana, Centro de Investigaciones Tropicales de la Universidad Veracruzana, Xalapa, Veracruz, Mexico

Birgit Schmook Departamento Conservación de la Biodiversidad, El Colegio de la Frontera Sur Unidad Chetumal, Chetumal, Quintana Roo, Mexico

Juan Carlos Serio-Silva Instituto Nacional de Ecología, A.C., Red de Biología y Conservación de Vertebrados, Xalapa, Veracruz, Mexico

Jorge Herrera Silveira Departamento de Recursos del Mar, Centro de Investigación y de Estudios Avanzados del IPN Unidad Mérida, Mérida, Yucatán, México

Nuria Torrescano-Valle Departamento Conservación de la Biodiversidad, El Colegio de la Frontera Sur Unidad Chetumal, Chetumal, Quintana Roo, Mexico

Mirna Valdez-Hernández Departamento Conservación de la Biodiversidad, El Colegio de la Frontera Sur Unidad Chetumal, Chetumal, Quintana Roo, Mexico

Alejandro Vega-Zepeda Departamento de Sistemática y Ecología Acuática, El Colegio de la Frontera Sur Unidad-Chetumal, Chetumal, Quintana Roo, Mexico

Rogel Villanueva-Gutierrez Departamento Conservación de la Biodiversidad, El Colegio de la Frontera Sur Unidad Chetumal, Chetumal, Quintana Roo, Mexico

Holger Weissenberger El Colegio de la Frontera Sur Unidad Chetumal, Chetumal, Quintana Roo, Mexico

Chapter 1
Introduction: Biodiversity and Conservation of the Yucatán Peninsula, Mexico

Gerald Alexander Islebe, Birgit Schmook, Sophie Calmé, and Jorge L. León-Cortés

Home to large tracts of biodiverse tropical forests, harboring valuable endemic species and carrying the spirit of its millennium Mayan culture, the Yucatán Peninsula is of high interest not only to ecologists, conservation biologists and social scientists, but to a wider audience interested in its biodiversity and conservation. The threats to biodiversity and the need for conservation are now of global societal concern, but for many tropical and subtropical countries or regions up-to-date and reliable information is not easily available. All of the authors of this book, most of them biologists, have been actively doing research in the Yucatán Peninsula for many years and possess an intimate knowledge of its natural resources. Many of the authors have linked research on biodiversity to questions of conservation by working closely with farmers and forest dwellers in rural communities. In this book we want to share this knowledge with our readers, hoping that it will help them to appreciate the incredible natural diversity and lively cultural heritage of the

G.A. Islebe (✉) • B. Schmook
El Colegio de la Frontera Sur, Unidad Chetumal, Avenida Centenario km 5.5, Col. Pacto Obrero Campesino, Apartado Postal 424, 77014 Chetumal, Quintana Roo, Mexico
e-mail: gislebe@ecosur.mx; bschmook@ecosur.mx

S. Calmé
El Colegio de la Frontera Sur, Unidad Chetumal, Avenida Centenario km 5.5, Col. Pacto Obrero Campesino, Apartado Postal 424, 77014 Chetumal, Quintana Roo, Mexico

Department of Biology, Université de Sherbrooke, 2500 Boulevard de l'Université, Sherbrooke, Quebec, Canada
e-mail: sophie.calme@usherbrooke.ca

J.L. León-Cortés
Conservación de la Biodiversidad, El Colegio de la Frontera Sur Unidad San Cristóbal, Carretera Panamericana & Periférico Sur S/N, 29000 San Cristóbal de las Casas, Chiapas, Mexico
e-mail: jleon@ecosur.mx

G.A. Islebe et al. (eds.), *Biodiversity and Conservation of the Yucatán Peninsula*,
DOI 10.1007/978-3-319-06529-8_1

Yucatán Peninsula and thereby provide tools for their conservation and food for thought.

In the 15 chapters we will not only present the most recent research on the ecology, distribution and use of plant communities and faunal assemblages, but also offer the reader an introduction to the peninsula's physical settings and its long history of human–environment interactions. These interactions are crucial for understanding the present structure of vegetation mosaics and species distribution, and they are also key to conservation efforts. Mexico is a country determined to demonstrate its conservation commitment to the rest of the world and the Yucatán Peninsula is a perfect example to show how these conservation efforts have played out on the ground. Approaches to conservation have changed over time and classic models of do-not-touch protected areas have given way to an interesting coexistence of state-sponsored protected areas (PA) that allow certain productive activities inside their borders. Activities include those that meet people's economic needs, traditional landscape management, and community-based natural resource management and conservation. However, the most important change of approach is that, increasingly, conservation advocates are receptive to considering humans as an integral part of the ecosystem, using and conserving it, and not only as a threat to biodiversity.

The unique features of the peninsula described above shape our consideration of a particular geographical area. From a biogeographical point of view, the Yucatán Peninsula covers the Mexican states of Campeche, Quintana Roo and Yucatán, as well as northern Guatemala and northern Belize. For this book we focus on the Mexican part, covering almost 145,000 km^2, but most of the information on biodiversity provided in the chapters applies to the wider region. However, land use policies and conservation strategies are quite different in the two neighboring countries of Guatemala and Belize, which is a main reason for limiting the geographical scope of the book to Mexico (Fig. 1.1).

The book is divided into four sections (parts) to guide the reader step by step, from the physical settings and environmental history (part I) to plant (part II) and fauna (part III) to conservation practice (part IV).

In part I we provide two chapters, the first one (Chap. 2) detailing the physical settings, including the geology and soils of the Yucatán Peninsula, which rests on a calcareous platform, and telling the complex environmental history of the last three millennia. The authors also give a comprehensive overview of the vegetation and highlight the climatic characteristics of the Yucatán Peninsula and analyze the consequences of on-going climate change. In Chap. 3 the authors describe vegetation types at the ecosystem scale. The most conspicuous biomes are tropical forests, which have been present since the end of the last ice age and have varied under changing climatic patterns and human influence.

The chapters in part II focus on plants and their environments at different scales, from the species to the community level. Two chapters analyze how plants have adapted to their environments (Chaps. 4 and 5). Chapter 4 provides phenological information for more than 2000 species, giving evidence of the close relationship between seasonality and flowering and fruiting. Given the particular climate of the

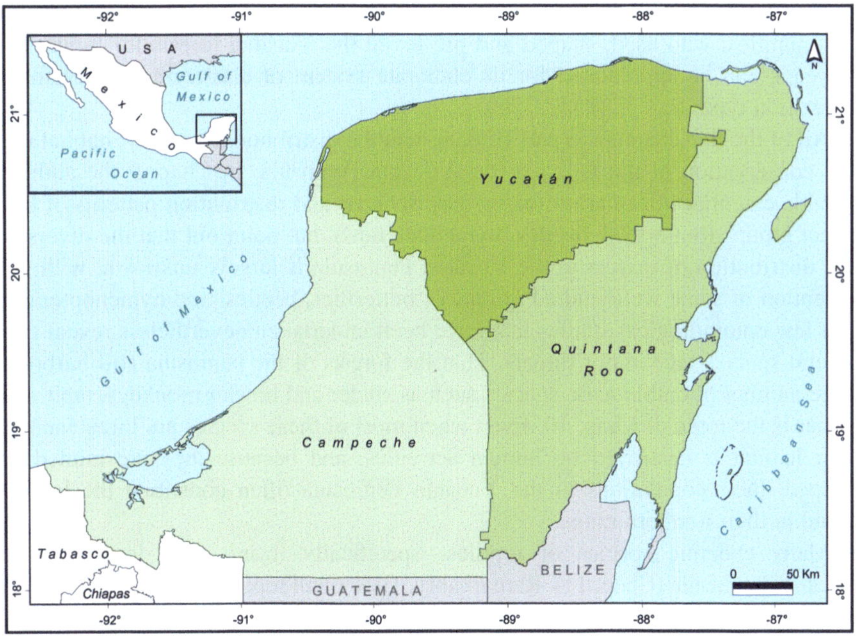

Fig. 1.1 The Yucatán Peninsula of Mexico, showing the states of Yucatán, Campeche and Quintana Roo

peninsula, with pronounced, but often varying, dry and wet seasons, a chapter on plant phenology (Chap. 5) is included so that the reader can better grasp climate–plant interactions at the individual, community and population level. Eco-physiological studies have elucidated the ability of plant species to withstand conditions of precipitation variability—especially drought conditions and poor soil conditions. Chapter 6 focuses on plant–bee interactions and the relevance of pollination by native and introduced bee species for conservation and production. Another important point raised by the authors is the importance of bees as bio-indicators of healthy environments. In Chap. 7 the authors show that large-scale natural disturbances, such as hurricanes and droughts, have influenced peninsular ecosystems. Dozens of hurricanes during the last 100 years have shaped the landscapes of the Yucatán Peninsula, which makes the region one of the most hurricane-affected geographical areas in the world. Human-induced disturbances, such as traditional shifting cultivation practices, among others, also impact ecosystems. Despite natural and human-induced disturbances at different temporal and spatial scales, ecosystems of the Yucatán Peninsula have been documented as relatively resilient. Given future climate change scenarios indicating increasingly dry conditions and ongoing land use changes and deforestation, this resilience might become compromised. Chapter 8 analyzes how humans use the forest ecosystems to extract timber and non-timber forest products, and temporarily deforest for shifting cultivation but simultaneously manage to conserve the forests.

This intricate interplay of use and conservation by mostly indigenous Mayan communities, who used, shaped and preserved the Yucatán forests for thousands of years, can be appreciated for its elaborate system of classifying successional vegetation types.

All of the five chapters in part III deal with the distribution, ecology, habitat use and conservation of the fauna of the Yucatán Peninsula. In Chap. 9 the authors provide essential information for species richness and distribution patterns of key insect groups (butterflies, beetles, hymenopterans), but point out that the diversity and distribution of insects in the Yucatán Peninsula is largely unknown, with the exception of some well-studied groups of butterflies, beetles, and hymenopterans. The few entomological studies that have been undertaken nevertheless reveal that several species face major threats. That the forests of the peninsula still harbor a large number of emblematic species such as spider and howler monkeys, tapir and jaguar is the topic of Chap. 10. Even when most of these species are threatened as their habitat is destroyed by human activities, and because they are hunted or trapped, their populations in the Yucatán Peninsula often constitute the largest found in their northern range.

Many endemic species of reptiles—specifically lizards—are found in the Yucatán Peninsula (Chap. 11). Remarkably, 19 % of all reptile species are endemic; albeit most of these species are threatened by habitat loss and degradation. Until now, most conservation efforts have concentrated on sea turtles, and to some degree on crocodiles. Birds (Chap. 12) are relatively well-studied organisms, but surprisingly little information exists on the 17 endemic and near-endemic species of the peninsula. The authors build on their knowledge of that group to provide new and detailed information on the natural history and status of these key species. They show that even though forest loss is an important factor in some species decline, the development of coastal areas and hurricanes are by far the most important threats to many species. Part III finishes with a chapter dedicated to showing how subsistence hunting remains a widespread activity in the forests of the Yucatán Peninsula and creates a strong link between people and their environment. Despite some species being overhunted, the authors argue that hunting is participating in forest conservation through its utilitarian and cultural value (Chap. 13).

Part IV of the book presents two chapters on conservation and management of forest and coastal ecosystems. Chapter 14 outlines lessons for conservation along coastal ecosystems. A new approach is presented that would guarantee ecosystem function and link key aspects of connectivity of coastal habitats, from mangroves and coastal lagoons to coral reefs. Evaluation of conservation efforts on the peninsula has occurred predominantly in forest ecosystems (Chap. 15). Forest management is part of the local culture in some *ejidos* of the Yucatán Peninsula. Efforts are undertaken to guarantee long term sustainability in logging and natural resources management. Changing land tenure systems are identified as threats to sustainable forest use and as a driver of land use change.

Taken together, the contributions in this volume provide a fresh view on the actual and potential conservation of biodiversity and resource management in the ancient Mayan lowlands of the Yucatán Peninsula. We anticipate that the research reported in the chapters will help researchers and conservation managers identify new directions for research on biodiversity, conservation, and cultural heritage. Clearly, some topics of investigation in the conservation of the Yucatán Peninsula remain underrepresented in this book and deserve deep consideration in the regional conservation agenda. These topics include: ecosystem functions and services, technological change, protected areas, marine ecosystems, freshwater ecosystems and organizational systems and processes (Sutherland et al. 2009). In a wider context, global environmental change might provide an important context for current biodiversity conservation research in the Yucatán Peninsula. Global change is leading to both long-term shifts in average conditions as well as potentially dramatic changes in environmental variation. Future research has to be concerned with understanding how systems will respond to such changes. We hope that the results presented in this book stimulate discussion and exciting new research.

Reference

Sutherland WJ, Adams WM, Aronson RB, Aveling R, Blackburn TM, Broad S, Ceballos G, Côte IM, Cowling RM, Da Fonseca AB, Dinerstein E, Ferraro PJ, Fleishman E, Gascon C, Hunter Jr M, Jutton J, Kareiva P, Kuria A, Macdonald DW, Mackinnon K, Madgwick FJ, Mascia MB, Mcneely J, Milner-Gulland EJ, Moon S, Morley CG, Nelson S, Osborn D, Pai M, Parsons ECM, Peck LS, Possingham H, Prior SV, Pullin AS, Rands RW, Ranganathan J, Redford KH, Rodriguez JP, Seymour F, Sobel J, Sodhi NS, Stott A, Vance-Borland K, Watkinson AR. One hundred questions of importance to the conservation of global biological diversity. Conserv Biol. 2009;23(3):557–67.

Part I
Physical Setting and Vegetation

Chapter 2
Physical Settings, Environmental History with an Outlook on Global Change

Nuria Torrescano-Valle and William J. Folan

Abstract The objective of this chapter is to offer a short but comprehensive overview of the Yucatán Peninsula, its karstic landscapes and soil types. In this chapter, we will also provide insights regarding Maya nomenclature associated with these physical characteristics. We include information on the hydrology of the Yucatán Peninsula (principally geohydrology) and regional climatic patterns (ITCZ, the North Atlantic and teleconnections). From this point of reference, we will explain the peculiarities of the peninsula, including a brief description of the origins of its vegetation distribution. Following this, we will present a paleoenvironmental and historic framework with a cultural focus, together with information on the uses and management of the natural resources by the Maya while explaining modifications of their environment. We will then include an analysis on global change based on data from the IPCC (Intergovernamental Panel on Climate Change) related to drivers of climatic change in the region such as deforestation, forest fires, hurricanes and changes in sea level. We will explain how these factors have influenced the loss of biodiversity and contributed to global change.

Keywords Geohydrology • Soils • Modern climate • Historical climate • Biogeography • Vegetation • Maya culture • Global change • Yucatán Peninsula

N. Torrescano-Valle (✉)
Departamento Conservación de la Biodiversidad, El Colegio de la Frontera Sur Unidad Chetumal, Avenida Centenario km 5.5, Apartado Postal 424, CP 77014 Chetumal, Quintana Roo, Mexico
e-mail: ntorresca@ecosur.mx

W.J. Folan
Centro de Investigaciones Históricas y Sociales, Universidad Autónoma de Campeche, Av. Universidad s/n entre Juan de la Barrera y Calle 20 Col. Buenavista, CP 24039 San Francisco de Campeche, Campeche, Mexico
e-mail: wijfolan@uacam.mx

G.A. Islebe et al. (eds.), *Biodiversity and Conservation of the Yucatán Peninsula*,
DOI 10.1007/978-3-319-06529-8_2

2.1 Geology and Hydrology

One of the most outstanding features of the Yucatán Peninsula (YP) is that it is a calcareous platform with great heterogeneity in its geological gradients, and the combination of geomorphological, climatological, hydrological, edaphic and ecological factors that have promoted the development of a particular type of biodiversity, resulting in a distinctive biogeographical unit. This unit corresponds to the Mexican States of Campeche, Quintana Roo and Yucatán, as well as small areas of Chiapas and Tabasco, the Department of the Peten in Guatemala and the northern part of Belize (Miranda 1958; Barrera 1962; Ibarra-Manríquez et al. 2002).

The YP had its origin with the breakup of the supercontinent of Pangaea during the Triassic, around 250 million years Before Present (My BP). During this period, the position of the YP was close to the Gulf of Mexico, Florida and North Africa. It was during the Jurassic and Cretaceous Periods that the YP experienced rotation that moved it in its present geographical location (Coney 1982; Burke 1988). The limestone, dolomite and anhydrite rocks that constitute the uppermost layer of the karstic platform of the YP are Mesozoic and Cenozoic (between 250 and 66 My BP) and rest upon 350 My BP crystalline and sedimentary rocks of the Paleozoic Era (Perry et al. 2003). The geologic origin of the YP, however, is frequently referred to as being recent (145-65 My BP), due to deep accumulations of carbonates and evaporates during the Cretaceous that were products of the shallow seas wherein the submerged YP was located. These deposits gave rise to the limestone, dolomites and gypsum that characterize the present day YP (Coney 1982; Bautista et al. 2005a).

The actual geomorphology and hydrology of the YP have resulted from various events that occurred near the end of the Cretaceous and during the Paleogene. The 10-km bolide impact that created the Chicxulub crater some 65 My BP, also gave rise to conditions that resulted in the ring of cenotes in northeastern YP, together with the movement of the Caribbean Plate (Back and Hanshaw 1970; Burke 1988; Hildebrand et al. 1991, 1995). During the Eocene (50-20 My BP), folding of rock strata with eastern and northeastern movements of the plate created the system of faults that form the major subterranean water channels that exist today. The Ticul Hills and the channel of the Río Hondo River were formed during the Miocene and Pliocene (20-4 My BP) epochs due to displacements of the Caribbean Plate in two directions (NW-SE and NE-SW, Perry et al. 2003; Bautista et al. 2005a). Figure 2.1 shows the principal geomorphological characteristics of the Yucatán Peninsula and the ages of the geological deposits. The karstic platform that forms the YP covers about 300,000 km^2, the most recent layers (dolomites and evaporites) are about 1500 m thick. The karstic aquifer of the YP is distributed by means of a system of caverns that is the largest in the world. The largest of these subterranean water reserves is situated on the eastern boundary of the Sian Ka'an Biosphere Reserve, which covers 165,000 km^2. This Biosphere Reserve is a World Heritage Site (UNESCO) and part of the Ramsar Convention on wetlands (www.unesco.org, www.ramsar.org). Another important water source to the north is associated with

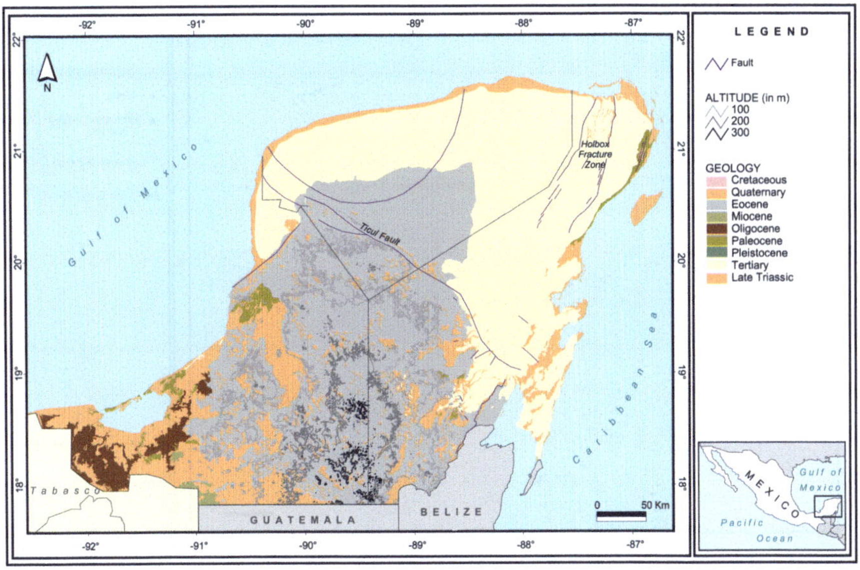

Fig. 2.1 Geology and altitude of Yucatán Peninsula

the ring of cenotes. In general terms, the system occurs across a network of faults (Figs. 2.1 and 2.2) and is supplied by rains that occur along an increasing gradient that runs from North to South (Perry et al. 2002; Gondwe et al. 2010; Bauer-Gottwein et al. 2011). The depth of phreatic aquifer varies considerably along the coast, it can be at less than 10 m depth while in the central region of the YP, it can be more than 30 m below the ground surface (Marín et al. 2005). In reality, this zone can be at the ground surface along the coast and at more than 80 m depth in the interior. Other major hydraulic features of the PY are *griekes*, i.e., structural cracks that are associated with karstic landscapes and with the pre-hispanic urban area of Edzna, Campeche. Matheny (1976) suggests that the Maya people in Edzna took advantage of these "canals" and modified the cracks by widening them close to domestic units (homes). They were thus described by Matheny et al. (1983) as canals that drained the city center, but Doolittle (2006) has suggested an alternative interpretation, viz., that the Edzna valley is a *polje* or a depression in the rock massif itself.

Only a few superficial bodies of water exist on the YP. In southern Campeche, the Palizada, Candelaria and the Champoton rivers form part of the watershed of the Usumacinta river, as well as several lagoons. Laguna Chichancanab is in Yucatán, bordering on the limits of Quintana Roo and the Hondo River in Belize and the system of the Bacalar Lagoons. Various lagoons are also associated with the ruins of Coba (Folan et al. 1983; Leyden et al. 1998). Although, numerous *aguadas* exist throughout of the YP. *Aguadas* are small bodies of water that are principally fed by rainwater; some are formed by topographic depressions that are at times lined with clay (González-Medrano 2004). These natural depressions are commonly seen

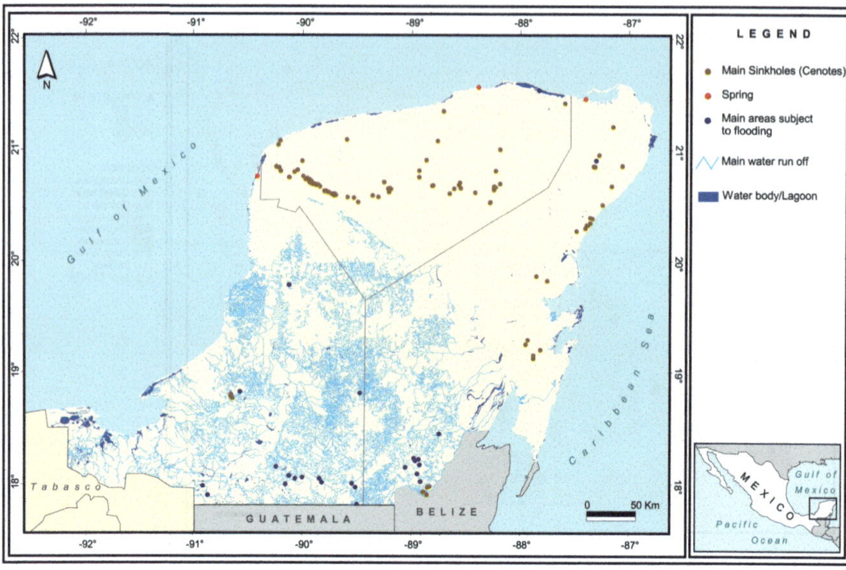

Fig. 2.2 Water bodies, main water run off (seasonal water currents) and Cenotes of Yucatán Peninsula. Map made from Red hidrográfica, 1:50000, 2.0 edition, INEGI 2010

within archaeological sites, their water storage capacities were reinforced by inhabitants of these Maya settlements. During our investigations (Domínguez-Carrasco et al. 1996) in the archaeological site of Calakmul, all of the aguadas that were tested by us were lined with either potsherds or limestone set in mortar.

Cenotes (sinkholes) are one of the most typical and recognizable karstic features of the YP, distinguished by their conical and circular forms. Mylroie et al. (1995) mention that cenotes are similar in formation (origin and form) to the Blue-Holes of the Bahamas and similar aquatic features in Florida. They are known for their cultural significance among the Maya. Their name is derived from the Maya term *tzonot*, which signifies a water-filled opening in the ground forming a subterranean cavern [Pearse (1936) is the primary source on cenote formation]. Cenotes are formed by rainfall, through the vertical dissolution of rocks (dolines) and by the dissolution of rocks in the subterranean zone. In the latter case, it is common for the surface rock or the cavern roof to collapse, thereby exposing the sinkhole at the surface. Cenotes are of varying diameter and depth due to the movement of subterranean water and the degree of dissolution of the rock. Steinich and Marín (1997) mapped over 7000 cenotes, but Bauer-Gottwein et al. (2011) have mentioned that it is difficult to know the total number of cenotes in the YP, due to large numbers that are likely hidden by dense vegetation and that are attributable to geologic processes, which occurred in the past. What is clear, however, is that most are found close to the northern and eastern coasts of the YP.

The coastal hydrology of the YP has experienced changes in sea levels, which were higher during the Pleistocene Epoch (2.6 My to 12 ky BP) as a result of glaciers and interglacial periods. During the last glacial maximum (LGM) when sea levels were lowest that the modern (ca. 130 m), the marine platform was exposed and subject to erosion and sedimentation of deltas, thereby initiating the development of coastal lagoons and systems of caves (Day et al. 2012). During the early Holocene Epoch, even lower sea levels than modern (ca. 3 m) allowed complete development of coastal lagoons and systems of caves through the accumulation of carbonates (Bautista et al. 2005a; Smart et al. 2006). The grottos or caves exhibit a great variety of geological features. Diverse deposits contain a complex mixture of cultural and fossil deposits, such as the cave of Balancanche, which is near Chichen Itza (Andrews 1970; Marín et al. 2005). On the northeastern coast, it is possible to find other karstic expressions such as semicircular caves that are associated with springs, resurgences and beaches created by the dissolution of carbonate rocks that are associated with aquifers.

Given the morphology and heterogeneity of the YP, Bautista et al. (2005b) devised a classification of geomorphological landscapes that were based on their morphogenetic, hydrologic and atmospheric surroundings. The resulting classification of 36 geomorphological landscapes emphasizes features of altitude, marshy plains, river currents, coastal areas, and the accumulation of quaternary sediments, together with high and low elevations on topographic relief and the degree of karstic evolution (Bautista et al. 2005b). Gates (1992), Domínguez-Carrasco et al. (2012), and Folan et al. (2000, 2014, 2015) have characterized the Karstic Mesoplano Calakmul to the south, as a bridge that connects the Peten of Guatemala (basin in northern) with the Yucatán Peninsula of Mexico. The bridge ranges in elevation from 80 to 400 m asl, and spans a distance of about 250 km. This bridge was very important during the Maya Preclassic (2000 BCE-250 CE) and Classic periods (250-950 CE), the relief and elevated regions were vital characteristics for the commercial relationships the Maya and construction of their impressive cityscapes.

2.2 Soils and Their Classification

The soils of the YP are characterized by their high calcium carbonate ($CaCO_3$) contents. Because of its coralline origins, the most common soil belongs to the Leptosols (FAO 2015). It is also possible to find Vertisols, Luvisols and Gleysols (FAO 2015). Leptosols are limited in depth either by layers of hard rock 25 cm below the soil surface, or by overlying material containing $CaCO_3$. In the latter instance, the $CaCO_3$ equivalent of the material is >40 % within 25 cm of the surface, or $<10\%$ (by mass) of the fine earth fraction (particles <2 mm in dia.) to a depth >75 cm, without diagnostic strata other than a mollic, ochric, umbric, yermic or vertic horizon. One of the most distinguishable features of all soil types in the YP is color, which is closely related to the nutrient and organic matter content (Bautista-Zuñiga et al. 2003, 2004; FAO 2015).

The soil layer is shallow across most of the YP platform. In some regions, it is practically absent, resulting in almost complete exposure of the bedrock. Evaporation, dilution, infiltration of subterranean water and the replacement of subterranean water, and the nature of limestone have a fundamental role in the formation of YP soils. In the northern YP, it is possible to observe this process, in that it is common to find two layers of calcite with low magnesium concentrations, where the upper layer is almost 3 m of impermeable *calcreta* with a layer of *sascab* (Maya, "white earth"; unconsolidated calcite) without cement. It is on these two Mio-Pliocene layers that a layer of quaternary soils have formed, which range from a few centimeters to no more than 1 m in thickness. In some zones, the *calcreta* has fractured and formed a mixture of materials (Perry et al. 2009). It is in the *aguadas* or karstic depressions that one can find large soil accumulations, the low permeability of which is associated with ejecta deposits originating from the Chicxulub meteorite impact. These *aguadas* are distinguished by their clay-like consistency and low hydraulic permeability (the emptying takes weeks or months). Ejecta deposits are also found in Southern Quintana Roo and Belize (Gondwe et al. 2010).

According to the classification of the World Reference Base for Soil Resources of the Food and Agriculture Organization of the United Nations (WRB/FAO/UNESCO), 13 Orders of soils can be found in the YP. These are Lithic Lepstosols, Rendiz-Leptosols, Cambisols, Luvisols, Vertisols, Gleysols, Regosols, Solonchaks, Phaeozems, Histosols, Solonetzs, Nitisols, and Castañozems. The types of soil that is most commonly found on the YP are Leptosols (Table 2.1). From of the fertility point of view, soils of the YP can be divided into two groups. One group of shallow black litosoles that somewhat cover rocky outcrops perhaps similar to those found in the ruins of Dzibilchaltun, Yucatán. And another are deep red rendzines in slightly low relief (Isphording 1984; Shang and Tiessen 2003). The basic composition of both soil types includes organic residues, limestone, amorphic metallic oxides, secondary minerals and some stratified minerals, such as illites, talc and chlorite (Isphording 1984).

The ancient agricultural activity carried out by the Maya in the YP is associated with their extensive knowledge of the soils and their fertility, as has been recognized by various authors (Barrera-Bassols and Zinck 2003; Barrera-Bassols and Toledo 2005; Barrera-Bassols et al. 2006). Barrera-Bassols and Toledo (2005), for instance, mention that there are currently more than 80 terms in Maya that are related to the characteristics of these soils, together with a classification of 30 types. Table 2.1 lists the principle soils that have been identified on the YP, their equivalents in Maya nomenclature, and their distributions.

Table 2.1 Soils types for Yucatan Península according WRB and Maya classification

Soil type WRB	Soil type Maya classification	Characteristics	Extension in YP
Leptosols	Chal'tun Stoney surface	Very thin soils less than 5 cm thick rich in organic material. Bedrock is located immediately below almost visible on the surface. These soils are not used for agriculture	North of the YP
Leptosols: Calcaric Leptosol LPca, Lithic Leptosol LPli	Tse'kel Rocky soils	These soils are dark brown, rich in gravel, calcareous rock and even bedrock. They are very thin measuring more or less 20 cm. They are rich in organic material with good drainage with pH somewhat alkaline. They are frequently used for agriculture. They can be associated with Chak lu'um Ak'alche and Puus lu'um	Principally in the northeast and northwest of the YP, but present in all the YP
Cambisols Leptosols: Lithic Leptosol LPli, Chromic Leptosol LPcr	Chak lu'um Chak means "red" and lu'um "earth", "red earth"	A red soil with little calcium carbonate. They include iron. The clay contains a mixture of haldisita montmorillonite. They are somewhat clay-like. They are up to 60 cm deep, but not rich in organic materials. It is rich in secondary minerals with a thickness that favors the growth or cultivated plants. This soil is associated with Puus lu'um, Tsk'el and K'ankab	On the line between Yucatan and Quintana Roo, some sites in the central and the southern part of Quintana Roo
Leptosols: Rendzic Leptosol LPrz	Puus lu'um Puus means "dust" and lu'um "earth", "loose soil with rocks"	The color is black to dark grey. It is 0.30–0.60 cm thick distributed over calcareous bedrock. It contains organic material and carbonates in a very fine soil with efficient drainage and a pH that is slightly alkaline. It is a fertile soil	Principally in the center, south and southeast of the YP

(continued)

Table 2.1 (continued)

Soil type WRB	Soil type Maya classification	Characteristics	Extension in YP
		utilized for agriculture. It is associated with Ak 'alche, Ya'ax hom, Tsek'el and Chak lu'um	
Leptosol: Rendzic Leptosol LPrz, Hyperskeletic Leptosol LPhk	Boox lu'um Boox means "dark or black" and lu'um "earth", "dark earth"	Similar to Puus lu'um, soils rich in organic material and carbonates, the surface stones are 0.5 cm–10 cm and retain moisture. In comparison to the chak lu'um soils, they are considered to be richer in micro nutrients, phosphorous and nitrates	Found in all the YP but they are reported principally in the north
Leptosol: Rendzic Leptosol LPrz	Chich lu'um Chich, small fragments of stone lu'um "earth"	Deep soils that retain a great quantity of humidity (water). They are dark in color and contain a low proportion of carbonates	Reported for the northern YP
Luvisol: Chromic Luvisol LVcr,	K'an kab k'an means "yellow" and kab "earth", "yellow earth"	These soils are found in Karstic valleys with a brownish red to a yellow color due to its chromic content. It is rich in clay due to the mixture of kaolinite-halosite, rich in iron oxides and aluminum with a slightly acidic pH. It is associated with Chak lu'um and Tsek'el and is considered suitable for agriculture	It is found in small zones of Yucatan, Campeche and Quintana Roo
Vertisols	Ya'ax hom Ya'ax means "Green"and hom "depression", "lowlands with ever-green vegetation"	It is a dark grey or reddish maroon very deep clay soil that does not contain rocks. The clay is montmormonite, the pH is slightly alkaline. These soils are found on soft calcareous rock with moderate, slow drainage. The internal drainage is moderate. In the low areas, there are processes of oxide reduction, associated with	It extends principally in the center and southwest of the YP frequently associated with the edges of bajos. Following the lepto soils, it is the soil type most amply extended. It is considered to be excellent for agriculture

(continued)

Table 2.1 (continued)

Soil type WRB	Soil type Maya classification	Characteristics	Extension in YP
		Puus'luum, K'ankab and Ak'alche	
Gleysols	Ak'alche Ak'al means "water body" and che "tree", "low lying terrains with seasonal inundation".	The upper level is dark grey or black and in accord with its depth reaching a grey or a green/grey horizon due to gleyzitation from a depth of 15 cm that occurs during a period of flooding from June to November. The pH is slightly acidic with an adhesive consistency caused by the montmorillonite clays that form fissures when they dry. The surface horizon is rich in organic material but in the lower part it is poorly represented. The internal and surface drainage is slow. It is associated with Ya'ax hom, Puus lu'um and Tsek'el	It extends toward the south central part of the YP. It supports the type of vegetation associated with bajos

References: Bautista et al. (2005a, b), Sánchez and Islebe (2002), Sedov et al. (2007), IUSS Work Group WRB (2007, 2014), INEGI (2003)

2.3 Modern and Historical Climate

The climate of the YP is characterized by its heterogeneity, particularly with respect to precipitation. The east coast and the basin of Laguna de Terminos (southwest) receive 1400 millimeters per year (mm/year) and decreases towards the northwest to 1000 mm/year (Fig. 2.3). The east coast is the most extensive humid region with 1200–1500 mm/year (Orellana et al. 2009; INEGI 2013). Two seasons are identified each year: a dry season, and a wet season, in the winter months occur incursions of Arctic air. Between August and September, a relative minimum of precipitation occurs, which is known as the midsummer drought (*canicula*). Temperatures vary by half a degree to a degree across the YP (25.5–26.75 °C). This is a synthetic and simplified description of the climate of the YP; however, climate is a dynamic and highly complex system that is controlled by the scales of time and space. Different climate patterns emerge over months, years, centuries, and millennia; moreover, local and regional climatic patterns act with different magnitudes. Also, it is important to identify the events that influence short (regional) and long distance

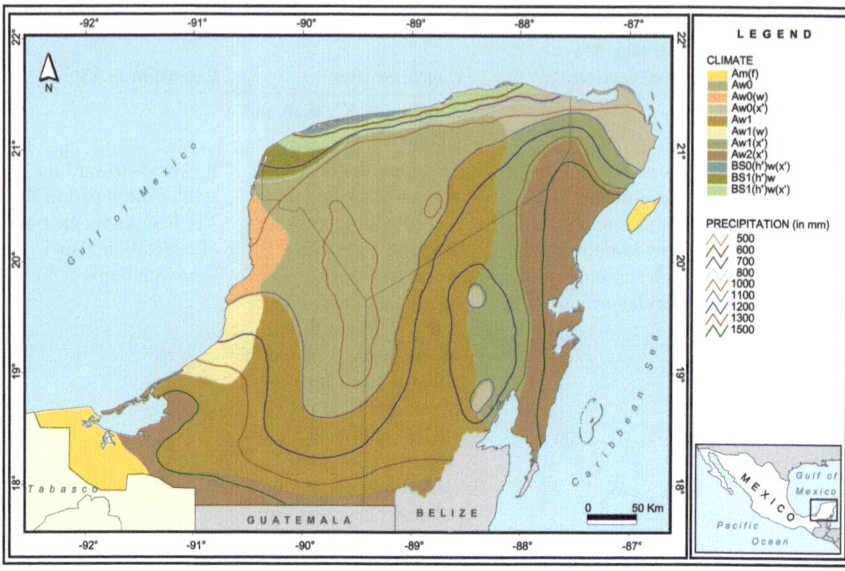

Fig. 2.3 Climate types and precipitation distribution across the Yucatán Peninsula. Map made from: Climas, 1:1000000, INEGI 2011

(global) like teleconnections (Gunn and Folan 1992; Sánchez-Santillán et al. 2006). The teleconnections are clime patterns that occur principally in North Atlantic (e.g. North Atlantic Oscillation) and South Pacific (e.g. El Niño/South Oscillation), derived from atmospheric and marine interactions. Which consist in oscillations and variations in Sea-Surface-Temperature (SST), atmospheric pressure and air masses.

To understand the climatic gradient in YP, it is necessary to know about the role of various factors in climatic behavior, such as temperature, rainfall, humidity, winds, cloudiness, and solar radiation, and geographic factors, such as latitude, altitude, relief, marine currents and the distribution of land and water. Given this, it is important to detail the origins of climate types in the YP and their histories. First one has to address the major interaction and connecting mechanisms between the atmosphere-sea-land that influence and determine precipitation. (1) Inter-tropical Convergence Zone (ITCZ), the so-called "Meteorologic equator," is a frontier between a dry pattern and wet pattern located 5°N of the equator, due to the asymmetry that is produced by changes in sea temperature. This frontier of low pressure and cloudiness that is caused by the convergence and connectivity of the hot and humid Trade Winds and Hadley Cell circulation, convective wind system that distributes humidity across tropical regions (Méndez and Magaña 2010).

The ITCZ determines patterns of precipitation across the YP during the summer months (May–October), when it migrates to the north. The zone also determines the occurrence of the dry season (November–April) when it migrates to the south (Gunn and Folan 2000; Stahle et al. 2012; Marshall et al. 2014). (2) Anomalies or changes in Sea-Surface-Temperature (SST), as well as changes in atmospheric pressure and air masses that occur in the South Pacific and North Atlantic influence provokes variation in the magnitude and occurrence of the teleconnections El Niño/ South Oscillation (ENSO) and the North Atlantic Oscillation (NAO), respectively. Climate is influenced by these anomalies, which affect the transport of the marine currents, such as the Meridional Overturning Circulation (MOC), and atmosphere transport by means of the Subtropical Jet Stream. The rains that are provided by the winter storms, which are called *nortes* (November–February), are the product of masses of cold air from Canada and the United States that interact with variability in the Pacific (ENSO and Tropical Jet Stream). The *canicula* or mid-summer drought that occurs between July and August is also associated with this variability (Gunn and Folan 1992; García 2003; Poveda et al. 2006; Stahle et al. 2012). These teleconnections determine droughts or wet conditions along the annual cycle (during a year), but can have influence for several years and decades (Méndez and Magaña 2010). (3) Other patterns that are similar to the "warm pools" (areas of water warmer than 28.5 °C with a interannual fluctuations and intensity significant) in the Eastern Pacific, and zones of high pressure in the Atlantic and Pacific, are the Atlantic cyclones or depressions that are directly related to distinct meteorological phenomena (Wang and Enfield 2001; García, 2003). In addition to atmosphere-marine patterns, latitude and elevation are fundamental geographic features that determine climate. Most of the YP is located within the tropics and has elevations that do not exceed 400 m between 17 and 21°N and 86–92°W. Latitudinal position assigns a tropical climate to the YP, while the low elevations offer almost direct access the hot and humid Trade Winds.

As a result of the different interactions between global, regional and local climatic factors, also of geographic factors; originate a gradient of temperature, humidity and precipitation across of YP. Each gradient or area is called climate type. The principal climates for the YP are: Most of the YP experiences Aw = tropical subhumid climate with summer rains and a dry winter. An Am = tropical humid climate with summer rains and a relatively dry winter, present in the south in the southeastern portion of the peninsula and the Island of Cozumel. A BS = with a short rainy season during the summer, between a very dry and subhumid, characterizes the northern coast. The positioning of the ITCZ above northern Yucatán results in a hot, dry type climate (BS) in this zone (Ibarra-Manríquez et al. 2002). These types of climate have nine subtypes related to local variations, according to Orellana et al. (2009). Figure 2.3 shows not only the climate types, but also the isotherms and isohyets of precipitation.

The YP climate has varied over the last 10,000 years (Holocene) due to the influence of solar activity on the diverse climatic patterns that have already been mentioned, eustatic changes in sea level, and human impacts (Coe 1994; Hodell et al. 1995; Torrescano-Valle and Islebe 2006). During the early Holocene, climate conditions were likely characterized by high levels of precipitation and by high levels in lakes. A subsequent dry period was observed in paleoenvironmental records from marshy areas, which mark a transition to tropical forest around 8600 years BP (Islebe et al. 1996; Whitmore et al. 1996; Leyden et al. 1998).

Consistent with diverse sources of information (proxies), it was during the Middle Holocene when the modern gradients of temperature and precipitation were established (Curtis et al. 1996; Leyden 2002; Torrescano-Valle and Islebe 2006). The coastline of the Mexican Caribbean was established prior to 3800 years BP, as were the levels of continental lakes (Hodell et al., 2000; Leyden 2002; Torrescano-Valle and Islebe 2006, 2012). The palaeoecological data for the Late Holocene show interference for climatic interpretations, due to relationship between landscape and Maya culture. The crop signal suggests the distribution of a large population during this period. The called "climatic deterioration" is manifested in reductions in precipitation and long periods of drought (various centuries) that are not comparable with the Early and Late Holocene (Gunn et al. 1994; Hodell et al. 1995; Leyden et al. 1996; Haug et al, 2003; Hodell et al. 2005; Müeller et al. 2009; Carrillo-Bastos et al. 2010). It is during the Late Holocene (ca. 3000–1000 years BP) that records indicate not only the end of the construction of Maya civic/ceremonial centers, but also the abandonment of great cities such as Calakmul (850 AD) and Tikal (830–850 AD). The contribution of environmental change, particularly the pronounced droughts of the ninth Century, which can be related to the demise of the Maya Late Classic, has been suggested and documented based on diverse studies from different sources (Folan 1981; Folan et al. 1983; Gunn et al. 1994; Haug et al. 2003; Hodell et al. 2005; Carrillo-Bastos et al. 2010; Medina-Elizalde and Rohling 2012).

2.4 Biogeography and Vegetation

Mexico is located between the Holarctic and Neotropical biogeographic realms, i.e., it is found between two regions with different environmental histories. The organisms that comprise these realms have differents histories of origin and distribution. The geographical location between North America and Meso America, is fundamental for the origin of its high biodiversity. The biogeographic analyses for the YP have been carried out for more than 70 years. Stand out the phytogeographic work of Lundell (1934), and the biogeographic work of Barrera (1962), based on endemic species of flora and fauna respectively.

Ibarra-Manríquez et al. (2002) mention the importance of the flora in establishing biogeographic regions, their contributions in the phytogeographic subdivision of the YP, has allowed to recognize as a biogeographic unit. Also mention during several decades many authors have recognized its standing as a geographical, climatic, geomorphologic and biogeographic unit, but the information was insufficient. The YP has been distinguished principally for by its high percentage of endemic tree species (72 species), together with its characteristic type of soils, climate, physiography, orography, hydrogeology and fauna. Data gathered from different various studies have demonstrated the underlying similarity in the floras of Tabasco, Chiapas, Campeche, Quintana Roo and Yucatán, together with northern Guatemala and Belize. This unity also shows a strong relation with

Mesoamerica. The floristic results are not surprising, given the close geographical proximity of these political entities; nor is a weak relationship with Antilles unexpected, given increasing distance from the rest of Mesoamerica (Ibarra-Manríquez et al. 2002; Espadas-Manrique et al. 2003).

In general, it is estimated that the floristic richness of the Mexican portion of the YP includes about 2300 species, which are distributed among 965 genera and 161 families that are either native or not found to exist in the wild. Only one monotypic genus is endemic in the Mexican collection, *Plagiolophus* (Asteraceae). The genera *Goldmanella* (Asteraceae) and *Asemnantha* (Rubiaceae) are also found in Belize and Guatemala; the first genus remains monotypic (*Goldmanella*), while the second has been subsumed into *Chiococca*.

As a biogeographic unit, the YP includes 203 endemic taxa. The Euphorbiaceae, Fabaceae and Orchidaceae have the largest number of species. The diversity of flora is low in proportion to the size of the YP territory, but endemism is high, given that the YP represents 5.7 % of the country of Mexico. The flora includes an Antillean component and two unique vegetation types: low-statured, low flooded semi-evergreen forests that are known as *bajos*, and flooded areas that are associated with mangroves known as *petenes* (Fernández-Concha et al. 2010).

Fernández-Concha et al. (2010) mention that the flora of the YP is still little known in biogeographic terms, given the very recent descriptions of *Attilaea abalak* E. Martinez & Ramos, a new genus and species in the Anacardiaceae (Martínez and Ramos-Alvarez 2007) and to new species, *Zephydranthes orellanae* Carneval, Duno & J.L. Tapia; (Carnevali et al. 2010) and *Hohenbergia mesoamericana* I. Ramírez, Carnevali & Cetzal (Ramírez-Morillo et al. 2010). Another notable aspect of the flora is the scarcity of gymnosperms (1) and native ferns (0). Taxonomic richness is considered to be low, considering that the YP covers a wide territory within the tropics. Fernández-Concha et al. (2010) briefly explore the causes for low taxonomic richness, including the peninsula's geological origins and historic processes. They also explore the hypothesis that was first proposed by Gentry (1988) and Whittaker and Field (2000), who detail the influence of latitude and water-energy balances. They further stress the importance of topographic relief (mountains, valleys, hillsides and ravines), when combined with seaward and windward exposure, in creating microsites that are not present in the YP.

According to the classification of Miranda and Hernandez-X (1963), which is based on physionomic and phenological criteria and numerous studies conducted in the region, the principal types of vegetation found in the YP are: low deciduous forest; low semi-deciduous forest, with columnar cacti; low and medium semi-deciduous forest; high, medium and low semi-evergreen forest; high evergreen forest; savannas; palm groves; mangroves; coastal dunes; *popales*; *tulles*; and reed beds. Table 2.2 describes these types of vegetation.

Table 2.2 Vegetation types for Yucatán Peninsula

Vegetation type	Description
Seasonally dry tropical forest. Low deciduous forest	It is distributed in the north and northwestern part of the PY. This forest sets on shallow superficial soils with large rock outcrops in the form of bedrock and large stony areas. It is extremely arid with 800 mm of rain a year and between 7 and 8 months of drought. The trees loose practically all their leaves during the dry season and grow to heights less than 10 m. The species that characterize this forest are: *Bursera simaruba*, *Caesalpinia gaumeri*, *Acacia pennatula*, *Metopium brownei*, *Gymnopodium floribundum*, *Jatropha gaumeri*, *Havardia albicans*, *Mimosa bahamensis*, *Alvaradora amorphoides*, *Sideroxylon obtusifolium*, *Caesalpinia yucatanensis*
Seasonally dry tropical forest with columnar cacti. Low deciduous forest	This is a variant of the seasonally dry, tropical forest characteristic of the northwestern part of the PY. It is developed on shallow soils with large rock outcrops in the form of bedrock and large stony areas in extreme aridity with 600-800 mm of rain per year and between 7 and 8 months of drought. The trees loose practically all their leaves during the dry season and reach less than 10 m in height. The large amount of rocky areas is notable and permits the formation of microniches. In addition to the species characteristic of this type of forest one can find the following endemic plants: *Mammillaria gaumeri*, *Beaucarnea pliagilis*, *Guaiacum sanctun*, *Pilosocereus gaumeri*, *Nopalea gaumeri*, *Nopalea inaperta* and *Pterocereus gaumeri*
Medium forest and low semi-deciduous forest	The canopy reaches a height of 8–25 m. Between 50 and 75 % of the trees lose their leaves during the dry period with precipitation between 1000 and 1200 m. The most common species are: *Vitex gaumeri*, *Brosimum alicastrum*, *Piscidia piscipula*, *Enterolobium cyclocarpum*, *Ceiba pentandra*, *Sideroxylon foetidissimum ssp. gaumeri*, *Caesalpinia gaumeri*, *Cedrela odorata*, *Alseis yucatanensis*, *Astronium graveolnes*, *Pseudobombax ellipticum*. The associations are dominated by: guayacán (*Guaiacum sanctum*), xu'ul (*Lonchocarpus yucatanensis*), ja'abin (*Piscidia piscipula*), jobillo (*Astronium graveolens*) and despeinada (*Beucarnea pliabilis*)
High and medium semi-evergreen forest	This type of forest is the one most amply distributed in the the PY. The height of the arboreal layer is 20–35 m. Around 25–50 % of the trees lose their leaves during the dry period of the year. The range of precipitation is between 1100 and 130 mm. The most dominant and characteristic specie is *Manilkara zapota*, other species are: *Brosimum alicastrum*, *Pimienta dioica*, *Lonchocarpus castilloi*, *Pouteria*

(continued)

Table 2.2 (continued)

Vegetation type	Description
	campechiana, *Swietenia macrophylla*, *Alseis yucatanensis*, *Zuelania guidonia*, *Cedrela odorata*, *Swartzia cubensis*. They are found associated with ramonales (*Brosimum alicastrum*), zapotales (*Manilkara zapota*), corozales (*Orbignya cohune*), pukteales (*Bucida buceras*), and the bayo (*Aspidosperma cruentus* and *A. megalocarpon*)
Low flooded semi-evergreen forest	This type of forest corresponds to what are known as "bajos" that develop in seasonally flooded areas with soils slightly permeable called "ak'alche" that are found all over the PY. Fifty percent of the trees lose their leaves during the dry season every year. Rainfall is between 1100 and 1300 mm. The canopy reaches heights of 8–10 m. The typical species are: *Cameraria latifolia*, *Acacia pringlei*, *Dalbergia glabra*, *Pisonia aculeata*, *Pithocellobium dulce* and *Pseudophoenix sargentii*. The presence of ferns and epiphytes are common. It is possible to find associations with: tintales (*Haematoxylon campechianum*), pukteales (*Bucida buceras*), chechenales (*Metopium brownei*), mukales (*Dalbergia glabra*), and tasistales (*Acoelorrhaphe wrightii*)
High evergreen forest	This forest is located in southern Campeche, northern Guatemala and Belize. The precipitation reaches between 1500 and 2000 mm. The height of the canopy reaches 30–60 m. The soils are rich in organic matter and are well drained. Vines and lianas are common as well as the following species: *Terminalia amazonia*, *Dialium guianensis*, *Vochysia hondurensis*, *Calophyllum brasiliense*, *Aspidosperma cruentum*, *Brosimum alicastrum*, *Pouteria campechiana*, *Licania platypus*, *Swietenia macrophylla*, *Manilkara zapota*, *Alseis yucatanensis* and *Zuelania guidonia*, as well as others
Savannas	These are found mixed with low flooded semi-evergreen forest. The herbaceous layer is dominant while the tree level does not reach more than 10 m high, giving a stunted effect. The density of the trees is variable. The dominant herbaceous species are the Poacea and Cyperacea that are resistant to fire. Southeastern Quintana Roo is characterized as being very humid with acidic soils where it's possible to find *Pinus caribaea*, *Myrsine cubana* and *Morella cerifera*. In southwestern Campeche, *Trachypogon* and *Curatella americana*, are common and probably of an introduced origin. In Guatemala and Belize are found: *Quercus oleoides*, *Pinus caribaea*. As characteristic tree species there are: *Byrsonima crassifolia*, *Acoelorraphe wrightii*, *Cresentia cujete*, and *Curatella americana*

(continued)

Table 2.2 (continued)

Vegetation type	Description
Pine forest	In Guatemala and Belize it is common to find these trees mixed in with savannas. The dominant species are *Pinus caribaea* and *Pinus oocarpa*. Another small zone of distribution exists in southeastern Quintana Roo
Palm groves	These are found in almost pure patches generally distributed in areas close to the east and west coast of the PY and areas close to lagoons and the Hondo River. The species that make up these groups are: amongst others: corozo (*Orbignya cohune*), tasiste (*Acoelorraphe wrightii*), jawuacte' (*Bactris mexicana*), xiat (*Chamaedora seifrizii*), chi'it (*Thrinax radiata*), huano (*Sabal yapa*), huano k'uum (*Cryosophila stauracantha*), kuka (*Pseudophoenix sargentii*), as well as others
Mangroves	They are found in all the coastal areas of the PY. There are variations including the height of the canopy that are directly related to the type of soils in which the communities are developed. The short mangroves do not grow to more than 1.5 m wherein are some zones of Campeche and Tabasco they can reach 10 m. The distribution and abundance of the specie is subject to flooding and salinity. The typical species are: *Rhizophora mangle*, *Avicennia germinans*, *Laguncularia racemosa* and *Conocarpus erectus*. In the Sian Ka'an reserve, Petenes and Celestun, one finds circular formations of mixed mangrove and forest that have been referred to as "Petenes" related to rising pools of fresh water. Species of medium subevergreen forests such as *Manilkara zapota*, *Gymnanthes lucida*, *Bravaisia berlanderiana*, *Sabal yapa*, as well as others make up these formations
Coastal Dunes	These dunes are found behind the beaches due to an association of halofita adapted to high concentrations of salt located in the coastal fringes restricted to sandy and rocky soils. The typical species are: *Ageratum littorale*, *Ambrosia hispida*, *Batis marítima*, *Chysobalanus icaco*, *Coccoloba uvifera*, *Suaeda linearis*, *Suriana marítima* and *Tournefortia gnaphalodes*
Popales, tulles and saibal	These are found under special soil and geomorphologic conditions with soils almost permanently saturated such as aguadas, rejolladas (ko'op) and cenotes. One could find formations of herbaceous dominance such as the Tulles (*Typha angustifolia*), popales (Poaceae), and saibals (*Cladium jamaicense*). Some variants include trees such as tasiste (*Acoelorraphe wrightii*), and cork

(continued)

Table 2.2 (continued)

Vegetation type	Description
	(*Annona glabra*) as well as cacao trees in Coba, Quintana Roo

References: Miranda and Hernandez-X (1963), Pennington and Sarukhán (1998), Sánchez et al. (1998), Martínez and Galindo-Leal (2002), Durán-García and Méndez-González (2010), Ek-Diaz (2011), Carnevali et al. (2010), Folan et al. (1983), Brokaw et al. (2011). For more detail see Chap. 3

2.5 Maya Culture and Landscape

One of the aspects most readily recognized in Maya culture is its adaptation and interaction with the physical surroundings, which is manifested in its long-term establishment (3000 years) and its sustainability over the centuries. Consequently, great population nuclei could develop, which according to various estimates (Folan et al. 2015), were of much greater extent and size than current population centers. Different types of cultivation techniques, forest management practices, and elaborate water distribution and retention techniques (Gates 1992; Folan et al. 2015), demonstrate the great understanding that the Maya had of their environment, together with an extensive knowledge of engineering. In spite of this degree of technological sophistication, the diverse periods of instability that the Maya experienced (Gunn et al. 1994) have been recognized as being closely linked to the variability of their environment and to other factors (Dunning and Beach 2000; Aimers and Iannone 2014).

One problem that has yet to be discussed is identifying the initial period of agriculture in the Maya Lowlands (Lohse 2010). Dating is complicated by the nature of the sediments in various sites that permit only poor preservation of evidence. Yet there is strong evidence that agriculture began 8000–6200 years BP on the watershed of Gulf of Mexico and in the Pacific Balsas River, where the archeological deposits have permitted better preservation (Pohl et al. 2007; Matsouka et al. 2002; Buckler and Stevens 2006; Sluyter and Dominguez 2006; Buckler and Stevens 2006). Recent studies that were carried out in northern Belize (Rosenswing et al. 2014) provide a possible date of ~6700 years BP for cultivation in the Maya Lowlands, while others have estimated more recent dates of 5400, 4600 and 4000 BP (Pohl et al. 1996; Wahl et al. 2007; Beach et al. 2009; Siemens 2011).

Maya subsistence is based on an integral use of resources. The most common system was slash-and-burn agriculture (shifting cultivation) and the use of terraces, such as those documented in the Peten Campechano (Turner 1983; Morales-López 1987), and a system of ridges (raised fields) in humid areas (Siemens 1983). Family orchards enhance the diet of the inhabitants with the production of fruit-bearing trees or vegetable gardens (Folan et al. 1983). The knowledge of vegetable resources was widespread; it is estimated that there were more than 300 plant species within this production system (Barrera-Bassols and Toledo 2005). Various authors recognize that forest agriculture was an important activity in diverse zones of the YP. One example is the characterization and classification of ten types of vegetation and six successional stages by the Maya. Another example is the diverse mono-specific patches of species, such as Maya nut or breadnut (*Brosimum alicastrum* Sw.), allspice (*Pimienta dioica*,[L.] Merr.) and cohune palm (*Attalea* [= *Orbingnia*] *cohune* Mart.), which do not correspond to the natural distributions of the forests that have been identified (Folan et al. 1979; Gómez-Pompa, 1987; Rico-Gray and García-Franco 1991). Other complimentary activities were the gathering of edible plants (in all types of tropical forest), hunting and trapping, raising of ocellated turkey (*Meleagris ocellata* Cuvier) apiculture and fishing.

The transformation of the landscape is a key to trying to understand the survival of large groups that developed principally during the Classic Period when Maya populations were at their highest levels. A population model that was proposed by Santley (1990) indicates demographic differences in various regions of the Maya area. This model corresponds with the process of regionalization proposed by Dunning and Beach (2000), based on their adaptive agricultural systems.

A clear example of landscape transformation is the pre-hispanic urban center of Calakmul, Campeche. A > 30 km^2 in area, Folan et al. (2000, 2015) mention that Calakmul was a larger metropolis than Tikal, with more than 6125 structures and some 120 stelas. Based on 30 years of investigation, Folan and his colleagues have argued that Calakmul's geographic location, its climate including precipitation, and its landscape including its relief and vegetation, provide many advantages. In principal, it maintained a strategic relationship in commerce and the production of goods such as worked shells from the Gulf of Mexico, the Caribbean and Pacific Ocean (Villanueva-García n.d.). The size and influence of the territory acquired by Calakmul was made possible by the support of many tributary and allied centers such as Oxpemul, La Muñeca and Becan, along with Cancuen (Demarest et al. 2011) and La Corona (Canuto and Barrientos 2013) to the south. It has also been recognized that it was through political relationship with major centers such as Copan, Edzna and Coba that Calakmul's stability was maintained for a long period of time. Both Gates (1992) and Folan et al. (2015) offer an environmental explanation for Calakmul's success. The discovery and analysis of the relief and geomorphological development of the karst in southern central Campeche and northern Guatemala suggests a region that has been named the Karst-Mesoplano, crossed by the Laberinto Bajo. The dynamic hydrology of the northern and southern regions of the Karstic Mesoplano are not the same, which favors significant differences in the productive nature of the soils and the processes of erosion. The north-central region of the Mesoplano is formed by an anticline, while the south is a syncline, as described by Gates (1992), Domínguez-Carrasco et al. (2012), and Folan et al. (2015). Given that different adaptations and migration from south to north, and vice-versa (Yucatecano), and from east to west (Cholano) (Josserand 1975) of the Karst Mesoplano occurred in different phases the Preclassic and Classic Periods, the management and control of soils was planned and efficient, thereby providing adaptations to climate variability, including extreme events such as droughts (Aimers and Iannone 2014). These advantages of Calakmul likely were key to its survival and political stability over many centuries.

In a manner similar to the Calakmul model, occupational and demographic models have been developed for other Maya regions that have been mentioned by Santley (1990). These models demonstrate variation in the adaptive responses of the inhabitants, with periods of stability and instability that were related to climatic and political changes. Soil fertility and control of erosion presented constant challenges, as demonstrated by terraces in several sites that are located up and down the anticline (Morales-López 1987). Without a doubt, however, the most valuable resource was water. The management and conservation of this resource was fundamental for Maya survival. Evidence of this is to be found in the architectural design of buildings, the network of canals and *chultunes* (cisterns) that captured and stored water, and the formation of *aguadas* (Domínguez-Carrasco and Folan 1996; Dunning and Beach 2000; Folan et al. 2014; Gunn et al. 2014; Lucero et al. 2014).

In addition to understanding the notion subsistence of the region, which is understood to be the minimum requirements that are required for population survival, but is a concept that is closely related to the use of land and natural resources, another line of major investigation has been to understand its cultural collapse through the perspective of the environment. On one hand, we have diverse paleo-environmental evidence (oxygen isotopes, pollen analysis, titanium content, and stalagmites) that has revealed reductions in precipitation that had resulted in droughts experienced over the last 3000 years, principally during the Late, Terminal and Postclassic Periods, where these events were most prolonged (Folan 1981; Gunn et al. 1994; Islebe et al. 1996; Medina-Elizalde et al. 2010; Aimers and Iannone 2014; Torrescano-Valle and Islebe 2015).

Has been strongly discussed about the magnitude of the environmental effects of deforestation carried out for milpa formation, cooking and lime production (for architecture) during nine centuries, principally during the Late Classic Period.

Effects of deforestation on the microclimate, hydrology, the evapotranspiration of bodies of water, such as *aguadas* and lagoons, the erosion of soils and the loss of soil moisture have been identified in diverse Maya sites (Beach et al. 2008). The analysis of sediments including geochemical analyses, sedimentation and aggradation rates, bulk density plus other characteristics, has revealed evidence of erosion caused by agricultural activity. However, it is recognized that there exists a strong and combined influence of climatic and environmental change in the paleoenvironmental records (Wahl et al. 2007; Beach et al. 2008; Aimers and Iannone 2014; Torrescano-Valle and Islebe 2015). The same line of investigation was pursued in demonstrated how subsistence agriculture was possible during the Classic Period (Scarborough et al. 2014).

Johnston (2003, 2006) analyzed fallow periods and the productivity of soils in the Maya Lowlands, concluding that the initial slash-and-burn system for preparing a *milpa* required long fallow times of 10–20 years, to be modified until arriving at a model called "prolonging the cultivation" (two continuous years of agriculture). This Johnston model was based on a revision of the conventional tropical-ecological model and the model of primitive agriculture evolution that was proposed by Boserup (1965). The lengthening of cultivation time brought about the gradual reduction of fallow times in cultivated areas. The *milperos* (agriculturalist, farmers) removed the undergrowth (slash) before it could produce seeds, while incorporating the debris into the soil to maintain moisture levels and nutrient availability. Under this modified system, increased time and energy was required on the part of the *milperos*. Which subsequently affected levels of consumption of the entire population.

2.6 An Outlook on Global Change

Understand the history of climate change in the Yucatán Peninsula, is the main objective of Paleoenvironmental studies. Identify the responses of vegetation and changes in water systems, helps to explain the implications for human subsistence. The establishment of environmental change scenarios, should cover the past, present and future. Most importantly are the implications for human existence, plus the management and conservation of natural resources. In short, we have to understand the past relationship between Maya culture and the environment to help us establish future scenarios of environmental and climatic change in the YP. - Human-influenced modifications of climatic patterns has been a strong theme of debate in recent decades. The fifth report of the IPCC (2013), which amasses the most recent scientific evidence related to observed climate change, confirms that human influence has affected the heat balance of the atmosphere and oceans, as well as the global water cycle, reduction in snow and ice, sea level rise, and changes of some extreme climates. Additionally, the report emphasizes the direct effects of human influence as trigger principal cause of climatic warming since the twentieth Century.

The scenarios and projections elaborated by scientists of the IPCC are based on paleoenvironmental and current data. These allow the attribution and differentiation of previous changes to human activities. The use of paleo-data by the IPCC provides a global benchmark against which future environmental changes in the YP can be predicted and assessed: increments of 1–2 °C are predicted for the mid-twenty-first Century. (from 2046 to 2065) and 1–4 °C increases for the end of the century (2081 and 2100). In the case of sea level, the IPCC estimates increases between 0.24 and 0.30 m for mid-century and 0.26–0.82 m for the end of the century. Changes in ocean circulation, particularly in the Atlantic Meridional Overturning Circulation (AMOC), are likely to diminish in velocity and magnitude, thereby affecting the interchange between cold and warm water currents. An increase in the frequency and magnitude of droughts is also projected. The frequencies of high level cyclones or hurricanes are projected, including extra-tropical events. The models also project a decrease of up to 12 % in precipitation levels by mid-century, as well as great increases in the force and the quantity of precipitation over short timespans, i.e., hours or days (torrential rains), rather than continuous rains last for several weeks.

Paleoenvironmental, historical and instrumental evidence have shown extreme climatic events in the YP, mainly hurricanes and droughts (see Chap. 7). Droughts have caused frequent human disasters in the YP and are well recorded. Instrument records only cover the last century (Méndez and Magaña 2010), but they provide key information for the development of models. Detailed historical records cover the last five centuries, which allow the development of drought time series (Mendoza et al. 2006, 2007). Paleoecological data provided evidence on different scales (millennial, century and decadal), mainly for the last 4000 years (Islebe et al. 1996; Hodell et al. 2001; Medina-Elizalde and Rohling 2012; Torrescano-Valle and Islebe 2015).

According to Méndez and Magaña (2010) protracted droughts were recorded in Mexico during the twentieth Century, which were related to SST anomalies, the Pacific Decadal Oscillation (PDO) and the multi-decadal Atlantic Oscillation (MDA). Those climatic events caused great economic and material damage, and the death and forced migration of many people (Delgadillo-Macías et al. 1999). Mendoza et al. (2006, 2007) analyzed the frequencies of drought between the sixteenth and nineteenth centuries. By using time series and the Palmer Drought Severity Index, they were able to identify a higher frequency of droughts during the eighteenth and nineteenth centuries compared to the sixteenth and seventeenth centuries. Particular dry years were 1650, 1782 and 1884. The data reveal some conspicuous cycles, principally with lengths of 1–7 years, but other cycles lasted 40 and 70 years.

Droughts showed a relation with the global pattern ENSO, and Atlantic Multidecadal Oscillation (AMO). The Little Ice Age occurred during the sixteenth and seventeenth centuries, which caused a considerable temperature decrease at higher latitudes, and the subsequent loss of farmland, famines, epidemics and the deaths of thousands of people. Similar human disasters were recorded for the YP for the same time period (Mendoza et al. 2007).

Paleoenvironmental records document a decrease in precipitation over the last 3000 years. These records from the late Classic Maya Period (800-1000 CE) are of particular interest, due to the possible role of diminished rainfall in the cultural collapse. Data from Lake Chichancanab and Lake Punta Laguna have yielded evidence of multiple droughts between 750 and 900 CE, with extreme events in 882, 986 and 1051 CE (Hodell et al. 1995; Curtis et al. 1996). Records from stalagmites in the northern YP estimate 40 % precipitation reductions (Medina-Elizalde et al. 2010; Medina-Elizalde and Rohling 2012), while fossil pollen suggests 20 % reductions in precipitation (Carrillo-Bastos et al. 2013).

Orellana et al. (2009) developed climate change scenarios and projections for the year 2020, based on climatic records between 1961 and 1990. GCM (HADCM3, GFDL-R30, CGCM2 and ECHAM4) were used to develop climate scenarios, based on recommendations by the IPCC. GCMs include data on atmosphere, ocean, ice cap and surface process interactions, and estimate variation in these components to develop scenarios under different conditions. Temperature scenarios yield no consensus, but all models agree with an increase in the range of 0.75–1.25 °C, with extreme temperatures to the Gulf of Mexico. Precipitation scenarios also provided no consensus, but show an amplitude of precipitations of lower precipitation (northern YP), and forecast a reduction in mean annual precipitation between 0 and 200 mm/year.

In short, the YP has experienced numerous episodes of drought over the last 2000 years, resulting in human disasters and strong economic losses. As we have mentioned these drought events are well documented. Based on historical and paleoenvironmental registers, and scenarios developed by Orellana et al. (2009). It could be assumed that those drought events will be the most likely challenge to human well being and biodiversity in the YP. The increase records and data to related with climate changes and environmental changes, it is urgent to decrease the vulnerability of the YP.

The IPCC not only reports projections of climate changes, but also a diverse range of risks and their effects: Increases in temperature of 1 °C are likely to place unique cultural systems and ecosystems at a high risk of loss. Many species with limited adaptive capacities may be lost. The extreme climatic events will put high-risk communities and human populations at a greater disadvantage in high-risk areas. Widespread displacement and migration of human populations that is caused by heat waves, droughts, extreme precipitation and flooding is projected in countries currently undergoing development, which we are currently witnessing in Africa and the Americas. There exists a high risk in terms of the loss of biodiversity and impacts on the global economy. Changes in the climate system and in ecosystems could be abrupt and irreversible. The latest reports of the IPCC (2014a, b) have included the themes of impacts, adaptation, vulnerability and mitigation. The objective of these reports has been to establish scenarios of change that are brought about with human intervention (mitigation) and without it for societies and for the environment over both short-and medium-periods of time. These point to the importance of government actions through their institutions and laws as well as the society in general. The least complex and generalized conclusion is that if we do not act rapidly on all levels, the changes that our climatic system, environmental, and as a result, our society, will experience a major synergy that will become less predictable with each passing day. Numerous publications (Hodell et al. 1995, 2000, 2005; Leyden et al. 1996, 1998; Leyden 2002; Haug et al. 2003; Beach et al. 2009; Medina-Elizalde et al. 2010; Medina-Elizalde and Rohling 2012; Aimers and Iannone 2014; IPCC 2013; Folan et al. 2014; Torrescano-Valle and Islebe 2015, among others) have prompted the need to look into the past to review and understand important changes that drove the collapse of great cultures, such as the Romans and the Maya (Tainter 2014), and establishing correspondences with modern data to predict the future.

References

Aimers J, Iannone G. The dynamics of ancient Maya developmental history. In: Iannone G, editor. The Great Maya droughts in cultural context. Case studies in resilience and vulnerability. Boulder, CO: University Press of Colorado; 2014. p. 21–50.

Andrews WE. Balankanche. Throne of the Tigre Priest. MARI Publication No. 32. New Orleans, LA: Tulane University; 1970. p. 182.

Back W, Hanshaw B. Comparison of chemical hydrogeology of the carbonate peninsulas of Florida and Yucatán. J Hydrol. 1970;10:330–68.

Barrera A. La península de Yucatán como provincia Biótica. Rev Soc Mex Hist Nat. 1962;23:71–105.

Barrera-Bassols N, Toledo VM. Ethnoecology of the Yucatec Maya: symbolism, Knowledge and Management of Natural Resources. J Lat Am Geogr. 2005;4(1):9–41.

Barrera-Bassols N, Zinck JA. Ethnopedology: a worldwide view on the soil knowledge of local people. Geoderma. 2003;111:171–95.

Barrera-Bassols N, Zinck JA, Van Ranst E. Local soil classification and comparison of indigenous and technical soil maps in a Mesoamerican community using spatial analysis. Geoderma. 2006;135:140–62.

Bauer-Gottwein P, Gondwe BRN, Charvet G, Marín LE. Review: the Yucatán Peninsula karst aquifer, Mexico. Hydrogeol J. 2011;19:507–24.

Bautista F, Palacio-Aponte G, Ortíz-Perez M, Batllori-Sampedro E, Castillo-González M. El Origen y el manejo Maya de las geoformas, suelos y aguas en la Península de Yucatán. In: Bautista F, Palacio G, editors. Caracterización y Manejo de los Suelos de la Península de Yucatán: Implicaciones Agropecuarias, Forestales y Ambientales. México: UAC, UADY, INE; 2005a. p. 21–32.

Bautista F, Batllori-Sampedro E, Palacio-Ponte G, Ortiz-Pérez M, Castillo-Gonzalez M. Integración del conocimiento actual sobre los paisajes geomorfológicos de la Península de Yucatán. In: Bautista F, Palacio G, editors. Caracterización y Manejo de los Suelos de la Península de Yucatán: Implicaciones Agropecuarias, Forestales y Ambientales. México: UAC, UADY, INE; 2005b. p. 33–58.

Bautista-Zuñiga F, Jiménez-Osornio J, Navarro-Alberto J, Manu A, Lozano R. Microrelieve y color del suelo como propiedades de diagnóstico en leptosoles cársticos. Terra Latinoamericana. 2003;21(1):1–11.

Bautista-Zuñiga F, Estrada-Medina H, Jiménez-Osornio JJM, González-Iturbe JA. Relación entre el relieve y unidades de suelo en zonas cársticas de Yucatán. Terra Latinoamericana. 2004;22 (3):243–54.

Beach T, Luzzadder-Beach S, Dunning N, Cook D. Human and natural impacts on fluvial and karst depressions of the Maya Lowlands. Geomorphology. 2008;101:308–31. doi:10.1016/j.geomorph.2008.05.019.

Beach T, Luzzadder-Beach S, Dunning N, Jones J, Lohse J, Guderjan T, Bozarth S, Millspaugh S, Bhattacharya T. A review of human and natural changes in Maya Lowland wetlands over the Holocene. Quat Sci Rev. 2009;28:1710–24. doi:10.1016/j.quascirev.2009.02.004.

Boserup E. The conditions of agricultural growth: the economics of agrarian change under population pressure. Chicago, IL: Aldine Publishing; 1965. p. 124.

Brokaw N, Bonilla N, Knapp S, MacVean AL, Ortiz JJ, Peña-Chocarro M, De Pöll E, Tun-Garrido J. Arboles del Mundo Maya. Natural History Museum, ProNatura Península de Yucatán, UADY, Fundación ProPetén, Universidad del Valle de Guatemal; 2011. p. 263.

Buckler ES, Stevens NM. Maize origins, domestication, and selection. In: Motley TJ, Zerega N, Cross H, editors. Darwin's harvest: new approaches to the origins, evolution, and conservation of crops. New York, NY: Columbia University Press; 2006. p. 67–90.

Burke K. Tectonic evolution of the Caribbean. Annu Rev Earth Planet Sci. 1988;16:201–30.

Canuto M, Barrientos T. Cinco años de investigaciones en La Corona: una adivinanza envuelta en un misterio dentro de un enigma. In: Arroyo B, Méndez-Salinas L, editors. XXVI Simposio de

Investigaciones Arqueológicas en Guatemala. Guatemala: Museo Nacional de Arqueología; 2013. p. 993–8.

Carnevali G, Duno R, Tapia JL, Ramírez IM. Reassessment of *Zephyranthes* (Amaryllidaceae) in the Yucatán Peninsula including a new species, *Z. orellanae*. J Torrey Bot Soc. 2010;137:39–48.

Carrillo-Bastos A, Islebe GA, Torrescano-Valle N, González NE. Holocene vegetation and climate history of central Quintana Roo, Yucatán Península, Mexico. Rev Palaeobot Palyno. 2010;160:189–96.

Carrillo-Bastos A, Islebe GA, Torrescano-Valle N. 3800 Years of quantitative precipitation reconstruction from the Northwest Yucatán Peninsula. PLoS One. 2013;8(12), e84333. doi:10.1371/journal.pone.0084333.

Coe MD. The Maya. 5th ed. London: Thames and Hudson; 1994. p. 224.

Coney PJ. Plate tectonic constraints on the biogeography of Middle America and The Caribbean Region. Ann Mo Bot Gard. 1982;69(3):432–43.

Curtis JH, Hodell DA, Brenner M. Climate variability on the Yucatán Peninsula (Mexico) during the past 3500 years, and implications for Maya Cultural Evolution. Quat Res. 1996;46:37–47.

Day JW, Gunn JD, Folan WJ, Yañez-Arancibia A, Horton BP. The influence of enhanced post-glacial coastal margin productivity on the emergence of complex societies. JICA. 2012;7:23–52.

Delgadillo-Macías J, Aguilar-Ortega T, Rodríguez-Velázquez D. Los aspectos económicos y sociales de El Niño. In: Magaña V, editor. Los Impactos de El Niño en México. México: Dirección General de Protección Civil, Secretaría de Gobernación; 1999. p. 181–212.

Demarest AA, Wolf M, Martínez H, Valle J, Barrientos T, Quiñones D, Rodas E, Luin W, Ajin J. Cancún, Ciudad de Agua y Pantanos: Algunos descubrimientos de la Temporada y el patrón General del Sitio Cancuen. In: Arroyo B, Paiz L, Linares A, Arroyave A, editors. XXIV Simposio de Investigaciones arqueológicos en Guatemala; 2011. p. 381–92.

Domínguez-Carrasco MR, Folan WJ. Calakmul, México: Aguadas, bajos, precipitación y asentamiento en el Peten Campechano. Memorias del IX Simposio de Arqueología Guatemalteca. Ministerio de Cultura y Deportes, Instituto de Antropología e Historia, Proyecto Tikal y Asociación Tikal, Guatemala, CA; 1996. p. 171–93.

Domínguez-Carrasco MR, Folan WJ, Gates G, González-Heredia R, Gunn JD, Morales-López A, Robichaux H, Volta B. Oxpemul, su altiplanicie Kárstica Ondulada-Calakmul, El Preclásico: 30 años en el corazón del Peten Campechano. In: Arroyo B, Paiz L, Mejía H, editors. XXV Simposio de Investigación Arqueológico en Guatemala, Ministro de Cultura y Deportes Instituto de Antropología e Historia, Ciudad de Guatemala, Guatemala; 2012. p. 887–901.

Doolittle WE. An epilogue and bibliographic supplement to canal irrigation in prehistoric Mexico: the sequence of technological change. J Mesoam Archaeol Res Lab. 2006;4:3–15. The University of Texas, Austin.

Dunning N, Beach T. Stability and instability in prehispanic Maya landscapes. In: Lentz DL, editor. Imperfect balance landscape transformations in the Precolumbian Americas. New York, NY: Columbia University Press; 2000. p. 179–202.

Durán-García R, Méndez-González ME. Selva baja caducifolia con cactáceas candelabriforme. In: Durán R, Méndez M, editors. Biodiversidad y desarrollo humano en Yucatán. México: CICY, PPD-FMAM, CONABIO, SEDUMA; 2010. p. 141–2.

Ek-Diaz A. Vegetación. In: Pozo C, Armijo-Canto N, Calmé S, editors. Riqueza Biológica de Quintana Roo. Un análisis para su conservación. Tomo I. ECOSUR, CONABIO. México: Gobierno Estado de Quintana Roo, PPD; 2011. p. 62–73.

Espadas-Manrique C, Durán R, Argáez J. Phytogeographic analysis of taxa endemic to the Yucatán Peninsula using geographic information systems, the domain heuristic method and parsimony analysis of endemicity. Divers Distrib. 2003;9:313–30.

FAO Food and Agricultural Organization of the United Nations. Land resources. 2015. http://www.fao.org/nr/land/soils/soil/wrb-soil-maps/classification-key/en/#c25341. Accessed Jan 2015.

Fernández-Concha GC, Tapia-Muñoz JL, Duno de Stefano R, Ramírez-Morillo I. Flora Ilustrada de la Península de Yucatán. Mexico: CICY; 2010. p. 326.

Folan WJ. CA Comments: in the Late Postclassic eastern frontier of Mesoamerica. Cultural innovation along the periphery by John W. Fox. Curr Anthropol. 1981;22(4):336–7.

Folan WJ, Fletcher LA, Kintz ER. Fruit, fiber, bark and resin: social organization of a Maya urban center. Science. 1979;204(4394):697–701.

Folan WJ, Gunn J, Eaton JD, Patch RW. Paleoclimatologic patterning in Southern Mesoamerica. J Field Archaeol. 1983;10(4):453–68.

Folan WJ, Faust B, Lutz W, Gunn JD. Social and environmental factors in the Classic Maya collapse. In: Lutz W, Prieto L, Sanderson W, editors. Population, development, and environment on the Yucatán Peninsula: from ancient Maya to 2030. Luxemburg: IIASA; 2000. p. 2–33.

Folan WJ, González-Heredia R, Domínguez-Carrasco MR, Florey FL. Calakmul, Oxpemul y Coba: Un patrón hidráulico en el Mesoplano y plano kárstico de la Península de Yucatán, México. Los Investigadores de la Cultura Maya. Universidad Autónoma de Campeche, Campeche, México. 2014; 22(2):307–24.

Folan WJ, Domínguez-Carrasco MR, Gunn JD, Morales-López A, González-Heredia R, Villanueva-García G, Torrescano-Valle N. Calakmul: power, perseverance, and persistence. In: Cucina A, editor. Archaeology and bioarchaeology of population movement among the prehispanic Maya. México: Springer; 2015. p. 37–50.

García E. Distribución de la precipitación en la República Mexicana. Bol Inst Geogr. 2003;50:67–76.

Gates G. Fisiografía, geología e hidrología. In: Folan WJ, Sánchez González MC, García-Ortega JM, editors. Programa de manejo Reserva de la Biosfera Calakmul. Campeche: Universidad Autónoma de Campeche, SEDUE; 1992.

Gentry AH. Changes in plant community diversity and floristic composition on environmental and geographical gradients. Ann Mo Bot Gard. 1988;75:1–34.

Gómez-Pompa A. On Maya silviculture. Mex Estud. 1987;3(1):1–17.

Gondwe BRN, Lerer S, Stisen S, Marín L, Rebolledo-Vieyra M, Merediz-Alonso G, Bauer-Gottwein P. Hydrogeology of the South-Eastern Yucatán Peninsula: new insights from wáter level measurements, geochemistry, geophysics and remote sensing. J Hydrol. 2010;389:1–17.

González-Medrano F. Las Comunidades Vegetales de México. 2nd ed. México: SEMARNAT, INE; 2004. p. 82.

Gunn JD, Folan WJ. Datos Climáticos, Apéndice 2, vol. 4. In: Folan WJ, García Ortega JM, Sánchez-González MC, editors. Programa de Manejo Reserva de la Biosfera Calakmul. PMRBC, 4 volumes. México: CIHS, UAC, SEDESOL; 1992.

Gunn JD, Folan WJ. Three rivers: Subregional variations in earth system impacts in the Southwestern Maya Lowlands (Candelaria, Usumacinta, and Champotón Watersheds). In: McIntosh RJ, Tainter JA, Keech MS, editors. The way the wind blows: climate, history and human action. New York, NY: Columbia University; 2000. p. 223–71.

Gunn JD, Folan WJ, Robichaux HR. Un análisis informativo sobre la descarga del sistema del Rio Candelaria en Campeche, México: Reflexiones acerca de los Paleoclimas que afectaron los antiguos sistemas Mayas en los sitios de Calakmul y El Mirador. In: Folan WJ, editor. Campeche Maya Colonial, Colección Arqueología. México: Universidad Autónoma de Campeche; 1994.

Gunn JD, Folan WJ, Isendahl C, Dominguez CMR, Faust BB, Volta B. 8 Calakmul: agent risk and sustainability in the Western Maya Lowlands. Archaeol Pap Am Anthropol Assoc. 2014;24:101–23. doi:10.1111/apaa.12032.

Haug GH, Günther D, Peterson LC, Sigman DM, Hughen KA, Aeschlimann B. Climate and the collapse of Maya Civilization. Science. 2003;299:1731–5.

Hildebrand AR, Penfield GT, Kring DA, Pilkington M, Camargo ZA, Jacobsen SB, Boynton WV. Chicxulub crater: a possible cretaceous/tertiary boundary impact cráter on the Yucatán Peninsula, Mexico. Geology. 1991;19:867–71.

Hildebrand AR, Pilkington M, Connors M, Ortíz-Aleman C, Chavez RE. Size and structure of Chickulub crater revealed by horizontal gravity gradients and cenotes. Nature. 1995;376:415–7.
Hodell DA, Curtis JH, Brenner M. Possible role of climatic change in the collapse of the Maya civilization. Nature. 1995;375:391–4.
Hodell DA, Brenner M, Curtis JH. Climate change in the Northern American Tropics and Subtropics since the last Ice Age: implications for environment and culture. In: Lentz DL, editor. Imperfect balance. Landscape transformations in the Precolumbian Americas. New York, NY: Columbia University Press; 2000. p. 13–38.
Hodell DA, Brenner M, Curtis JH, Guilderson T. Frequency in the Maya lowlands. Science. 2001;292:1367–70.
Hodell DA, Brenner M, Curtis JH. Terminal classic drought in the northern Maya Lowlands inferred from multiple sediment cores in Lake Chichancanab (México). Quat Sci Rev. 2005;24:1413–27.
Ibarra-Manríquez G, Villaseñor JL, Durán R, Meave J. Biogeographical analysis of the tree flora of the Yucatán Peninsula. J Biogeogr. 2002;29:17–29.
INEGI. Mapa Digital de México. 2003. http://gaia.inegi.org.mx/mdm6/?v=bGF0OjIzLjMyMDA4LGxvbjotMTAyLjE0NTY1LHo6MSxsOmM0MTY=
INEGI. Mapa Digital Red Hidrográfica, 1:50000, 2.0 edición. 2010. http://www.inegi.org.mx/geo/contenidos/topografia/regiones_hidrograficas.aspx
INEGI. Mapa Digital de Climas, 1:1000000. 2011. http://www.inegi.org.mx/geo/contenidos/recnat/clima/infoescala.aspx
INEGI. Mapa Precipitación México. 2013. http://www.inegi.org.mx/sistemas/buzon/buzon.aspx?c=2652&s=inegi
IPCC. Climate change 2013: the physical science basis. In: Stocker TF, Qin D, Plattner GK, Tignor M, Allen SK, Boschung J, Nauels A, Xia Y, Bex V, Midgley PM, editors. Contribution of working group I to the fifth assessment report of the intergovernmental panel on climate change. Cambridge: Cambridge University Press; 2013. p. 1535.
IPCC. Climate change 2014: impacts, adaptation, and vulnerability. Part A: global and sectoral aspects. In: Field CB, Barros VR, Dokken DJ, Mach MD, Mastrandrea TE, Bilir M, Chatterjee KL, Ebi YO, Estrada KJ, Genova RC, Girma B, Kissel ES, Levy AN, MacCracken S, Mastrandrea PR, White LL, editors. Contribution of working group II to the fifth assessment report of the intergovernmental panel on climate change. Cambridge, UK: Cambridge University Press; 2014a. p. 1132.
IPCC. Climate change 2014: mitigation of climate change. In: Edenhofer O, Pichs-Madruga R, Sokona Y, Farahani E, Kadner S, Seyboth K, Adler A, Baum I, Brunner S, Eickemeier P, Kriemann B, Savolainen J, Schlömer S, von Stechow C, Zwickel T, Minx JC, editors. Contribution of working group III to the fifth assessment report of the intergovernmental panel on climate change. Cambridge, UK: Cambridge University Press; 2014b.
Islebe GA, Hooghiemstra H, Brenner M, Curtis JH, Hodell DA. A Holocene vegetation history from lowland Guatemala. The Holocene. 1996;6(3):265–71.
Isphording WC. The clay of Yucatán Mexico: a contrast in genesis. In: Singer A, Galan E, editors. Palygorskite-Sepiolite: Occurrences Genesis and Uses. Amsterdam: Elsevier; 1984. p. 59–73.
IUSS Grupo de Trabajo WRB. Base Referencial Mundial del Recurso Suelo. Primera actualización 2007, Informes sobre Recursos Mundiales de Suelos No. 103. Roma: FAO; 2007. p 117.
IUSS Working Group WRB. World reference base for soil resources 2014. International soil classification system for naming soils and creating legends for soil maps. World soil resources reports No. 106. Rome: FAO; 2014. p. 182.
Johnston KJ. The intensification of pre-industrial cereal agriculture in the tropics: Boserup, cultivation lengthening, and the Classic Maya. J Anthropol Archaeol. 2003;22:126–61. doi:10.1016/S0278-4165(03)00013-8.

Johnston KJ. La intensificación de la agricultura Maya Clásica. In: Laporte JP, Arroyo B, Mejía H, editors. XIX Simposio de Investigaciones Arqueológicas en Guatemala 2005. Guatemala: Museo Nacional de Arqueología y Etnología; 2006. p. 1090–100.

Josserand JK. Archaeological and linguistic correlations for Mayan prehistory. Actas del XLI Congreso Internacional de Américas, México. 1975;1:504–10.

Leyden BW. Pollen evidence for climatic variability and cultural disturbance in the Maya Lowlands. Anc Mesoam. 2002;13:85–101.

Leyden B, Brenner M, Whitmore T, Curtis JH, Piperno DR, Dahlin BH. A record of long-and short-term climatic variation from Northwest Yucatán: Cenote San José Chulchacá. In: Fedick SL, editor. The managed mosaic: ancient Maya agriculture and resource use. Salt Lake City: University of Utah Press; 1996. p. 30–50.

Leyden BW, Brenner B, Dahlin BH. Cultural and climatic history of Cobá, a lowland Maya city in Quintana Roo, Mexico. Quat Res. 1998;49:111–22.

Lohse JC. Archaic origins of the Lowland Maya. Lat Am Antiq. 2010;21(3):312–52.

Lucero LJ, Fedick SL, Dunning NP, Lentz DL, Scarborough VL. 3 Water and landscape: ancient Maya settlement decisions. Archeol Pap Am Anthropol Assoc. 2014;24:30–42. doi:10.1111/apaa.12027.

Lundell CL. Preliminary sketch of the phytogeography of the Yucatán Peninsula. Contrib Am Archaeol. 1934;12:257–321.

Marín L, Pacheco-Ávila JG, Méndez-Ramos R. Hidrogeología de la Península de Yucatán. In: Jiménez B, Marín L, editors. El agua en México vista desde la Academia. México: Academia Mexicana de Ciencias; 2005. p. 159–76.

Marshall J, Donohoe A, Ferreira D, McGee D. The ocean's role in setting the mean position of the Inter-Tropical Convergence Zone. Clim Dyn. 2014;42:1967–79.

Martínez E, Galindo-Leal C. La Vegetación de Calakmul, Campeche, México: Clasificación, Descripción y Distribución. Bol Soc Bot México. 2002;71:7–32.

Martínez E, Ramos-Alvarez CH. Un nuevo género de Anacardiaceae de la Península de Yucatán. Acta Botanica Hungarica. 2007;49(3-4):353–8.

Matheny RT. Maya Lowlands hydraulic systems. Science. 1976;193:639–46.

Matheny RT, Gurr DL, Forsyth DW, Hauck FR. Investigations at Edzná Campeche, Mexico. In: Thayer DD, editors. Papers of the New World Archaeological Foundation, Number 46, Vol. I, Part 1. Utah: Brigham Young University Provo; 1983. p. 239.

Matsouka Y, Vigouroux Y, Goodman MM, Sanchez GS, Buckler E, Doebley J. A single domestication for maize shown by multilocus microsatellite genotyping. PNAS. 2002;99(9):6080–4. doi:10.1073/pnas.052125199.

Medina-Elizalde M, Rohling EJ. Collapse of classic Maya civilization related to modest reduction in precipitation. Science. 2012;335:956–9.

Medina-Elizalde M, Burns SJ, Lea DW, Asmerom Y, von Gunten L, Polyak V, Vuille M, Karmalkar A. High resolution stalagmite climate record from the Yucatán Peninsula spanning the Maya terminal classic period. Earth Planet Sci Lett. 2010;298:255–62.

Méndez M, Magaña V. Regional aspects of prolonged meteorological droughts over Mexico and Central America. Am Meteorol Soc. 2010;23:1175–88. doi:10.1175/2009JCLI3080.1.

Mendoza B, Velasco V, Jáuregui E. A study of historical droughts in Southeastern Mexico. J Clim. 2006;19:2916–34.

Mendoza B, García-Acosta V, Velasco V, Jáuregui E, Díaz-Sandoval R. Frequency and duration of historical droughts from the 16th to the 19th centuries in the Mexican Maya lands, Yucatán Peninsula. Clim Chang. 2007;83:151–68. doi:10.1007/s10584-006-9232-1.

Miranda F. Estudios acerca de la vegetación. In: Beltran E, editor. Los recursos naturales del sureste y su aprovechamiento, segundo tomo. Mexico: Instituto Mexicano de Recursos Naturales Renovables AC; 1958. p. 215–71.

Miranda F, Hernández-Xolocotzi E. Tipos de vegetación de México y su clasificación. Bol Soc Bot Méx. 1963;28:29–179.

Morales-López A. Arqueología de salvamento en la Nueva Carretera a Calakmul, Municipio de Champotón, Campeche. Información. 1987;12:75–109.

Müeller AD, Islebe GA, Hillesheim M, Hillesheim MB, Grzesik DA, Anselmetti FS, Ariztegui D, Brenner M, Curtis JH, Hodell DA, Venz KA. Climate drying and associated forest decline in the lowlands of Northern Guatemala during the late Holocene. Quat Res. 2009;71:133–41.

Mylroie JE, Carew JL, Moore AI. Blue holes: definition and genesis. Carbonates Evaporites. 1995;10(2):225–33.

Orellana R, Espadas C, Conde C, Gay C. Atlas. Escenarios de Cambio Climático en la Península de Yucatán. México: CICY; 2009. p. 1–7.

Pearse AS. Parasites from Yucatán. Carnegie Inst Wash Publ. 1936;457:45–9.

Pennington TD, Sarukhan J. Arboles Tropicales de México. Manual para la Identificación de las Principales Especies. UNAM/FCE; 1998. p. 498.

Perry E, Velazquez-Oliman G, Marin L. The hydrogeochemistry of the karst aquifer system of the Northern Yucatán Peninsula, Mexico. Int Geol Rev. 2002;44:191–221.

Perry E, Velazquez-Oliman G, Socki RA. Hidrogeology of the Yucatán Peninsula. In: Gómez-Pompa A, Allen MF, Fedick SL, Jimenez-Osornio JJ, editors. The Lowland Maya: three millenia at human-wildland interface. Binghamton, NY: Haworth Press; 2003. p. 115–38.

Perry E, Paytan A, Pedersen B, Velazquez-Oliman G. Groundwater geochemistry of the Yucatán Peninsula, Mexico: constraints on stratigraphy and hydrogeology. J Hydrol. 2009;367:27–40.

Pohl MED, Pope KO, Jones JG, Jacob JS, Piperno DR, deFrance DS, Lentz DL, Gifford JA, Danforth ME, Josserand JK. Early agriculture in the Maya Lowlands. Lat Am Antiq. 1996;7 (4):355–72.

Pohl MED, Piperno DR, Pope KO, Jones JG. Microfossil evidence for pre-Columbian maize dispersals in the neotropics from San Andrés, Tabasco, México. PNAS. 2007;104(16):6870–5. doi:10.1073/pnas.0701425104.

Poveda G, Waylen PR, Pulwarty RS. Annual and inter-annual variability of the present climate in northern South America and Southern Mesoamerica. Palaeogeogr Palaeoclimatol Palaeoecol. 2006;234:3–27.

Ramírez-Morillo IM, Carnevali G, Cetzal-Ix W. *Hohenbergia mesoamericana* (Bromeliaceae), first record of the genus for Mesoamerica. Rev Mex Biodivers. 2010;81(1):21–6.

Rico-Gray V, García-Franco JG. The Maya and the vegetation of the Yucatán Peninsula. J Ethnobiol. 1991;11(1):135–42.

Rosenswing RM, Pearsall DM, Masson MA, Culleton BJ, Kennett DJ. Archaic period settlement and subsistence in the Maya Lowlands: new starch grain and lithic data from Freshwater Creek, Belize. J Archaeol Sci. 2014;41:308–21.

Sánchez SO, Islebe GA, Herrera EP. Flora representativa de Quintana Roo. In: Xacur J, editor. Enciclopedia de Quintana Roo, Vols. 1–10. 1998.

Sánchez-Sánchez O, Islebe GA. Tropical forest communities in Southeastern Mexico. Plant Ecol. 2002;158:183–200.

Sánchez-Santillán N, Signoret-Poillon M, Garduño-López R. La Oscilación del Atlántico Norte: un fenómeno que incide en la variabilidad climática de México. Ing Inv Tec VII(2) Abril-Junio. 2006.

Santley RS. Demographic archaeology in the Maya Lowlands. In: Culbert TP, Rice DS, editors. Precolumbian population history in the Maya Lowlands. Albuquerque: University of New Mexico Press; 1990. p. 325–43.

Scarborough VL, Valdez F. 9 The alternative economy: resilience in the face of complexity from the Eastern lowlands. Archeol Pap Am Anthropol Assoc. 2014;24:124–41. doi:10.1111/apaa.12033.

Sedov S, Solleiro-Rebolledo E, Fedick SL, Gama-Castro J, Palacios-Mayorga S, Vallejo-Gómez E. Soil genesis in relation to landscape evolution and ancient sustainable land use in the Northeastern Yucatán Peninsula, Mexico. Atti Soc Tosc Sci Nat Mem. Serie A. 2007;112:115–26.

Shang C, Tiessen H. Soil organic C sequestration and stabilization in karstic soils of Yucatán. Biogeochemistry. 2003;62:177–96.

Siemens AH. Wetland Agriculture in pre-hispanic Mesoamerica. Am Geogr Soc. 1983;73 (2):166–81. http://www.jstor.org/stable/214642.

Siemens AH. Subidas y Bajadas: Desafíos en la investigación del factor agua en la ecología humana del mundo Maya. Paper read at the XXI Encuentro Internacional: Los Investigadores de la Cultura Maya. Universidad Autónoma de Campeche, Campeche, México; 2011.

Sluyter A, Dominguez G. Early maize (Zea mays L.) cultivation in Mexico: dating sedimentary pollen records and its implications. PNAS. 2006;103(4):1147–51.

Smart PL, Beddows PA, Coke J, Doerr S, Smith S, Whitaker FF. Cave development on the Caribbean coast of the Yucatán Peninsula, Quintana Roo, Mexico. In: Harmon RS, Wicks C, editors. Perspectives on Karst geomorphology, hydrology, and geochemistry-A tribute volume to Derek C. Ford and William B. White: Geological Society of America Special Paper 404; 2006. p. 105–28.

Stahle DW, Burnette DJ, Villanueva-Diaz J. Pacific and Atlantic influences on Mesoamerican climate over the past millennium. Clim Dyn. 2012;39:1431–46. doi:10.1007/s00382-011-1205-z.

Steinich B, Marín LE. Determination of flow characteristics in the aquifer of the Northwestern Peninsula of Yucatán, Mexico. J Hydrol. 1997;191(1–4):315–31.

Tainter JA. 14 Collapse and sustainability: Rome, the Maya, and the modern world. Archeol Pap Am Anthropol Assoc. 2014;24:201–14. doi:10.1111/apaa.12038.

Torrescano-Valle N, Islebe GA. Tropical forest and mangrove history from southeastern Mexico: a 5000 yr pollen record and implications for sea level rise. Veg Hist Archaeobot. 2006;15:191–5.

Torrescano-Valle N, Islebe GA. Mangroves of Southeastern Mexico: palaecology and conservation. Open Geogr J. 2012;5:6–15.

Torrescano-Valle N, Islebe GA. Holocene paleoecology, climate history and human influence in the southwestern Yucatán Peninsula. Rev Palaeobot Palynol. 2015. doi:10.1016/j.revpalbo.2015.03.003.

Turner II BL. Once beneath the forest: prehistoric terracing in the Rio Bec Region of the Maya Lowlands. Dellplain Latin American studies, no. 13. Boulder, CO: Westview Press; 1983.

Villanueva-García G. Las conchas y caracoles de Calakmul. Manuscript on file. México: Instituto Nacional de Antropología e Historia; n.d.

Wahl D, Byrne R, Schreiner T, Hansen R. Palaeolimnological evidence of late-Holocene settlement and abandonment in the Mirador Basin, Peten, Guatemala. The Holocene. 2007;17 (6):813–20.

Wang C, Enfield DB. The tropical Western Hemisphere warm pool. Geophys Res Lett. 2001;28 (8):1635–8.

Whitmore TJ, Brenner M, Curtis J, Dahlin BH, Leyden BW. Holocene climatic and human influences on lakes of the Yucatán Peninsula, Mexico: an interdisciplinary paleolimnological approach. The Holocene. 1996;6:273–87.

Whittaker RJ, Field R. Tree species richness modelling: an approach of global applicability? Oikos. 2000;89(2):399–402.

Chapter 3
Distribution of Vegetation Types

Gerald Alexander Islebe, Odilón Sánchez-Sánchez, Mirna Valdéz-Hernández, and Holger Weissenberger

Abstract This chapter presents an overview of the vegetation types of the Yucatán Peninsula. Mean annual precipitation, terrain and soil characteristics explain the large-scale distribution of forest types in the Yucatán Peninsula. Tropical (high) forest is found in areas with >800 mm of mean annual precipitation, while dry (low) tropical forest is distributed in areas with <800 mm of mean annual precipitation. Tropical forest has canopy heights of more than 30 m and mostly presents three well-defined vegetative strata. The distribution of tropical forest is mainly in the central, eastern and southern parts of the Yucatán Peninsula. Dry tropical forest has canopy heights up to 20 m, and is widely distributed in the Yucatán Peninsula in different successional stages. Other woody vegetation types include mangroves, *petenes* and pine savannas. Open vegetation types include coastal dunes, marsh, and savanna vegetation. Disturbed tropical and dry tropical forest of all successional stages covers more than 7.4 million ha at present, and requires detailed management plans to maintain future ecosystems benefits.

Keywords Tropical forest • Distribution • Precipitation • Succession • Mangroves • Coastal dunes • Marsh vegetation

3.1 Introduction

Although the Yucatán Peninsula is characterized as an extensive limestone plateau with no significant altitudinal variation, a relatively large number of vegetation types can be identified. Vegetation types are mainly distributed along a north-to-south

G.A. Islebe (✉) • M. Valdéz-Hernández • H. Weissenberger
El Colegio de la Frontera Sur Unidad Chetumal, Avenida Centenario km 5.5, Apartado Postal 424, CP 77014 Chetumal, Quintana Roo, Mexico
e-mail: gislebe@ecosur.mx; mavaldez@ecosur.mx; hweissenberger@ecosur.mx

O. Sánchez-Sánchez
Universidad Veracruzana, Centro de Investigaciones Tropicales de la Universidad Veracruzana, Ex Hacienda Lucas Martín, Privada de Araucarias s/n, Col. Periodistas, 91019 Xalapa, Veracruz, Mexico
e-mail: sasodil@yahoo.com.mx

G.A. Islebe et al. (eds.), *Biodiversity and Conservation of the Yucatán Peninsula*, DOI 10.1007/978-3-319-06529-8_3

gradient of precipitation and according to different soil types, as well as precipitation variability in an east to west direction (Miranda 1958; Wright 1967). Tropical forest vegetation can be classified into two major groups with stand-specific associations, namely high and low forest. Other forest types, like mangrove forest, are found along coastal areas of the peninsula. Aquatic and subaquatic vegetation is present in open seasonally flooded areas locally known as *sabanas*, swamps, lakes and other water bodies.

The climate and hence the vegetation of the Yucatán Peninsula is determined by a series of factors: the lack of major orographic variation, the Tropic of Cancer, the influence of the Bermuda High, the presence of jet streams, tropical cyclones during the rainy season, cold fronts during the winter months, and the presence of a warm ocean current in the Yucatán Channel (Orellana et al. 2003). The driest region is located in the northwestern Yucatán Peninsula, in the Sisal-Progreso region, where mean annual precipitation is around 500 mm. The isohyet of 1000 mm runs east to west from northern Campeche, Yucatán to northern Quintana Roo. Toward the southern part of the Gulf of Mexico, precipitation increases rapidly up to 2000 mm; to the south of Quintana Roo, isohyets of around 1400 mm are found. Most of the precipitation falls between May and November, while the period between December and April is considered dry. Cold fronts between December and February are locally known as *nortes*, and provide occasional winter precipitation.

Tropical forest is found as the potential natural vegetation in areas with > 800 mm of annual precipitation, mainly in the central, eastern and southern Yucatán Peninsula, in the states of Campeche and Quintana Roo. High forest as a major vegetation type presents floristic and structural variations and covers around 70 % of the peninsula before significant human-induced landscape transformation, which began in the twentieth century (Rogan et al. 2011; Schmook et al. 2011).

Tropical dry forest is generally found in areas with generally less than 800 mm average precipitation, on a variety of soil types, including Gleysols to Vertisols, and as a potential natural vegetation type, covers large parts of Quintana Roo, Campeche and Yucatán. A unique vegetation type is low flooded forest, covering low-lying areas in Quintana Roo and Campeche. Species of this vegetation type withstand several months of flooding, as well as dry conditions. *Petenes*, also a unique vegetation type, are forest islands with fresh water inlets surrounded by marsh vegetation.

Recent human impact has caused profound changes in vegetation (Rico-Gray and García-Franco 1991) due to over-exploitation of timber species and changes in land use (Rueda 2010; Snook and Negreros 2004). Many plant species have specific ethnobotanical uses and values, ranging from medicinal to religious purposes (Barrera 1962; La Torre-Cuadros and Islebe 2003; Andersen et al. 2005). The Yucatán Peninsula is recognized to have an exceptional degree and detail of traditional ecological knowledge and utilization of local plants.

3.2 Forest Types and Distribution

The first systematic descriptions of the vegetation of the Yucatán Peninsula are provided by Miranda (1958), Miranda and Hernández-X (1964) and Barrera (1962). Earlier work by Lundell (1934, 1937) compiled botanical and ecological aspects of the vegetation of northern Guatemala and the Yucatán Peninsula. The work of Lundell was impressive, as he also included ethnobotanical and economic factors in his reports, mainly from observations and data collection in Campeche, Yucatán and the Petén province of northern Guatemala. Lundell's botanical collections can be seen in Chicago's Field Museum of Natural History (http://collections.mnh.si.edu/search/botany/). Lundell (1937) divided the vegetation in two categories, upland and lowland, which correspond to specific soil characteristics. The upland type included broad-leaved forest, also known as "high forest" or (seasonally) evergreen tropical forest, the soils of which featured a distinct organic layer. Lowland vegetation types were characterized by Lundell as forest with clayey soils originating in erosion from the uplands. This vegetation type is locally known as *akalché*.

In the classic work "Vegetación de México", Rzedowski (1978) described species composition and physiognomy of the major vegetation types of the Yucatán Peninsula. Other regional vegetation studies are from Sánchez-Sánchez and Islebe (2002), analyzing plant communities along precipitation gradients, and Barber and colleagues (1999), presenting phytosociological units. Ibarra-Manríquez and co-authors (1995) analyzed the phytogeographical relationships of the trees of the Yucatán Peninsula and concluded that the strongest ecological and botanical affinities exist with other Mesoamerican locations, while affinities with the Caribbean region are much weaker.

Distribution of vegetation types are presented in two maps (Figs. 3.1 and 3.2) based on inventories carried out in 2009, 2010 and 2011 (Inegi 2013). Vegetation types are separated on the two maps into forest (zonal) vegetation, and vegetation related to water bodies (azonal). Area covered and geographical distribution is given in Table 3.1. Disturbed dry tropical forest is the largest category with more than 3.9 million ha, followed by disturbed tropical forest with over 3.5 million ha. Following the data of Inegi, tropical and dry tropical forests together occupy less than 240,000 ha. However, it should be noted that disturbed tropical and dry tropical forests include all successional stages, from 5 to 30 or more years of age.

Vegetation related to water bodies with fluctuating water tables covers more than 2,474,244 ha in the Yucatán Peninsula. Most of these areas are covered by *Typha*- and Cyperaceae-dominated swamps and disturbed low flooded forest. Mangrove forests cover more than 400,000 ha along the coasts of the peninsula. However increased deforestation rates of nearly 1 % per year (Hirales-Cota et al. 2010; Sánchez-Sánchez et al. 2009), specifically along the Caribbean coast, are threatening this ecosystem.

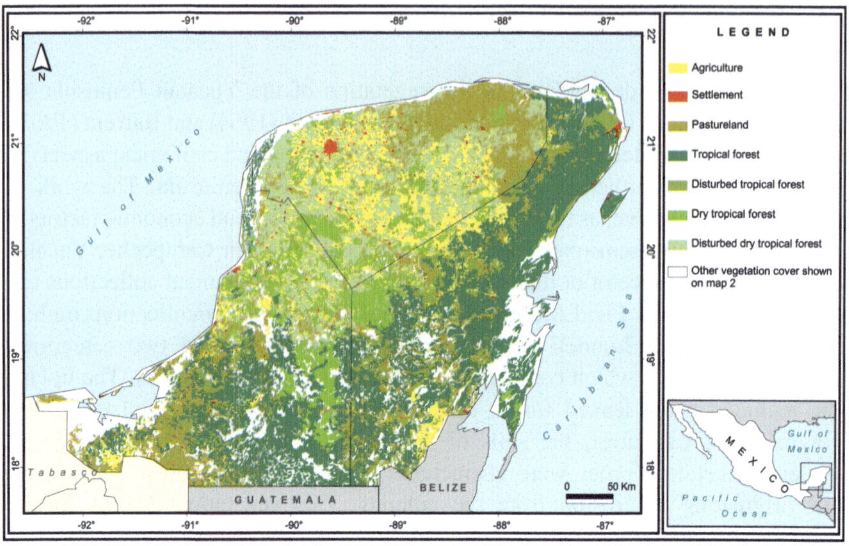

Fig. 3.1 Distribution of main vegetation types

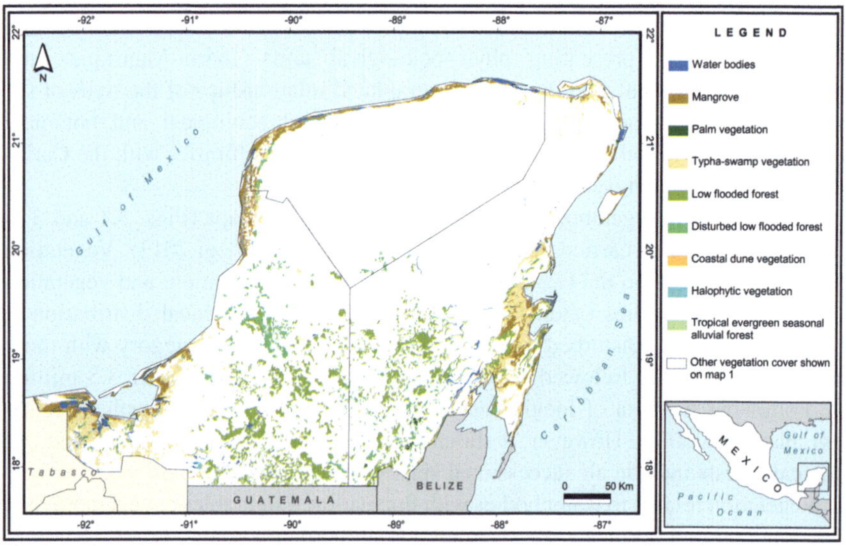

Fig. 3.2 Distribution of vegetation related to water bodies

3.2.1 Tropical Forest

Tropical forest or high forest (Selva alta and *selva mediana subperennifolia* in the Spanish literature) is widely distributed in the Yucatán Peninsula

Table 3.1 Main vegetation types, land use and areas covered 2011

Main vegetation type and land use	Cover (ha)[a]	Distribution
Agriculture	724,880	PY
Human settlement	149,358	PY
Pastureland	1,561,832	N, C
Tropical forest	1,356,095	E, C, S
Disturbed tropical forest	3,638,819	E, C, S
Dry tropical forest	92,221	W, C, N, S
Disturbed dry tropical forest	3,907,388	W, C, N, S
Water bodies	286,045	
Mangrove	400,124	E, N, W
Palm vegetation	6831	E
Typha-swamp vegetation	516,159	E, N, W
Low flooded forest	513,673	E, N, W
Disturbed low flooded forest	656,360	E, N, W
Coastal dune vegetation	8525	E, N, W
Halophytic vegetation	14,392	E, W
Tropical evergreen seasonal alluvial forest	57,176	E
Disturbed mangrove	15,673	E, N, W
Disturbed tropical evergreen seasonal alluvial forest	5339	E
Undefined vegetation types	500,000	PY

PY Yucatán Peninsula, *N* North, *C* Central Yucatán Peninsula, *E* East, *S* South, *W* West
[a]Cover of all data is based on Inegi (2013)

(Sánchez-Sánchez 2000). This vegetation type is characterized by 25–50 % of the trees composing the high forest losing their leaves during the dry season. Mean annual precipitation values are above 1200 mm. Mean annual temperature in these forests is above 20 °C, while mean minimum annual temperatures are 11 °C. Trees have canopy heights of more 30 m and three well-defined strata can generally be observed (Martínez and Galindo-Leal 2002).

The dominant tree is *Manilkara zapota* (Sapotaceae), with a wide distribution in southern Mexico, extending into northern Belize and Guatemala. Tropical forest is common in southern-central Campeche and most parts of Quintana Roo. Characteristic and commercially valuable canopy species include *Swietenia macrophylla* (Meliaceae), *Swartzia cubensis* (Leguminosae), *Pimenta dioica* (Myrtaceae), *Aspidosperma megalocarpon* (Apocynaceae), *Caesalpinia gaumeri* (Leguminosae), among many others (Snook and Negreros-Castillo 2004). An upper shrub layer consists mainly of the species *Ficus spp.* (Moraceae), *Bursera simaruba* (Burseraceae), *Sickingia salvadorensis* (Rubiaceae), *Chlorophora tinctoria* (Moraceae), *Enterolobium cyclocarpum* (Leguminosae), *Sapindus saponaria* (Sapindaceae) and *Acrocomia mexicana* (Arecaceae). High forest consists of different plant communities, in which *Manilkara zapota-Coccothrinax readii* is the dominant (Sánchez-Sánchez and Islebe 2002).

Another widely distributed plant community of the high forest is *Vitex gaumeri-Caesalpinia gaumeri*. This community has as characteristic species *Acacia gaumeri*

(Leguminosae), *Byrsonima bucidaefolia* (Malphigiaceae), *Alseis yucatanensis* (Rubiaceae), *Spondias mombin* (Anacardiaceae), *Diospyros verae-crucis* (Ebenaceae) and *Caesalpinia gaumeri* (Leguminosae). An additional characteristic feature of this plant community are the stone slabs, which can cover up to 45 % of the soil surface (Sánchez-Sánchez and Islebe 2002). In southern Quintana Roo, where the wettest conditions are found, the characteristic community is *Trichilia glabra-Brosimum alicastrum-Attalea cohune*. This community is a dense forest with the following characteristic subdominant species: *Chrysophyllum mexicanum* (Sapotaceae), *Piper sempervirens* (Piperaceae), *Chlorophora tinctorea* (Moraceae), and *Swartzia cubensis* (Leguminosea).

The *Hampea trilobata-Metopium brownei-Bursera simaruba* community is widely distributed in the central Yucatán Peninsula and includes many characteristic taxa of secondary forests (Sánchez-Sánchez and Islebe 2002) like *Nectandra coriacea* (Lauraceae), *Sabal yapa* (Arecaceae), *Bauhinia divaricata* (Leguminosea), *Lysiloma latisiliquum* (Leguminosae) and *Bauhinia jenningsii* (Leguminosae).

Soils of the high forest are generally shallow, but rich in organic material and of Lithosol-Rendzina or even Luvisol and Vertisol types (Flores 1977, FAO 1988; Wright 1967), depending on the region. In the southern Yucatán Peninsula, soils are mostly Vertisols, while in the central Yucatán Peninsula soils are of the Lithosol-Rendzina type.

Climbers are common in these forests, and characteristic species are *Paullina sp.* (Sapindaceae) and *Cardiospermum corindum* (Sapindaceae). At local terrain depressions in high forests, pure stands of *Attalea cohune* (Arecaceae) can be found (Quero 1992). These local vegetative communities consist of one distinctive upper layer 20 m in height, and an understory herbaceous layer, typically with 90 % cover. At similar conditions of deficient drainage, pure stands of *Sabal mauritiiformis* (Arecaceae) can be found in southern Quintana Roo. These palm communities have total heights of nearly 25 m. High forest has nearly disappeared in the Yucatán Peninsula due to excessive timber extraction, cattle farming and development of rice plantations in southern Quintana Roo over the last 50 years (Sánchez-Sánchez et al. 2007; Turner et al. 2001). Old growth forest plots have estimated values of 103.5 ± 4.4 Mg ha^{-1} AGB (above ground biomass) (Urquiza-Haas et al. 2007). However, higher values of 190 Mg ha^{-1} AGB are additionally reported from the same authors (Urquiza-Haas et al. 2007). Basal area estimates of tropical forest are highly variable: values between 11.9 and 45.0 m^2 ha^{-1} are reported (Urquiza-Haas et al. 2007; Dickinson et al. 2000; Ceccon et al. 2002; González-Iturbe et al. 2002; Lawrence and Foster 2002; La Torre-Cuadros and Islebe 2003; White and Hood 2004). Most likely, differences are due to the sampled vegetation type and numerical methods applied.

Selva mediana subcaducifolia (*semi-deciduous medium dry forest*) can be found in the states of Yucatán, Campeche and in some parts of northern Quintana Roo, although much of its original vegetation has been removed, mostly for logging, cattle and agriculture (Zamora-Crescencio et al. 2008). Canopy heights are usually from 20 up to 35 m. Fifty to 75 % of all trees lose their leaves during the dry season.

Soils are shallow with rocky outcrops with a very thin organic upper layer, mostly less than 7 cm (Sánchez-Sánchez and Islebe 2002). These soils are classified as Lithosols and Luvisols (Zamora-Crescencio et al. 2008). Characteristic species include *Brosimum alicastrum* (Moraceae), *Vitex gaumeri* (Verbenaceae), *Byrsonima* (Malphigiaceae), *Lysiloma latisiliquum*, among others (Dzib-Castillo et al. 2014; Zamora-Crescencio 2003). Epiphytes can be found in these forests.

3.2.2 Dry Forest

Low seasonally deciduous forest, or *selva baja caducifolia* in Spanish, has canopy heights of less than 15 m, and occurs in areas with annual precipitation varying between 600 and 800 mm or rarely more, located in the states of Yucatán, Campeche and Quintana Roo. Nearly 100 % of tree species drop their leaves during the dry season. Characteristic species are *Lysiloma bahamensis* (Leguminosae), *Baeucarnea pliabilis* (Nolinaceae), *Gymnopodium floribundum* (Polygonaceae), *Cassia alata* (Leguminosae), *Acacia milleriana* (Leguminosae), *Mimosa bahamensis* (Leguminosae), *Diospyros anisandra* (Ebenaceae), *Pseudophoenix sargentii* (Arecaceae) and *Piscidia piscipula* (Leguminosae) (Valdez-Hernández et al. 2014). Those species can also grow directly on eroded limestone. On a community level *Sebastiania adenophora-Plumeria obtusa var.cericifolia-Agave angustifolia* was identified as dominant in the eastern Yucatán Peninsula (Sánchez-Sánchez and Islebe 2002).

Seasonal flooded low forest (*Selva baja inundable*) occurs in areas that are geographically similar to those of high forest (Miranda and Hernández 1964), but is related to *sabana* areas. These forests are found in central and southern Quintana Roo, as well as northern Campeche and Yucatán. Soils with poor drainage are characteristic of these dry forests, and soils can withstand high water levels during long periods. The areas of low drainage are locally known as *akalchés*, which are periodically flooded during the rainy season. Several tree species can be found, the most conspicuous being *Byrsonima crassifolia* (Malphigiaceae), *Chrysobalanus icaco* (Sapotaceae), *Curatella americana* (Dilleniaceae), *Crescentia cujete* (Bignoniaceae) and *Hyperbaena winzerlingii* (Menispermaceae) (Cortés-Castelán and Islebe 2002; Díaz-Gallegos et al. 2002). Canopy heights rarely surpass 10 m. Some of those low forests can form pure stands, as in the case of *Haematoxylon campechianum* (Fabaceae) (common name *tasistal*), *Bucidas buceras* (Combretaceae) (common name *pukte*) and *Metopium brownei* (Anacardiaceae) (Miranda and Hernández 1964). There is no distinctive herbaceous layer present, mainly due to the prevalence of seasonal flooding. If undergrowth is present, species belonging to the Poaceae and Cyperaceae families can be found, like *Eleocharis*. Epiphytes of Orchidaceae and Piperaceae are quite common, such as *Encyclia alata, Peperomia* spp., and climbers belonging to *Dahlbergia* (Leguminosae). Deforestation rates for low forest have been estimated around 1.2 % per year in Calakmul for the last 30 years (Schmook et al. 2011).

In the northwestern part of the peninsula seasonally dry tropical forest (low deciduous forest) with columnar cacti can be found (Miranda and Hernández 1964). It is restricted to areas of 600–800 mm of annual precipitation, with no precipitation occurring during a 7- to 8-month period. The presence of rocky areas is conspicuous, and the soils are described as Lithosols (Thien et al. 1982). Characteristic endemic species of this vegetation type are: *Mammillaria gaumeri*, *Beaucarnea pliabilis*, *Guaiacum sanctum*, *Pilosocereus gaumeri*, *Nopalea gaumeri*, *Nopalea inaperta* and *Pterocereus gaumeri*.

3.2.3 Mangrove Vegetation

Mangroves can be found along the entire coast of the Yucatán Peninsula. Four mangrove species are distributed within different types of vegetative association: *Rhizophoramangle* (Rhizophoraceae), *Conocarpus erectus* (Combretaceae), *Avicennia germinans* (Acanthaceae), and *Laguncularia racemosa* (Combretaceae). *R. mangle* forests form nearly pure stands along the coast, while *A. germinans* and *L. racemosa* are found in mixed stands depending on the degree of soil salinity (Sánchez-Sánchez et al. 1991). The same authors distinguish four different mangrove communities. *C. erectus* is distributed on higher grounds with Gleysols, and can withstand variation in water table depths. Canopy heights of mangrove forests do not surpass 6–12 m, and consist of one predominant tree layer. The edges of mangrove areas are mostly dominated by *Acrostichum danaeifolium* (Pteridophyta), *Bravaisia tubiflora* (Acanthaceae), *Cladium jamaicensis* and other Cyperaceae (Islebe and Sánchez 2002; Torrescano-Valle and Islebe 2012). In southern Quintana Roo, dwarf *Rhizophora mangle* stands can be found with heights of only up to 1 m (Valdéz-Hernandez and Islebe 2011). Soils of mangrove communities are mostly sandy with clayey material, but can develop deep organic horizons.

3.2.4 Peten Vegetation

Peten vegetation is unique to the Yucatán Peninsula. It takes the form of closed woody vegetation islands consisting of a mosaic of mangroves and tropical forest species surrounded by salt marshes and mangroves (Durán 1987). Mangrove species can occasionally be found mixed with common tropical forest tree species. The most prominent and largest petenes are those of the biosphere reserve Los Petenes in Campeche, but these formations are also present on a smaller scale in the eastern part of Quintana Roo (Sian Ka'an Biosphere). Canopy heights may reach 12 m. Soils are rich in organic material, deep and slightly saline. Characteristic species are *Conocarpus erectus* (Combretaceae), *Metopium brownei* (Anacardiaceae), *Thrinax radiata* (Arecaceae), *Bucida buceras* (Combretaceae), *H. campechianum* (Leguminosae), among others.

3.3 Open Vegetation Types

Open vegetation types, including sabana, marsh and coastal dune vegetation are well defined and widely distributed in the Yucatán Peninsula. The coastal dune type is found along the shoreline of the Yucatán Peninsula (Espejel 1984, 1986, 1987; Moreno-Casasola and Espejel 1986; Flores and Espejel 1986; Torres et al 2010), and coastal plant communities are distributed along distinctive ecological gradients, such as salinity (Islebe, unpublished). Espejel (1987) identified 237 species for coastal dunes, and cosmopolitan families like Poaceae, Asteraceae and Leguminosae are the most common. Low non-woody plants (<30 cm high) like *Ambrosia hispida* (Asteraceae), *Canavalia rosea* (Leguminosae), *Distichlis spicata* (Poaceae), *Ipomoea pes-caprae* (Convolvulaceae), *Suriana maritima* (Surianaceae) and *Ernodea littoralis* (Rubiaceae), among other species, are found close to the sea. A transition occurs to higher elevation plant communities with woody taxa like *Coccoloba uviferae* (Polygonaceae), *Pouteria campechana* (Sapotaceae), *Cordia sebestena* (Boraginaceae) and *Chrysobalanus icaco* (Chrysobalanaceae). Woody taxa reach up to 6 m high, and sometimes a low shrub layer is present (Sánchez-Sánchez et al. 1991). Plant species from coastal dunes show typical morphological adaptations such as thicker leaves and glandules (like *S. maritima* and *E. littoralis*). Altitudinal differences between the highest and lowest parts of the dune along the coasts of Quintana Roo averages 5 m. Near-shore vegetation includes creeping species like *I. pres-caprae,* while dunes with woody taxa show the development of early shallow soil formation with thin organic layers. Coastal dunes are one of the most threatened vegetation types of the Yucatán Peninsula, with less than 8500 ha remaining, as coastal dunes are used to build hotels and tourism infrastructure (Lapointe 2011). Coastal dune vegetation is mostly delimited by the presence of *Thrinax radiata* palm vegetation, or by swamp like vegetation with abundant Cyperaceae and a transition to tropical forest types. Marsh vegetation is influenced by a high level of carbonate dissolved in water and soil, and is relatively species-poor relative to its biogeographical position. *Typha dominguensis, Cyperus jamaicensis, Scirpus* and *Carex* spp. are the main taxa found in marsh environments and are locally named *tulares* or *popales.* Marsh vegetation is mainly distributed along the eastern and western margins of the peninsula, though marshes cover large areas of the northern Yum Balam reserve. Their floristic composition is determined by nutrient availability, phosphate limitation, and the influence of salinity. *T. dominguensis* dominates in nutrient-rich conditions (Rejmankova et al. 1996).

In southern Quintana Roo and reaching further into Belize and parts of Guatemala, relic *Pinus caribaea* forest can be found (Macario-Mendoza et al. 1998). It is an open savanna-like vegetation type on sandy soils, with taxa from the Poaceae dominating the herbaceous layer. Following Chavelas (1981) it is the only pine species occurring at sea level in Mesoamerica, with an original distribution running from the Yucatán Peninsula through Belize to Nicaragua, along coastal areas.

3.4 Human and Natural Impact

Human impact on the vegetation of the region has occurred over the last four millennia (Rico-Gray and García-Franco 1991; Islebe et al. 1996). The earliest corn cultivation in the Yucatán Peninsula and its subsequent impact on vegetation is recorded from 2000 BC on (Aragón-Moreno et al. 2012; Carrillo-Bastos et al. 2010; Torrescano-Valle and Islebe 2015), although it could date back as much as 5000 years, as evidence from Northern Belize suggests. The use of agroforestry is part of ancient Mayan cultural knowledge and is well established among present-day farmers (Barrera 1962). The sustainable use of natural resources has been documented in many ethnobotanical studies (Barrera 1962; Andersen et al. 2005). Forest sustainability is incorporated into agroforestry practices like shifting cultivation, depending on the crops and areas cultivated, as well as cultivation intensity (Schmook et al. 2011; Valdez-Hernández et al. 2014). Fire is an important factor, and has been used since early Maya culture to clear land (Islebe et al. 1996). For the last 4000 years, and until the present day it is, intentionally or unintentionally, used to modify the landscape at different scales (López-Portillo et al. 1990). According to soil and paleoecological data from central Peten in northern Guatemala, full recovery of tropical forest vegetation from fire can take up to 80 years (Müller et al. 2010). However, using physiognomic criteria and plant species composition data, secondary vegetation cannot be distinguished from primary vegetation after 30 years (Miranda 1958; Sánchez-Sánchez and Islebe 2002). Traditional Mayan nomenclature makes no linguistic distinction between secondary vegetation of more than 30 years and primary forest (Barrera et al. 1976), but the first seral stages are well defined in traditional ecological knowledge (Sánchez-Sánchez et al. 2007). Conversely, Lawrence and Foster (2002) report that biomass from 25 year old secondary forests was only 40 % that of mature forests.

Hurricanes as natural hazards are a major determinant of forest structure in the Yucatán Peninsula (Tanner et al. 1991; Sánchez-Sánchez and Islebe 1999; Ramírez-Barajas et al. 2012; Rogan et al. 2011; Vandecar et al. 2011). The coastal and immediate interior vegetation of Quintana Roo is especially vulnerable to hurricanes as natural disturbances, since coral reefs and mangroves are severely affected and reduced in geographical extent. Hurricane damage to vegetation includes tree snapping (complete or partial, by branches), uprooting, defoliation, tree mortality, strong flooding and eventually, weeks after the event, intensive fires, due to the high quantities of accumulated woody and foliar debris on the ground (López-Portillo et al. 1990; Rodríguez-Trejo et al. 2011). During the last century, more than 100 hurricanes have hit the coast of Quintana Roo and the Yucatán Peninsula, damaging mangroves and other forest types (Islebe et al. 2009). The hurricanes causing the most severe damage were Gilbert in 1989 (Sánchez-Sánchez and Islebe 1999), Isidor in 2003, Emily in 2005 and Hurricane Dean (Islebe et al. 2009; Ramírez-Barajas et al. 2012), which damaged more than 2 million ha of forests in 2007 (CONAFOR, Ramírez-Barajas et al. 2012). For a high forest in northern Quintana Roo, Sánchez-Sánchez and Islebe (1999) estimated 4.5 t/ha of fallen biomass during Hurricane Gilbert. For more details, please see Chaps. 6 and 7.

3.5 Floristic Diversity

Floristic lists of the Yucatán Peninsula include the works of Standley (1930), Cabrera and Sousa (1983) for Quintana Roo, Sosa and colleagues (1985), Duran and co-authors (1998), Martínez and Galindo-Leal (2002), Gutierrez-Baez (2003) for Campeche, and Carnevali and co-authors (2010). The latter authors recognize some 3000 taxa. Some checklists on a local or regional scale are available, e.g., for the El Eden reserve in northern Quintana Roo, see Schultz (2005) and for Cozumel, Téllez and Cabrera (1987). Endemic species of the Yucatán Peninsula are listed in Durán and colleagues (1998), which follows Ibarra-Manríquez and co-authors (1995) in summing up 54 species, most of them with Caribbean phytogeographical affinities.

More than 500 tree species have been identified for the Yucatán Peninsula (based on the CIQRO-Herbarium data base, 2015). The most species-rich woody families are Leguminosae, Euphorbiaceae and Rubiaceae. Including all plant life forms, Fabaceae, Poaceae, Asteraceae and Orchidaceae (Carnevali et al. 2010) are the most represented families. The genera with most species are *Ipomoea, Croton, Euphorbia* and *Cyperus,* following Carnevali and colleagues (2010).

3.6 Outlook for Future Vegetation Studies

Given the threat to all vegetation types, several actions should be taken to preserve large connected forest tracts, specifically in the core conservation areas of Calakmul, central Quintana Roo, Campeche and the Sian Ka'an Biosphere reserve. Conservation efforts are being made to guarantee genetic connectivity and habitat range for biodiversity, such as the biological corridor of Sian Ka'an-Calakmul. The effectiveness of these efforts must be analyzed in future years, as the areas containing this corridor are also used for the construction of roads and rural infrastructure. Floristic and structural parameters should be monitored; forests should be locally recognized as ecosystem service providers to boost local income; and plant demographic studies should be undertaken to establish scientifically based rules for long term sustainable timber extraction (Valdez-Hernández et al. 2014). Scientific knowledge of natural succession should be included in management plans to avoid excessive unnecessary exploitation of major timber species with specific basal area parameters. The last 30 years of history has shown that the *ejido* (communally owned land) system has worked well (Porter-Bolland et al. 2013), though not perfectly, to guarantee the conservation of major forest tracts on a large regional scale in the central and southern Yucatán Peninsula. Although most of the remaining forests are secondary, at different successional stages, they cover more than 3.5 million ha, and still harbor a high and valuable degree of biodiversity (Ramírez-Barajas et al. 2012) and supply rural communities with income and use-value (Rico-Gray and García-Franco 1991). The following research should be encouraged in the future: understanding the succession of the different vegetation types; improvement and application of restoration ecology; accurate and precise assessment of the carbon stocks of distinct forest types; and the eco-physiology of characteristic tree species of Yucatán forests. In the case of mangrove forests, some species are currently protected by national laws. Continuous development of the tourism industry along coastal areas makes effective conservation almost impossible, however, as ecosystem functioning is altered with the construction of roads, drainage and other infrastructure.

Acknowledgments We are grateful to the Consejo Nacional de Ciencia y Tecnología (Conacyt) for funding our projects and to El Colegio de la Frontera Sur for access to all facilities provided over the years.

References

Anderson EN, Canul JC, Dzib A, Flores SG, Islebe GA, Medina Tzuc F, Sánchez-Sánchez O, Valdez P. Las plantas de los Mayas. Etnobotánica en Quintana Roo, México. Conabio and Ecosur; 2005.

Aragón-Moreno AA, Islebe GA, Torrescano-Valle N. A ~3800-yr, high-resolution record of vegetation and climate change on the north coast of the Yucatán Peninsula. Rev Palaeobot Palynol. 2012;178:35–42.

Barber A, Tun J, Crespo MB. A new approach of the bioclimatology and potential vegetation of the Yucatán Peninsula, Mexico. Phytocoenologia. 1999;31:1–31.

Barrera A. La Península de Yucatán como provincia biótica. Revista de la Sociedad Mexicana de Historia Natural. 1962;23:71–105.

Barrera AM, Barrera A, López F. Nomenclatura Etnobotánica Maya. Colección científica INAH, México; 1976.

Cabrera E, Sousa M. Listado florístico preliminar: 25–50. In: Sian Ka'an. Puerto Morelos, Q.Roo: CIQRO. SEDUE; 1983.

Carnevali Fernández-Concha G, Tapia Muñoz JL, Duno de Stefano R, Ramírez Morillo IM. Flora ilustrada de la península de Yucatán. CICY. 2010.

Carrillo-Bastos A, Islebe GA, Torrescano Valle N. Holocene vegetation and climate history of central Quintana Roo, Yucatán Península, Mexico. Rev Palaeobot Palynol. 2010;160:189–96.

Ceccon E, Olmsted I, Vazquez-Yanes C, Campo J. Vegetation and soil properties in two tropical dry forests of differing regeneration status in Yucatán. Agrociencia. 2002;36:621–31.

Chavelas PJ. El Pinus caribaea Morelet en el estado de Quintana Roo, México. INIF-SARH, México. Nota Técnica No. 10, 7; 1981.

Cortés-Castelán JC, Islebe GA. Influencia de factores ambientales en la distribución de especies arbóreas en las selvas del sureste de México. Rev Biol Trop. 2002;53:115–33.

Díaz-Gallegos J, Castillo-Acosta Y, García-Gil G. Distribución espacial y estructura arbórea de la selva baja subperennifolia en un Ejido de la Reserva de la Biosfera Calakmul, Campeche, México. Universidad y Ciencia. 2002;35:11–27.

Dickinson MB, Whigham DF, Hermann SM. Tree regeneration in felling and natural treefall disturbances in a semideciduous tropical forest in Mexico. Forest Ecol Manage. 2000;134: 137–51.

Durán R. Descripción y análisis de la estructura y composición de la vegetación de los petenes del Noroeste de Campeche, México. Biótica. 1987;12:181–98.

Durán R, Trejo-Torres JC, Ibarra-Manríquez G. Endemic phytotaxa of the Peninsula of Yucatán. Harv Pap Bot. 1998;3:263–314.

Dzib-Castillo B, Chanatásig-Vaca C, González-Valdivia NA. Structure and composition of two tree communities of tropical deciduous and subdeciduous forests in Campeche, Mexico. Revista Mexicana de Biodiversidad. 2014;85:167–78.

Espejel I. La vegetación de las dunas costeras de la Península de Yucatán. I. Norte de Yucatán. Biotica. 1984;9:183–210.

Espejel I. La vegetación de las dunas costeras de la Península de Yucatán, II. Sian Ká an. Quintana Koo, México. Biotica. 1986;11:7–24.

Espejel I. Phytogeographic relationships of the Yucatán Peninsula. J Biogeogr. 1987;14:499–519.

FAO. Soil map of the world with revised legend. World Soil Resources Report 60. Rome: FAO; 1988.

Flores JS, Espejel I. Etnoflora Yucatánense. Mapa de la Vegetación de la Península de Yucatán. 2. Xalapa, México: INIREB; 1986.

Flores DA. Los suelos de la Península de Yucatán y sus posibilidades agropecuarias. Reunión nacional de reconstrucción de la infraestructura agrícola. Mérida, México: Secretaría de Agricultura y Recursos Hidráulicos; 1977.

González-Iturbe JA, Olmsted I, Tun-Dzul FA. Tropical dry forest recovery after long term Henequen (sisal, *Agave fourcroydes* Lem.) plantation in northern Yucatán, Mexico. For Ecol Manag. 2002;167:67–82.

Gutierrez-Baez C. Listado florístico actualizado de Campeche. RN/C/09; 2003.

Hirales-Cota M, Espinoza-Avalos J, Schmook B, Ruiz-Luna A, Ramos-Reyes R. Agentes de deforestación de manglar en Mahahual-Xcalak, Quintana Roo, sureste de México. Ciencias marinas. 2010;36:147–59.

Ibarra-Manríquez G, Villaseñor JL, Durán R. Riqueza de especies y endemismo del componente arbóreo de la Península de Yucatán, México. Bol Sociedad Bot México. 1995;57:49–77.

Inegi. Uso de suelo y vegetación, Serie V., Escala 1:250,000. Mexico. 2013.

Islebe GA, Sánchez O. History of late Holocene vegetation at Quintana Roo, Mexico. Plant Ecol. 2002;160:187–92.

Islebe GA, Hooghiemstra H, Brenner M, Hodell DA, Curtis JA. A Holocene vegetation history from lowland Guatemala. The Holocene. 1996;6:265–71.

Islebe GA, Torrescano-Valle N, Valdez-Hernández M, Tuz-Novelo M, Weissenberger H. Efectos del impacto del huracán Dean en la vegetación del sureste de Quintana Roo. México: Foresta Veracruzana. 2009;11:1–6.

Lapointe P. Le tourisme de nature :un moyen de conserver l'écosystème côtier de la Costa Maya. Master thesis. University of Sherbrooke. 2011.

La Torre-Cuadros MD, Islebe GA. Traditional ecological knowledge and use of vegetation in southeastern Mexico: a case study from Solferino, Quintana Roo. Biodivers Conserv. 2003;12: 2455–76.

Lawrence D, Foster D. Changes in forest biomass, litter dynamics and soils following shifting cultivation in southern Mexico: an overview. Interciencia. 2002;27:400–8.

López-Portillo J, Keyes MR, González A, Cano EC, Sánchez O. Los incendios de Quintana Roo: catástrofe ecológica o evento periódico? Ciencia y Desarrollo. 1990;91:43–57.

Lundell CL. Preliminary sketch of the phytogeography of the Yucatán Peninsula. Contrib Am Archaeol. 1934;12:257–21.

Lundell CL. The vegetation of Petén, Publ. 478. Washington, D.C.: Carnegie Institution; 1937.

Macario-Mendoza PA, Torres-Pech SA, Cabrera Cano EG. Estructura y Composición de una Comunidad con Pinus caribaea var. hondurensis (Sénecl.) Barr. y Golf., en el Estado de Quintana Roo, México. Caribb J Sci. 1998;34:50–7.

Martínez E, Galindo-Leal C. La vegetación de Calakmul, México: clasificación, descripción y distribución. Bol Soc Bot Méx. 2002;71:7–32.

Miranda F. Estudios acerca de la vegetación. In: Los recursos naturales del sureste y su aprovechamiento. II parte. Imernat. 1958.

Miranda F, Hernández XE. Los tipos de vegetación de México y su clasificación. Bol Soc Bot Méx. 1964;28:29–179.

Moreno-Casasola P, Espejel I. Classification and ordination of coastal dune vegetation along the Gulf and Caribbean Sea of Mexico. Vegetatio. 1986;66:147–82.

Müller AD, Islebe GA, Anselmetti FS, Ariztegui D, Brenner M, Hodell DA, Hajdas I, Hamann Y, Haug GH, Kennett DJ. Recovery of the forest ecosystem in the tropical lowlands of northern Guatemala after disintegration of Classic Maya polities. Geology. 2010;38(6):523–6.

Orellana R, Islebe GA, Espadas C. Presente, pasado y futuro de los climas de la península de Yucatán. In: Colunga-García Marín P, Larqué-Saavedra A, editors. Naturaleza y sociedad en el área maya. Mexico: Amc&CICY; 2003.

Porter-Bolland L, García-Frapolli E, Sánchez-Gonzaléz MC. Local perceptions of conservation initiatives in the Calakmul region. In: Porter-Bolland L et al., editors. Community action for conservation: Mexican experiences. New York, NY: Springer; 2013.

Quero H. Las palmas silvestres de la península de Yucatán. Publ. No. 10. Mexico: UNAM, Instituto de Biología; 1992.

Ramírez-Barajas P, Islebe GA, Calme S. Impact of hurricane Dean (2007) on game species of the Selva Maya, Mexico. Biotropica. 2012;44:402–11.

Rejmankova E, Pope KO, Post R, Maltby E. Herbaceous Wetlands of the Yucatán Peninsula: Communities at Extreme Ends of Environmental Gradients. Int Revue ges Hydrobiol Hydrogr. 1996;81:223–52. doi:10.1002/iroh.19960810208.

Rico-Gray V, García-Franco G. The Maya and the vegetation of the Yucatán Peninsula. J Ethnobiol. 1991;11:135–42.

Rodríguez-Trejo DA, Tchikoue H, Cíntora-González C, Contreras-Aguado R, Rosa-Vázquez A. Modelaje del peligro de incendio forestal en las zonas afectadas por el huracán Dean. Agrociencia. 2011;45(5):593–608.

Rogan J, Schneider L, Christman Z, Millones M, Lawrence D, Schmook B. Hurricane disturbance mapping using MODIS EVI data in the southeastern Yucatán, Mexico. Remote Sens Lett. 2011;2:259–67.

Rueda X. Understanding deforestation in the southern Yucatán: insights from a sub–regional, multi–temporal analysis. Reg Environ Chang. 2010;10:175–89.

Rzedowski J. Vegetación de México. Ed. Limusa., Mexico City. 1978.

Sánchez-Sánchez O. Análisis estructural de la selva del jardín botánico. In: y Sánchez O, Islebe GA, editors. El Jardín Botánico Dr Alfredo Barrera Marín, fundamento y estudios particulares. El Colegio de la Frontera Sur and CONABIO. 2000.

Sánchez-Sánchez O, Cabrera-Cano E, Torres-Pech S, Herrera-Escudero P, Serralta-Peraza L, Salazar-Gómez C. Vegetación. In: Camarena-Luhrs T, Salazar-Vallejo S, editors. Estudios ecológicos preliminares de la zona sur de Quintana Roo. CIQRO. 1991.

Sánchez-Sánchez O, Islebe GA. Hurricane Gilbert and structural changes in a tropical forest of south-eastern Mexico. Glob Ecol Biogeogr. 1999;8:29–38.

Sánchez-Sánchez O, Islebe GA. Tropical forest communities of southeastern Mexico. Plant Ecol. 2002;158:183–200.

Sánchez-Sánchez O, Islebe GA, Valdéz-Hernández M. Flora arbórea y caracterización de gremios ecológicos en distintos estados sucesionales de la selva mediana de Quintana Roo. Foresta Veracruzana. 2007;9:17–26.

Sánchez-Sánchez O, Islebe GA, Valdez-Hernández M. Vegetación costera del Santuario del Manatí, In Espinoza-Avalos J, Islebe GA, Hernández-Arana H, editors. El sistema ecológico de la bahía de Chetumal/Corozal: costa occidental del Mar. ECOSUR. 2009.

Schmook B, Dickson RP, Sangermano F, Vadjunec JM, Eastman JR, Rogan J. A step-wise land-cover classification of the tropical forests of the Southern Yucatán, Mexico. Int J Remote Sens. 2011;32:1139–64.

Schultz GP. Vascular flora of the El Eden Ecological Reserve, Quintana Roo, Mexico. J Torrey Bot Soc. 2005;132:311–22.

Snook LK, Negreros-Castillo P. Regenerating mahogany (Swietenia macrophylla King) on clearings in Mexico's Maya forest: the effects of clearing method and cleaning on seedling survival and growth. Forest Ecol Manage. 2004;189:143–60.

Sosa V, Flores JS, Rico-Gray V, Lira R, Ortiz JJ. Lista florística y sinonimia maya. Etnoflora Yucatánense. 1985;1:1–225.
Standley PC. Flora of Yucatán. Chicago, IL: Field Museum of Natural History; 1930.
Tanner EVJ, Kapos V, Healey JR. Hurricane effects on forest ecosystems in the Caribbean. Biotropica. 1991;23:513–21.
Téllez O, Cabrera EF. Flórula de la Isla de Cozumel. Listados florísticos de México. 1987;6:1–34.
Thien L, Bradburn AS, Welden AL. The woody vegetation of Dzibilchaltun. A Mayan archaeological site in Northwest Yucatán, Mexico. Middle Am Res Inst Occasional Pap. 1982;5:1–24.
Torres W, Méndez M, Dorantes A, Durán R. Estructura, composición y diversidad del matorral de duna costera en el litoral yucateco. Bol Soc Bot Méx. 2010;86:37–51.
Torrescano-Valle N, Islebe GA. Mangroves of Southeastern Mexico: palaeoecology and conservation. Open Geogr J. 2012;5:6–15.
Torrescano-Valle N, Islebe GA. Holocene paleoecology, climate history and human influence in the southwestern Yucatán Peninsula. Rev Palaeobot Palynol. 2015;217:1–8.
Turner II BL, Villar SC, Foster D, Geoghegan J, Keys E, Klepeis P, Lawrence D, Mendoza PM, Manson S, Ogneva-Himmelberger Y, Plotkin AB, Perez Salicrup D, Chowdhury RR, Savitsky B, Schneider L, Schmook B, Vance C. Deforestation in the Southern Yucatán Peninsular Region: an integrative approach. Forest Ecol Manage. 2001;154:353–70.
Urquiza-Haas T, Dolman PM, Peres CA. Regional scale variation in forest structure and biomass in the Yucatán Peninsula, Mexico: effects of forest disturbance. Forest Ecol Manage. 2007;247:80–90.
Valdez-Hernández M, Islebe GA. Tipos de vegetación de Quintana Roo. In: Pozo C, editor. Riqueza biológica de Quintana Roo. Un análisis para su conservación, Tomo 2. Mexico: Ecosur and Conabio; 2011. p. 32–6.
Valdez-Hernández M, Sánchez-Sánchez O, Islebe GA, Snook L, Negreiros-Castillo P. Recovery and early succession after experimental disturbance in a seasonally dry tropical forest in Mexico. Forest Ecol Manage. 2014;334:331–43.
Vandecar KL, Lawrence D, Richards D, Schneider L, Rogan J, Schmook B, Wilbur H. High mortality for rare species following hurricane disturbance in the Southern Yucatán. Biotropica. 2011;43:676–84. doi:10.1111/j.1744-7429.2011.00756.x.
White DA, Hood CS. Vegetation patterns and environmental gradients in tropical dry forests of the Northern Yucatán Peninsula. J Veg Sci. 2004;15:151–60. doi:10.1111/j.1654-1103.2004.tb02250.x.
Wright ACS. El reconocimiento de los suelos en la Península de Yucatán, México. Informe final. FAO. 1967.
Zamora-Crescencio P. Contribución al estudio florístico y descripción de la vegetación del municipio de Tenabo, Campeche, México. Polibotánica. 2003;15:1–40.
Zamora-Crescencio P, Garcia-Gil G, Flores-Guido SF, Ortiz JJ. Estructura y composición florística de la selva mediana subcaducifolia en el sur del Estado de Yucatán. Polibotánica. 2008;26:39–66.

Part II
Plants and Environment

Chapter 4
Vegetative and Reproductive Plant Phenology

Mirna Valdez-Hernández

Abstract An analysis of phenological patterns of plant species of the Yucatán Peninsula is presented. Vegetative phenology is analyzed on community level, based on vegetation types of the region. Spatial differences of precipitation availability play an important role in foliar behavior at community level. For some species patterns of foliar phenophases are detailed. Phenophase of leaf-fall is determined by the length and intensity of the dry season in deciduous species. While in evergreen or leaf-exchanging species the leaf renewal occurs during the rainy season. 2170 species are presented with flowering season and fructification. Those data were obtained by reviewing herbarium data. Most of flowering and fructification happens during the dry season, but both have a large number of species presenting flowering and fructification during the whole year. Plant phenology has not been intensively studied in the Yucatán Peninsula, but it is necessary to foster its study to understand cycles and development of plants, which help to elucidate responses and adaptation to species to climate change, interaction with fauna, and sound management plans for non-timber forest resources.

Keywords Phenological patterns • Phenophases • Leaf-fall • Flowering • Fructification • Dry tropical forests • Vegetal physionomy • Evergreen species • Deciduous species • Water stress

4.1 Introduction

Plant species as primary producers depend directly on the physical environment (light, water, nutrients) to obtain their resources and achieve a proper development from germination to growth and reproduction. This development involves a series of cyclical steps like leaf, flower and fruit production. The study of these biological

M. Valdez-Hernández (✉)
Herbarium, El Colegio de la Frontera Sur Unidad Chetumal, Avenida Centenario km 5.5, 77014 Chetumal, Quintana Roo, Mexico
e-mail: mavaldez@ecosur.mx

G.A. Islebe et al. (eds.), *Biodiversity and Conservation of the Yucatán Peninsula*,
DOI 10.1007/978-3-319-06529-8_4

cycles is known as phenology. Phenological observations have been made since man uses plants.

In particular, in the Yucatán Peninsula, records of corn production since the last 4000 years area are available. Additional to corn, ancient Maya cultivated sow squash and depended on species like: *Brosimum alicastrum, Manilkara zapota, Byrsonima* sp, *Amaranthus* sp, *Pouteria* sp; to complement their food sources in times of shortage (Anderson et al. 2003; Goldstein and Hageman 2010; Anderson 2010). Additionally to fruits and seeds used for direct consumption, the flowering of *Gymnopodium floribundum* and *Tagetes erecta*, among other species, maintained honey production from native bees known as melipones (Anderson and Medina-Tzuc 2005).

We can assume the existence of empirical phenological observations of all used species for management and use. Due to those observations the ancient Maya understood the relation of plants with the environment in face of different climatic scenarios (Gunn et al. 1995). For example, in the case of corn, the Yucatec Maya selected corn varieties with different cycles, one with a short cycle with a duration of 7–10 weeks and a long cycle with a duration between 12 and 16 weeks (Tuxill et al. 2010). The duration of the crop cycle is important, because in the Yucatán Peninsula during the rainy season, an midsummer drought known as "canícula" affects cultivars (Holston 1986). Given that the planting of corn happens at the beginning of the rainy season, this short drought can coincide temporally with the development of spikes or during the growth of the corn grains in the cob, stages of higher demand of water (Tuxill et al. 2010). For this reason, the Maya farmers planted their milpas mixing different corn varieties in the same plot. The phenology of corn varieties with a short cycle are known as *nal t'eel*, *x-t'uup nal*, and *x-mejen nal*; and assure the development of cobs before the dry period. The varieties with the long cycle are called *X-nuuk nal, Ts'íit bakal, Bek'ech bakal*; and guarantee a higher production of grain in the case of good rainy season (Tuxill et al. 2010).

The pronounced high seasonality during in the rainy season is a characteristic of the climate of the Yucatán Peninsula. Therefore, vegetation is classified as seasonal dry tropical forests (SDTF). Those forests are characterized by an extreme seasonality in the rainy season, which causes a drought of 4–6 months, causing seasonal changes in the physionomy in the vegetation (Borchert 1994; Bullock and Solis-Magallanes 1990; Eamus and Prior 2001; Singh and Kushwaha 2006; Valdez-Hernández et al. 2010). The main phenophases like leaf-fall, leaf-flushing, flowering, and fruiting are closely linked to water availability (Borchert 1994; Bullock and Solis-Magallanes 1990; Eamus and Prior 2001; Singh and Kushwaha 2006; Valdez-Hernández et al. 2010). The phenophases of leaf-flushing, flowering, expansion and development of fruits are triggered by the start of the rainy season (Daubenmire 1972; Borchert 1994; Bullock and Solis-Magallanes 1990; Eamus and Prior 2001; Do et al. 2005), while in most of the species the leaf-fall is related to the start of the dry season (Borchert 1994; Sayer and Newbery 2003; Valdez-Hernández et al. 2010). The water status of the plants can be determined by intrinsic factors like foliar area, wood density and size of the roots, modified by the phenological response between individuals and species (Borchert 1994, 1998; Valdez-Hernández et al. 2010).

The phenology of the vegetation of the Yucatán Peninsula follow the trends of SDTF with some important environmental specifics, viz (1) a high intra and inter annual variability in precipitation (Márdero et al. 2012), (2) a high spatial variability in precipitation, with a gradient between 600 mm annual precipitation in the northwest to 1400 mm in the southeast of the peninsula (Orellana et al. 2009), (3) High infiltration of the precipitation given the karstic origin of the peninsula, generating an extensive network of underground rivers and sinkholes, known as "cenotes" (Bauer-Gottwein et al. 2011). The environmental characteristics previously described can influence and create spatio-temporal differences in inter and intra specific phenological cycles. However, the patterns of vegetative and reproductive phenology of the species distributed in the Yucatán Peninsula are poorly studied.

This chapter the information is organized into three sections, the first part covers the vegetative phenology, the second part reproductive phenology and finally a conclusion and perspective in the study of phenology of the Yucatán Peninsula.

4.2 Vegetative Phenology

One of the most conspicuous features in the physionomy of the vegetation of the Yucatán Peninsula is the seasonal variability in foliar cover (Fig. 4.1). This feature is based on the vegetation classification of Miranda and Hernández-X (1963), one of the most used vegetation classification system in the Yucatán Peninsula and Mexico.

This system uses the deciduous character of the vegetation, recognizing two groups: deciduous forests (more than 75 % of the species drop their leaves in the dry season) and evergreen forests (less than 25 % drop their leaves in the dry season). Besides, in each group there are subdivisions which consider intermedia categories (sub-deciduous and semi-evergreen).

This classification system allows us to understand on a large scale the foliar phenology of the vegetation. According to the INEGI vegetation map (2013, Chap. 3, Fig. 3.1), the state of Quintana Roo and the southern part of the state of Campeche, are mainly covered by evergreen and semi-evergreen forests. The state of Yucatán and the northern part of Campeche present deciduous or sub-deciduous forests. This distribution is closely related to the precipitation gradient recorded from the northeast to the southeast of the peninsula, in an interval between 600 and 1400 mm (Orellana et al. 2009). However, it is necessary to study of factors which determine foliar phenology in much more detail. In these forests, nearly 2400 species of vascular plants are distributed (Duno et al. 2011). Some of the most frequent species including *Bursera simaruba*, *Metopium brownei*, *Piscidia piscipula* and different taxa of the Fabaceae family, are indistinctively present in deciduous and evergreen forest, which indicates that the deciduous character of the

Fig. 4.1 Changes in the physionomy of a dry tropical forest, archaeological reserve Dzibilchaltún, Yucatán. (**a**) rainy season (June–October), (**b**) dry season (March–May)

vegetation is influenced by environmental conditions causing a differential response of foliar phenology in the taxa.

Foliar phenology has been poorly studied in the Yucatán Peninsula, there is only one study which documents changes in foliar cover, in a limited number of conspecific species. In this study, Valdez-Hernández et al. (2010) analyzed the effect of water availability in the phenology of five representative species of deciduous forests in the north of the Yucatán Peninsula. Considering that water access could be limited by the seasonality of the precipitation, the distance of the individuals to a *cenote* as water source, and physiological and morphological characteristics like the water potential and wood density. According to a 2 year follow up study on foliation three patterns were identified. Deciduous hard-wood species (*Acacia gaumeri*, *Apoplanesia paniculata*) with leaf-fall in the dry warm season (March–May) and leaf-flush at the beginning of the rainy season (May–June). Deciduous species with soft-wood (*Bursera simaruba)* with leaf-fall and leaf-flush in the dry-temperate season (November–February). Leaf-exchanging species (*Diospyros cuneata* and *Gymnopodium floribundum*) present less than 50 % of foliar cover in the dry season (December–May) reducing while water stress increases. Leaf-exchanging happens in the mid rainy season (June–July), when precipitation is constant. In all foliar patterns, a micro climate effect of water availability was observed, as individuals close to a *cenote* all species retained their leaves for a longer time periods. In leaf-exchanging species, their capacity to retain leaves during the dry season could indicate access to underground water.

In Belize, with a similar vegetation and floristic composition, foliar phenology was studied by Sayer and Newbery (2003). These authors studied the effect of diametric size in foliar phenology of ten arboreal species, considering two categories of diameter and soil depth. In deciduous hard-wood species (*Gymnopodium floribundum*, *Esenbeckia pentaphylla*, *Metopium brownei*), with increasing diameter the foliation occurs earlier. While in the deciduous species with soft-wood (*Bursera simaruba*, *Jatropha gaumeri*, *Gliricidia sepium*) the diameter size had no influence on foliation, which is related with its capacity to store water in the stem.

From both previously mentioned studies the attention is drawn to the fact that *G. floribundum* can behave as deciduous species or leaf-exchanging species, depending on micro environmental conditions of water availability. A key factor related with its capacity of response to environmental variability is synchrony of foliation. A low intraspecific synchrony can indicate a high plasticity and therefore a higher response capacity of changing environments (Borchert et al. 2002). In Yucatán, it was observed that species present a low synchrony in the phenophases of leaf-fall and leaf-flush (Valdez-Hernández et al. 2010). Leaves have a lifespan, which varies strongly among species and is defined by several factors. There is a direct relation of useful life of the leaves with energetic costs involved in their production (Santiago and Wrigth 2007). An analysis of physiological adaptations of some species is presented in Chap. 5.

4.3 Reproductive Phenology

Flowering and fructification periods of many plant species from the Yucatán Peninsula are well identified. In rural communities it is well known when to visit the forests to harvest Chico zapote (*Manilkara zapota*) or Nance (*Byrsonima crassifolia*). People know when bees visit flowers of Dzizilche (*Gymnopodium floribundum*) or Tajonal (*Tagetes erecta*). Despite this well-known empirical knowledge, most studies of reproductive phenology have focused on a reduced number of species (Parra-Tabla and Vargas 2004; Salinas-Peba and Parra-Tabla 2007; Valdez-Hernández et al. 2010). There are few works documenting floral phenology of a higher number of species (Porter-Bolland 2003). Studies of reproductive phenology of species of plants with no direct human use are rare.

In this section the reproductive phenology of 2170 species collected in the Yucatán Peninsula (Table 4.1) are presented. To establish patterns of flowering and fructification a revision of herbarium specimens was undertaken. The months of flowering and fructification were obtained from the recorded information in the data base of the CIQR herbarium, and by physical examination of the specimens to record reproductive structures. The revision aimed at identifying flowering patterns of 8225 herbarium specimens belonging to 2202 species. To analyze patterns of fructification 6235 specimens were considered, belonging to 1032 species.

Reproductive phenology at the community level presents a seasonal behavior (Fig. 4.2a). Flowering occurs in two main periods, one in the dry season (March–May) with nearly 500 species in each month and another more prominent period at the end of the rainy season (October–November) with 550 species each month.

Table 4.1 Flowering and fruiting obtained for 2170 species collected in the Yucatán Peninsula

Family	Specie	J	F	M	A	M	J	J	A	S	O	N	D
Acanthaceae	*Aphelandra deppeana* - Schltdl. & Cham.												
Acanthaceae	*Aphelandra scabra* - (Vahl) Sm.												
Acanthaceae	*Avicennia germinans* - (L.) L.												
Acanthaceae	*Bactris balanoidea* - (Oerst.) H. Wendl.												
Acanthaceae	*Bactris mexicana* - Mart.												
Acanthaceae	*Blechum brownei* - Juss.												
Acanthaceae	*Bravaisia berlandieriana* - (Nees) T. F. Daniel												
Acanthaceae	*Bravaisia integerrima* - (Spreng.) Standl.												
Acanthaceae	*Bravaisia tubiflora* - Hemsl.												
Acanthaceae	*Dicliptera assurgens* - (L.) Juss.												
Acanthaceae	*Dicliptera sciadephora* - Donn. Sm.												
Acanthaceae	*Elytraria bromoides* - Oerst.												
Acanthaceae	*Elytraria imbricata* - (Vahl) Pers.												
Acanthaceae	*Jacobinia spicigera* - (Schltdl.) L. H. Bailey												
Acanthaceae	*Jacquemontia agrestia* - (Mart. ex Choisy) Meisn.												
Acanthaceae	*Jacquemontia havanensis* - (Jacq.) Urb.												
Acanthaceae	*Jacquemontia nodiflora* - (Desr.) G. Don												
Acanthaceae	*Jacquemontia pentantha* - (Jacq.) G. Don												
Acanthaceae	*Jacquemontia polyantha* - (Schltdl. & Cham.) Hallier f.												
Acanthaceae	*Jacquemontia sphaerostigma* - (Cav.) Rusby												
Acanthaceae	*Jacquemontia tamnifolia* - (L.) Griseb.												
Acanthaceae	*Jacquemontia verticillata* - (L.) Urb.												
Acanthaceae	*Justicia breviflora* - (Nees) Rusby												
Acanthaceae	*Justicia campechiana* - Standl. ex Lundell												
Acanthaceae	*Justicia carthagenensis* - Jacq.												
Acanthaceae	*Justicia chiapensis* - Brandegee												
Acanthaceae	*Justicia ensiflora* - (Standl.) D. N. Gibson												
Acanthaceae	*Justicia furcata* - Jacq.												
Acanthaceae	*Justicia metallica* - Lindau												
Acanthaceae	*Justicia spicigera* - Schltdl.												
Acanthaceae	*Odontonema callistachyum* - (Schltdl. & Cham.) Kuntze												
Acanthaceae	*Pseuderanthemum alatum* - (Nees) Radlk.												
Acanthaceae	*Pseuderanthemum verapazense* - Donn. Sm.												
Acanthaceae	*Pseudobombax ellipticum* - (Kunth) Dugand												
Acanthaceae	*Pseudolmedia oxyphyllaria* - Donn. Sm.												
Acanthaceae	*Pseudophoenix sargentii* - H. Wendl. ex Sarg.												
Acanthaceae	*Ruellia inundata* - Kunth												
Acanthaceae	*Ruellia nudiflora* - (Engelm. & A. Gray) Urb.												
Acanthaceae	*Ruellia paniculata* - L.												
Acanthaceae	*Russelia acuminata* - Carlson												
Acanthaceae	*Russelia campechiana* - Standl.												
Acanthaceae	*Russelia equisetiformis* - Schltdl. & Cham.												
Acanthaceae	*Russelia sarmentosa* - Jacq.												
Acanthaceae	*Spathacanthus hahnianus* - Baill.												
Acanthaceae	*Tetramerium nervosum* - Nees												
Acanthaceae	*Thunbergia alata* - Bojer ex Sims												
Acanthaceae	*Thunbergia grandiflora* - Roxb.												
Achatocarpaceae	*Achatocarpus nigricans* - Triana												
Actinidiaceae	*Saurauia pringlei* - Rose												
Actinidiaceae	*Saurauia scabrida* - Hemsl.												
Actinidiaceae	*Saurauia serrata* - DC.												
Actinidiaceae	*Saurauia villosa* - DC.												
Adoxaceae	*Sambucus mexicana* - C. Presl ex DC.												
Adoxaceae	*Viburnum acutifolium* - Benth.												
Adoxaceae	*Viburnum hartwegii* - Benth.												
Aizoaceae	*Sesuvium portulacastrum* - (L.) L.												
Aizoaceae	*Trianthema portulacastrum* - L.												
Alismataceae	*Echinodorus andrieuxii* - (Hook. & Arn.) Small												
Alismataceae	Echinodorus berteroi - (Spreng.) Fassett												
Alismataceae	*Echinodorus nymphaeifolius* - (Griseb.) Buchenau												
Alismataceae	*Echinodorus ovalis* - C. Wright												
Alismataceae	*Sagittaria lancifolia* - L.												
Amaranthaceae	*Achyranthes aspera* - L.												
Amaranthaceae	*Alternanthera lanceolata* - (Benth.) Schinz												
Amaranthaceae	*Alternanthera mexicana* - Moq.												
Amaranthaceae	*Alternanthera ramosissima* - (Mart.) Chodat												
Amaranthaceae	*Amaranthus caudatus* - L.												
Amaranthaceae	*Amaranthus chihuahuensis* - S. Watson												
Amaranthaceae	*Amaranthus cruentus* - L.												
Amaranthaceae	*Amaranthus greggii* - S. Watson												
Amaranthaceae	*Amaranthus hybridus* - L.												
Amaranthaceae	*Amaranthus spinosus* - L.												

(continued)

Table 4.1 (continued)

Family	Specie	J	F	M	A	M	J	J	A	S	O	N	D
Amaranthaceae	*Atriplex pentandra* - (Jacq.) Standl.												
Amaranthaceae	*Blutaparon vermiculare* - (L.) Mears												
Amaranthaceae	*Celosia argentea* - L.												
Amaranthaceae	*Celosia nitida* - Vahl												
Amaranthaceae	*Celosia virgata* - Jacq.												
Amaranthaceae	*Chamissoa altissima* - (Jacq.) Kunth												
Amaranthaceae	*Chenopodium ambrosioides* - L.												
Amaranthaceae	*Gomphrena globosa* - L.												
Amaranthaceae	*Gomphrena serrata* - L.												
Amaranthaceae	*Iresine calea* - (Ibáñez) Standl.												
Amaranthaceae	*Iresine canescens* - Humb. & Bonpl. ex Willd.												
Amaranthaceae	*Iresine celosia* - L.												
Amaranthaceae	*Iresine flavescens* - Humb. & Bonpl. ex Willd.												
Amaranthaceae	*Iresine nigra* - Uline & W. L. Bray												
Amaranthaceae	*Iresine pringlei* - S. Watson												
Amaranthaceae	*Philoxerus vermicularis* - (L.) R. Br.												
Amaranthaceae	*Salicornia bigelovii* - Torr.												
Amaranthaceae	*Suaeda linearis* - (Elliott) Moq.												
Amaranthaceae	*Teloxys ambrosioides* - (L.) W. A. Weber												
Amaryllidaceae	*Allium schoenoprasum* - L.												
Amaryllidaceae	*Hymenocallis littoralis* - (Jacq.) Salisb.												
Amaryllidaceae	*Sprekelia formosissima* - (L.) Herb.												
Amaryllidaceae	*Zephyranthes tubispatha* - (L'Her.) Herb.												
Anacardiaceae	*Anacardium occidentale* - L.												
Anacardiaceae	*Astronium graveolens* - Jacq.												
Anacardiaceae	*Mangifera indica* - L.												
Anacardiaceae	*Metopium brownei* - (Jacq.) Urb												
Anacardiaceae	*Mosquitoxylum jamaicense* - Krug & Urb.												
Anacardiaceae	*Rhus schiedeana* - Schltdl.												
Anacardiaceae	*Spondias mombin* - L.												
Anacardiaceae	*Spondias purpurea* - L.												
Anacardiaceae	*Tapirira mexicana* - Marchand												
Annonaceae	*Annona cherimola* - Mill.												
Annonaceae	*Annona diversifolia* - Saff.												
Annonaceae	*Annona glabra* - L.												
Annonaceae	*Annona muricata* - L.												
Annonaceae	*Annona primigenia* - Standl. & Steyerm.												
Annonaceae	*Annona reticulata* - L.												
Annonaceae	*Annona squamosa* - L.												
Annonaceae	*Guatteria amplifolia* - Triana & Planch.												
Annonaceae	*Guatteria galeottiana* - Baill.												
Annonaceae	*Malmea depressa* - (Baill.) R. E. Fr.												
Annonaceae	*Oxandra lanceolata* - (Sw.) Baill.												
Annonaceae	*Sapranthus campechianus* - (Kunth) Standl.												
Annonaceae	*Sapranthus microcarpus* - (Donn. Sm.) R. E. Fr.												
Annonaceae	*Xylopia frutescens* - Aubl.												
Apiaceae	*Coriandrum sativum* - L.												
Apiaceae	*Donnellsmithia cordata* - (J. M. Coult. & Rose) Mathias & Constance												
Apiaceae	*Donnellsmithia guatemalensis* - J. M. Coult. & Rose												
Apiaceae	*Donnellsmithia juncea* - (Humb. & Bonpl. ex Spreng.) Mathias & Constance												
Apiaceae	*Eryngium bonplandianum* - Kunth												
Apiaceae	*Micropleura renifolia* - Lag.												
Apiaceae	*Spananthe paniculata* - Jacq.												
Apocynaceae	*Allamanda cathartica* - L.												
Apocynaceae	*Asclepias curassavica* - L.												
Apocynaceae	*Asclepias jaliscana* - B. L. Rob.												
Apocynaceae	*Asclepias oenotheroides* - Schltdl. & Cham.												
Apocynaceae	*Asclepias similis* - Hemsl.												
Apocynaceae	*Aspidosperma megalocarpon* - Müll. Arg.												
Apocynaceae	*Blepharodon mucronatum* - (Schltdl.) Decne.												
Apocynaceae	*Cameraria latifolia* - L.												
Apocynaceae	*Catharanthus roseus* - (L.) G. Don												
Apocynaceae	*Cryptostegia grandiflora* - R. Br												
Apocynaceae	*Cynanchum angustifolium* - Pers.												
Apocynaceae	*Cynanchum foetidum* - (Cav.) Kunth												
Apocynaceae	*Cynanchum schlechtendalii* - (Decne.) Standl. & Steyerm.												
Apocynaceae	*Cynanchum trichophyllum* - L. O. Williams												
Apocynaceae	*Echites tuxtlensis* - Standl.												
Apocynaceae	*Echites umbellatus* - Jacq.												
Apocynaceae	*Echites yucatanensis* - Millsp. ex Standl.												
Apocynaceae	*Gonolobus barbatus* - Kunth												
Apocynaceae	*Gonolobus chloranthus* - Schltdl.												

(continued)

Table 4.1 (continued)

Family	Specie	J	F	M	A	M	J	J	A	S	O	N	D
Apocynaceae	*Gonolobus ctenophorus* - (S. F. Blake) Woodson				■								
Apocynaceae	*Gonolobus stenanthus* - (Standl.) Woodson							■					
Apocynaceae	*Gonolobus xanthotrichus* - Brandegee							■					
Apocynaceae	*Lochnera rosea* - (L.) Rchb. ex Endl.			■	░		■	■		■		■	
Apocynaceae	*Mandevilla hirsuta* - (Rich.) K. Schum.					■				░			
Apocynaceae	*Mandevilla subsagittata* - (Ruiz & Pav.) Woodson	■	▒	■	■	■	■	▒	■	▒	▒	▒	▒
Apocynaceae	*Mandevilla torosa* - (Jacq.) Woodson			■	■								
Apocynaceae	*Mandevilla tubiflora* - (M. Martens & Galeotti) Woodson				■								
Apocynaceae	*Marsdenia coulteri* - Hemsl.					■							
Apocynaceae	*Marsdenia gualanensis* - Donn. Sm.	■				░		░	■				■
Apocynaceae	*Marsdenia macrophylla* - (Humb. & Bonpl. ex Schult.) E. Fourn.	■											
Apocynaceae	*Marsdenia maculata* - Hook.											■	
Apocynaceae	*Matelea belizensis* - (Lundell & Standl.) Woodson				■			■	▒	■	■	■	
Apocynaceae	*Matelea campechiana* - (Standl.) Woodson					░		■	■				
Apocynaceae	*Matelea cordata* - (Brandegee) Woodson					■							
Apocynaceae	*Matelea crassifolia* - Woodson	■		░		■							
Apocynaceae	*Matelea magnifolia* - (Pittier) Woodson											■	
Apocynaceae	*Matelea quirosii* - (Standl.) Woodson			■									
Apocynaceae	*Matelea velutina* - (Schltdl.) Woodson			■				■	■			■	■
Apocynaceae	*Matelea yucatanensis* - (Standl.) Woodson	■						░	■				
Apocynaceae	*Metastelma schlechtendalii* - Decne.	■	■	■	■	■	■		■			■	
Apocynaceae	*Nerium oleander* - L.				■	■	■						
Apocynaceae	*Oxypetalum cordifolium* - (Vent.) Schltr.	▒			░				■				░
Apocynaceae	*Pentalinon andrieuxii* - (Müll. Arg.) B. F. Hansen & Wunderlin	░	▒	■	▒	▒	■	░	░	■	▒	■	░
Apocynaceae	*Plumeria obtusa* - L.	░	■	■	■	■	■	■		■	■	░	■
Apocynaceae	*Plumeria rubra* - L.			■	■	■					■	░	
Apocynaceae	*Prestonia mexicana* - A. DC.			■									
Apocynaceae	*Rauvolfia heterophylla* - Willd. ex Roem. & Schult.							░	▒				
Apocynaceae	*Rauvolfia ligustrina* - Willd. ex Roem. & Schult.							░		■			
Apocynaceae	*Rauvolfia tetraphylla* - L.	▒	■	▒	▒	▒	▒	▒	░	░	░	▒	
Apocynaceae	*Rhabdadenia biflora* - (Jacq.) Müll. Arg.			■		░	■	■	■		■		
Apocynaceae	*Sarcostemma bilobum* - Hook. & Arn.	░		░				■	■				
Apocynaceae	*Sarcostemma clausum* - (Jacq.) Schult.	■				■	■	■	■		░	▒	▒
Apocynaceae	*Sarcostemma odoratum* - (Hemsl.) R. W. Holm								■				
Apocynaceae	*Stapelia gigantea* - N. E. Br.												■
Apocynaceae	*Stemmadenia donnell-smithii* - (Rose) Woodson					░		■	■				■
Apocynaceae	*Stemmadenia obovata* - K. Schum.							■					
Apocynaceae	*Tabernaemontana alba* - Mill.					▒	■	▒	■	■	■		
Apocynaceae	*Tabernaemontana amygdalifolia* - Jacq.	■	░	░		░		■	■	■	■	■	
Apocynaceae	*Tabernaemontana arborea* - Rose							■					
Apocynaceae	*Tabernaemontana chrysocarpa* - S. F. Blake			■		■	░	░			■	░	
Apocynaceae	*Tabernaemontana divaricata* - (L.) R. Br. ex Roem. & Schult.			■				■	■				
Apocynaceae	*Thevetia ahouai* - (L.) A. DC.		■		▒	▒	░	░	▒	■		■	
Apocynaceae	*Thevetia gaumeri* - Hemsl.	▒	■	▒	▒	▒	▒	▒	░	▒	▒	▒	■
Apocynaceae	*Thevetia peruviana* - (Pers.) K. Schum.		■	■	▒	░	■	░	░	■			■
Apocynaceae	*Tonduzia longifolia* - (A. DC.) Markgr.		■			■							
Apocynaceae	*Urechites andrieuxii* - Müll. Arg.		■	░		■	■	■	░	░	░	■	
Apocynaceae	*Vallesia antillana* - Woodson						■	░					
Aquifoliaceae	*Ilex mexicana* - (Turcz.) Black ex Hemsl.												■
Aquifoliaceae	*Ilex pringlei* - Standl.									■			
Araceae	*Monstera deliciosa* - Liebm.										■		
Araceae	*Pistia stratiotes* - L.							■			■		■
Araceae	*Syngonium angustatum* - Schott			░									
Araceae	*Syngonium podophyllum* - Schott			░									
Araliaceae	*Dendropanax arboreus* - (L.) Decne. & Planch.						■		░	▒	▒	▒	
Araliaceae	*Hydrocotyle bonariensis* - Lam.					■					■		
Araliaceae	*Oreopanax capitatus* - (Jacq.) Decne. & Planch.				■								
Araliaceae	*Schefflera actinophylla* - (Endl.) Harms					■				■			
Arecaceae	*Acoelorraphe wrightii* - (Griseb. & H. Wendl.) H. Wendl. ex Becc.				■	■	▒			■		░	
Arecaceae	*Acrocomia mexicana* - Karw ex Mart.	■					░				░	░	
Arecaceae	*Anthurium schlechtendalii* - Kunth					■	░	▒					
Arecaceae	*Chamaedorea neurochlamys* - Burret		░					■				■	
Arecaceae	*Chamaedorea seifrizii* - Burret		■	░			■	■				■	■
Arecaceae	*Coccothrinax readii* - H. J. Quero	░		░		■		■	■		■	░	░
Arecaceae	*Cocos nucifera* - L.						░		■				
Arecaceae	*Cryosophila argentea* - Bartlett											░	
Arecaceae	*Desmoncus quasillarius* - Bartlett			■	░	▒		■	░				
Arecaceae	*Sabal gretherae* - H. J. Quero								■				
Arecaceae	*Sabal mexicana* - Mart.			▒	▒								
Arecaceae	*Sabal yapa* - C. Wright ex Becc.			░			▒						
Arecaceae	*Thrinax radiata* - Lodd. ex Schult. & Schult. f.	░			░	■			░		■	░	
Arecaceae	*Veitchia merrillii* - (Becc.) H. E. Moore		■	■									

(continued)

Table 4.1 (continued)

Family	Specie	J	F	M	A	M	J	J	A	S	O	N	D
Aristolochiaceae	*Aristolochia grandiflora* - Sw.			■									
Aristolochiaceae	*Aristolochia maxima* - Jacq.			■			■	■					
Aristolochiaceae	*Aristolochia pentandra* - Jacq.		■									■	
Aristolochiaceae	*Aristolochia trilobata* - L.					░							
Asparagaceae	*Agave angustifolia* - Haw.	░		■		░							■
Asparagaceae	*Asparagus densiflorus* - (Kunth) Jessop										■		
Asparagaceae	*Asparagus plumosus* - Baker				░							■	
Asparagaceae	*Asparagus sprengeri* - Regel							░	■				
Asparagaceae	*Dracaena americana* - Donn. Sm.									░	▒		
Asparagaceae	*Echeandia macrocarpa* - Greenm.										■		
Asparagaceae	*Furcraea cahum* - Trel.			░									
Asparagaceae	*Furcraea cubensis* - Baker						■						
Asparagaceae	*Furcraea quicheensis* - Trel.			■									
Asparagaceae	*Milla biflora* - Cav.						■						
Asparagaceae	*Sansevieria trifasciata* - Prain						■						
Asparagaceae	*Sansevieria zeylanica* - (L.) Willd.	░						■					
Asparagaceae	*Yucca lacandonica* - Gómez-Pompa & J. Valdés			■									
Asparagaceae	*Yucca schidigera* - Roezl ex Ortgies									■			
Asparagaceae	*Maianthemum paniculatum* - (M. Martens & Galeotti) LaFrankie							■					
Asteraceae	*Acmella filipes* - (Greenm.) R. K. Jansen							■					
Asteraceae	*Acmella pilosa* - R. K. Jansen		■		■	■							
Asteraceae	*Aldama dentata* - La Llave	■										■	
Asteraceae	*Artemisia ludoviciana* - (Willd. ex Spreng.) D. D. Keck	■			■							■	
Asteraceae	*Aster bullatus* - Klatt												■
Asteraceae	*Aster novi-belgii* - L.												■
Asteraceae	*Aster subulatus* - Michx.	■										■	
Asteraceae	*Baccharis dioica* - Vahl	■											
Asteraceae	*Baccharis heterophylla* - Kunth								■				■
Asteraceae	*Baccharis trinervis* - Pers.	■											
asteraceae	*Chromolaena lundellii* - R. M. King & H. Rob.			■				■					
Asteraceae	*Chrysanthemum parthenium* - (L.) Bernh.			■									
Asteraceae	*Cirsium horridulum* - Michx.	■			■								
Asteraceae	*Cirsium subcoriaceum* - (Less.) Sch. Bip.					■							
Asteraceae	*Clibadium arboreum* - Donn. Sm.					■						■	■
Asteraceae	*Conyza bonariensis* - (L.) Cronquist										■	■	■
Asteraceae	*Conyza canadensis* - (L.) Cronquist			■	■				░			■	■
Asteraceae	*Conyza schiedeana* - (Less.) Cronquist								■				
Asteraceae	*Coreopsis mutica* - DC.										■		
Asteraceae	*Cosmos caudatus* - Kunth											■	
Asteraceae	*Cosmos sulphureus* - Cav.										■	▒	
Asteraceae	*Dahlia coccinea* - Cav.	■											
Asteraceae	*Dahlia imperialis* - Roezl ex Ortgies		■								■		
Asteraceae	*Dahlia pteropoda* - Sherff								■				
Asteraceae	*Delilia biflora* - (L.) Kuntze	░		░		■				░	■	▒	░
Asteraceae	*Eclipta prostrata* - (L.) L.						■	■			■		
Asteraceae	*Egletes liebmannii* - Shinners					■		■					■
Asteraceae	*Elephantopus mollis* - Kunth				■								
Asteraceae	*Emilia fosbergii* - Nicolson		■				■		■				
Asteraceae	*Erechtites hieraciifolius* - (Fisch. ex Spreng.) Griseb.		■		■	■				■	■		
Asteraceae	*Erigeron karvinskianus* - DC.								■				■
Asteraceae	*Eupatorium albicaule* - Sch. Bip. ex Klatt	■	■	■	■	░		■	■	■	■	■	■
Asteraceae	*Eupatorium araliifolium* - Less.	■											
Asteraceae	*Eupatorium breedlovei* - (R. M. King & H. Rob.) B. L. Turner											■	■
Asteraceae	*Eupatorium campechense* - B. L. Rob.		■	■						■	■	■	░
Asteraceae	*Eupatorium daleoides* - (DC.) Hemsl.	■	■				■		■	■			
Asteraceae	*Eupatorium espinosarum* - A. Gray	■								■			
Asteraceae	*Eupatorium galeottii* - B. L. Rob.												■
Asteraceae	*Eupatorium ligustrinum* - DC.				■								
Asteraceae	*Eupatorium morifolium* - Mill.	■											
Asteraceae	*Eupatorium nubigenum* - Benth.												■
Asteraceae	*Eupatorium odoratum* - L.	▒	■	■	■	■	■	■	■	■	■	▒	■
Asteraceae	*Eupatorium pinabetense* - B. L. Rob.								■				
Asteraceae	*Eupatorium pycnocephalum* - Less.			■			■	■	■	■	■		■
Asteraceae	*Eupatorium schultzii* - Schnittsp.												■
Asteraceae	*Eupatorium sordidum* - Less.										■		
Asteraceae	*Flaveria linearis* - Lag.		■	■	▒	■			■	▒	■	■	
Asteraceae	*Flaveria trinervia* - (Spreng.) C. Mohr										■	■	
Asteraceae	*Galinsoga quadriradiata* - Ruiz & Pav.					■							
Asteraceae	*Gnaphalium attenuatum* - DC.				■					■			
Asteraceae	*Gnaphalium lavandulifolium* - Willd.											■	
Asteraceae	*Gochnatia hypoleuca* - (S. F. Blake) Cabrera											■	
Asteraceae	*Goldmanella sarmentosa* - (Greenm.) Greenm.		■								■		

(continued)

Table 4.1 (continued)

Family	Specie	J	F	M	A	M	J	J	A	S	O	N	D
Asteraceae	*Gymnolaena oaxacana* - (Greenm.) Rydb.										■		
Asteraceae	*Harleya oxylepis* - (Benth.) S. F. Blake									■			
Asteraceae	*Hymenostephium microcephalum* - (Less.) S. F. Blake				■								
Asteraceae	*Hymenostephium pilosulum* - S. F. Blake				■								
Asteraceae	*Isocarpha oppositifolia* - (L.) Cass.	■										■	
Asteraceae	*Lactuca intybacea* - Jacq.			░			■		■	■	░	░	
Asteraceae	*Lagascea mollis* - Cav.							■			■		
Asteraceae	*Lasianthaea fruticosa* - (L.) K. M. Becker		■	■		■		■		■		■	■
Asteraceae	*Lasianthaea guatemalensis* - (Donn. Sm.) B. L. Turner			■		■		■					■
Asteraceae	*Liabum discolor* - (Hook. & Arn.) Benth. & Hook. f. ex Hemsl.			■		■							
Asteraceae	*Melampodium costaricense* - Stuessy				■								
Asteraceae	*Melampodium divaricatum* - (Rich.) DC.	■			■						■	■	
Asteraceae	*Melampodium gracile* - Less.							■				■	
Asteraceae	*Melampodium perfoliatum* - (Cav.) Kunth											■	
Asteraceae	*Melanthera aspera* - (Jacq.) Steud. ex Small	■		■				■	■	■			
Asteraceae	*Melanthera nivea* - (L.) Small	■				■			■				▓
Asteraceae	*Mikania cordifolia* - (L. f.) Willd.	■	■	■	■		■					■	
Asteraceae	*Mikania houstoniana* - (L.) B. L. Rob.	■	■										
Asteraceae	*Mikania micrantha* - Kunth	▓			■								■
Asteraceae	*Milleria quinqueflora* - L.					■	■	■		░	░	■	
Asteraceae	*Montanoa atriplicifolia* - (Pers.) Sch. Bip.	▓	▓		■		■		■				■
Asteraceae	*Montanoa grandiflora* - Alamán ex DC.												■
Asteraceae	*Montanoa mollissima* - Brongn. ex Groenl.					■							
Asteraceae	*Montanoa tomentosa* - (Sch. Bip. ex K. Koch) V. A. Funk		■										
Asteraceae	*Neurolaena lobata* - (L.) Cass.	■		■		■	■	■			■	■	■
Asteraceae	*Otopappus curviflorus* - (R. Br.) Hemsl.			■									
Asteraceae	*Otopappus scaber* - S. F. Blake	■								■			
Asteraceae	*Parthenium hysterophorus* - L.	■	■	■	■	■	■	■	■	■		■	■
Asteraceae	*Parthenium schottii* - Greenm. ex Millsp. & Chase		■										
Asteraceae	*Pluchea odorata* - (L.) Cass.	■			■			■	■				
Asteraceae	*Pluchea purpurascens* - (Sw.) DC.	■				■							
Asteraceae	*Pluchea symphytifolia* - (Mill.) Gillis	▓	■	■	▓	▓	■	■		■			■
Asteraceae	Polymnia maculata - Cav.				■	■	■				■	■	■
Asteraceae	Porophyllum punctatum - (Mill.) S. F. Blake	■	■	■		■	■			■	▓	■	
Asteraceae	*Ratibida latipalearis* - E. L. Richards	■											
Asteraceae	*Sanvitalia procumbens* - Lam.							■					
Asteraceae	*Schistocarpha eupatorioides* - (Fenzl) Kuntze		■	■							■		
Asteraceae	*Schistocarpha longiligula* - Rydb.		■								■		
Asteraceae	*Schistophragma pusillum* - Benth.							■		▓	■		
Asteraceae	*Sclerocarpus divaricatus* - (Benth.) Benth. & Hook. f. ex Hemsl.					■			■				
Asteraceae	*Senecio chenopodioides* - Kunth	■		■	■			■	■	■	■		■
Asteraceae	*Senecio confusus* - Britten					■		■			■		
Asteraceae	*Senecio roseus* - Sch. Bip.					■							
Asteraceae	*Senecio schaffneri* - Sch. Bip. ex Klatt				■						■		
Asteraceae	*Sigesbeckia jorullensis* - Kunth				■								
Asteraceae	*Simsia eurylepis* - S. F. Blake					■							
Asteraceae	*Simsia foetida* - (Cav.) S. F. Blake			■									
Asteraceae	*Simsia villasenorii* - D. M. Spooner					■							
Asteraceae	*Sonchus oleraceus* - L.	■		■	■							■	
Asteraceae	*Spilanthes oppositifolia* - (Lam.) D'Arcy											■	
Asteraceae	*Spilanthes pilosa* - R. K. Jansen											■	
Asteraceae	*Spiracantha cornifolia* - Kunth				■								
Asteraceae	*Stevia ephemera* - Grashoff								■				
Asteraceae	*Synedrella nodiflora* - (L.) Gaertn.		■									■	
Asteraceae	*Tagetes erecta* - L.	■				■	■	■	■			■	
Asteraceae	*Tagetes filifolia* - Lag.									■			
Asteraceae	*Tagetes lucida* - Cav.				■								
Asteraceae	*Tagetes patula* - L.	■		■	■				■	■		■	
Asteraceae	*Tithonia calva* - (B. L. Rob. & Greenm.) S. F. Blake												■
Asteraceae	*Tithonia diversifolia* - (Hemsl.) A. Gray		░	■	■			■				░	■
Asteraceae	*Tithonia rotundifolia* - (Mill.) S. F. Blake	■	■	■									
Asteraceae	*Tithonia tubiformis* - (Jacq.) Cass.		■								■		
Asteraceae	*Tridax procumbens* - L.					■					■	▓	
Asteraceae	*Trixis inula* - Crantz			▓				■					
Asteraceae	*Vernonia argyropappa* - H. Buek								■				
Asteraceae	*Vernonia aschenborniana* - S. Schauer		■	■									
Asteraceae	*Vernonia canescens* - Kunth	■			■	■			■			■	
Asteraceae	*Vernonia cinerea* - (L.) Less.	■		■			■				■	■	
Asteraceae	*Vernonia ctenophora* - Gleason									■	■	■	
Asteraceae	*Vernonia deppeana* - Less.			■									
Asteraceae	*Vernonia jonesii* - B. L. Turner							■					
Asteraceae	*Vernonia jucunda* - Gleason				■								

(continued)

Table 4.1 (continued)

Family	Specie	J	F	M	A	M	J	J	A	S	O	N	D
Asteraceae	*Vernonia karvinskiana* - DC.												
Asteraceae	*Vernonia leiocarpa* - DC.												
Asteraceae	*Vernonia oaxacana* - Sch. Bip. ex Klatt												
Asteraceae	*Vernonia polypleura* - S. F. Blake												
Asteraceae	*Vernonia pooleae* - B. L. Turner												
Asteraceae	*Vernonia salcifolia* - (Mart.) Less.												
Asteraceae	*Vernonia tortuosa* - (L.) S. F. Blake												
Asteraceae	*Viguiera dentata* - (Cav.) Spreng.												
Asteraceae	*Viguiera tenuis* - A. Gray												
Asteraceae	*Wedelia acapulcensis* - (Greenm.) Strother												
Asteraceae	*Wedelia calycina* - Rich.												
Asteraceae	*Wedelia fertilis* - McVaugh												
Asteraceae	*Wedelia hispida* - Kunth												
Asteraceae	*Wedelia parviceps* - S. F. Blake												
Asteraceae	*Wedelia trilobata* - (L.) Hitchc.												
Asteraceae	*Ageratum gaumeri* - B. L. Rob.												
Asteraceae	*Ageratum littorale* - A. Gray												
Asteraceae	*Ageratum maritimum* - Kunth												
Asteraceae	*Ambrosia cumanensis* - Kunth												
Asteraceae	*Ambrosia hispida* - Pursh												
Asteraceae	*Baltimora recta* - L.												
Asteraceae	*Bidens alba* - (L.) DC.												
Asteraceae	*Bidens odorata* - Cav.												
Asteraceae	*Bidens pilosa* - L.												
Asteraceae	*Bidens squarrosa* - Kunth												
Asteraceae	*Borrichia arborescens* - (L.) DC.												
Asteraceae	*Borrichia frutescens* - (L.) DC.												
Asteraceae	*Calea integrifolia* - (DC.) Hemsl.												
Asteraceae	*Calea jamaicensis* - (L.) L.												
Asteraceae	*Calea megacephala* - B. L. Rob. & Greenm.												
Asteraceae	*Calea peckii* - B. L. Rob.												
Asteraceae	*Calea ternifolia* - Kunth												
Asteraceae	*Calea urticifolia* - (Mill.) DC.												
Balsaminaceae	*Impatiens balsamina* - L.												
Basellaceae	*Anredera vesicaria* - (Lamb.) C. F. Gaertn.												
Bataceae	*Batis maritima* - L.												
Begoniaceae	*Begonia glabra* - Aubl.												
Begoniaceae	*Begonia heracleifolia* - Schltdl. & Cham.												
Begoniaceae	*Begonia ludicra* - A. DC.												
Begoniaceae	*Begonia manicata* - Brongn. ex F. Cels												
Begoniaceae	*Begonia pinetorum* - A. DC.												
Begoniaceae	*Begonia semiovata* - Liebm.												
Betulaceae	*Ostrya virginiana* - (Mill.) K. Koch												
Bignoniaceae	*Adenocalymma fissum* - Loes.												
Bignoniaceae	*Adenocalymma inundatum* - Mart. ex DC.												
Bignoniaceae	*Amphilophium paniculatum* - (L.) Kunth												
Bignoniaceae	*Amphitecna latifolia* - (Mill.) A. H. Gentry												
Bignoniaceae	*Amphitecna tuxtlensis* - A. H. Gentry												
Bignoniaceae	*Arrabidaea chica* - (Humb. & Bonpl.) B. Verl.												
Bignoniaceae	*Arrabidaea corallina* - (Jacq.) Sandwith												
Bignoniaceae	*Arrabidaea floribunda* - (Kunth) Loes.												
Bignoniaceae	*Arrabidaea mollissima* - (Kunth) Bureau & K. Schum.												
Bignoniaceae	*Arrabidaea patellifera* - (Schltdl.) Sandwith												
Bignoniaceae	*Arrabidaea podopogon* - (DC.) A. H. Gentry												
Bignoniaceae	*Arrabidaea pubescens* - (L.) A. H. Gentry												
Bignoniaceae	*Arrabidaea verrucosa* - (Standl.) A. H. Gentry												
Bignoniaceae	*Ceratophytum tetragonolobum* - (Jacq.) Sprague & Sandwith												
Bignoniaceae	*Crescentia cujete* - L.												
Bignoniaceae	*Cydista aequinoctialis* - (L.) Miers												
Bignoniaceae	*Cydista diversifolia* - (Kunth) Miers												
Bignoniaceae	*Cydista heterophylla* - Seibert												
Bignoniaceae	*Cydista potosina* - (K. Schum. & Loes.) Loes.												
Bignoniaceae	*Godmania aesculifolia* - (Kunth) Standl.												
Bignoniaceae	*Macfadyena unguis-cati* - (L.) A. H. Gentry												
Bignoniaceae	*Mansoa hymenaea* - (DC.) A. H. Gentry												
Bignoniaceae	*Mansoa verrucifera* - (Schltdl.) A. H. Gentry												
Bignoniaceae	*Melloa quadrivalvis* - (Jacq.) A. H. Gentry												
Bignoniaceae	*Paragonia pyramidata* - (Rich.) Bureau												
Bignoniaceae	*Parmentiera aculeata* - (Kunth) L. O. Williams												
Bignoniaceae	*Parmentiera millspaughiana* - L. O. Williams												
Bignoniaceae	*Pithecoctenium crucigerum* - (L.) A. H. Gentry												
Bignoniaceae	*Pseudocatalpa caudiculata* - (Standl.) A. H. Gentry												

(continued)

Table 4.1 (continued)

Family	Specie	J	F	M	A	M	J	J	A	S	O	N	D
Bignoniaceae	*Tabebuia chrysantha* - G. Nicholson	■			■		■	■		■			
Bignoniaceae	*Tabebuia rosea* - (Bertol.) A. DC.			▓									
Bignoniaceae	*Tecoma capensis* - (Thunb.) Lindl.							■					
Bignoniaceae	*Tecoma stans* - (L.) Juss. ex Kunth	■	▓	▓	▓	■		▓	■	▓	■	■	
Bignoniaceae	*Tectona grandis* - L. f.						■						
Bignoniaceae	*Tephrosia belizensis* - Lundell			■									
Bignoniaceae	*Tephrosia cinerea* - (L.) Pers.		░	▓	░	░		▓	▓	▓	▓	▓	■
Bignoniaceae	*Tephrosia lanata* - M. Martens & Galeotti			■									
Bignoniaceae	*Tephrosia pachypoda* - Riley							■					
Bignoniaceae	*Tephrosia vogelii* - Hook. f.							■					
Bignoniaceae	*Teramnus uncinatus* - (L.) Sw.							░					■
Bignoniaceae	*Tynanthus guatemalensis* - Donn. Sm.							■				░	
Bixaceae	*Amoreuxia palmatifida* - DC.					░			░				
Bixaceae	*Bixa orellana* - L.	░	■	▓	▓	░	▓			░	▓	░	
Bixaceae	*Cochlospermum vitifolium* - (Willd.) Spreng.			■		■		■					
Boraginaceae	*Antiphytum heliotropioides* - DC.						■						
Boraginaceae	*Borago officinalis* - L.						■						
Boraginaceae	*Bourreria andrieuxii* - (DC.) Hemsl.				■								
Boraginaceae	*Bourreria huanita* - (Lex.) Hemsl.								■	░			
Boraginaceae	*Bourreria ovata* - Miers				■		■						
Boraginaceae	*Bourreria oxyphylla* - Standl.	■	▓	▓	░	■	▓	■		■		░	
Boraginaceae	*Bourreria pulchra* - (Millsp.) Greenm.		░	■	▓	▓			■	■	■	■	
Boraginaceae	*Bourreria spathulata* - (Miers) Hemsl.										■		
Boraginaceae	*Cordia alliodora* - (Ruiz & Pav.) Oken			░	■				■				
Boraginaceae	*Cordia curassavica* - (Jacq.) Roem. & Schult.	■		■	■	■	■	▓	▓	░	▓	░	
Boraginaceae	*Cordia cylindrostachya* - (Ruiz & Pav.) Roem. & Schult.								■	■	░		
Boraginaceae	*Cordia dentata* - Poir.					▓					░	▓	
Boraginaceae	*Cordia diversifolia* - Pav. ex DC.					■		■	░				
Boraginaceae	*Cordia dodecandra* - DC.	■	░	■		░	░		░	■	■	■	■
Boraginaceae	*Cordia gerascanthus* - L.	■	■	▓				■	■			■	
Boraginaceae	*Cordia globosa* - (Jacq.) Kunth		■	░		░	■		■	▓		░	
Boraginaceae	*Cordia pringlei* - B. L. Rob.												■
Boraginaceae	*Cordia sebestena* - L.		■			■	▓	■	■	░	▓		
Boraginaceae	*Cordia spinescens* - L.			■	■				■				■
Boraginaceae	*Cordia stellata* - Greenm.						■						
Boraginaceae	*Cordia stellifera* - I. M. Johnst.			░	░	░	■	■					
Boraginaceae	*Cordia stenoclada* - I. M. Johnst.						■						
Boraginaceae	*Ehretia tinifolia* - L.	░	░	░	░	▓	░	▓	■		■	▓	
Boraginaceae	*Ehretia viscosa* - Fernald							■					
Boraginaceae	*Hackelia mexicana* - (Schltdl. & Cham.) I. M. Johnst.				■								
Boraginaceae	*Heliotropium angiospermum* - Murray		▓	▓	▓	■	▓	▓	■	■	■	■	░
Boraginaceae	*Heliotropium curassavicum* - L.			░	■	░		■			■		
Boraginaceae	*Heliotropium indicum* - L.	■		░								░	■
Boraginaceae	*Heliotropium procumbens* - Mill.			■	■	■	■	▓		■			
Boraginaceae	*Heliotropium ternatum* - Vahl	■	■	■	■	▓	■		▓	■	■	■	■
Boraginaceae	*Nama jamaicensis* - L.			■	■								
Boraginaceae	*Rochefortia lundellii* - Camp							░					
Boraginaceae	*Tournefortia acutiflora* - M. Martens & Galeotti					■							
Boraginaceae	*Tournefortia glabra* - L.		■	■		■	▓	▓	░	▓	░	░	
Boraginaceae	*Tournefortia gnaphalodes* - (L.) R. Br. ex Roem. & Schult.		░	░		▓					▓	░	
Boraginaceae	*Tournefortia hirsutissima* - L.		■	░			■	■			▓		
Boraginaceae	*Tournefortia laurifolia* - Vent.									░			
Boraginaceae	*Tournefortia maculata* - Jacq.								░				
Boraginaceae	*Tournefortia volubilis* - L.	■		░	░		░	░		▓	▓	░	■
Boraginaceae	*Wigandia caracasana* - Kunth		■										
Boraginaceae	*Macromeria discolor* - Benth.			■									
Brassicaceae	*Brassica integrifolia* - (H. West) Rupr.						■						
Brassicaceae	*Cakile edentula* - (Bigelow) Hook.			░	■					■	■		
Brassicaceae	*Cakile lanceolata* - (Willd.) O. E. Schulz			■		▓	■	■	■		▓	░	■
Brassicaceae	*Lepidium virginicum* - L.	░		■				░	■	▓	░	▓	
Bromeliaceae	*Aechmea bracteata* - (Sw.) Griseb.						■	░		■	■		
Bromeliaceae	*Aechmea tillandsioides* - (Mart. ex Schult. & Schult. f.) Baker		■								■		
Bromeliaceae	*Ananas comosus* - (L.) Merr.						■						
Bromeliaceae	*Bromelia alsodes* - H. St. John					░							
Bromeliaceae	*Bromelia alsodes* - H. St. John	■											
Bromeliaceae	*Bromelia plumieri* - (E. Morren) L. B. Sm.				░								
Bromeliaceae	*Catopsis nutans* - (Sw.) Griseb.		░										
Bromeliaceae	*Tillandsia balbisiana* - Schult. f.				░								
Bromeliaceae	*Tillandsia bulbosa* - Hook.	■											
Bromeliaceae	*Tillandsia fasciculata* - Sw.		■										
Bromeliaceae	*Tillandsia festucoides* - Brongn. ex Mez					░							
Bromeliaceae	*Tillandsia guatemalensis* - L. B. Sm.		■										

(continued)

Table 4.1 (continued)

Family	Specie	J	F	M	A	M	J	J	A	S	O	N	D
Burseraceae	*Bursera longipes* - (Rose) Standl.								░				■
Burseraceae	*Bursera schlechtendalii* - Engl.					░	░	░			░		
Burseraceae	*Bursera simaruba* - (L.) Sarg.	░	▒	▒	▒	▒	▒	▒	▒	▒	░	▒	▒
Burseraceae	*Protium copal* - (Schltdl. & Cham.) Engl.				░	▒	░	░			■	■	
Buxaceae	*Buxus bartlettii* - Standl.		■	■	░	░	▒	░	░	▒	■	■	░
Cabombaceae	*Cabomba aquatica* - Aubl.										■	■	
Cabombaceae	*Cabomba palaeformis* - Fassett										■		
Cactaceae	*Acanthocereus tetragonus* - (L.) Hummelinck											░	
Cactaceae	*Epiphyllum phyllanthus* - (L.) Haw.				░					■	░	■	
Cactaceae	*Hylocereus undatus* - (Haw.) Britton & Rose											░	
Cactaceae	*Nopalea cochenillifera* - (L.) Salm-Dyck			■							▒		
Cactaceae	*Nopalea gaumeri* - Britton & Rose			▒	░	▒	░					■	■
Cactaceae	*Nopalea inaperta* - Schott ex Griffiths					░							
Cactaceae	*Opuntia dillenii* - (Ker Gawl.) Haw.			■							░		
Cactaceae	*Selenicereus donkelaarii* - (Salm-Dyck) Britton & Rose						▒	■					
Cactaceae	*Selenicereus testudo* - (Karw. ex Zucc.) Buxb.											■	
Cactaceae	*Stenocereus laevigatus* - (Salm-Dyck) Buxb.							■					
Calophyllaceae	*Mammea americana* - L.									■			
Campanulaceae	*Centropogon cordifolius* - Benth.	■	■					■		■			
Campanulaceae	*Laurentia longiflora* - (L.) Peterm.		■	■		░							
Campanulaceae	*Lobelia berlandieri* - A. DC.	■											
Campanulaceae	*Lobelia laxiflora* - Kunth			■									
Cannabaceae	*Celtis iguanaea* - (Jacq.) Sarg.		■			■	░	░		▒	░	■	
Cannabaceae	*Celtis trinervia* - Lam.						■	░					
Cannabaceae	*Trema micrantha* - (L.) Blume	■	■	▒		░	▒	▒		▒	░	░	░
Cannaceae	*Canna edulis* - Ker Gawl.				■							■	
Cannaceae	*Canna generalis* - L. H. Bailey & E. Z. Bailey				■								
Cannaceae	*Canna indica* - L.		▒	▒	■						▒	▒	
Capparaceae	*Capparis baducca* - L.										░		
Capparaceae	*Capparis cynophallophora* - L.				░	■	░	░		░			
Capparaceae	*Capparis flexuosa* - (L.) L.	■					░	░	▒		■		
Capparaceae	*Capparis incana* - Kunth	■	■	■	▒	▒	▒	▒	▒	▒	▒	▒	
Capparaceae	*Capparis pachaca* - (C. Wright ex Radlk.) H. H. Iltis						░						
Capparaceae	*Crateva tapia* - L.		▒	░				░	░				
Caricaceae	*Carica papaya* - L.		■	░		▒		■	▒		■	░	
Caricaceae	*Carica pennata* - Heilborn		■										
Casuarinaceae	*Casuarina equisetifolia* - L.		░		░	░	░	░					
Celastraceae	*Celastrus vulcanicola* - Donn. Sm.			■									
Celastraceae	*Crossopetalum gaumeri* - (Loes.) Lundell	■	░	░	░	░		░	░	░	■	▒	■
Celastraceae	*Crossopetalum rhacoma* - Crantz		░										
Celastraceae	*Elaeodendron trichotomum* - (Turcz.) Lundell			░						░			
Celastraceae	*Elaeodendron xylocarpum* - (Vent.) DC.				░		░	░					■
Celastraceae	*Gyminda latifolia* - (Sw.) Urb.				░								
Celastraceae	*Hemiangium excelsum* - (Kunth) A. C. Sm.											■	
Celastraceae	*Hippocratea volubilis* - L.			░									
Celastraceae	*Maytenus guatemalensis* - Lundell			░		░	░	░	░				
Celastraceae	*Maytenus phyllanthoides* - Benth.								■				
Celastraceae	*Maytenus schippii* - Lundell												■
Celastraceae	*Rhacoma crossopetalum* - L.						░			▒			
Celastraceae	*Rhacoma gaumeri* - (Loes.) Standl.				▒	■		░		░	░		■
Celastraceae	*Rhacoma puberula* - (Lundell) Standl. & Steyerm.			░									
Celastraceae	*Wimmeria obtusifolia* - Standl.								■				
Celastraceae	*Zinowiewia rubra* - Lundell											■	
Chrysobalanaceae	*Chrysobalanus icaco* - L.				░	▒	▒	■			▒		■
Chrysobalanaceae	*Hirtella americana* - L.				░	■	░						
Chrysobalanaceae	*Licania hypoleuca* - Benth.	■											
Cleomaceae	*Cleome gynandra* - L.	■									░		
Cleomaceae	*Cleome serrata* - Jacq.			░								■	
Cleomaceae	*Cleome speciosa* - Raf.	■	▒			■					▒		
Cleomaceae	*Cleome spinosa* - Jacq.										■		
Cleomaceae	*Cleome viscosa* - L.										■		
Clethraceae	*Clethra oleoides* - L. O. Williams									■	■		
Clusiaceae	*Clusia salvinii* - Donn. Sm.									░		■	░
Combretaceae	*Bucida buceras* - L.		■	■		▒	░		■	■		■	■
Combretaceae	*Bucida spinosa* - Jenn.	■		■		■	▒				░	■	
Combretaceae	*Bucida wigginsiana* - Miranda			■									
Combretaceae	*Combretum fruticosum* - (Loefl.) Stuntz.	▒	▒	▒								■	
Combretaceae	*Combretum laxum* - Jacq.					■			■				■
Combretaceae	*Conocarpus erectus* - L.	■	■	▒	▒	▒		░	▒	■	▒	▒	
Combretaceae	*Laguncularia racemosa* - (L.) C. F. Gaertn.					■	░	▒	▒		■		■
Combretaceae	*Terminalia catappa* - L.	░		░	■	■	■					░	
Commelinaceae	*Gibasis geniculata* - (Jacq.) Rohweder									■			

(continued)

Table 4.1 (continued)

Family	Specie	J	F	M	A	M	J	J	A	S	O	N	D
Commelinaceae	*Tinantia erecta* - (Jacq.) Schltdl.				■								
Commelinaceae	*Tradescantia monosperma* - Brandegee							■					
Commelinaceae	*Tradescantia zebrina* - Bosse								■				
Commelinaceae	*Commelina diffusa* - Burm. f.					■		░			■		
Commelinaceae	*Commelina elegans* - Kunth						■						
Commelinaceae	*Commelina erecta* - L.				■	■	▒					■	■
Connaraceae	*Rourea glabra* - Kunth						░	░		■	░	■	
Convolvulaceae	*Cuscuta tinctoria* - Mart. ex Engelm.						■						
Convolvulaceae	*Evolvulus alsinoides* - (L.) L.		■							▒		■	
Convolvulaceae	*Evolvulus sericeus* - Sw.		■										
Convolvulaceae	*Ipomoea alba* - L.	■			■	■		■	■	■		■	
Convolvulaceae	*Ipomoea anisomeres* - B. L. Rob. & Bartlett										■		
Convolvulaceae	*Ipomoea batatas* - (L.) Lam.						■				■		
Convolvulaceae	*Ipomoea carnea* - Jacq.	▒	■		▒			■	▒	■	■	■	■
Convolvulaceae	*Ipomoea clavata* - (G. Don) Ooststr. ex J. F. Macbr.	░											
Convolvulaceae	*Ipomoea contrerasii* - L. O. Williams	■											
Convolvulaceae	*Ipomoea crinicalyx* - S. Moore	■	■		▒	■			■	■	■	▒	■
Convolvulaceae	*Ipomoea fistulosa* - Mart. ex Choisy	■											
Convolvulaceae	*Ipomoea hederifolia* - L.	▒				░		■		░	▒	▒	■
Convolvulaceae	*Ipomoea heterodoxa* - Standl. & Steyerm.		░			■		■	■	■			░
Convolvulaceae	*Ipomoea imperati* - (Vahl) Griseb.	■									■	■	
Convolvulaceae	*Ipomoea indica* - (Burm.) Merr.	■	■	■	░					░	■	■	■
Convolvulaceae	*Ipomoea jalapa* - (L.) Pursh									░			
Convolvulaceae	*Ipomoea lutea* - Hemsl.												■
Convolvulaceae	*Ipomoea nil* - (L.) Roth	■		■		■		■	■	■		▒	■
Convolvulaceae	*Ipomoea orizabensis* - (Pelletan) Ledeb. ex Steud.	■											
Convolvulaceae	*Ipomoea pes-caprae* - (L.) R. Br.		■								■		
Convolvulaceae	*Ipomoea quamoclit* - L.	■									░		
Convolvulaceae	*Ipomoea sagittata* - Poir.								■			■	
Convolvulaceae	*Ipomoea steerei* - (Standl.) L. O. Williams			■									
Convolvulaceae	*Ipomoea stolonifera* - (Cirillo) J. F. Gmel.						■						
Convolvulaceae	*Ipomoea trifida* - (Kunth) G. Don		■	■	■								
Convolvulaceae	*Ipomoea triloba* - L., 1753	■					■	■		■	■		
Convolvulaceae	*Ipomoea tuxtlensis* - House	■	■	■				■	▒	■		■	■
Convolvulaceae	*Ipomoea umbraticola* - House					■		■	■			■	
Convolvulaceae	*Ipomoea violacea* - L.	■			■	■		░			░	░	
Convolvulaceae	*Itzaea sericea* - (Standl.) Standl. & Steyerm.												░
Convolvulaceae	*Merremia aegyptia* - (L.) Urb.	░	■	■	■	■		■	■		▒	▒	
Convolvulaceae	*Merremia cissoides* - (Lam.) Hallier f.	■		■		■					■		
Convolvulaceae	*Merremia quinquefolia* - (L.) Hallier f.			■									
Convolvulaceae	*Merremia tuberosa* - (L.) Rendle	░	░								■		░
Convolvulaceae	*Merremia umbellata* - (L.) Hallier f.		■	■		■				■	■		
Convolvulaceae	*Operculina pinnatifida* - (Kunth) O'Donell	■			■				░	■			
Convolvulaceae	*Rivea corymbosa* - (L.) Hallier f.		■										
Convolvulaceae	*Rivina humilis* - L.	▒	■	▒	■	▒	▒	▒	▒	▒	▒	▒	■
Convolvulaceae	*Turbina corymbosa* - (L.) Raf.	▒	▒	■	■				■		■		■
Coriariaceae	*Coriaria ruscifolia* - L.											■	
Crassulaceae	*Kalanchoe blossfeldiana* - Poelln.					■							
Crassulaceae	*Kalanchoe laciniata* - (L.) DC.			■		■							
Crassulaceae	*Kalanchoe pinnata* - (Lam.) Pers.			░									
Cucurbitaceae	*Apodanthera undulata* - A. Gray				■								
Cucurbitaceae	*Cayaponia attenuata* - (Hook. & Arn.) Cogn.				■					■			
Cucurbitaceae	*Cayaponia racemosa* - (Mill.) Cogn.	▒	░	░									
Cucurbitaceae	*Cionosicys excisus* - (Griseb.) C. Jeffrey	░			■		■	■		▒	▒	▒	░
Cucurbitaceae	*Cionosicys macranthus* - (Pittier) C. Jeffrey		░			■							
Cucurbitaceae	*Citrullus lanatus* - (Thunb.) Matsum. & Nakai					■	■				■		■
Cucurbitaceae	*Cucumis melo* - L.								■	░		░	
Cucurbitaceae	*Cucumis sativus* - L.			■		░			░				
Cucurbitaceae	*Cucurbita maxima* - Duchesne ex Lam.		░									■	
Cucurbitaceae	*Cucurbita mixta* - Pangalo		■			▒					■		
Cucurbitaceae	*Cucurbita moschata* - Duchesne	░						░	■	■	░		
Cucurbitaceae	*Cucurbita pepo* - L.					■		■				░	
Cucurbitaceae	*Cyclanthera multifoliola* - Cogn.				■								
Cucurbitaceae	*Hanburia parviflora* - Donn. Sm.								■				
Cucurbitaceae	*Ibervillea millspaughii* - (Cogn.) C. Jeffrey				░	░							
Cucurbitaceae	*Lagenaria siceraria* - (Molina) Standl.							■	■		■	░	
Cucurbitaceae	*Luffa cylindrica* - M. Roem.										■		
Cucurbitaceae	*Melothria pendula* - L.	░	▒	░	░	▒	■	▒	░		▒		
Cucurbitaceae	*Microsechium helleri* - (Peyr.) Cogn.	■						■					
Cucurbitaceae	*Momordica charantia* - L.		▒		░	■	▒	░	▒		░	▒	
Cucurbitaceae	*Psiguria triphylla* - (Miq.) C. Jeffrey		■									■	
Cucurbitaceae	*Rytidostylis carthaginensis* - Kuntze						■				░		

(continued)

Table 4.1 (continued)

Family	Specie	J	F	M	A	M	J	J	A	S	O	N	D
Cucurbitaceae	*Sechium edule* - (Jacq.) Sw.												
Cucurbitaceae	*Sicydium tamnifolium* - (Kunth) Cogn.												
Cucurbitaceae	*Tecunumania quetzalteca* - Standl. & Steyerm.												
Cunoniaceae	*Weinmannia pinnata* - L.												
Cyperaceae	*Carex donnell-smithii* - L. H. Bailey												
Cyperaceae	*Cladium jamaicense* - Crantz												
Cyperaceae	*Cyperus articulatus* - L.												
Cyperaceae	*Cyperus elegans* - L.												
Cyperaceae	*Cyperus esculentus* - L.												
Cyperaceae	*Cyperus hermaphroditus* - (Jacq.) Standl.												
Cyperaceae	*Cyperus ligularis* - L.												
Cyperaceae	*Cyperus macrocephalus* - Liebm.												
Cyperaceae	*Cyperus ochraceus* - Vahl												
Cyperaceae	*Cyperus odoratus* - L.												
Cyperaceae	*Cyperus planifolius* - Rich.												
Cyperaceae	*Cyperus polystachyos* - Rottb.												
Cyperaceae	*Cyperus rotundus* - L.												
Cyperaceae	*Cyperus squarrosus* - L.												
Cyperaceae	Cyperus thyrsiflorus - Jungh.												
Cyperaceae	*Eleocharis caribaea* - (Rottb.) S. F. Blake												
Cyperaceae	*Eleocharis cellulosa* - Torr.												
Cyperaceae	*Eleocharis flavescens* - (Poir.) Urb.												
Cyperaceae	*Eleocharis geniculata* - (L.) Roem. & Schult.												
Cyperaceae	*Eleocharis interstincta* - (Vahl) Roem. & Schult.												
Cyperaceae	*Fimbristylis cymosa* - R. Br.												
Cyperaceae	*Fimbristylis spadicea* - (L.) Vahl												
Cyperaceae	*Fuirena camptotricha* - C. Wright												
Cyperaceae	*Fuirena simplex* - Vahl												
Cyperaceae	*Scleria lithosperma* - (L.) Sw.												
Cyperaceae	*Scleria melaleuca* - Rchb. ex. Schltdl. & Cham.												
Cyperaceae	*Scleria mexicana* - (Liebm.) Boeck.												
Cyperaceae	*Scleria microcarpa* - Nees ex Kunth												
Cyperaceae	*Torulinium odoratum* - (L.) S. S. Hooper												
Dichapetalaceae	*Dichapetalum donnell-smithii* - Engl.												
Dilleniaceae	*Curatella americana* - L.												
Dilleniaceae	*Tetracera mollis* - Standl.												
Dilleniaceae	*Tetracera volubilis* - L.												
Dilliniaceae	*Davilla kunthii* - A. St.-Hil.												
Dioscoreaceae	*Dioscorea alata* - L.												
Dioscoreaceae	*Dioscorea aspera* - Humb. & Bonpl. ex Willd.												
Dioscoreaceae	*Dioscorea bartlettii* - C. V. Morton												
Dioscoreaceae	*Dioscorea bernoulliana* - Prain & Burkill												
Dioscoreaceae	*Dioscorea carionis* - Prain & Burkill												
Dioscoreaceae	*Dioscorea chiapasensis* - Matuda												
Dioscoreaceae	*Dioscorea composita* - Hemsl.												
Dioscoreaceae	*Dioscorea convolvulacea* - Schltdl. & Cham.												
Dioscoreaceae	*Dioscorea cyanisticta* - Donn. Sm.												
Dioscoreaceae	*Dioscorea densiflora* - Hemsl.												
Dioscoreaceae	*Dioscorea floribunda* - M. Martens & Galeotti												
Dioscoreaceae	*Dioscorea galeottiana* - Kunth												
Dioscoreaceae	*Dioscorea gaumeri* - R. Knuth												
Dioscoreaceae	*Dioscorea liebmannii* - Uline												
Dioscoreaceae	*Dioscorea matagalpensis* - Uline												
Dioscoreaceae	*Dioscorea pilosiuscula* - Bertero ex Spreng.												
Dioscoreaceae	*Dioscorea polygonoides* - Humb. & Bonpl. ex Willd.												
Dioscoreaceae	*Dioscorea spiculiflora* - Hemsl.												
Dioscoreaceae	*Dioscorea urceolata* - Uline												
Ebenaceae	*Diospyros anisandra* - S. F. Blake												
Ebenaceae	*Diospyros bumelioides* - Standl.												
Ebenaceae	*Diospyros cuneata* - Standl.												
Ebenaceae	*Diospyros nicaraguensis* - (Standl.) Standl.												
Ebenaceae	*Diospyros verae-crucis* - (Standl.) Standl.												
Ericaceae	*Cavendishia crassifolia* - (Benth.) Hemsl.												
Ericaceae	*Gaultheria esculenta* - Corcoran & Breckon ex Luteyn												
Eriocaulaceae	*Eriocaulon seemannii* - Moldenke												
Euphorbiaceae	*Acalypha alopecuroidea* - Jacq.												
Euphorbiaceae	*Acalypha arvensis* - Poepp. & Endl.												
Euphorbiaceae	*Acalypha diversifolia* - Jacq.												
Euphorbiaceae	*Acalypha ferdinandii* - K. Hoffm.												
Euphorbiaceae	*Acalypha hirsuta* - Hochst. ex A. Rich.												
Euphorbiaceae	*Acalypha hispida* - Burm. f.												
Euphorbiaceae	*Acalypha leptopoda* - Müll. Arg.												

(continued)

Table 4.1 (continued)

Family	Specie	J	F	M	A	M	J	J	A	S	O	N	D
Euphorbiaceae	*Acalypha radians* - Torr.												
Euphorbiaceae	*Acalypha seleriana* - Greenm.												
Euphorbiaceae	*Acalypha setosa* - A. Rich.												
Euphorbiaceae	*Acalypha unibracteata* - Müll. Arg.												
Euphorbiaceae	*Acalypha vagans* - Cav.												
Euphorbiaceae	*Acalypha villosa* - Jacq.												
Euphorbiaceae	*Acalypha wilkesiana* - Müll. Arg.												
Euphorbiaceae	*Adelia barbinervis* - Schltdl. & Cham.												
Euphorbiaceae	*Adelia oaxacana* - (Müll. Arg.) Hemsl.												
Euphorbiaceae	*Alchornea latifolia* - Sw.												
Euphorbiaceae	*Argythamnia guatemalensis* - Müll. Arg.												
Euphorbiaceae	*Caperonia castaneifolia* - (L.) A. St.-Hil.												
Euphorbiaceae	*Caperonia palustris* - (L.) A. St.-Hil.												
Euphorbiaceae	*Chamaesyce blodgettii* - (Engelm. ex Hitchc.) Small												
Euphorbiaceae	*Chamaesyce buxifolia* - (Lam.) Small												
Euphorbiaceae	*Chamaesyce dioeca* - (Kunth) Millsp.												
Euphorbiaceae	*Chamaesyce glomerifera* - Millsp.												
Euphorbiaceae	*Chamaesyce hirta* - (L.) Millsp.												
Euphorbiaceae	*Chamaesyce hyssopifolia* - (L.) Small												
Euphorbiaceae	*Cnidoscolus aconitifolius* - (Mill.) I. M. Johnst.												
Euphorbiaceae	*Cnidoscolus chayamansa* - McVaugh												
Euphorbiaceae	*Cnidoscolus multilobus* - (Pax) I. M. Johnst.												
Euphorbiaceae	*Cnidoscolus souzae* - McVaugh												
Euphorbiaceae	*Codiaeum variegatum* - (L.) Rumph. ex A. Juss.												
Euphorbiaceae	*Croton ameliae* - Lundell												
Euphorbiaceae	*Croton arboreus* - Millsp.												
Euphorbiaceae	*Croton argenteus* - L.												
Euphorbiaceae	*Croton campechianus* - Standl.												
Euphorbiaceae	*Croton chichenensis* - Lundell												
Euphorbiaceae	*Croton ciliatoglandulifer* - Ortega												
Euphorbiaceae	*Croton cortesianus* - Kunth												
Euphorbiaceae	*Croton flavens* - L.												
Euphorbiaceae	*Croton glabellus* - L.												
Euphorbiaceae	*Croton glandulosepalus* - Millsp.												
Euphorbiaceae	*Croton humilis* - L.												
Euphorbiaceae	*Croton jutiapensis* - Croizat												
Euphorbiaceae	*Croton lobatus* - L.												
Euphorbiaceae	*Croton lundellii* - Standl.												
Euphorbiaceae	*Croton malvaviscifolius* - Millsp.												
Euphorbiaceae	*Croton miradorensis* - Müll. Arg.												
Euphorbiaceae	*Croton niveus* - Jacq.												
Euphorbiaceae	*Croton ortholobus* - Müll. Arg.												
Euphorbiaceae	*Croton peraeruginosus* - Croizat												
Euphorbiaceae	*Croton perobtusus* - Lundell												
Euphorbiaceae	*Croton punctatus* - Jacq.												
Euphorbiaceae	*Croton reflexifolius* - Kunth												
Euphorbiaceae	*Croton yucatanensis* - Lundell												
Euphorbiaceae	*Dalechampia heteromorpha* - Pax & K. Hoffm.												
Euphorbiaceae	*Dalechampia scandens* - L.												
Euphorbiaceae	*Dalechampia schippii* - Standl.												
Euphorbiaceae	*Ditaxis guatemalensis* - (Müll. Arg.) Pax & K. Hoffm.												
Euphorbiaceae	*Enriquebeltrania crenatifolia* - (Miranda) Rzed.												
Euphorbiaceae	*Euphorbia buxifolia* - Lam.												
Euphorbiaceae	*Euphorbia cozumelensis* - Millsp.												
Euphorbiaceae	*Euphorbia cyathophora* - Murray												
Euphorbiaceae	*Euphorbia dentata* - Michx.												
Euphorbiaceae	*Euphorbia dioeca* - Kunth												
Euphorbiaceae	*Euphorbia francoana* - Boiss.												
Euphorbiaceae	*Euphorbia glomerifera* - (Millsp.) L. C. Wheeler												
Euphorbiaceae	*Euphorbia graminea* - Jacq.												
Euphorbiaceae	*Euphorbia heterophylla* - L.												
Euphorbiaceae	*Euphorbia hirta* - L.												
Euphorbiaceae	*Euphorbia hypericifolia* - L.												
Euphorbiaceae	*Euphorbia hyssopifolia* - L.												
Euphorbiaceae	*Euphorbia ocymoidea* - L.												
Euphorbiaceae	*Euphorbia prostrata* - Aiton												
Euphorbiaceae	*Euphorbia pteroneura* - A. Berger												
Euphorbiaceae	*Euphorbia pulcherrima* - Willd. ex Klotzsch												
Euphorbiaceae	*Euphorbia schlechtendalii* - Boiss.												
Euphorbiaceae	*Gymnanthes lucida* - Sw.												
Euphorbiaceae	*Hevea brasiliensis* - (Willd. ex A. Juss.) Müll. Arg.												
Euphorbiaceae	*Hura polyandra* - Baill.												

(continued)

Table 4.1 (continued)

Family	Specie	J	F	M	A	M	J	J	A	S	O	N	D
Euphorbiaceae	*Jatropha curcas* - L.												
Euphorbiaceae	*Jatropha gaumeri* - Greenm.												
Euphorbiaceae	*Jatropha multifida* - L.												
Euphorbiaceae	*Manihot aesculifolia* - (Kunth) Pohl												
Euphorbiaceae	*Manihot carthagenensis* - (Jacq.) Müll. Arg.												
Euphorbiaceae	*Manihot esculenta* - Crantz												
Euphorbiaceae	*Omphalea oleifera* - Hemsl.												
Euphorbiaceae	*Pedilanthus deamii* - Millsp.												
Euphorbiaceae	*Pedilanthus nodiflorus* - Millsp.												
Euphorbiaceae	*Pedilanthus tithymaloides* - (L.) Poit.												
Euphorbiaceae	*Ricinus communis* - L.												
Euphorbiaceae	*Sapindus saponaria* - L.												
Euphorbiaceae	*Sapium nitidum* - (Monach.) Lundell												
Euphorbiaceae	*Sapium oligoneurum* - K. Schum. & Pittier												
Euphorbiaceae	*Sebastiania adenophora* - Pax & K. Hoffm.												
Euphorbiaceae	*Sebastiania longicuspis* - Standl.												
Euphorbiaceae	*Sebastiania pavoniana* - (Müll. Arg.) Müll. Arg.												
Euphorbiaceae	*Tetrorchidium rotundatum* - Standl.												
Euphorbiaceae	*Tragia nepetifolia* - Cav.												
Euphorbiaceae	*Tragia yucatanensis* - Millsp.												
Fabaceae	*Abrus precatorius* - L.												
Fabaceae	*Acacia angustissima* - (Mill.) Kuntze												
Fabaceae	*Acacia cedilloi* - L. Rico												
Fabaceae	*Acacia collinsii* - Saff.												
Fabaceae	*Acacia cornigera* - (L.) Willd.												
Fabaceae	*Acacia dolichostachya* - S. F. Blake												
Fabaceae	*Acacia farnesiana* - (L.) Willd.												
Fabaceae	*Acacia gaumeri* - S. F. Blake												
Fabaceae	*Acacia gentlei* - Standl.												
Fabaceae	*Acacia glomerosa* - Benth.												
Fabaceae	*Acacia pennatula* - (Schltdl. & Cham.) Benth.												
Fabaceae	*Acacia pringlei* - Rose												
Fabaceae	*Acacia riparia* - Kunth												
Fabaceae	*Acacia riparioides* - (Britton & Rose) Standl.												
Fabaceae	*Acacia villosa* - (Sw.) Willd.												
Fabaceae	*Aeschynomene americana* - L.												
Fabaceae	*Aeschynomene fascicularis* - Schltdl. & Cham.												
Fabaceae	*Aeschynomene purpusii* - Brandegee												
Fabaceae	*Aeschynomene sensitiva* - Sw.												
Fabaceae	*Albizia guachapele* - (Kunth) Dugand												
Fabaceae	*Albizia lebbeck* - (L.) Benth.												
Fabaceae	*Albizia tomentosa* - (Micheli) Standl.												
Fabaceae	*Amicia zygomeris* - DC.												
Fabaceae	*Andira inermis* - (W. Wright) Kunth ex DC.												
Fabaceae	*Apoplanesia paniculata* - C. Presl												
Fabaceae	*Astragalus astragatus* - Sessé & Moc.												
Fabaceae	*Ateleia gummifera* - (Bertero & DC.) D. Dietr.												
Fabaceae	*Bauhinia divaricata* - L.												
Fabaceae	*Bauhinia erythrocalyx* - Wunderlin												
Fabaceae	*Bauhinia herrerae* - (Britton & Rose) Standl. & Steyerm.												
Fabaceae	*Bauhinia jenningsii* - P. Wilson												
Fabaceae	*Bauhinia microstachya* - (Raddi) J. F. Macbr.												
Fabaceae	*Bauhinia rubeleruziana* - Donn. Sm.												
Fabaceae	*Bauhinia subrotundifolia* - Cav.												
Fabaceae	*Bauhinia ungulata* - L.												
Fabaceae	*Bauhinia variegata* - L.												
Fabaceae	*Brongniartia lupinoides* - (Kunth) Taub.												
Fabaceae	*Brongniartia sousae* - Dorado												
Fabaceae	*Caesalpinia bonduc* - (L.) Roxb.												
Fabaceae	*Caesalpinia crista* - L.												
Fabaceae	*Caesalpinia gaumeri* - Greenm.												
Fabaceae	*Caesalpinia major* - (Medik.) Dandy & Exell												
Fabaceae	*Caesalpinia mollis* - (Kunth) Spreng.												
Fabaceae	*Caesalpinia platyloba* - S. Watson												
Fabaceae	*Caesalpinia pulcherrima* - (L.) Sw.												
Fabaceae	*Caesalpinia vesicaria* - L.												
Fabaceae	*Caesalpinia violacea* - (Mill.) Standl.												
Fabaceae	*Caesalpinia yucatanensis* - Greenm.												
Fabaceae	*Cajanus cajan* - (L.) Millsp.												
Fabaceae	*Calliandra belizensis* - (Britton & Rose) Standl.												
Fabaceae	*Calliandra calothyrsus* - Meisn.												
Fabaceae	*Calliandra capillata* - Benth.												

(continued)

Table 4.1 (continued)

Family	Specie	J	F	M	A	M	J	J	A	S	O	N	D
Fabaceae	*Calliandra cookii* - (Britton & Rose) Standl.												
Fabaceae	*Calliandra formosa* - (Kunth) Benth.												
Fabaceae	*Calliandra houstoniana* - (Mill.) Standl.												
Fabaceae	*Calliandra magdalenae* - (Bertero ex DC.) Benth.												
Fabaceae	*Calliandra portoricensis* - (Jacq.) Benth.												
Fabaceae	*Calliandra tergemina* - (L.) Benth.												
Fabaceae	*Calopogonium caeruleum* - (Benth.) C. Wright												
Fabaceae	*Canavalia brasiliensis* - Mart. ex Benth.												
Fabaceae	*Canavalia ensiformis* - (L.) DC.												
Fabaceae	*Canavalia hirsuta* - (M. Martens & Galeotti) Standl.												
Fabaceae	*Canavalia rosea* - (Sw.) DC.												
Fabaceae	*Canavalia villosa* - Benth.												
Fabaceae	*Cassia diphylla* - L.												
Fabaceae	*Cassia fistula* - L.												
Fabaceae	*Cassia grandis* - L. f.												
Fabaceae	*Cassia javanica* - L.												
Fabaceae	*Cassia leiophylla* - Vogel												
Fabaceae	*Cassia leptocarpa* - Benth.												
Fabaceae	*Cercis canadensis* - L.												
Fabaceae	*Chaetocalyx scandens* - (L.) Urb.												
Fabaceae	*Chamaecrista diphylla* - (L.) Greene												
Fabaceae	*Chamaecrista glandulosa* - (L.) Greene												
Fabaceae	*Chamaecrista nictitans* - (L.) Moench												
Fabaceae	*Chamaecrista rufa* - (M. Martens & Galeotti) Britton & Rose												
Fabaceae	*Chamaecrista zygophylloides* - (Taub.) H. S. Irwin & Barneby												
Fabaceae	*Coursetia caribaea* - (Jacq.) Lavin												
Fabaceae	*Coursetia glandulosa* - A. Gray												
Fabaceae	*Coursetia mollis* - B. L. Rob. & Greenm.												
Fabaceae	*Cracca caribaea* - (Jacq.) Benth.												
Fabaceae	*Crotalaria acapulcensis* - Hook. & Arn.												
Fabaceae	*Crotalaria cajanifolia* - Kunth												
Fabaceae	*Crotalaria eriocarpa* - Benth.												
Fabaceae	*Crotalaria incana* - L.												
Fabaceae	*Crotalaria maypurensis* - Kunth												
Fabaceae	*Crotalaria nayaritensis* - Windler												
Fabaceae	*Crotalaria pumila* - Ortega												
Fabaceae	*Crotalaria purdiana* - H. Senn												
Fabaceae	*Crotalaria retusa* - L.												
Fabaceae	*Crotalaria sagittalis* - L.												
Fabaceae	*Crotalaria verrucosa* - L.												
Fabaceae	*Dalbergia glabra* - (Mill.) Standl.												
Fabaceae	*Dalea bicolor* - Humb. & Bonpl. ex Willd.												
Fabaceae	*Dalea carthagenensis* - (Jacq.) J. F. Macbr.												
Fabaceae	*Dalea scandens* - (Mill.) R. T. Clausen												
Fabaceae	*Delonix regia* - (Bojer ex Hook.) Raf.												
Fabaceae	*Desmanthus virgatus* - (L.) Willd.												
Fabaceae	*Desmodium barbatum* - (L.) Benth.												
Fabaceae	*Desmodium conzattii* - Greenm.												
Fabaceae	*Desmodium distortum* - (Aubl.) J. F. Macbr.												
Fabaceae	*Desmodium glabrum* - (Mill.) DC.												
Fabaceae	*Desmodium incanum* - DC.												
Fabaceae	*Desmodium intortum* - (Mill.) Urb.												
Fabaceae	*Desmodium metallicum* - (Rose & Standl.) Standl.												
Fabaceae	*Desmodium molle* - (Vahl) DC.												
Fabaceae	*Desmodium orbiculare* - Schltdl.												
Fabaceae	*Desmodium procumbens* - (Mill.) Hitchc.												
Fabaceae	*Desmodium scorpiurus* - (Sw.) Desv.												
Fabaceae	*Desmodium tortuosum* - (Sw.) DC.												
Fabaceae	*Desmodium triflorum* - (L.) DC.												
Fabaceae	*Dialium guianense* - (Aubl.) Sandwith												
Fabaceae	*Dioclea wilsonii* - Standl.												
Fabaceae	*Diphysa carthagenensis* - Jacq.												
Fabaceae	*Diphysa floribunda* - Peyr.												
Fabaceae	*Diphysa macrophylla* - Lundell												
Fabaceae	*Enterolobium cyclocarpum* - (Jacq.) Griseb.												
Fabaceae	*Eriosema simplicifolium* - (DC.) G. Don												
Fabaceae	*Erythrina berteroana* - Urb.												
Fabaceae	*Erythrina crista-galli* - L.												
Fabaceae	*Erythrina folkersii* - Krukoff & Moldenke												
Fabaceae	*Erythrina herbacea* - Krukoff & Barneby												
Fabaceae	*Erythrina horrida* - DC.												
Fabaceae	*Erythrina standleyana* - Krukoff												

(continued)

Table 4.1 (continued)

Family	Specie	J	F	M	A	M	J	J	A	S	O	N	D
Fabaceae	*Erythroxylum areolatum* - L.												
Fabaceae	*Erythroxylum bequaertii* - Standl.												
Fabaceae	*Erythroxylum confusum* - Britton												
Fabaceae	*Erythroxylum havanense* - Jacq.												
Fabaceae	*Erythroxylum rotundifolium* - Lunan												
Fabaceae	*Erythroxylum tabascense* - Britton												
Fabaceae	*Galactia spiciformis* - Torr. & A. Gray												
Fabaceae	*Galactia striata* - (Jacq.) Urb.												
Fabaceae	*Gliricidia ehrenbergii* - (Schltdl.) Rydb.												
Fabaceae	*Gliricidia sepium* - Kunth ex Steud.												
Fabaceae	*Haematoxylum campechianum* - L.												
Fabaceae	*Harpalyce arborescens* - A. Gray												
Fabaceae	*Harpalyce formosa* - DC.												
Fabaceae	*Harpalyce rupicola* - Donn. Sm.												
Fabaceae	*Havardia albicans* - (Kunth) Britton & Rose												
Fabaceae	*Havardia platyloba* - (Bertero ex DC.) Britton & Rose												
Fabaceae	*Indigofera jamaicensis* - Spreng.												
Fabaceae	*Indigofera lespedezioides* - Kunth												
Fabaceae	*Indigofera suffruticosa* - Mill.												
Fabaceae	*Inga oerstediana* - Benth. ex Seem.												
Fabaceae	*Inga punctata* - Willd.												
Fabaceae	*Inga sapindoides* - Willd.												
Fabaceae	Inga thibaudiana - DC.												
Fabaceae	Inga vera - Willd.												
Fabaceae	*Lens esculenta* - Moench												
Fabaceae	*Leucaena esculenta* - (Moc. & Sessé ex DC.) Benth.												
Fabaceae	*Leucaena leucocephala* - (Lam.) de Wit												
Fabaceae	*Leucaena shannonii* - Donn. Sm.												
Fabaceae	*Lonchocarpus castilloi* - Standl.												
Fabaceae	*Lonchocarpus guatemalensis* - Benth.												
Fabaceae	*Lonchocarpus hondurensis* - Benth.												
Fabaceae	*Lonchocarpus longistylus* - Pittier												
Fabaceae	*Lonchocarpus luteomaculatus* - Pittier												
Fabaceae	*Lonchocarpus punctatus* - Kunth												
Fabaceae	*Lonchocarpus robustus* - Pittier												
Fabaceae	*Lonchocarpus rugosus* - Benth.												
Fabaceae	*Lonchocarpus violaceus* - Benth.												
Fabaceae	*Lonchocarpus xuul* - Lundell												
Fabaceae	*Lonchocarpus yucatanensis* - Pittier												
Fabaceae	*Lysiloma acapulcense* - (Kunth) Benth.												
Fabaceae	*Lysiloma bahamense* - Benth.												
Fabaceae	*Lysiloma divaricatum* - (Jacq.) J. F. Macbr.												
Fabaceae	*Lysiloma latisiliquum* - (L.) Benth.												
Fabaceae	*Machaerium falciforme* - Rudd												
Fabaceae	*Macroptilium atropurpureum* - (Moc. & Sessé ex DC.) Urb.												
Fabaceae	*Macroptilium lathyroides* - (L.) Urb.												
Fabaceae	*Mimosa albida* - Humb. & Bonpl. ex Willd.												
Fabaceae	*Mimosa bahamensis* - Benth.												
Fabaceae	*Mimosa hondurana* - Britton												
Fabaceae	*Mimosa langlassei* - Micheli												
Fabaceae	*Mimosa pigra* - L.												
Fabaceae	*Mimosa pudica* - L.												
Fabaceae	*Mimosa somnians* - Humb. & Bonpl. ex Willd.												
Fabaceae	*Mimosa tenuiflora* - (Willd.) Poir.												
Fabaceae	*Mucuna argyrophylla* - Standl.												
Fabaceae	*Mucuna pruriens* - (L.) DC.												
Fabaceae	*Neptunia plena* - (L.) Benth.												
Fabaceae	*Nissolia fruticosa* - Jacq.												
Fabaceae	*Oxyrhynchus trinervius* - (Donn. Sm.) Rudd												
Fabaceae	*Pachyrhizus erosus* - (L.) Urb.												
Fabaceae	*Pachyrhizus vernalis* - R. T. Clausen												
Fabaceae	*Parkinsonia aculeata* - L.												
Fabaceae	*Peltophorum inerme* - Naves ex Fern.-Vill.												
Fabaceae	*Phaseolus coccineus* - L.												
Fabaceae	*Phaseolus leptophyllus* - G. Don												
Fabaceae	*Phaseolus lunatus* - L.												
Fabaceae	*Phaseolus tuerckheimii* - Donn. Sm.												
Fabaceae	*Phaseolus vulgaris* - L.												
Fabaceae	*Piptadenia flava* - (Spreng. ex DC.) Benth.												
Fabaceae	*Piscidia piscipula* - (L.) Sarg.												
Fabaceae	*Pithecellobium albicans* - (Kunth) Benth.												
Fabaceae	*Pithecellobium arboreum* - (L.) Urb.												

(continued)

Table 4.1 (continued)

Family	Specie	J	F	M	A	M	J	J	A	S	O	N	D
Fabaceae	*Pithecellobium cognatum* - (Schltdl.) Benth.									■			
Fabaceae	*Pithecellobium donnell-smithii* - (Britton & Rose) Standl.						░						
Fabaceae	*Pithecellobium dulce* - (Roxb.) Benth.	░	░	░		■			■		■	■	■
Fabaceae	*Pithecellobium keyense* - Britton	░	▓	■	░	▓		▓	■		■	■	■
Fabaceae	*Pithecellobium lanceolatum* - (Humb. & Bonpl. ex Willd.) Benth.		░						░				
Fabaceae	*Pithecellobium leucospermum* - Brandegee		░		░	■							
Fabaceae	*Pithecellobium mangense* - (Jacq.) J. F. Macbr.	░	░										
Fabaceae	*Pithecellobium platylobum* - (Bertero ex DC.) Urb.	■				░							
Fabaceae	*Pithecellobium recordii* - (Britton & Rose) Standl.			■								■	
Fabaceae	*Pithecellobium seleri* - Harms				░								
Fabaceae	*Pithecellobium stevensonii* - (Standl.) Standl. & Steyerm.				■					░			
Fabaceae	*Platymiscium pinnatum* - (Jacq.) Dugand										■		
Fabaceae	*Platymiscium yucatanum* - Standl.					░	░					■	
Fabaceae	*Prosopis chilensis* - (Molina) Stuntz.						░						
Fabaceae	*Prosopis juliflora* - (Sw.) DC.	■	■	░	░		░				■	■	
Fabaceae	*Rhynchosia longeracemosa* - M. Martens & Galeotti	░	░	▓		■		■		■		■	■
Fabaceae	*Rhynchosia minima* - (L.) DC.	░	░	░	▓	░		░	■	▓	■	▓	▓
Fabaceae	*Rhynchosia swartzii* - (Vail) Urb.		░										
Fabaceae	*Rhynchosia yucatanensis* - Grear												░
Fabaceae	*Rhynchospora aristata* - Boeck.							■					
Fabaceae	*Rhynchospora cephalotes* - (L.) Vahl	■				░							
Fabaceae	*Rhynchospora colorata* - (L.) H. Pfeiff.									■			
Fabaceae	*Rhynchospora floridensis* - (Britton) H. Pfeiff.		■		■	■	■	■	░	▓	■	░	■
Fabaceae	*Rhynchospora globularis* - (Chapm.) Small											■	
Fabaceae	*Rhynchospora holoschoenoides* - (Rich.) Herter	▓					░	░	░	■			
Fabaceae	*Rhynchospora nervosa* - (Vahl) Boeck.			■									
Fabaceae	*Rhynchospora polyphylla* - (Vahl) Vahl							■	■				
Fabaceae	*Senna alata* - (L.) Roxb.		▓	■					■				
Fabaceae	*Senna atomaria* - (L.) H. S. Irwin & Barneby	░	░		░	■		▓	▓	░	░		
Fabaceae	*Senna cobanensis* - (Britton) H. S. Irwin & Barneby									░	■		░
Fabaceae	*Senna fruticosa* - (Mill.) H. S. Irwin & Barneby							■					
Fabaceae	*Senna hayesiana* - (Britton & Rose) H. S. Irwin & Barneby	░										░	
Fabaceae	*Senna hirsuta* - (L.) H. S. Irwin & Barneby	░	░								░		
Fabaceae	*Senna obtusifolia* - (L.) H. S. Irwin & Barneby				■			■		■	▓	░	
Fabaceae	*Senna occidentalis* - (L.) Link	■	■	■		░	▓	■		■	▓	░	
Fabaceae	*Senna pallida* - (Vahl) H. S. Irwin & Barneby	▓	▓	▓	▓	■	■	■		░	▓	░	■
Fabaceae	*Senna papillosa* - (Britton & Rose) H. S. Irwin & Barneby								■				
Fabaceae	*Senna pendula* - (Humb. & Bonpl. ex Willd.) H. S. Irwin & Barneby			■					■			■	
Fabaceae	*Senna peralteana* - (Kunth) H. S. Irwin & Barneby		░	■	■	░	■	░	░	░	░	░	░
Fabaceae	*Senna polyantha* - (Moc. & Sessé ex Collad.) H. S. Irwin & Barneby	■											
Fabaceae	*Senna quinquangulata* - (Rich.) H. S. Irwin & Barneby												■
Fabaceae	*Senna racemosa* - (Mill.) H. S. Irwin & Barneby		▓	▓	░		■	▓	■	■	▓	■	
Fabaceae	*Senna undulata* - (Benth.) H. S. Irwin & Barneby	░	░	▓	■	■	■			■	▓	■	░
Fabaceae	*Senna uniflora* - (Mill.) H. S. Irwin & Barneby		■	■	■	▓		■	▓	■	▓		
Fabaceae	*Senna villosa* - (Mill.) H. S. Irwin & Barneby	■	▓	■	■			■		░	▓	▓	▓
Fabaceae	*Sesbania emerus* - (Aubl.) Urb.	░		■			■	■	■		░	░	
Fabaceae	*Sesbania grandiflora* - (L.) Pers.					░						░	
Fabaceae	*Sesbania herbacea* - (Mill.) McVaugh				■								
Fabaceae	*Sesbania macrocarpa* - Muhl. ex Raf.						■						
Fabaceae	*Sophora tomentosa* - L.			░	■	■	■			■	░	░	░
Fabaceae	*Stizolobium pruriens* - (L.) Medik.		■				░					░	
Fabaceae	*Stizophyllum riparium* - (Kunth) Sandwith.	■			■	░	░	▓		░		▓	
Fabaceae	*Stylosanthes hamata* - (L.) Taub.											■	
Fabaceae	*Stylosanthes humilis* - Kunth				■							■	
Fabaceae	*Sutherlandia frutescens* - (L.) R. Br.									■	■		
Fabaceae	*Swartzia cubensis* - (Britton & P. Wilson) Standl.			░	░	░	░				■		
Fabaceae	*Sweetia panamensis* - Benth.						■						
Fabaceae	*Tamarindus indica* - L.	■	■	■		▓		░			░	▓	
Fabaceae	*Vigna candida* - (Vell.) Maréchal, Mascherpa & Stainier											■	
Fabaceae	*Vigna elegans* - (Piper) Maréchal, Mascherpa & Stainier		░					■			■	▓	■
Fabaceae	*Vigna linearis* - (Benth.) Maréchal, Mascherpa & Stainier						■						
Fabaceae	*Vigna luteola* - (Jacq.) Benth.	■							■		░	■	
Fabaceae	*Vigna speciosa* - (Kunth) VerDC.		■					■					
Fabaceae	*Vigna spectabilis* - (Standl.) A. Delgado								■				
Fabaceae	*Vigna unguiculata* - (L.) Walp.		■	░				■	■	░		░	
Fabaceae	*Vigna vexillata* - (L.) A. Rich.											■	
Fabaceae	*Zapoteca formosa* - (Kunth) H. M. Hern.	■		■	■	■			░	▓	░	░	▓
Fabaceae	*Zapoteca lambertiana* - (G. Don) H. M. Hern.						■						
Fabaceae	*Zapoteca mollis* - (Standl.) H. M. Hern.		░										
Fabaceae	*Zapoteca portoricensis* - (Jacq.) H. M. Hern.												■
Fabaceae	*Zapoteca ravenii* - H. M. Hern.		■										
Fabaceae	*Zapoteca tetragona* - (Willd.) H. M. Hern.			■									

(continued)

Table 4.1 (continued)

Family	Specie	J	F	M	A	M	J	J	A	S	O	N	D
Fabaceae	*Zygia paucijugata* - (Lundell) L. Rico			■									
Fabaceae	*Zygia recordii* - Britton & Rose				■								
Fabaceae	*Zygia stevensonii* - (Standl.) Record	■	■	■				░	░				
Fabaceae	*Centrosema angustifolium* - (Kunth) Benth.					▦							
Fabaceae	*Centrosema galeottii* - Fantz				░								
Fabaceae	*Centrosema plumieri* - (Turpin ex Pers.) Benth.	░								■	■	■	
Fabaceae	*Centrosema pubescens* - Benth.										■	■	
Fabaceae	*Centrosema schottii* - (Millsp.) K. Schum.	▦	░	░							■	▦	▦
Fabaceae	*Centrosema virginianum* - (L.) Benth.	░	░					░		▦	■	▦	░
Gentianaceae	*Centaurium quitense* - (Kunth) B. L. Rob.				■		■						
Gentianaceae	*Coutoubea spicata* - Aubl.						■					■	
Gentianaceae	*Eustoma exaltatum* - (L.) Salisb. ex G. Don			▦	░	■		■	■	■		■	■
Gentianaceae	*Leiphaimos parasitica* - Schltdl. & Cham.				■								
Gentianaceae	*Lisianthius axillaris* - (Hemsl.) Kuntze	░	░		■	▦			■	■	■	▦	
Gentianaceae	*Lisianthius saponarioides* - Schltdl. & Cham.		■			■				■			
Gentianaceae	*Lisianthius viscidiflorus* - B. L. Rob.					■							
Gentianaceae	*Voyria parasitica* - (Schltdl. & Cham.) Ruyters & Maas		■	■									
Geraniaceae	*Geranium albidum* - Rydb. ex Hanks & Small			■									
Gesneriaceae	*Achimenes grandiflora* - (Schltdl.) DC.							■					
Gesneriaceae	*Achimenes pedunculata* - Benth.							■					
Gesneriaceae	*Achimenes warszewicziana* - (Regel) H. E. Moore						■						
Gesneriaceae	*Alloplectus tetragonus* - (Hanst.) Hanst.		■										
Gesneriaceae	*Columnea erythrophaea* - Decne. ex Houllet								■				
Gesneriaceae	*Columnea schiedeana* - Schltdl.								■				
Gesneriaceae	*Gasteranthus pansamalanus* - (Donn. Sm.) Wiehler	■											
Gesneriaceae	*Kohleria spicata* - (Kunth) Oerst.									■			
Gesneriaceae	*Moussonia deppeana* - (Schltdl. & Cham.) Hanst.										■		
Gesneriaceae	*Smithiantha cinnibarina* - (Linden) Kuntze				■								
Gesneriaceae	*Drymonia strigosa* - (Oerst.) Wiehler							■					
Goodeniaceae	*Scaevola plumieri* - (L.) Vahl			■		■		▦	░	■	▦	░	
Heliconiaceae	*Heliconia spissa* - Griggs												■
Hernandiaceae	*Gyrocarpus jatrophifolius* - Domin		■										
Hydroleaceae	*Hydrolea spinosa* - L.	■										░	
Hypericaceae	*Hypericum formosum* - Kunth									■			
Icacinaceae	*Ottoschulzia pallida* - Lundell							▦	░	░			
Iridaceae	*Alophia silvestris* - (Loes.) Goldblatt		■										
Iridaceae	*Cipura campanulata* - Ravenna			■			░	░		░	░	▦	
Iridaceae	*Cipura paludosa* - Aubl.								░				
Iridaceae	*Sisyrinchium arizonicum* - Rothr.	■											
Iridaceae	*Tritonia crocosmiiflora* - G. Nicholson												■
Krameriaceae	*Krameria cuspidata* - C. Presl											■	
Lamiaceae	*Aegiphila monstrosa* - Moldenke					░							
Lamiaceae	*Aegiphila valerioi* - Standl.						■					■	
Lamiaceae	*Callicarpa acuminata* - Kunth	▦		■	■	▦	■	▦	▦	▦	░	▦	
Lamiaceae	*Clerodendrum ligustrinum* - (Jacq.) R. Br.			■	■					■			
Lamiaceae	*Clerodendrum philippinum* - Schauer											■	
Lamiaceae	*Clerodendrum speciosissimum* - C. Morren	■										■	
Lamiaceae	*Cornutia grandifolia* - (Schltdl. & Cham.) Schauer			■									■
Lamiaceae	*Cornutia pyramidata* - L.	■	■	■	■	■	■	▦	▦	■		■	■
Lamiaceae	*Hyptis albida* - Kunth	■						■					
Lamiaceae	*Hyptis capitata* - Jacq.	▦	■			■			■	■	■		
Lamiaceae	*Hyptis mutabilis* - (Rich.) Briq.											■	
Lamiaceae	*Hyptis pectinata* - (L.) Poit.									■	■		
Lamiaceae	*Hyptis suaveolens* - (L.) Poit.											■	
Lamiaceae	*Hyptis tomentosa* - Poit.												■
Lamiaceae	*Hyptis urticoides* - Kunth					■							
Lamiaceae	*Hyptis verticillata* - Jacq.						■						
Lamiaceae	*Leonotis nepetifolia* - (L.) R. Br.						░				▦		
Lamiaceae	*Mentha pulegium* - L.	■											
Lamiaceae	*Monarda fistulosa* - L.										■		
Lamiaceae	*Ocimum basilicum* - L.	■			■					■		■	■
Lamiaceae	*Ocimum micranthum* - Willd.	■		▦	░	■			▦		▦	▦	░
Lamiaceae	*Ocimum selloi* - Benth.					■							
Lamiaceae	*Salvia candicans* - M. Martens & Galeotti										■		
Lamiaceae	*Salvia chiapensis* - Fernald				■								
Lamiaceae	*Salvia cinnabarina* - M. Martens & Galeotti									■			
Lamiaceae	*Salvia coccinea* - Buc'hoz ex Etl.	░	▦	■	■						■	■	
Lamiaceae	*Salvia fernaldii* - Standl.		░									■	
Lamiaceae	*Salvia gracilis* - Benth.							■					
Lamiaceae	*Salvia hyptoides* - M. Martens & Galeotti				■		▦		■	■			
Lamiaceae	*Salvia infuscata* - Epling								■				
Lamiaceae	*Salvia karwinskii* - Benth.	■											

(continued)

Table 4.1 (continued)

Family	Specie	J	F	M	A	M	J	J	A	S	O	N	D
Lamiaceae	*Salvia lasiantha* - Benth.							■	■				
Lamiaceae	*Salvia lavanduloides* - Kunth	■						■				■	
Lamiaceae	*Salvia mexicana* - L.			■									
Lamiaceae	*Salvia micrantha* - Vahl				■								
Lamiaceae	*Salvia misella* - Kunth	■			■								■
Lamiaceae	*Salvia nitida* - (M. Martens & Galeotti) Benth.						■						
Lamiaceae	*Salvia occidentalis* - Sw.		■						■				
Lamiaceae	*Salvia polystachya* - Cav.	■				■							
Lamiaceae	*Salvia purpurea* - Cav.	■			■							■	
Lamiaceae	*Salvia riparia* - Kunth									■			
Lamiaceae	*Salvia rubiginosa* - Benth.							■	■				
Lamiaceae	*Salvia thyrsiflora* - Benth.						■						
Lamiaceae	*Salvia tiliifolia* - Vahl				■								
Lamiaceae	*Salvia townsendii* - Fernald						■						
Lamiaceae	*Salvia vitifolia* - Benth.											■	■
Lamiaceae	*Salvia xalapensis* - Benth.											■	
Lamiaceae	*Satureja macrostema* - (Moc. & Sessé ex Benth.) Briq.						■						
Lamiaceae	*Teucrium vesicarium* - Mill.										░	■	■
Lamiaceae	*Vitex gaumeri* - Greenm.	■		▓	■	▓	▓	▓	▓	▓	▓	■	■
Lamiaceae	*Vitex mollis* - Kunth											■	
Lamiaceae	*Vitex trifolia* - L.											░	■
Lamiaceae	*Scutellaria gaumeri* - Leonard									■			
Lamiaceae	*Scutellaria seleriana* - Loes.												■
Lauraceae	*Aiouea inconspicua* - van der Werff										■		
Lauraceae	*Cassytha filiformis* - L.		▓	▓		▓		■		░	░	■	■
Lauraceae	*Licaria peckii* - (I. M. Johnst.) Kosterm.				░		░		░		■		
Lauraceae	*Litsea glaucescens* - Kunth			■				■	■				
Lauraceae	*Nectandra coriacea* - (Sw.) Griseb.	■	■	■	■	▓	▓	░	▓	░	░	■	
Lauraceae	*Nectandra cuspidata* - Nees & Mart.					■							
Lauraceae	*Nectandra rudis* - C. K. Allen						■						
Lauraceae	*Nectandra salicifolia* - (Kunth) Nees		■	■	■	■	▓	▓	░	▓	▓	▓	■
Lauraceae	*Ocotea cernua* - (Nees) Mez											■	
Lauraceae	*Ocotea dendrodaphne* - Mez									■			
Lauraceae	*Ocotea helicterifolia* - (Meisn.) Hemsl.						■		■			■	
Lauraceae	*Ocotea truncata* - Lundell										■		
Lauraceae	*Persea americana* - Mill.		░	▓	░	░	■	■		■	■	■	░
Lauraceae	*Persea ferruginea* - Mez									■			
Lauraceae	*Persea liebmanni* - Mez			■									
Lauraceae	*Persea schiedeana* - Nees	■				░							
Lentibulariaceae	*Utricularia gibba* - L.											■	
Lentibulariaceae	*Utricularia purpurea* - Walter											■	
Loasaceae	*Gronovia scandens* - L.			■								■	■
Loasaceae	*Mentzelia aspera* - L.	▓		■				■					
Loasaceae	*Mentzelia hispida* - Willd.					■							
Loganiaceae	*Cynoctonum mitreola* - (L.) Britton								■				
Loganiaceae	*Spigelia anthelmia* - L.	■		■	■		░		░	▓		▓	■
Loganiaceae	*Spigelia humboldtiana* - Cham. & Schltdl.											░	
Loganiaceae	*Strychnos nigricans* - Progel			■								░	
Loganiaceae	*Strychnos panamensis* - Seem.					■							
Loganiaceae	*Strychnos tabascana* - Sprague & Sandwith												░
Loranthaceae	*Oryctanthus cordifolius* - (C. Presl) Urb.			■									
Loranthaceae	*Psittacanthus americanus* - (L.) Mart.		░			■					■	■	░
Loranthaceae	*Psittacanthus calyculatus* - (DC.) G. Don		░										■
Loranthaceae	*Psittacanthus mayanus* - Standl. & Steyerm.	■		■		■			░	■	■		
Loranthaceae	*Struthanthus cassythoides* - Millsp. ex Standl.	■		■		■	░		▓	░		▓	■
Loranthaceae	*Struthanthus densiflorus* - (Benth.) Standl.											■	
Lythraceae	*Cuphea aequipetala* - Cav.												■
Lythraceae	*Cuphea calophylla* - Cham. & Schltdl.	■											
Lythraceae	*Cuphea carthagenensis* - (Jacq.) J. F. Macbr.										■		
Lythraceae	*Cuphea hyssopifolia* - Kunth			■	■	■		■		■		■	
Lythraceae	*Cuphea utriculosa* - Koehne	▓				■		■	■	■	■	■	■
Lythraceae	*Lagerstroemia indica* - L.			■	■			■		■	■	■	■
Lythraceae	*Lawsonia inermis* - L.						░				■		
Lythraceae	*Punica granatum* - L.		■	■				░		░	■		░
Lythraceae	*Rotala ramosior* - (L.) Koehne								■				
Malpighiaceae	*Banisteriopsis cornifolia* - (Kunth) C. B. Rob.		■										
Malpighiaceae	*Bunchosia glandulosa* - (Cav.) DC.		░			░	░	░		░	░		
Malpighiaceae	*Bunchosia lanceolata* - Turcz.			■				░	■	░			
Malpighiaceae	*Bunchosia lindeniana* - A. Juss.	■	■			■					■	■	■
Malpighiaceae	*Bunchosia palmeri* - S. Watson						■						
Malpighiaceae	*Bunchosia swartziana* - Griseb.		░		░	░		░	▓		░		░
Malpighiaceae	*Byrsonima bucidaefolia* - Standl.	■	■	■				▓		■		■	

(continued)

Table 4.1 (continued)

Family	Specie	J	F	M	A	M	J	J	A	S	O	N	D
Malpighiaceae	*Byrsonima crassifolia* - (L.) Kunth		■	▓	░	▓	▓	■	░	■	▓	▓	░
Malpighiaceae	*Callaeum malpighioides* - (Turcz.) D. M. Johnson										■		
Malpighiaceae	*Gaudichaudia albida* - Schltdl. & Cham.		■										
Malpighiaceae	*Heteropterys beecheyana* - A. Juss.	░	░								░	▓	░
Malpighiaceae	*Heteropterys brachiata* - (L.) DC.			■				■			■	■	
Malpighiaceae	*Heteropterys cotinifolia* - A. Juss.										■		
Malpighiaceae	*Heteropterys laurifolia* - (L.) A. Juss.											■	
Malpighiaceae	*Heteropterys lindeniana* - A. Juss.	■						■		■		■	
Malpighiaceae	*Hintonia octomera* - (Hemsl.) Bullock	■		■	░				■	■	▓	■	■
Malpighiaceae	*Hintonia standleyana* - Bullock											■	
Malpighiaceae	*Hiraea obovata* - Huber	■										■	
Malpighiaceae	*Hiraea reclinata* - Jacq.	■		■								■	
Malpighiaceae	*Malpighia emarginata* - DC.									■			
Malpighiaceae	*Malpighia glabra* - L.	■		▓	▓			▓	░	▓	■	░	▓
Malpighiaceae	*Malpighia lundellii* - C. V. Morton		■			■			■				
Malpighiaceae	*Mascagnia hiraea* - (Gaertn.) Fawc. & Rendle				■								
Malpighiaceae	*Mascagnia sinemariensis* - (Aubl.) Griseb.				■								
Malpighiaceae	*Stigmaphyllon cordatum* - Rose			■									
Malpighiaceae	*Stigmaphyllon dichotomum* - (L.) Griseb.			■									
Malpighiaceae	*Stigmaphyllon ellipticum* - (Kunth) A. Juss.			░	░	■		░					
Malpighiaceae	*Stigmaphyllon lindenianum* - A. Juss.			▓	▓	▓					■	■	
Malpighiaceae	*Tetrapterys schiedeana* - Schltdl. & Cham.		■	■	■			░	▓	▓	▓	▓	▓
Malpighiaceae	*Tetrapterys seleriana* - Nied.	■		■		■	■	░	■		■	■	
Malvaceae	*Abelmoschus esculentus* - (L.) Moench	░		■								░	
Malvaceae	*Abutilon andrieuxii* - Hemsl.	■											
Malvaceae	*Abutilon barrancae* - M. E. Jones											■	
Malvaceae	*Abutilon permolle* - (Willd.) Sweet	░	▓	■	▓	■	■		■		■	■	▓
Malvaceae	*Abutilon tridens* - Standl. & Steyerm.	■											
Malvaceae	*Abutilon trisulcatum* - (Jacq.) Urb.											░	░
Malvaceae	*Abutilon umbellatum* - (L.) Sweet	▓		■			■					░	■
Malvaceae	*Anoda cristata* - (L.) Schltdl.				■						■	■	■
Malvaceae	*Ayenia ardua* - Cristóbal									░			
Malvaceae	*Ayenia fasciculata* - Millsp.		■				■		■				
Malvaceae	*Ayenia magna* - L.								■		░		■
Malvaceae	*Ayenia ovata* - Hemsl.									■			
Malvaceae	*Ayenia pusilla* - L.	░		■					■		░		■
Malvaceae	*Ayenia standleyi* - Cristóbal	■								░			
Malvaceae	*Bakeridesia gaumeri* - (Standl.) D. M. Bates		■	░								■	
Malvaceae	*Bakeridesia notolophium* - (A. Gray) Hochr.											■	
Malvaceae	*Bakeridesia yucatana* - (Standl.) D. M. Bates		■		░				■			■	
Malvaceae	*Byttneria aculeata* - (Jacq.) Jacq.		■				■	■				▓	▓
Malvaceae	*Byttneria catalpifolia* - Jacq.							■					
Malvaceae	*Ceiba aesculifolia* - (Kunth) Britten & Baker f.	■		░	░	░	■	░	■		■	■	■
Malvaceae	*Ceiba pentandra* - (L.) Gaertn.											■	
Malvaceae	*Ceiba schottii* - Britten & Baker f.					▓	■	▓				■	
Malvaceae	*Cienfuegosia yucatanensis* - Millsp.						■					░	
Malvaceae	*Corchorus aestuans* - L.									■		░	
Malvaceae	*Corchorus orinocensis* - Kunth					▓					░		
Malvaceae	*Corchorus siliquosus* - L.	░	▓	▓	▓	░	■				■	░	░
Malvaceae	*Gossypium barbadense* - L.			■									■
Malvaceae	*Gossypium gossypioides* - (Ulbr.) Standl.											■	
Malvaceae	*Gossypium hirsutum* - L.	░	░	■	▓	▓	■	■	■	░	■	▓	
Malvaceae	*Guazuma ulmifolia* - Lam.	▓	░	■	■	▓	▓	■	■	▓	▓	▓	░
Malvaceae	*Hampea longipes* - Miranda				■								
Malvaceae	*Hampea rovirosae* - Standl.									■	■		■
Malvaceae	*Hampea trilobata* - Standl.	▓		■	■	■	▓	░		▓	▓	▓	▓
Malvaceae	*Helicteres baruensis* - Jacq.		░	░	▓	░	■	░	▓	▓	▓	▓	░
Malvaceae	*Helicteres guazumafolia* - Kunth	■			■		■			░		■	
Malvaceae	*Heliocarpus appendiculatus* - Turcz.				■								
Malvaceae	*Heliocarpus donnellsmithii* - Rose	▓	▓	■	■	■	■					■	░
Malvaceae	*Heliocarpus pallidus* - Rose				■								
Malvaceae	*Heliocarpus tomentosus* - Turcz.	■											
Malvaceae	*Herissantia crispa* - (L.) Brizicky	░											▓
Malvaceae	*Hibiscus mutabilis* - L.											■	
Malvaceae	*Hibiscus rosa-sinensis* - L.					■	■	■		■		■	
Malvaceae	*Hibiscus sabdariffa* - L.								■			▓	
Malvaceae	*Hibiscus tiliaceus* - L.				■					■			
Malvaceae	*Hibiscus tubiflorus* - DC.											■	
Malvaceae	*Hibiscus uncinellus* - DC.			■		■							
Malvaceae	*Kearnemalvastrum lacteum* - (Aiton) D. M. Bates				■	■							■
Malvaceae	*Luehea seemannii* - Triana & Planch.				■								
Malvaceae	*Luehea speciosa* - Willd.	░	░	▓	■	■	░			■		▓	■

(continued)

Table 4.1 (continued)

Family	Specie	J	F	M	A	M	J	J	A	S	O	N	D
Malvaceae	*Malachra alceifolia* - Jacq.										░		
Malvaceae	*Malachra capitata* - (L.) L.			░									
Malvaceae	*Malachra fasciata* - Jacq.					■							
Malvaceae	*Malvastrum corchorifolium* - (Desr.) Britton ex Small	░	░	■			■		■		■	■	
Malvaceae	*Malvastrum coromandelianum* - (L.) Garcke	░									■		░
Malvaceae	*Malvaviscus arboreus* - Cav.	░	▒	▒	▒	▒	■	■	░	■	▒	▒	■
Malvaceae	*Malvaviscus urticifolius* - (C. Presl) Fryxell	■											
Malvaceae	*Melochia hirsuta* - Cav.								■				
Malvaceae	*Melochia lupulina* - Sw.				■			■		■			
Malvaceae	*Melochia nodiflora* - Sw.	░	░	■									
Malvaceae	*Melochia pyramidata* - L.	░	■			■	▒	░	■	▒			░
Malvaceae	*Melochia tomentosa* - L.		■	■	▒		■		■	■	■	▒	
Malvaceae	*Ochroma lagopus* - Sw.					■							
Malvaceae	*Ochroma pyramidale* - (Cav. ex Lam.) Urb.				■								
Malvaceae	*Pachira aquatica* - Aubl.	■					■		■		■		
Malvaceae	*Pavonia fryxellii* - Krapov.								■				
Malvaceae	*Pavonia paniculata* - Cav.				■								
Malvaceae	*Pavonia sidifolia* - Kunth				■								
Malvaceae	*Physodium corymbosum* - C. Presl						■						
Malvaceae	*Quararibea funebris* - (La Llave) Vischer		■		■		■		░	░		░	
Malvaceae	*Robinsonella cordata* - Rose & Baker f.						■						
Malvaceae	*Robinsonella lindeniana* - (Turcz.) Rose & Baker f.						■						
Malvaceae	*Robinsonella samaricarpa* - Fryxell				■								
Malvaceae	*Sida acuta* - Burm. f.	▒			■		▒			■	░	░	
Malvaceae	*Sida cordifolia* - L.	░			■								
Malvaceae	*Sida glabra* - Mill.				■								
Malvaceae	*Sida glutinosa* - Comm. ex Cav.	░		░									
Malvaceae	*Sida haenkeana* - C. Presl				■		■						
Malvaceae	*Sida malvacea* - Vell.						■						
Malvaceae	*Sida rhombifolia* - L.	░	■	■		■	■	■	■		■	▒	
Malvaceae	*Sida spinosa* - L.						■						
Malvaceae	*Sida urens* - L.						■						
Malvaceae	*Theobroma cacao* - L.						■					░	
Malvaceae	*Thespesia populnea* - (L.) Sol. ex Corrêa			▒	■		░					░	■
Malvaceae	*Triumfetta grandiflora* - Vahl									■			
Malvaceae	*Triumfetta lappula* - L.			░									
Malvaceae	*Triumfetta semitriloba* - Jacq.	▒	░		■	■	■	■	■		■		■
Malvaceae	*Waltheria americana* - L.	■	■	▒	■	▒	■	■	░	■	▒	■	
Malvaceae	*Waltheria indica* - L.					■					■		░
Malvaceae	*Wissadula amplissima* - (L.) R. E. Fr.	■		■						░		░	░
Marantaceae	*Maranta arundinacea* - L.			■					░	░		■	
Marantaceae	*Maranta gibba* - Sm.										░		
Marantaceae	*Thalia geniculata* - L.				■							░	
Marantaceae	*Calathea macrocephala* - K. Schum										■		
Marcgraviaceae	*Souroubea loczyi* - (V. A. Rich.) de Roon							■					
Melastomataceae	*Acisanthera quadrata* - Pers.						■						
Melastomataceae	*Arthrostemma ciliatum* - Pav. ex D. Don					■		■					
Melastomataceae	*Conostegia icosandra* - (Sw. ex Wikstr.) Urb.			■									
Melastomataceae	*Conostegia superba* - D. Don ex Naudin			■									
Melastomataceae	*Conostegia xalapensis* - (Bonpl.) D. Don ex DC.	■		■				■					
Melastomataceae	*Miconia ciliata* - (Rich.) DC.					░							
Melastomataceae	*Miconia fulvostellata* - L. O. Williams			■									
Melastomataceae	*Miconia glaberrima* - (Schltdl.) Naudin						■				■		
Melastomataceae	*Miconia guatemalensis* - Cogn.											■	
Melastomataceae	*Miconia impetiginosa* - DC.							■					
Melastomataceae	*Miconia lacera* - (Bonpl.) Naudin							■		■			
Melastomataceae	*Miconia minutiflora* - (Bonpl.) DC.		■										
Melastomataceae	*Miconia oligotricha* - (DC.) Naudin											■	
Melastomataceae	*Miconia pinetorum* - Naudin									■			
Melastomataceae	*Miconia prasina* - (Sw.) DC.			░									
Melastomataceae	*Miconia triplinervis* - Ruiz & Pav.											■	
Melastomataceae	*Mouriri gleasoniana* - Standl.							■					
Meliaceae	*Cedrela odorata* - L.			■	■	■		■	■	░		░	
Meliaceae	*Guarea glabra* - Vahl	■		■					■				
Meliaceae	*Melia azederach* - L.				■	■	▒	■		■		░	
Meliaceae	*Swietenia humilis* - Zucc.				■								
Meliaceae	*Swietenia macrophylla* - King			▒		■				▒		░	
Meliaceae	*Trichilia americana* - (Sessé & Moc.) T. D. Penn.						▒	░					
Meliaceae	*Trichilia arborea* - C. DC.					■							
Meliaceae	*Trichilia cuneata* - Radlk.			░					░				
Meliaceae	*Trichilia havanensis* - Jacq.		■									░	
Meliaceae	*Trichilia hirta* - L.			░	■	■			■		▒		

(continued)

Table 4.1 (continued)

Family	Specie	J	F	M	A	M	J	J	A	S	O	N	D
Meliaceae	*Trichilia martiana* - C. DC.												
Meliaceae	*Trichilia minutiflora* - Standl.												
Meliaceae	*Trichilia moschata* - Sw.												
Meliaceae	*Trichilia pallida* - Sw.												
Menispermaceae	*Cissampelos pareira* - L.												
Menispermaceae	*Cissampelos tropaeolifolia* - DC.												
Menispermaceae	*Hyperbaena axilliflora* - (Griseb.) Urb.												
Menispermaceae	*Hyperbaena mexicana* - Miers												
Menispermaceae	*Hyperbaena winzerlingi* i - Standl.												
Menyanthaceae	*Nymphoides indica* - (L.) Kuntze												
Moraceae	*Artocarpus communis* - J. R. Forst. & G. Forst.												
Moraceae	*Brosimum alicastrum* - Sw.												
Moraceae	*Castilla elastica* - Sessé ex Cerv.												
Moraceae	*Chlorophora tinctoria* - (L.) Gaudich.												
Moraceae	*Dorstenia contrajerva* - L.												
Moraceae	*Dorstenia drakena* - L.												
Moraceae	*Dorstenia lindeniana* - Bureau												
Moraceae	*Ficus benjamina* - L.												
Moraceae	*Ficus carica* - L.												
Moraceae	*Ficus citrifolia* - Mill.												
Moraceae	*Ficus cotinifolia* - Kunth												
Moraceae	*Ficus goldmanii* - Standl.												
Moraceae	*Ficus lapathifolia* - (Liebm.) Miq.												
Moraceae	*Ficus lyrata* - Warb.												
Moraceae	*Ficus maxima* - Mill.												
Moraceae	*Ficus oblongifolia* - D. Don												
Moraceae	*Ficus obtusifolia* - Kunth												
Moraceae	*Ficus ovalis* - (Liebm.) Miq.												
Moraceae	*Ficus padifolia* - Kunth												
Moraceae	*Ficus petenensis* - Lundell												
Moraceae	*Ficus religiosa* - L.												
Moraceae	*Ficus tecolutensis* - (Liebm.) Miq.												
Moraceae	*Ficus trigonata* - L.												
Moraceae	*Trophis chorizantha* - Standl.												
Moraceae	*Trophis racemosa* - (L.) Urb.												
Muntingiaceae	*Muntingia calabura* - L.												
Musaceae	*Musa acuminata* - Colla												
Musaceae	*Musa balbisiana* - Colla												
Musaceae	*Musa paradisiaca* - L.												
Musaceae	*Musa sapientum* - L.												
Myricaceae	*Myrica cerifera* - L.												
Myrtaceae	*Calyptranthes chytraculia* - (L.) Sw.												
Myrtaceae	*Calyptranthes millspaughii* - Urb.												
Myrtaceae	*Calyptranthes pallens* - Griseb.												
Myrtaceae	*Calyptranthes schiediana* - O. Berg												
Myrtaceae	*Chamguava schippi* i - (Standl.) Landrum												
Myrtaceae	*Eugenia aeruginea* - DC.												
Myrtaceae	*Eugenia axillaris* - (Sw.) Willd.												
Myrtaceae	*Eugenia biflora* - (L.) DC.												
Myrtaceae	*Eugenia buxifolia* - Lam.												
Myrtaceae	*Eugenia capuli* - (Schltdl. & Cham.) Hook. & Arn.												
Myrtaceae	*Eugenia karwinskyana* - O. Berg												
Myrtaceae	*Eugenia laevis* - O. Berg]												
Myrtaceae	*Eugenia mayana* - Standl.												
Myrtaceae	*Eugenia oerstediana* - O. Berg												
Myrtaceae	*Eugenia origanoides* - O. Berg												
Myrtaceae	*Eugenia rhombea* - (O. Berg) Krug & Urb.												
Myrtaceae	*Eugenia tikalana* - Lundell												
Myrtaceae	*Eugenia winzerlingii* - Standl.												
Myrtaceae	*Eugenia yucatanensis* - Standl.												
Myrtaceae	*Mosiera contrerasii* - (Lundell) Landrum												
Myrtaceae	*Myrcia splendens* - (Sw.) DC.												
Myrtaceae	*Myrcianthes fragrans* - (Sw.) McVaugh												
Myrtaceae	*Pimenta dioica* - (L.) Merr.												
Myrtaceae	*Psidium guajava* - L.												
Myrtaceae	*Psidium sartorianum* - (O. Berg) Nied.												
Myrtaceae	*Ugni myricoides* - (Kunth) O. Berg												
Nyctaginaceae	*Boerhavia caribaea* - Jacq.												
Nyctaginaceae	*Boerhavia erecta* - L.												
Nyctaginaceae	*Bougainvillea buttiana* - Holttum & Standl.												
Nyctaginaceae	*Bougainvillea glabra* - Choisy												
Nyctaginaceae	*Commicarpus scandens* - (L.) Standl.												

(continued)

Table 4.1 (continued)

Family	Specie	J	F	M	A	M	J	J	A	S	O	N	D
Nyctaginaceae	*Guapira linearibracteata* - (Heimerl) Lundell												
Nyctaginaceae	*Mirabilis jalapa* - L.												
Nyctaginaceae	*Neea choriophylla* - Standl.												
Nyctaginaceae	*Neea fagifolia* - Heimerl												
Nyctaginaceae	*Neea psychotrioides* - Donn. Sm.												
Nyctaginaceae	*Okenia hypogaea* - Schltdl. & Cham.												
Nyctaginaceae	*Oxybaphus comatus* - (Small) Weath.												
Nyctaginaceae	*Pisonia aculeata* - L.												
Nyctaginaceae	*Torrubia linearibracteata* - (Heimerl) Standl.												
Nymphaeaceae	*Nymphaea ampla* - (Salisb.) DC.												
Ochnaceae	*Ouratea lucens* - (Kunth) Engl.												
Ochnaceae	*Ouratea nitida* - (Sw.) Engl.												
Oleaceae	*Chionanthus domingensis* - Lam.												
Oleaceae	*Jasminum officinale* - L.												
Oleaceae	*Jasminum sambac* - (L.) Aiton												
Onagraceae	*Fuchsia arborescens* - Sims												
Onagraceae	*Fuchsia encliandra* - (Lindl.) Breedlove												
Onagraceae	*Fuchsia macrophylla* - I. M. Johnst.												
Onagraceae	*Fuchsia paniculata* - Lindl.												
Onagraceae	*Hauya elegans* - (Hemsl.) P. H. Raven & Breedlove												
Onagraceae	*Ludwigia octovalvis* - (Jacq.) P. H. Raven												
Onagraceae	*Ludwigia sedoides* - (Bonpl.) H. Hara												
Orchidaceae	*Arpophyllum giganteum* - Hartw. ex Lindl.												
Orchidaceae	*Arpophyllum laxiflorum* - Pfitzer												
Orchidaceae	*Bletia purpurea* - (Lam.) DC.												
Orchidaceae	*Brassavola cucullata* - (L.) R. Br.												
Orchidaceae	*Brassavola nodosa* - (L.) Lindl.												
Orchidaceae	*Brassavola venosa* - Lindl.												
Orchidaceae	*Calanthe calanthoides* - (A. Rich. & Galeotti) Hamer & Garay												
Orchidaceae	*Campylocentrum tyrridion* - Garay & Dunst.												
Orchidaceae	*Cohniella ascendens* - (Lindl.) Christenson												
Orchidaceae	*Dendrophylax porrectus* - (Rchb. f.) Carlsward & Whitten												
Orchidaceae	*Dichaea graminoides* - (Sw.) Lindl.												
Orchidaceae	*Dichaea squarrosa* - Lindl.												
Orchidaceae	*Dichromanthus cinnabarinus* - (La Llave & Lex.) Garay												
Orchidaceae	*Elleanthus caricoides* - Nash												
Orchidaceae	*Elleanthus cynarocephalus* - (Rchb. f.) Rchb. f.												
Orchidaceae	*Encyclia alata* - (Bateman.) Schltr.												
Orchidaceae	*Encyclia bractescens* - (Lindl.) Hoehne												
Orchidaceae	*Encyclia guatemalensis* - (Klotzsch) Dressler & G. E. Pollard												
Orchidaceae	*Encyclia nematocaulon* - (A. Rich.) Acuña												
Orchidaceae	*Epidendrum cardiophorum* - Schltr.												
Orchidaceae	*Epidendrum chlorocorymbos* - Schltr.												
Orchidaceae	*Epidendrum flexuosum* - G. Mey.												
Orchidaceae	*Epidendrum radicans* - Pav. ex Lindl.												
Orchidaceae	*Epidendrum raniferum* - Lindl.												
Orchidaceae	*Epidendrum scriptum* - A. Rich. & Galeotti												
Orchidaceae	*Gongora truncata* - Lindl.												
Orchidaceae	*Habenaria macroceratitis* - Willd.												
Orchidaceae	*Habenaria quinqueseta* - (Michx.) Eaton												
Orchidaceae	*Harrisella porrecta* - (Rchb. f.) Fawc. & Rendle												
Orchidaceae	*Lycaste deppei* - (G. Lodd.) Lindl.												
Orchidaceae	*Maxillaria pulchra* - (Schltr.) L. O. Williams ex Correll												
Orchidaceae	*Maxillaria uncata* - Lindl.												
Orchidaceae	*Myrmecophila christinae* - Carnevali & Gómez-Juárez												
Orchidaceae	*Nidema boothii* - (Lindl.) Schltr.												
Orchidaceae	*Notylia barkeri* - Lindl.												
Orchidaceae	*Notylia orbicularis* - A. Rich. & Galeotti												
Orchidaceae	*Oncidium ascendens* - Lindl.												
Orchidaceae	*Oncidium ensatum* - Lindl.												
Orchidaceae	*Oncidium sphacelatum* - Lindl.												
Orchidaceae	*Ornithocephalus inflexus* - Lindl.												
Orchidaceae	*Platystele stenostachya* - (Rchb. f.) Garay												
Orchidaceae	*Pleurothallis broadwayi* - Ames												
Orchidaceae	*Pleurothallis tribuloides* - (Sw.) Lindl.												
Orchidaceae	*Polystachya foliosa* - (Hook.) Rchb. f.												
Orchidaceae	*Ponera glomerata* - Correll												
Orchidaceae	*Ponera striata* - Lindl.												
Orchidaceae	*Prosthechea boothiana* - (Lindl.) W. E. Higgins												
Orchidaceae	*Prosthechea ochracea* - (Lindl.) W. E. Higgins												
Orchidaceae	*Prosthechea pterocarpa* - (Lindl.) W. E. Higgins												
Orchidaceae	*Prosthechea pygmaea* - (Hook.) W. E. Higgins												

(continued)

Table 4.1 (continued)

Family	Specie	J	F	M	A	M	J	J	A	S	O	N	D
Orchidaceae	*Prosthechea radiata* - (Lindl.) W. E. Higgins												
Orchidaceae	*Prosthechea varicosa* - (Bateman ex Lindl.) W. E. Higgins												
Orchidaceae	*Restrepia xanthophthalma* - Rchb. f.												
orchidaceae	*Restrepiella ophiocephala* - (Lindl.) Garay & Dunst.												
Orchidaceae	*Rhyncholaelia digbyana* - (Lindl.) Schltr.												
Orchidaceae	*Rhynchostele cordata* - (Lindl.) Soto Arenas & Salazar												
Orchidaceae	*Rhynchostele stellata* - (Lindl.) Soto Arenas & Salazar												
orchidaceae	*Sacoila lanceolata* - (Aubl.) Garay]												
Orchidaceae	*Schomburgkia tibicinis* - (Bateman) Bateman												
Orchidaceae	*Vanilla insignis* - Ames												
Orchidaceae	*Isochilus carnosiflorus* - Lindl.												
Orchidaceae	*Isochilus major* - Schltdl. & Cham.												
Orobanchaceae	*Buchnera pusilla* - Kunth												
Orobanchaceae	*Gerardia maritima* - Benth.												
Orobanchaceae	*Seymeria laciniata* - (M. Martens & Galeotti) Standl.												
Oxalidaceae	*Oxalis frutescens* - (Kunth) Lourteig												
Oxalidaceae	*Oxalis latifolia* - Kunth												
Oxalidaceae	*Oxalis yucatanensis* - (Rose) L. Riley												
Papaveraceae	*Argemone mexicana* - L.												
Passifloraceae	*Passiflora alata* - Curtis												
Passifloraceae	*Passiflora biflora* - Lam.												
Passifloraceae	*Passiflora capsularis* - L.												
Passifloraceae	*Passiflora coriacea* - Juss.												
Passifloraceae	*Passiflora foetida* - L.												
Passifloraceae	*Passiflora obovata* - Killip												
Passifloraceae	*Passiflora palmeri* - Killip												
Passifloraceae	*Passiflora pulchella* - Kunth												
Passifloraceae	*Passiflora serratifolia* - L.												
Passifloraceae	*Passiflora suberosa* - L.												
Passifloraceae	*Passiflora yucatanensis* - Killip												
Passifloraceae	*Turnera diffusa* - Willd. ex Schult.												
Passifloraceae	*Turnera odorata* - Rich.												
Passifloraceae	*Turnera ulmifolia* - L.												
Pedaliaceae	*Sesamum indicum* - L.												
Pentaphylacaceae	*Cleyera integrifolia* - (Benth.) Choisy												
Pentaphylacaceae	*Freziera grisebachii* - Krug & Urb.												
Pentaphylacaceae	*Ternstroemia tepezapote* - Schltdl. & Cham.												
Phrymaceae	*Leucocarpus perfoliatus* - (Kunth) Benth.												
Phrymaceae	*Mimulus pachystylus* - A. L. Grant												
Phyllanthaceae	*Astrocasia tremula* - (Griseb.) G. L. Webster												
Phyllanthaceae	*Margaritaria nobilis* - L. f.												
Phyllanthaceae	*Phyllanthus acidus* - (L.) Skeels												
Phyllanthaceae	*Phyllanthus acuminatus* - Vahl												
Phyllanthaceae	*Phyllanthus carolinensis* - Walter												
Phyllanthaceae	*Phyllanthus glaucescens* - Kunth												
Phyllanthaceae	*Phyllanthus grandifolius* - L.												
Phyllanthaceae	*Phyllanthus mocinianus* - Baill.												
Phyllanthaceae	*Phyllanthus niruri* - L.												
Phyllanthaceae	*Phyllanthus nobilis* - (L. f.) Müll. Arg.												
Phytolaccaceae	*Petiveria alliacea* - L.												
Phytolaccaceae	*Phytolacca icosandra* - L.												
Picramniaceae	*Alvaradoa amorphoides* - Liebm.												
Picramniaceae	*Picramnia andicola* - Tul.												
Picramniaceae	*Picramnia antidesma* - Sw.												
Picramniaceae	*Picramnia sphaerocarpa* - Planch.												
Picramniaceae	*Picramnia teapensis* - Tul.												
Picramniaceae	*Picramnia tetramera* - Turcz.												
Pinaceae	*Pinus caribaea* - Morelet												
Piperaceae	*Peperomia asarifolia* - Schltdl. & Cham.												
Piperaceae	*Peperomia crassiuscula* - Millsp.												
Piperaceae	*Peperomia glutinosa* - Millsp.												
Piperaceae	*Peperomia obtusifolia* - (L.) A. Dietr.												
Piperaceae	*Peperomia pellucida* - (L.) Kunth												
Piperaceae	*Peperomia pereskiifolia* - (Jacq.) Kunth												
Piperaceae	*Piper aduncum* - L.												
Piperaceae	*Piper amalago* - L.												
Piperaceae	*Piper auritum* - Kunth												
Piperaceae	*Piper gaumeri* - Trel.												
Piperaceae	*Piper jacquemontianum* - Kunth												
Piperaceae	*Piper marginatum* - Jacq.												
Piperaceae	*Piper patulum* - Bertol.												
Piperaceae	*Piper psilorhachis* - C. DC.												

(continued)

Table 4.1 (continued)

Family	Specie	J	F	M	A	M	J	J	A	S	O	N	D
Piperaceae	*Piper sempervirens* - (Trel.) Lundell							░		░			
Piperaceae	*Piper umbellatum* - L.									■			
Piperaceae	*Piper yucatanense* - C. DC.					■	▓		▓	░	▓		
Piperaceae	*Piper yzabalanum* - C. DC.										■		
Piperaceae	*Pothomorphe umbellata* - (L.) Miq.		■			■							
Piriquetaceae	*Piriqueta cistoides* - G. Mey. ex Steud.		░							░	■		
Plantaginaceae	*Angelonia angustifolia* - Benth.		▓			■	■	▓	■	░	■	■	
Plantaginaceae	*Bacopa lacertosa* - Standl.									■			
Plantaginaceae	*Bacopa monnieri* - (L.) Wettst.								■			■	
Plantaginaceae	*Bacopa procumbens* - (Mill.) Greenm.											■	
Plantaginaceae	*Lophospermum atrosanguineum* - Zucc.											■	
Plantaginaceae	*Mecardonia procumbens* - (Mill.) Small								■				
Plantaginaceae	*Penstemon gentianoides* - (Kunth) Poir.											■	
Plantaginaceae	*Penstemon hartwegii* - Benth.											■	
Plantaginaceae	*Penstemon hidalgensis* - Straw									■			
Plantaginaceae	*Plantago major* - L.			▓		▓							
Plantaginaceae	*Scoparia dulcis* - L.					░		■		▓			■
Plantaginaceae	*Stemodia durantifolia* - (L.) Sw.	■									■		
Plantaginaceae	*Stemodia maritima* - L.			░							■		
Plumbaginaceae	*Plumbago auriculata* - Lam.									■			
Plumbaginaceae	*Plumbago scandens* - L.	▓	■		■			■		■	▓	▓	
Poaceae	*Aegopogon cenchroides* - Humb. & Bonpl. ex Willd.				■								
Poaceae	*Andropogon glomeratus* - (Walter) Britton, Sterns & Poggenb.								░	■	■	■	
Poaceae	*Aristida ternipes* - Cav.												░
Poaceae	*Arundinella berteroniana* - (Schult.) Hitchc. & Chase							■					
Poaceae	*Arundinella deppeana* - Nees ex Steud.									■			
Poaceae	*Arundo donax* - L.										■		
Poaceae	*Bouteloua triaena* - (Trin. ex Spreng.) Scribn.										■		
Poaceae	*Brachiaria fasciculata* - (Sw.) Parodi							░					
Poaceae	*Cenchrus brownii* - Roem. & Schult.	░											
Poaceae	*Cenchrus echinatus* - L.			░					▓		░		
Poaceae	*Cenchrus incertus* - M. A. Curtis				░					░	▓	░	■
Poaceae	*Cenchrus pilosus* - Kunth								░	■			
Poaceae	*Chloris ciliata* - Sw.								░		■		
Poaceae	*Chloris virgata* - Sw.			░								■	
Poaceae	*Dactyloctenium aegyptium* - (L.) Willd.				░			▓	■			▓	■
Poaceae	*Dichanthelium dichotomum* - (L.) Gould					■							
Poaceae	*Digitaria insularis* - (L.) Fedde			■				■				■	░
Poaceae	*Echinochloa colona* - (L.) Link			■									
Poaceae	*Eleusine indica* - (L.) Gaertn.		■					░				■	
Poaceae	*Eragrostis ciliaris* - (L.) R. Br.									■	▓		
Poaceae	*Eragrostis maypurensis* - (Kunth) Steud.								■				
Poaceae	*Eragrostis prolifera* - (Sw.) Steud.		■						░			■	
Poaceae	*Eragrostis tenuifolia* - (A. Rich.) Hochst. ex Steud.							■					
Poaceae	*Hyparrhenia rufa* - (Nees) Stapf			▓									░
Poaceae	*Ichnanthus lanceolatus* - Scribn. & J. G. Sm.				■	■					░		
Poaceae	*Ichnanthus pallens* - (Sw.) Munro ex Benth.											■	▓
Poaceae	*Ichnanthus tenuis* - (J. Presl & C. Presl) Hitchc. & Chase			■									
Poaceae	*Imperata brasiliensis* - Trin.						■						
Poaceae	*Ischaemum latifolium* - (Spreng.) Kunth								■				
Poaceae	*Lasiacis divaricata* - (L.) Hitchc.	▓	▓	░			░	░	▓	░	░	░	░
Poaceae	*Lasiacis grisebachii* - (Nash) Hitchc.		░		■				■		░	░	
Poaceae	*Lasiacis nigra* - Davidse											░	
Poaceae	*Lasiacis rugelii* - (Griseb.) Hitchc.				■						░		
Poaceae	*Lasiacis ruscifolia* - (Kunth) Hitchc.	░		░				░	▓	▓	░	░	░
Poaceae	*Lasiacis sloanei* - (Griseb.) Hitchc.							■	■	░	▓	▓	
Poaceae	*Leptochloa mucronata* - (Michx.) Kunth							░				■	
Poaceae	*Leptochloa virgata* - (L.) P. Beauv.							■			░		
Poaceae	*Olyra glaberrima* - Raddi						■	■	■		░		
Poaceae	*Olyra latifolia* - L.											■	
Poaceae	*Olyra yucatana* - Chase			■									
Poaceae	*Oplismenus burmannii* - (Retz.) P. Beauv.								■				
Poaceae	*Oplismenus hirtellus* - (L.) P. Beauv.	■		▓						■			
Poaceae	*Panicum amarum* - Elliott							░				░	
Poaceae	*Panicum bartlettii* - Swallen	▓		░	░	■					▓	░	
Poaceae	*Panicum cayoense* - Swallen						■						
Poaceae	*Panicum frondescens* - G. Mey.						■		■				
Poaceae	*Panicum hirsutum* - Sw.											■	
Poaceae	*Panicum ichnanthoides* - E. Fourn.										■		
Poaceae	*Panicum maximum* - Jacq.	░											
Poaceae	*Panicum pilosum* - Sw.								■				
Poaceae	*Panicum sucosum* - Hitchc. & Chase		■								■		

(continued)

Table 4.1 (continued)

Family	Specie	J	F	M	A	M	J	J	A	S	O	N	D
Poaceae	*Panicum tenerum* - Beyr. ex Trin.		■			■							
Poaceae	*Panicum trichoides* - Sw.										░		
Poaceae	*Panicum tuerckheimii* - Hack.									■			
Poaceae	*Panicum virgatum* - L.						■					■	
Poaceae	*Paspalidium chapmanii* - (Vasey) R. W. Pohl								░				
Poaceae	*Paspalum clavuliferum* - C. Wright									░			
Poaceae	*Paspalum millegrana* - Schrad.			■									
Poaceae	*Paspalum virgatum* - L.	■	■		▦				■		░		
Poaceae	*Pennisetum ciliare* - (L.) Link				■					■			
Poaceae	*Pennisetum purpureum* - Schumach.				■					■	░	■	
Poaceae	*Phragmites australis* - (Cav.) Trin. ex Steud.			■		■					■		
Poaceae	*Phragmites communis* - Trin.							■					
Poaceae	*Rhipidocladum bartlettii* - (McClure) McClure								■				
Poaceae	*Rhipidocladum racemiflorum* - (Steud.) McClure					■							
Poaceae	*Saccharum officinarum* - L.	▦							■				
Poaceae	*Schizachyrium condensatum* - (Kunth) Nees											░	
poaceae	*Setaria gracilis* - Kunth												░
poaceae	*Setaria grisebachii* - E. Fourn.				■								
poaceae	*Setaria paniculifera* - (Steud.) E. Fourn. ex Hemsl.				■					■	■		
poaceae	*Setaria parviflora* - (Poir.) Kerguélen	■			■	▦			■				
poaceae	*Setariopsis auriculata* - (E. Fourn.) Scribn.		■	■			░		░	░	■	■	
Poaceae	*Sorghastrum incompletum* - (J. Presl) Nash		■										
Poaceae	*Sorghastrum nudipes* - Nash										■		
Poaceae	*Sorghastrum setosum* - (Griseb.) Hitchc.		■							■			
Poaceae	*Sorghum bicolor* - (L.) Moench			░							■	■	░
Poaceae	*Sorghum halepense* - (L.) Pers.		■	░							■	■	■
Poaceae	*Spartina spartinae* - (Trin.) Merr. ex Hitchc.		■						■	░			
Poaceae	*Sporobolus buckleyi* - Vasey							░					
Poaceae	*Sporobolus domingensis* - (Trin.) Kunth								■		■		
Poaceae	*Sporobolus virginicus* - (L.) Kunth						■	▦	■		░		
Poaceae	*Streptochaeta spicata* - Schrad. ex Nees		■										
Poaceae	*Streptogyna americana* - C. E. Hubb.		■										
Poaceae	*Tripsacum laxum* - Nash, 1909												■
Poaceae	*Urochloa fasciculata* - (Sw.) R. D. Webster		■				▦	░	░		■	▦	■
Polygalaceae	*Polygala aparinoides* - Hook. & Arn.	■										■	
Polygalaceae	*Polygala floribunda* - Benth.	■		■								■	
Polygalaceae	*Polygala jamaicensis* - Chodat					■		■					
Polygalaceae	*Polygala paniculata* - L.								■		■	■	
Polygalaceae	*Securidaca diversifolia* - (L.) S. F. Blake			░		░							
Polygonaceae	*Antigonon cinerascens* - M. Martens & Galeotti									■			
Polygonaceae	*Antigonon leptopus* - Hook. & Arn.	■	■			■				▦		■	■
Polygonaceae	*Coccoloba acapulcensis* - Standl.	■				■	░		░				
Polygonaceae	*Coccoloba acuminata* - Kunth						■						
Polygonaceae	*Coccoloba barbadensis* - Jacq.	░									░		
Polygonaceae	*Coccoloba belizensis* - Standl.									░			
Polygonaceae	*Coccoloba caracasana* - Meisn.							■					
Polygonaceae	*Coccoloba cozumelensis* - Hemsl.	░	▦	▦	■	░		▦	▦	░	▦	▦	░
Polygonaceae	*Coccoloba diversifolia* - Jacq.	░		░		░	░		░	■	░		
Polygonaceae	*Coccoloba floribunda* - (Benth.) Lindau			■				■				■	
Polygonaceae	*Coccoloba spicata* - Lundell	■	■	■	■	▦	░	░	░			■	░
Polygonaceae	*Coccoloba swartzii* - Meisn.			■								■	
Polygonaceae	*Coccoloba uvifera* - (L.) L.		■	■	▦	▦	■	░		░	■	▦	
Polygonaceae	*Gymnopodium floribundum* - Rolfe	■	░	▦	▦	░	▦	■	■	■	■	■	▦
Polygonaceae	*Neomillspaughia emarginata* - (H. Gross) S. F. Blake	■	■	▦	■	▦	■	▦	░	▦	▦	▦	▦
Polygonaceae	*Podopterus mexicanus* - Bonpl.	■		░	░								
Polygonaceae	*Polygonum acre* - Lam.	■											
Polygonaceae	*Polygonum hydropiperoides* - Michx.					■							
Polygonaceae	*Polygonum punctatum* - Elliott											■	
Polygonaceae	*Ruprechtia chiapensis* - Lundell										■		
Pontederiaceae	*Eichhornia crassipes* - (Mart.) Solms									■			
Portulacaceae	*Portulaca oleracea* - L.		■	■	■	▦	░			■	■		
Portulacaceae	*Portulaca pilosa* - L.			■	░	▦	■	▦		■		■	
Primulaceae	*Ardisia compressa* - Kunth							■		■			
Primulaceae	*Ardisia escallonioides* - Schltdl. & Cham.	■	░	▦	▦		░	■		■		░	░
Primulaceae	*Ardisia sexpartita* - Lundell	■											
Primulaceae	*Ardisia tuerckheimii* - Donn. Sm.		■										
Primulaceae	*Gentlea venosissima* - (Ruiz & Pav.) Lundell	■											
Primulaceae	*Jacquinia aurantiaca* - W. T. Aiton			■	▦	■		▦	░	■		░	
Primulaceae	*Jacquinia flammea* - Millsp. ex Mez					░	░						
Primulaceae	*Jacquinia macrocarpa* - Cav.	░	░			▦		■	■	░	░	░	
Primulaceae	*Jacquinia pungens* - A. Gray	■	■			■							
Primulaceae	*Jacquinia schippii* - Standl.			■									

(continued)

Table 4.1 (continued)

Family	Specie	J	F	M	A	M	J	J	A	S	O	N	D
Primulaceae	*Parathesis cubana* - (A. DC.) Molinet & M. Gómez	░		■			░			▓	▓	▓	▓
Primulaceae	*Parathesis donnell-smithii* - Mez							■					
Primulaceae	*Parathesis melanosticta* - (Schltdl.) Hemsl.							■					
Primulaceae	*Parathesis serrulata* - (Sw.) Mez								■				
Primulaceae	*Rapanea myricoides* - (Schltdl.) Lundell				■								
Primulaceae	*Samolus ebracteatus* - Kunth							░			■		
Psilotaceae	*Psilotum nudum* - (L.) P. Beauv.					░							
Putranjivaceae	*Drypetes lateriflora* - (Sw.) Krug & Urb.			░	░	▓	░		░				
Ranunculaceae	*Clematis dioica* - L.	░				■		■				■	
Ranunculaceae	*Clematis grossa* - Benth.						■		■				
Ranunculaceae	*Clidemia dentata* - Pav. ex D.Don		■			░							
Ranunculaceae	*Clidemia octona* - (Bonpl.) L. O. Williams	■		░						■			
Ranunculaceae	*Clidemia petiolaris* - (Schltdl. & Cham.) Schltdl. ex Triana								░				
Resedaceae	*Forchhammeria trifoliata* - Radlk.				░	░	░						
Rhamnaceae	*Colubrina arborescens* - (Mill.) Sarg.	░	░	■	■	░	■	■	■			■	░
Rhamnaceae	*Colubrina asiatica* - (L.) Brongn.	■				■					░	■	░
Rhamnaceae	*Colubrina elliptica* - (Sw.) Brizicky & W. L. Stern	░		░				■				░	░
Rhamnaceae	*Colubrina greggii* - S. Watson	░	■	▓		░	░	■	■	▓			░
Rhamnaceae	*Colubrina guatemalensis* - Standl.				░					■			
Rhamnaceae	*Gouania eurycarpa* - Standl.	░		■						■			
Rhamnaceae	*Gouania lupuloides* - (L.) Urb.	▓	▓	▓	■	■	■	■			■	▓	░
Rhamnaceae	*Gouania polygama* - (Jacq.) Urb.						■						
Rhamnaceae	*Karwinskia humboldtiana* - (Willd. ex Roem. & Schult.) Zucc.			▓		■		░	░		░	░	
Rhamnaceae	*Krugiodendron ferreum* - (Vahl) Urb.						░		░	▓		░	■
Rhamnaceae	*Rhamnus capreifolia* - Schltdl.							■					
Rhamnaceae	*Rhamnus discolor* - (Donn. Sm.) Rose						■						
Rhamnaceae	*Sageretia elegans* - (Kunth) Brongn.					■			░	░			
Rhamnaceae	*Ziziphus mauritiana* - Lam.			■		▓	■			░			
Rhamnaceae	*Ziziphus yucatanensis* - Standl.				■		■	░					░
Rhizophoraceae	*Rhizophora mangle* - L.	■		■		▓		░					
Rosaceae	*Agrimonia gryposepala* - Wallr.			■									
Rosaceae	*Alchemilla orbiculata* - Ruiz & Pav.									■			
Rosaceae	*Crataegus pubescens* - (C. Presl) C. Presl		■										
Rosaceae	*Holodiscus argenteus* - (L. f.) Maxim.					■				■	■	■	■
Rosaceae	*Prunus brachybotrya* - Zucc.								■				
Rosaceae	*Prunus rhamnoides* - Koehne								■				
Rosaceae	*Rubus coriifolius* - Liebm.									■			
Rosaceae	*Rubus irasuensis* - Liebm.	■											
Rubiaceae	*Alibertia edulis* - (Rich.) A. Rich. ex DC.	░											
Rubiaceae	*Allenanthus hondurensis* - Standl.	■											
Rubiaceae	*Alseis yucatanensis* - Standl.			░	░		░	▓	▓	▓	▓	░	■
Rubiaceae	*Amaioua corymbosa* - Kunth		■										
Rubiaceae	*Antirhea lucida* - (Sw.) Benth. & Hook. f.	■											
Rubiaceae	*Appunia guatemalensis* - Donn. Sm.					░							
Rubiaceae	*Asemnantha pubescens* - Hook. f.	■	░		■	■	■	▓		░	▓	▓	■
Rubiaceae	*Blepharidium mexicanum* - Standl.					■			■				
Rubiaceae	*Borreria laevis* - (Lam.) Griseb.										■		
Rubiaceae	*Borreria verticillata* - (L.) G. Mey.	■	■	░	■	■		■	░		■	■	■
Rubiaceae	*Bouvardia longiflora* - (Cav.) Kunth				■								
Rubiaceae	*Bouvardia ternifolia* - (Cav.) Schltdl.				■								
Rubiaceae	*Chiococca alba* - (L.) Hitchc.	░	■	■	▓	▓	■	▓	▓	▓	▓	▓	▓
Rubiaceae	*Chiococca phaenostemon* - Schltdl.				■								
Rubiaceae	*Chomelia protracta* - (Bartl. ex DC.) Standl.					■							
Rubiaceae	*Coccocypselum herbaceum* - Aubl.											■	
Rubiaceae	*Coffea arabica* - L.											■	
Rubiaceae	*Cosmocalyx spectabilis* - Standl.								░			▓	
Rubiaceae	*Coutarea hexandra* - (Jacq.) K. Schum.									░			
Rubiaceae	*Crusea calocephala* - DC.			■		■		■					
Rubiaceae	*Crusea hispida* - (Mill.) B. L. Rob.					■							
Rubiaceae	*Deppea grandiflora* - Schltdl.		■										
Rubiaceae	*Erithalis fruticosa* - L.			■	■	▓	░	▓	▓	■	░	░	■
Rubiaceae	*Ernodea littoralis* - Sw.			░	░	▓		░		■			
Rubiaceae	*Exostema caribaeum* - (Jacq.) Roem. & Schult.		▓	■	■	■		■	▓	▓	▓	░	
Rubiaceae	*Exostema mexicanum* - A. Gray			░		■	■		░		■	░	■
Rubiaceae	*Faramea occidentalis* - (L.) A. Rich.					■				■			
Rubiaceae	*Faramea suerrensis* - (Donn. Sm.) Donn. Sm.												■
Rubiaceae	*Genipa vulcanicola* - Standl.					■							
Rubiaceae	*Gonzalagunia panamensis* - (Cav.) K. Schum.										■		
Rubiaceae	*Gonzalagunia tacanensis* - Lundell												
Rubiaceae	*Guettarda combsii* - Urb.		■	■	■	■	▓	▓	▓	▓	■	■	
Rubiaceae	*Guettarda elliptica* - Sw.						■	░	░	▓	░	░	░
Rubiaceae	*Guettarda gaumeri* - Standl.	■	░					░		░	░	■	■

(continued)

Table 4.1 (continued)

Family	Specie	J	F	M	A	M	J	J	A	S	O	N	D
Rubiaceae	*Guettarda macrosperma* - Donn. Sm.												
Rubiaceae	*Hamelia axillaris* - Sw.												
Rubiaceae	*Hamelia calycosa* - Donn. Sm.												
Rubiaceae	*Hamelia patens* - Jacq.												
Rubiaceae	*Hillia tetrandra* - Sw.												
Rubiaceae	*Hoffmannia ghiesbreghtii* - (Lem.) Hemsl.												
Rubiaceae	*Hoffmannia nicotianifolia* - (M. Martens & Galeotti) L. O. Williams												
Rubiaceae	*Hoffmannia pittieri* - Standl.												
Rubiaceae	*Isertia haenkeana* - DC.												
Rubiaceae	*Ixora acuminata* - Thwaites												
Rubiaceae	*Ixora coccinea* - L.												
Rubiaceae	*Ixora finlaysoniana* - Wall. ex G. Don												
Rubiaceae	*Lindenia rivalis* - Benth.												
Rubiaceae	*Machaonia lindeniana* - Baill.												
Rubiaceae	*Manettia reclinata* - L.												
Rubiaceae	*Morinda citrifolia* - L.												
Rubiaceae	*Morinda panamensis* - Seem.												
Rubiaceae	*Morinda royoc* - L.												
Rubiaceae	*Morinda yucatanensis* - Greenm.												
Rubiaceae	*Moringa oleifera* - Lam.												
Rubiaceae	*Oldenlandia corymbosa* - L.												
Rubiaceae	*Oldenlandia xestosperma* - B. L. Rob. & Greenm.												
Rubiaceae	*Palicourea padifolia* - (Willd. ex Roem. & Schult.) C. M. Taylor & Lorence												
Rubiaceae	*Posoqueria latifolia* - (Rudge) Roem. & Schult.												
Rubiaceae	*Psychotria chiapensis* - Standl.												
Rubiaceae	*Psychotria costivenia* - Griseb.												
Rubiaceae	*Psychotria erythrocarpa* - Schltdl.												
Rubiaceae	*Psychotria fruticetorum* - Standl.												
Rubiaceae	*Psychotria furcata* - DC.												
Rubiaceae	*Psychotria graciliflora* - Benth.												
Rubiaceae	*Psychotria grandis* - Sw.												
Rubiaceae	*Psychotria horizontalis* - Sw.												
Rubiaceae	*Psychotria limonensis* - K. Krause												
Rubiaceae	*Psychotria lundellii* - Standl.												
Rubiaceae	*Psychotria microdon* - (DC.) Urb.												
Rubiaceae	*Psychotria nervosa* - Sw.												
Rubiaceae	*Psychotria papantlensis* - (Oerst.) Hemsl.												
Rubiaceae	*Psychotria psychotriifolia* - (Seem.) Standl.												
Rubiaceae	*Psychotria pubescens* - Sw.												
Rubiaceae	*Psychotria tenuifolia* - Sw.												
Rubiaceae	*Psychotria trichotoma* - M. Martens & Galeotti												
Rubiaceae	*Psychotria veracruzensis* - Lorence & Dwyer												
Rubiaceae	*Rachicallis americana* - (Jacq.) Hitchc.												
Rubiaceae	*Randia aculeata* - L.												
Rubiaceae	*Randia albonervia* - Brandegee												
Rubiaceae	*Randia longiloba* - Hemsl.												
Rubiaceae	*Randia obcordata* - S. Watson												
Rubiaceae	*Randia truncata* - Greenm. & C. H. Thomps.												
Rubiaceae	*Richardia scabra* - L.												
Rubiaceae	*Rondeletia albida* - Lundell												
Rubiaceae	*Rondeletia amoena* - (Planch.) Hemsl.												
Rubiaceae	*Rondeletia buddleioides* - Benth.												
Rubiaceae	*Rondeletia leucophylla* - Kunth												
Rubiaceae	*Rondeletia ovandensis* - Lundell												
Rubiaceae	*Rondeletia salicifolia* - Dwyer & M. V. Hayden												
Rubiaceae	*Rondeletia stachyoidea* - Donn. Sm.												
Rubiaceae	*Rondeletia stenosiphon* - Hemsl.												
Rubiaceae	*Rondeletia strigosa* - (Benth.) Hemsl.												
Rubiaceae	*Rondeletia tacanensis* - Lundell												
Rubiaceae	*Simira salvadorensis* - (Standl.) Steyerm.												
Rubiaceae	*Spermacoce tetraquetra* - A. Rich.												
Rutaceae	*Amyris attenuata* - Standl.												
Rutaceae	*Amyris elemifera* - L.												
Rutaceae	*Amyris sylvatica* - Jacq.												
Rutaceae	*Casimiroa tetrameria* - Millsp.												
Rutaceae	*Citrus aurantifolia* - (Christm.) Swingle												
Rutaceae	*Citrus aurantium* - L.												
Rutaceae	*Citrus limetta* - Risso												
Rutaceae	*Citrus limon* - (L.) Burm. f.												
Rutaceae	*Citrus reticulata* - Blanco												
Rutaceae	*Citrus sinensis* - (L.) Osbeck												
Rutaceae	*Esenbeckia berlandieri* - Baill.												

(continued)

Table 4.1 (continued)

Family	Specie	J	F	M	A	M	J	J	A	S	O	N	D
Rutaceae	*Esenbeckia pentaphylla* - (Macfad.) Griseb.												
Rutaceae	*Esenbeckia yaaxhokob* - Lundell												
Rutaceae	*Murraya paniculata* - (L.) Jack												
Rutaceae	*Pilocarpus racemosus* - Vahl												
Rutaceae	*Ruta chalepensis* - L.												
Rutaceae	*Ruta graveolens* - L.												
Rutaceae	*Zanthoxylum belizense* - Lundell												
Rutaceae	Zanthoxylum caribaeum - Lam.												
Rutaceae	*Zanthoxylum elegantissimum* - (Engl.) P. Wilson												
Rutaceae	*Zanthoxylum fagara* - (L.) Sarg.												
Rutaceae	*Zanthoxylum pringlei* - S. Watson												
Rutaceae	*Zanthoxylum procerum* - Donn. Sm.												
Salicaceae	*Casearia commersoniana* - Cambess.												
Salicaceae	*Casearia corymbosa* - Kunth												
Salicaceae	*Casearia emarginata* - C. Wright ex Griseb.												
Salicaceae	*Casearia nitida* - (L.) Jacq.												
Salicaceae	*Casearia subsessiliflora* - Lundell												
Salicaceae	*Casearia sylvestris* - Sw.												
Salicaceae	*Laetia thamnia* - L.												
Salicaceae	*Lunania mexicana* - Brandegee												
Salicaceae	*Salix chilensis* - Molina												
Salicaceae	*Samyda yucatanensis* - Standl.												
Salicaceae	*Xylosma anisophylla* - Standl.												
Salicaceae	*Xylosma flexuosa* - (Kunth) Hemsl.												
Salicaceae	*Zuelania guidonia* - (Sw.) Britton & Millsp.												
Santalaceae	*Phoradendron calyculatum* - Trel.												
Santalaceae	*Phoradendron flavescens* - Nutt.												
Santalaceae	*Phoradendron quadrangulare* - (Kunth) Griseb.												
Santalaceae	*Phoradendron yucatanum* - Trel.												
Sapindaceae	*Allophylus camptostachys* - Radlk.												
Sapindaceae	*Allophylus cominia* - (L.) Sw.												
Sapindaceae	*Allophylus psilospermus* - Radlk.												
Sapindaceae	*Billia hippocastanum* - Peyr.												
Sapindaceae	*Blighia sapida* - K. D. Koenig												
Sapindaceae	*Cardiospermum corindum* - L.												
Sapindaceae	*Cardiospermum grandiflorum* - Sw.												
Sapindaceae	*Cardiospermum halicacabum* - L.												
Sapindaceae	*Cupania belizensis* - Standl.												
Sapindaceae	*Cupania dentata* - DC.												
Sapindaceae	*Cupania glabra* - Sw.												
Sapindaceae	*Cupania macrophylla* - Mart.												
Sapindaceae	*Dodonaea viscosa* - Jacq.												
Sapindaceae	*Exothea diphylla* - (Standl.) Lundell												
Sapindaceae	*Exothea paniculata* - (Juss.) Radlk.												
Sapindaceae	*Matayba oppositifolia* - (A. Rich.) Britton												
Sapindaceae	*Melicoccus bijugatus* - Jacq.												
Sapindaceae	*Paullinia clavigera* - Schltdl.												
Sapindaceae	*Paullinia costata* - Schltdl. & Cham.												
Sapindaceae	*Paullinia cururu* - L.												
Sapindaceae	*Paullinia fuscescens* - Kunth												
Sapindaceae	*Paullinia pinnata* - L.												
Sapindaceae	*Paullinia tomentosa* - Jacq.												
Sapindaceae	*Serjania adiantoides* - Radlk.												
Sapindaceae	*Serjania atrolineata* - C. Wright												
Sapindaceae	*Serjania caracasana* - (Jacq.) Willd.												
Sapindaceae	*Serjania goniocarpa* - Radlk.												
Sapindaceae	*Serjania macrocarpa* - Croat												
Sapindaceae	*Serjania mexicana* - (L.) Willd.												
Sapindaceae	*Serjania pterarthra* - Standl.												
Sapindaceae	*Serjania racemosa* - Schumach.												
Sapindaceae	*Serjania rhombea* - Radlk.												
Sapindaceae	*Serjania triquetra* - Radlk.												
Sapindaceae	*Serjania yucatanensis* - Standl.												
Sapindaceae	*Talisia floresii* - Standl.												
Sapindaceae	*Talisia oliviformis* - (Kunth) Radlk.												
Sapindaceae	*Thinouia tomocarpa* - Standl.												
Sapindaceae	*Thouinia canescens* - Radlk.												
Sapindaceae	*Thouinia paucidentata* - Radlk.												
Sapindaceae	*Thouinidium insigne* - (Brandegee) Radlk.												
Sapindaceae	*Urvillea ulmacea* - Kunth												
Sapotaceae	*Bumelia obtusifolia* - Humb. ex Roem. & Schult.												
Sapotaceae	*Chrysophyllum cainito* - L.												

(continued)

Table 4.1 (continued)

Family	Specie	J	F	M	A	M	J	J	A	S	O	N	D
Sapotaceae	*Chrysophyllum mexicanum* - Brandegee ex Standl.												
Sapotaceae	*Dipholis salicifolia* - (L.) A. DC.												
Sapotaceae	*Manilkara zapota* - (L.) P. Royen												
Sapotaceae	*Pouteria amygdalina* - (Standl.) Baehni												
Sapotaceae	*Pouteria campechiana* - (Kunth) Baehni												
Sapotaceae	*Pouteria glomerata* - (Miq.) Radlk.												
Sapotaceae	*Pouteria mammosa* - (L.) Cronquist												
Sapotaceae	*Pouteria reticulata* - (Engl.) Eyma												
Sapotaceae	*Pouteria sapota* - (Jacq.) H. E. Moore & Stearn												
Sapotaceae	*Sideroxylon americanum* - (Mill.) T. D. Penn.												
Sapotaceae	*Sideroxylon celastrinum* - (Kunth) T. D. Penn.												
Sapotaceae	*Sideroxylon foetidissimum* - Jacq.												
Sapotaceae	*Sideroxylon gaumeri* - Pittier												
Sapotaceae	*Sideroxylon obtusifolium* - (Roem. & Schult.) T. D. Penn.												
Sapotaceae	*Sideroxylon salicifolium* - (L.) Lam.												
Schoepfiaceae	*Schoepfia schreberi* - J. F. Gmel.												
Scrophulariaceae	*Alonsoa caulialata* - Ruiz & Pav.												
Scrophulariaceae	*Buddleja americana* - L.												
Scrophulariaceae	*Buddleja crotonoides* - A. Gray												
Scrophulariaceae	*Buddleja megalocephala* - Donn. Sm.												
Scrophulariaceae	*Buddleja nitida* - Benth.												
Scrophulariaceae	*Buddleja sessiliflora* - Kunth												
Scrophulariaceae	*Capraria biflora* - L.												
Simaroubaceae	*Simarouba glauca* - DC.												
Siparunaceae	*Siparuna andina* - (Tul.) A. DC.												
Siparunaceae	*Siparuna pauciflora* - (Beurl.) A. DC.												
Smilacaceae	*Smilax luculenta* - Killip & C. V. Morton												
Smilacaceae	*Smilax mexicana* - Griseb. ex Kunth												
Smilacaceae	*Smilax mollis* - Humb. & Bonpl. ex Willd.												
Smilacaceae	*Smilax moranensis* - M. Martens & Galeotti												
Smilacaceae	*Smilax spinosa* - Mill.												
Solanaceae	*Athenaea cernua* - Donn. Sm.												
Solanaceae	*Brunfelsia americana* - L.												
Solanaceae	*Brunfelsia nitida* - Benth.												
Solanaceae	*Cestrum diurnum* - L.												
Solanaceae	*Cestrum dumetorum* - Schltdl.												
Solanaceae	*Cestrum nocturnum* - L.												
Solanaceae	*Cestrum racemosum* - Ruiz & Pav.												
Solanaceae	*Cestrum scandens* - Vahl												
Solanaceae	*Datura candida* - (Pers.) Saff.												
Solanaceae	*Datura inoxia* - Mill.												
Solanaceae	*Datura metel* - L.												
Solanaceae	*Datura stramonium* - L.												
Solanaceae	*Lycianthes armentalis* - J. L. Gentry												
Solanaceae	*Lycianthes geminiflora* - (M. Martens & Galeotti) Bitter												
Solanaceae	*Lycianthes heteroclita* - (Sendtn.) Bitter												
Solanaceae	*Lycianthes hypoleuca* - Standl.												
Solanaceae	*Lycianthes lenta* - (Cav.) Bitter												
Solanaceae	*Lycianthes quichensis* - (J. M. Coult. & Donn. Sm.) Bitter												
Solanaceae	*Lycianthes sideroxyloides* - (Schltdl.) Bitter												
Solanaceae	*Lycianthes virgata* - (Lam.) Bitter												
Solanaceae	*Lycium carolinianum* - Walter												
Solanaceae	*Lycopersicon esculentum* - Mill.												
Solanaceae	*Nicandra physalodes* - (L.) Gaertn.												
Solanaceae	*Nicotiana glauca* - Graham												
Solanaceae	*Nicotiana tabacum* - L.												
Solanaceae	*Physalis arborescens* - L.												
Solanaceae	*Physalis cinerascens* - (Dunal) Hitchc.												
Solanaceae	*Physalis lagascae* - Roem. & Schult.												
Solanaceae	*Physalis maxima* - Mill.												
Solanaceae	*Physalis philadelphica* - Lam.												
Solanaceae	*Physalis pubescens* - L.												
Solanaceae	*Schwenckia americana* - L.												
Solanaceae	*Solandra guttata* - D. Don												
Solanaceae	*Solandra maxima* - (Sessé & Moc.) P. S. Green												
Solanaceae	*Solandra nitida* - Zuccagni												
Solanaceae	*Solanum adhaerens* - Willd. ex Roem. & Schult.												
Solanaceae	*Solanum amazonium* - Ker Gawl.												
Solanaceae	*Solanum americanum* - Mill.												
Solanaceae	*Solanum aphyodendron* - S. Knapp												
Solanaceae	*Solanum appendiculatum* - Dunal												
Solanaceae	*Solanum asperum* - Rich.												

(continued)

Table 4.1 (continued)

Family	Specie	J	F	M	A	M	J	J	A	S	O	N	D
Solanaceae	*Solanum campechiense* - L.												
Solanaceae	*Solanum chiapasense* - K. E. Roe												
Solanaceae	*Solanum chrysotrichum* - Schltdl.												
Solanaceae	*Solanum diaboli* - Standl. & L. O. Williams												
Solanaceae	*Solanum diphyllum* - L.												
Solanaceae	*Solanum donianum* - Walp.												
Solanaceae	*Solanum douglasii* - Dunal												
Solanaceae	*Solanum erianthum* - D. Don												
Solanaceae	*Solanum hirtum* - Vahl												
Solanaceae	*Solanum hispidum* - Pers.												
Solanaceae	*Solanum lanceifolium* - Jacq.												
Solanaceae	*Solanum lanceolatum* - Cav.												
Solanaceae	*Solanum mammosum* - L.												
Solanaceae	*Solanum myrianthum* - Britton ex Rusby												
Solanaceae	*Solanum nigrescens* - M. Martens & Galeotti												
Solanaceae	*Solanum nigricans* - M. Martens & Galeotti												
Solanaceae	*Solanum nigrum* - L.												
Solanaceae	*Solanum nudum* - Dunal												
Solanaceae	*Solanum pubigerum* - Dunal												
Solanaceae	*Solanum refractum* - Hook. & Arn.												
Solanaceae	*Solanum roei* - Ugent & H. H. Iltis												
Solanaceae	*Solanum schlechtendalianum* - Walp.												
Solanaceae	*Solanum seaforthianum* - Andrews												
Solanaceae	*Solanum torvum* - Sw.												
Solanaceae	*Solanum tridynamum* - Dunal												
Solanaceae	*Solanum tuerckheimii* - Greenm.												
Solanaceae	*Solanum umbellatum* - Mill.												
Solanaceae	*Solanum verbascifolium* - L.												
Solanaceae	*Solanum wendlandii* - Hook. f.												
Solanaceae	*Capsicum annuum* - L.												
Solanaceae	*Capsicum chinense* - Jacq.												
Solanaceae	*Capsicum frutescens* - L.												
Solanaceae	*Capsicum rhomboideum* - (Dunal) Kuntze												
Solanaceae	*Jaltomata procumbens* - (Cav.) J. L. Gentry												
Solanaceae	*Juanulloa mexicana* - (Schltdl.) Miers												
Staphyleaceae	*Turpinia insignis* - (Kunth) Tul.												
Staphyleaceae	*Turpinia occidentalis* - (Sw.) G. Don												
Styracaceae	*Styrax argenteus* - C. Presl												
Surianaceae	*Suriana maritima* - L.												
Symplocaceae	*Symplocos limoncillo* - Bonpl.												
Talinaceae	*Talinum paniculatum* - (Jacq.) Gaertn.												
Talinaceae	*Talinum triangulare* - (Jacq.) Willd.												
Thymelaeaceae	*Daphnopsis americana* - (Mill.) J. R. Johnst.												
Tropaeolaceae	*Tropaeolum emarginatum* - Turcz.												
Typhaceae	*Typha domingensis* - Pers.												
Typhaceae	*Typha latifolia* - L.												
Urticaceae	*Cecropia peltata* - L.												
Urticaceae	*Myriocarpa longipes* - Liebm.												
Urticaceae	*Pilea mexicana* - Wedd.												
Urticaceae	*Pilea microphylla* - (L.) Liebm.												
Urticaceae	*Pilea riparia* - Donn. Sm.												
Urticaceae	*Urera baccifera* - (L.) Gaudich. ex Wedd.												
Urticaceae	*Urera caracasana* - (Jacq.) Gaudich. Ex Griseb.												
Verbenaceae	*Bouchea nelsonii* - Grenzeb												
Verbenaceae	*Bouchea prismatica* - (L.) Kuntze												
Verbenaceae	*Citharexylum caudatum* - L.												
Verbenaceae	*Citharexylum crassifolium* - Greenm.												
Verbenaceae	*Citharexylum donnell-smithii* - Greenm.												
Verbenaceae	*Citharexylum hexangulare* - Greenm.												
Verbenaceae	*Citharexylum hirtellum* - Standl.												
Verbenaceae	*Citharexylum schottii* - Greenm.												
Verbenaceae	*Duranta repens* - L.												
Verbenaceae	*Ghinia curassavica* - Moldenke												
Verbenaceae	*Lantana camara* - L.												
Verbenaceae	*Lantana canescens* - Kunth												
Verbenaceae	*Lantana hirta* - Graham												
Verbenaceae	*Lantana involucrata* - L.												
Verbenaceae	*Lantana microcephala* - A. Rich.												
Verbenaceae	*Lantana urticifolia* - Mill.												
Verbenaceae	*Lantana velutina* - M. Martens & Galeotti												
Verbenaceae	*Lippia alba* - (Mill.) N.E. Br. ex Britton & P. Wilson												
Verbenaceae	*Lippia controversa* - Moldenke												

(continued)

Table 4.1 (continued)

Family	Specie	J	F	M	A	M	J	J	A	S	O	N	D
Verbenaceae	*Lippia dulcis* - Trevis.							F				F	
Verbenaceae	*Lippia graveolens* - Kunth		F	F				F		F	F	F	F
Verbenaceae	*Lippia myriocephala* - Schltdl. & Cham.					F		F			F		
Verbenaceae	*Lippia nodiflora* - (L.) Michx.				F	F	F				F	F	
Verbenaceae	*Lippia nutans* - B. L. Rob. & Greenm.						F				F		
Verbenaceae	*Lippia stoechadifolia* - (L.) Kunth	F		F				F	F		F	F	
Verbenaceae	*Lippia strigulosa* - M. Martens & Galeotti				F	F	Fr			F	F		
Verbenaceae	*Lippia substrigosa* - Turcz.							F					
Verbenaceae	*Lippia umbellata* - Cav.											F	
Verbenaceae	*Petrea volubilis* - L.	F				Fr		F	F	F	F	F	F
Verbenaceae	*Phyla nodiflora* - (L.) Greene		F			F		F					
Verbenaceae	*Phyla scaberrima* - (A. Juss. ex Pers.) Moldenke						F				F		
Verbenaceae	*Phyla stoechadifolia* - (L.) Small					F							
Verbenaceae	*Priva aspera* - Kunth									F			
Verbenaceae	*Priva lappulacea* - (L.) Pers.	Fr		F	F		F/Fr	Fr	Fr	Fr		F	F
Verbenaceae	*Stachytarpheta angustifolia* - (Mill.) Vahl	F				F		F			F/Fr	F	
Verbenaceae	*Stachytarpheta cayennensis* - (Rich.) Vahl	F		F	F	F	F			F		F	
Verbenaceae	*Stachytarpheta frantzii* - Pol.	F	F	F		F/Fr	F	F		F	F	Fr	F
Verbenaceae	*Stachytarpheta guatemalensis* - Moldenke					F			F				
Verbenaceae	*Stachytarpheta jamaicensis* - (L.) Vahl	F		F		F	F	F				F	F
Verbenaceae	*Stachytarpheta miniacea* - Moldenke			F		F					F		
Verbenaceae	*Verbena carolina* - L.								F			F	
Verbenaceae	*Verbena laciniata* - (L.) Briq.			F									
Verbenaceae	*Verbena litoralis* - Kunth		F	F								F	
Verbenaceae	*Verbena longifolia* - M. Martens & Galeotti												F
Verbenaceae	*Verbesina chiapensis* - B. L. Rob. & Greenm.								F				
Verbenaceae	*Verbesina crocata* - (Cav.) Less.							F					
Verbenaceae	*Verbesina gigantea* - Jacq.		F	F	F		F				F	F	Fr
Verbenaceae	*Verbesina gigantoides* - B. L. Rob.											F	
Verbenaceae	*Verbesina neriifolia* - Hemsl.	F											
Verbenaceae	*Verbesina perymenioides* - Sch. Bip. ex Klatt	F											
Verbenaceae	*Verbesina turbacensis* - Kunth					F		F					
Violaceae	*Corynostylis arborea* - (L.) S. F. Blake			F									
Violaceae	*Hybanthus calceolaria* - (L.) Schulze-Menz			F									
Violaceae	*Hybanthus longipes* - (Dowell) Standl.												F
Violaceae	*Hybanthus thiemei* - (Donn. Sm.) C. V. Morton					F							
Violaceae	*Hybanthus yucatanensis* - Millsp.		F	Fr	F	F/Fr		Fr	F	F			
Violaceae	*Rinorea guatemalensis* - (S. Watson) Bartlett				Fr			F/Fr			Fr		
Violaceae	*Rinorea hummelii* - Sprague								Fr		F		
Violaceae	*Viola scandens* - Humb. & Bonpl. ex Roem. & Schult.									F			
Violaceae	*Orthion subsessile* - (Standl.) Steyerm. & Standl.					F							
Vitaceae	*Cissus biformifolia* - Standl.												Fr
Vitaceae	*Cissus cacuminis* - Standl.									F			
Vitaceae	*Cissus erosa* - Rich.					F			Fr	Fr			
Vitaceae	*Cissus gossypiifolia* - Standl.			F	F			F/Fr	F	Fr	F/Fr	F	F/Fr
Vitaceae	*Cissus microcarpa* - Vahl				F						Fr		
Vitaceae	*Cissus rhombifolia* - Vahl	F		F			F					F/Fr	
Vitaceae	*Cissus sicyoides* - L.	F	F	F	F	F	F/Fr		Fr	F/Fr	F/Fr	F/Fr	F
Vitaceae	*Cissus trifoliata* - (L.) L.		F			Fr	F/Fr				F	Fr	
Vitaceae	*Cissus verticillata* - (L.) Nicholson & C. E. Jarvis			F									
Vitaceae	*Vitis tiliifolia* - Humb. & Bonpl. ex Roem. & Schult.					F			F/Fr	Fr	Fr		
Vochysiaceae	*Vochysia guatemalensis* - Donn. Sm.					F							
Winteraceae	*Drimys granadensis* - L. f.							F		F			
Ximeniaceae	*Ximenia americana* - L.			F	F	Fr	F			Fr			
Zamiaceae	*Zamia loddigesii* - Miq.										Fr		
Zingiberaceae	*Alpinia purpurata* - (Vieill.) K. Schum.	F											
Zingiberaceae	*Hedychium coronarium* - J. König				F					F		F	
Zingiberaceae	*Hedychium gardnerianum* - Roscoe					F	F						
Zingiberaceae	*Zingiber officinale* - Roscoe												F
Zygophyllaceae	*Guaiacum coulteri* - A. Gray					F							
Zygophyllaceae	*Guaiacum sanctum* - L.	Fr	F	F	F/Fr	F/Fr	F/Fr	Fr	Fr	Fr		F	F
Zygophyllaceae	*Kallstroemia brachystylis* - Vail									Fr			
Zygophyllaceae	*Kallstroemia maxima* - (L.) Hook. & Arn.				F	F	F/Fr	F		F		Fr	
Zygophyllaceae	*Kallstroemia mexicana* - Hook. & Arn.					F							
Zygophyllaceae	*Tribulus cistoides* - L.	F	F/Fr	Fr	F	F	F	Fr	F/Fr			F/Fr	

Grey area: flowering; dotted area: fruiting; grey-dotted: flowering and fruiting

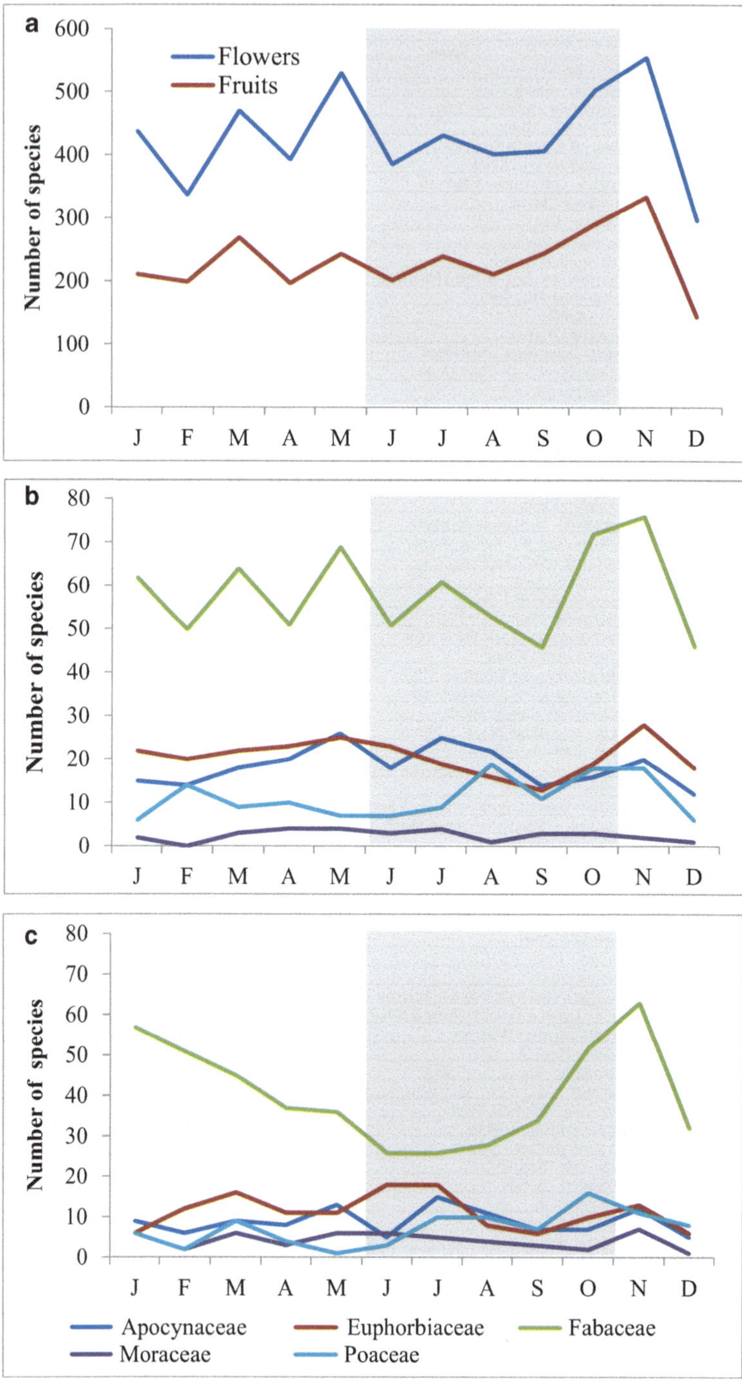

Fig. 4.2 (**a**) Patterns of flowering and fruiting of 2170 plant species distributed in the Yucatán Peninsula. (**b**) Flowering of representative families of the flora of the Yucatán Peninsula. (**c**) Fructification of representative families of the flora of the Yucatán Peninsula. The *grey area* indicates the rainy season

Fructification shows a similar behavior, recording two peaks of fructification, one peak in the dry season (March) and the most important one at the end of the rainy season (October–November). Every month at least 15 % of the species present flowering and/or fructification.

The importance of the flowering in the dry season was reported by Porter-Bolland (2003), who found that at least 30 % of the melliferous flora of Campeche flowers in the dry season. It is possible that the seasonality of the flowering is related with foliar patterns and the water status of the plants. Leaf-exchanging species (*Diospyros cuneata* and *Gymnopodium floribundum)* and deciduous soft-wood species (*Bursera simaruba*) flower in the dry season, which could be related to their capacity to retain a water reservoir in their stem, or obtaining water from underground sources (Valdez-Hernández et al. 2010). While in deciduous hard-wood species (*Apoplanesia panniculata* y *Acacia gaumeri*) flowering and fructification occurs in the rainy season, when there is sufficient water to meet the demands caused by flowering, fruiting and simultaneous foliation (Valdez-Hernández et al. 2010). In *Manilkara zapota*, an evergreen hard-wood species, Salinas-Peba and Parra-Tabla (2007) report that in forests and homegardens in Yucatán, the maximum flowering and fructification peak is found in the rainy season (June–September), although flowering and fruiting is scarce occurring throughout the year. Similar behavior has been reported in dry tropical forests (Pavón and Briones 2001; Borchert et al. 2004; Singh and Kushwaha 2006).

Patterns of reproductive phenology per family for five representative families are shown in Fig. 4.2b, c. The phenological patterns show that Fabaceae have a strong seasonality. Even this family presents flowering during the whole year (Fig. 4.2b), there is a conspicuous increase at the end of the rainy season (October–November). Fructification occurs mainly in dry season from November to February (Fig. 4.2c). A similar pattern but less obvious is shown by Poaceae. The families Apocynaceae, Moraceae and Euphorbiaceae show a constant reproductive phenology. Although no studies exist about reproductive phenology on a family level in the region, it is known that in some species, flowering and fructification can be controlled by abiotic conditions. Favorable environmental conditions can speed up or increase flowering and fructification. An increase in water and nutrient availability can be observed near permanent water sources (Valdez-Hernández et al. 2010), in managed home gardens (Salinas-Peba and Parra-Tabla 2007) and due to sporadic precipitation in the dry season known as *cabañuelas* (Porter-Bolland 2003).

4.4 Conclusions and Perspectives

The study of phenology requires monitoring of the same individuals duringover long time periods. For the Yucatán Peninsula phenological studies cover 2 years (Parra-Tabla and Vargas 2004; Valdez-Hernández et al. 2010), some studies just considered few months, as monitoring is done for specific phenophase (Salinas-

Peba and Parra-Tabla 2007). Short term studies, limit the possibility to distinguish changes in seasonality of phenological patterns related to environmental changes. The understanding of these relationships is essential in a seasonal environmental like the Yucatán Peninsula. Mainly, because of the increase in seasonality due to the effects of climate change (Orellana et al. 2009).

Changes in the periodicity of the phenology of species by environmental changes, causes change at community level. Intense and long drought events, hurricanes and fires of different intensity are frequent in the Yucatán Peninsula (Boose et al. 2003). Those factors cause changes in vegetation structure which range from defoliation or even decoupling of individuals (Sánchez and Islebe 1999; Islebe et al. 2009). Frequency and intensity of those events can modify phenological patterns as an adaptive response to new conditions. Valdez-Hernández et al. (2010) reported that an early rainy season in 2005, induced earlier leave production in two Fabaceae species (*Acacia gaumeri* and *Apoplanesia panniculata*). Availability of resources in the dry season provoked an explosion of caterpillars, which consumed all palatable leaves of Fabaceae (Campo and Dirzo 2003), causing that these plants produce new leaves at expense of low flowering and fructification.

The effect of continuous disturbances in phenology at population, species and community level is unknown. In the Yucatán Peninsula, phenology can be related to a variety of factors like water availability, wood density, number and duration of reproductive events (Valdez-Hernández et al. 2010), interspecific interactions like specificity of pollinators (Parra-Tabla and Vargas 2004; Salinas-Peba and Parra-Tabla 2007).

A complete monitoring of phenological patterns is required with the highest number of possible species, which allows prediction of behavior of plant species in terms of abundance and distribution. Also it is necessary to analyze the capacity of plant communities to keep fauna of herbivores and frugivorous (Scott and Martin 1984; Montiel et al. 2011). Finally, knowledge on phenology brings knowledge on proper management for non-timber forest resources (Salinas-Peba and Parra-Tabla 2007).

Acknowledgements I appreciate the funding of the following projects, CONABIO EC08, SEP-CONACYT 177842.

References

Anderson EN, Cauich-Canul J, Dzib A, Flores-Guido S, Islebe G, Medina-Tzuc F, Sánchez-Sánchez O, Valdez-Chale P. Those who bring the flowers: Maya ethnobotany in Quintana Roo, México. Mexico: ECOSUR; 2003.

Anderson EN. Food and feasting in the zona Maya of Quintana Roo. In: Staller JE, Carrasco MD, editors. Pre-Columbian foodways: interdisciplinary approaches to food, culture, and markets in ancient Mesoamerica. New York, NY: Springer; 2010.

Anderson EN, Medina-Tzuc F. Animals and the Maya in Southeast Mexico. Tucson, AZ: University of Arizona Press; 2005.

Bauer-Gottwein P, Gondwe BRN, Charvet G, Marín LE, Rebolledo-Vieyra M, Merediz-Alonso G. Review: the Yucatán Peninsula karst aquifer, Mexico. Hydrogeol J. 2011;19:507–24.

Boose ER, Foster D, Barker Plotkin A, Hall B. Geographical and historical variation in hurricanes across the Yucatán Peninsula. In: Gomez-Pompa A, Allen M, Fedick SL, Jimenez-Osornio JJ, editors. The lowland Maya: three millennia at the human–wildland interface. New York, NY: Haworth Press; 2003.

Borchert R. Soil and stem water storage determine phenology and distribution of tropical dry forest trees. Ecology. 1994;75(5):1437–49.

Borchert R. Response of tropical trees to rainfall seasonality and its long-term changes. Clim Change. 1998;39:381–93.

Borchert R, Rivera G, Hagnauer W. Modification of vegetative phenology in a tropical semi-deciduous forest by abnormal drought and rain. Biotropica. 2002;34:27–39.

Borchert R, Meyer SA, Felger RS, Porter-Bolland L. Environmental control of flowering periodicity in Costa Rican and Mexican tropical dry forests. Global Ecol Biogeogr. 2004;13:409–25.

Bullock SH, Solis-Magallanes A. Phenology of canopy trees of a tropical deciduous forest in Mexico. Biotropica. 1990;22(1):22–35.

Campo J, Dirzo R. Leaf quality and herbivory response to soil nutrient addition in secondary tropical dry forests of Yucatán, Mexico. J Trop Ecol. 2003;19:525–30.

Daubenmire R. Phenology and other characteristics of tropical semi-deciduous forest in north-eastern Costa Rica. J Ecol. 1972;60:147–70.

Do FC, Goudiaby VA, Gimenez O, Diagne AL, Diouf M, Rocheteau A, Akpo LE. Environmental influence on canopy phenology in the dry tropics. For Ecol Manage. 2005;215:319–28.

Duno R, Ramírez Morillo IM, Tapia-Muñoz JL, Carnevali G. Plantas vasculares de Quintana Roo. In: Pozo C, editor. Riqueza biológica del estado de Quintana Roo: un análisis para su conservación (tomo 2). ECOSUR-CONABIO-Gobierno del Estado de Quintana Roo, Programa de Pequeñas Donaciones (PNUD), México; 2011.

Eamus D, Prior L. Ecophysiology of trees of seasonally dry tropics: comparisons among phenologies. Adv Ecol Res. 2001;32:113–97.

Goldstein DJ, Hageman JB. Power plants: paleobotanical evidence of rural feasting in late classic Belize. In: Staller JE, Carrasco MD, editors. Pre-Columbian foodways: interdiscipilanary approaches to food, culture, and markets in ancient Mesoamerica. New York, NY: Springer; 2010.

Gunn JD, Folan WJ, Robichaux HR. A landscape analysis of the Candelaria watershed in Mexico: insights paleoclimates affecting upland horticulture in the southern Yucatán peninsula semi-karts. Geoarchaeology. 1995;3(1):3–42.

Holston J. Geofact or artifact? Canicular drought in northern Yucatán. Anthropos. 1986;81:696–702.

Inegi. Uso de suelo y vegetación, Serie V, Escala 1:250,000. Mexico. 2013.

Islebe GA, Torrescano-Valle N, Valdez-Hernández M, Tuz-Novelo M, Weissenberger H. Efectos del impacto del huracán Dean en la vegetación del sureste de Quintana Roo, México. Foresta Veracruzana. 2009;11:1–6.

Márdero S, Nickl E, Schmook B, Schneider L, Rogan J, Christman Z, Lawrence D. Sequías en el sur de la península de Yucatán: análisis de la variabilidad anual y estacional de la precipitación. Investigaciones Geográficas. 2012;78:19–33.

Miranda F, Hernández-X E. Los tipos de vegetación de México y su clasificación. Bol Soc Bot Méx. 1963;23(8):29–47.

Montiel S, Estrada A, Leon P. Reproductive seasonality of fruit-eating bats in northwestern Yucatán, Mexico. Acta Chiropterologica. 2011;13:139–45.

Orellana R, Espadas C, Conde C, Gay C. Atlas escenarios de cambio climático en la Península de Yucatán. Mérida, México: Centro de Investigación Científica de Yucatán; 2009.

Parra-Tabla V, Vargas CF. Phenology and phenotypic natural selection on the flowering time of a deceit-pollinated tropical orchid, *Myrmecophila christinae*. Ann Bot. 2004;94:243–50.

Pavón NP, Briones O. Phenological patterns of nine perennial plants in an intertropical semi-arid Mexican scrub. J Arid Environ. 2001;49:265–77.
Porter-Bolland L. La apicultura y el paisaje maya. Estudio sobre la fenología de floración de las especies melíferas y su relación con el ciclo apícola en La Montaña, Campeche, México. Estudios Mexicanos. 2003;19:303–30.
Salinas-Peba L, Parra-Tabla V. Phenology and pollination of *Manilkara zapota* in forest and homegardens. Forest Ecol Manag. 2007;248:136–42.
Sánchez O, Islebe GA. Hurricane Gilbert and structural changes in a tropical forest in South-eastern Mexico. Global Ecol Biogeogr. 1999;8:29–38.
Santiago LS, Wright SJ. Leaf functional traits of tropical forest plants in relation to growth form. Funct Ecol. 2007;21(1):19–27.
Sayer EJ, Newbery DM. The role of tree size in the leafing phenology of a seasonally dry tropical forest in Belize, Central America. J Trop Ecol. 2003;19:539–48.
Scott PE, Martin RF. Avian consumers of *Bursera, Ficus*, and *Ehretia* fruit in Yucatán. Biotropica. 1984;16:319–23.
Singh KP, Kushwaha CP. Diversity of flowering and fruiting phenology of trees in a tropical deciduous forest in India. Ann Bot. 2006;97:265–76.
Tuxill J, Arias-Reyes L, Latournerie L, Cob V, Jarvis DI. All maize is not equal: maize variety choices and Mayan foodways in rural Yucatán, Mexico. In: Staller JE, Carrasco MD, editors. Pre-Columbian foodways: interdisciplinary approaches to food, culture, and markets in ancient Mesoamerica. New York, NY: Springer; 2010.
Valdez-Hernández M, Andrade JL, Jackson PC, Rebolledo-Vieyra M. Phenology of five tree species of a tropical dry forest in Yucatán, Mexico: effects of environmental and physiological factors. Plant Soil. 2010;329:155–71.

Chapter 5
Physiological Ecology of Vascular Plants

Mirna Valdez-Hernández, Claudia González-Salvatierra, Casandra Reyes-García, Paula C. Jackson, and José Luis Andrade

Abstract The Yucatán Peninsula has opposing gradients of precipitation and evaporation from north to south: as precipitation increases, evaporation rates decrease. Also, due to bedrock composed primarily of limestone, the area presents high porosity and rainfall infiltrates quickly, resulting in no superficial runoff. Natural disturbances such as hurricanes and fires are also common. The interaction of these factors has created a mosaic of environmental conditions that has given rise to a series of physiological adaptations in the plant species of the area. This chapter focuses mainly on the morphophysiological responses and adaptations of native plant species of the Yucatán to natural conditions, and on how plants respond to environmental factors at the level of the individual, species, population and/or functional groups. In this chapter, species were grouped according to their metabolic pathway (C_3 species or crassulacean acid metabolism species (CAM)), and discussed based on adaptations to limiting resources: water, nutrients, temperature, and light. For C_3 species only trees were included, and studies incorporated in this section spanned from plant establishment, growth, water use and water relations, alternative water sources, and carbon and nutrient flow. For the section on CAM the focus was on the factors that affect CAM plants in the microenvironments in which they occur. Other issues analyzed include plant physiological responses to natural

M. Valdez-Hernández (✉)
Herbarium, El Colegio de la Frontera Sur Unidad Chetumal, Avenida Centenario km. 5.5, 77014 Chetumal, Quintana Roo, Mexico
e-mail: mavaldez@ecosur.mx

C. González-Salvatierra
Ciencias Ambientales, Instituto Potosino de Investigación Científica y Tecnológica, Camino a la Presa San Jose 2055, Col. Lomas 4ta sección, 78216 San Luis Potosí, San Luis Potosí, Mexico
e-mail: claudia.gonzalez@ipicyt.edu.mx

C. Reyes-García • J.L. Andrade
Unidad de Recursos Naturales, Centro de Investigación Científica de Yucatán, Calle 43 No. 130, Col. Chuburná de Hidalgo, 97200 Mérida, Yucatán, Mexico
e-mail: creyes@cicy.mx; andrade@cicy.mx

P.C. Jackson
Department of Ecology, Evolution and Organismal Biology, Kennesaw State University, 370 Paulding Avenue Bldg.12, Rm.308, Kennesaw 30144, GA, USA
e-mail: pjackson@kennesaw.edu

G.A. Islebe et al. (eds.), *Biodiversity and Conservation of the Yucatán Peninsula*,
DOI 10.1007/978-3-319-06529-8_5

disturbances, the potential impact of climate change on plant populations, and gaps in information as well as additional perspectives of study.

Keywords Tropical dry forests • Water stress • Microenvironments • Precipitation gradient • Photosynthesis • Carbon stable isotopes • Light stress • Ecophysiology • Epiphytes • CAM

5.1 Introduction

The floristic composition of the woody vegetation of the Yucatán Peninsula resembles that of other dry forests in the Caribbean and Mesoamerica that have limestone as parent material (White and Hood 2004). In the Yucatán Peninsula, differences in rainfall and temperature play a major role in floristic change, both directly through climate effects on vegetation, and indirectly through climate effects on other factors such as pH, organic matter, and soil depth (White and Hood 2004). Due to lime dissolution, rock fractures, cavities, and soil pockets are sometimes formed in the soil/bedrock profile (Querejeta et al. 2006, 2007). These fractures and soil pockets may modify water retention of different soil layers (Estrada-Medina et al. 2013), which may influence the phenology of the local vegetation (Valdez-Hernández et al. 2010).

Vegetation types found in the Yucatán Peninsula include extensive areas of mangrove, particularly along the West and East coasts (Espejel 1987). One particular type of mangrove forest stands out on the West coast around springs, easily identified as large domes from the air (called 'petenes'; Zaldivar-Jimenez et al. 2010). Trees of up to 6 m in height are found growing in the extreme conditions of coastal sand dunes, with annual rainfall between 300 and 400 mm and average annual temperatures of 30 °C (Torres et al. 2010; Flores Guido et al. 2010). Further inland, parallel to the coast, seasonally dry tropical forests are found, with an annual rainfall from 700 to 1000 mm, a dry season (<100 mm per month) of 4–7 months, and shallow soils (≤ 30 cm) (Valdez-Hernández et al. 2010; Estrada-Medina et al. 2013). In these seasonally dry tropical forests, trees are taller than those on the sand dunes, reaching about 12 m, and legumes represent the main tree family (Flores Guido et al. 2010; Reyes-García et al. 2012a).

In the central part of the Yucatán Peninsula, plant communities vary in composition according to precipitation and soil depth (See Fig. 5.1 for rainfall patterns). Seasonally dry tropical forests are found in areas with rainfall between 900 and 1100 mm annually, dry seasons of up to 6 months, and shallow soils. Here the canopy reaches between 15 and 25 m, and between 50 and 75 % of the trees are leafless in the dry season. Trees in the Fabaceae, Sapotaceae, and Boraginaceae families are among the most common (Cabrera et al. 1982; Sánchez and Islebe 2002). In areas with an annual rainfall of 1000–1500 mm, where the carbonate rock layer is naturally eroded, communities with a higher proportion of evergreen species are found (50–75 %). Here the canopy also reaches between 15 and 25 m. Common tree families include Fabaceae, Sapotaceae, and Moraceae (Cabrera et al. 1982; Sánchez and Islebe 2002). In the south, where precipitation is higher

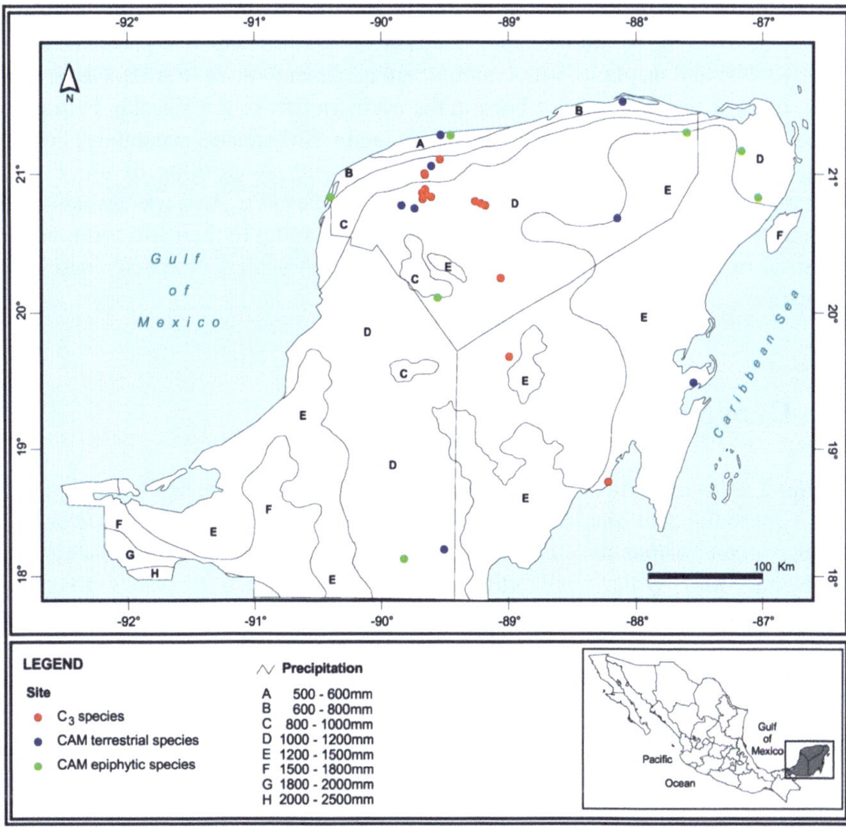

Fig. 5.1 Location of studies on plant physiological ecology in the Yucatán Peninsula (courtesy of H. Weissenberger, El Colegio de la Frontera Sur, Mexico)

(>1500 mm per year), soils are deeper and relatively rich in organic material, and have efficient drainage; the forest is composed mainly of evergreen species (>75 %) and the canopy reaches about 25 m. Common tree families include Sapotaceae, Moraceae and Arecaceae (Cabrera et al. 1982; Sánchez and Islebe 2002).

Hurricanes are a recurring disturbance on the Yucatán Peninsula. Their frequency and intensity varies across the region, but tends to be highest in the northeast tip of the peninsula (Boose et al. 2003). The long history of disturbance, both natural (hurricanes and fires) and anthropogenic (fires, agriculture, and more recently, cattle ranching), has led plants of the Yucatán to have a series of characteristics and adaptations, many of which allow for a rapid recovery after disturbance (Boose et al. 2003; Valdez-Hernández et al. 2014, see Chap. 7).

The particular characteristics of the vegetation of the Yucatán Peninsula, and the continuous disturbances that occur there, makes their native plants an exceptional and remarkable subject for physiological studies under natural conditions. However, only 2 % (50 species) of the total vascular plant species estimated for the peninsula (2400 species; Duno et al. 2011) have been reported in physiological studies. Most of these studies have focused on species with C_3 photosynthesis (mainly trees) and on plants with crassulacean acid metabolism (CAM). Geographically, most of the studies have been in the northern part of the Yucatán Peninsula (Fig. 5.1). In this chapter we will discuss C_3 and CAM species separately. For C_3 species the focus will be on changes in the floristic composition of the woody vegetation across the Yucatán Peninsula and on studies of C_3 tree species native to the peninsula. CAM species will be considered according to their life form, either terrestrial or epiphytic and the focus will also be on studies of species native to Yucatán.

5.2 C_3 Species

In natural areas of north central Yucatán, important woody species (stems greater than 3 cm main stem diameter) include *Caesalpinia gaumeri*, *Bursera simaruba*, *Gymnopodium floribundum*, *Thouinia paucidentata* and *Piscidia piscipula* (White and Hood 2004). Water availability appears to influence tree species diversity, which increases following increase in water availability from north to south (Flores Guido et al. 2010). Tree richness and stem density are also correlated to nutrient availability (Ceccon et al. 2002); tree diversity also appears to be related to successional age, with older areas generally presenting greater diversity of tree species.

Research on C_3 species in the Yucatán has focused mainly on tree species and of the estimated 500 tree species in the peninsula (Ibarra-Manríquez et al. 1995), only about 15 species (or 3 %) have been studied. These studies have focused mainly on the response of a few morphophysiological parameters (e.g. wood density, water potential, water-use efficiency, among others) to water deficit; on the identification of the source of water used by trees, and on the community level effects of climate change and other disturbances on the ecology of the area. The main features and parameters that have been studied for trees in the Yucatán Peninsula are presented in Table 5.1.

5.2.1 Responses to Water Stress

As might be expected in a tropical dry forest where water is an important determinant of plant success, research on the early stages of development of deciduous species has found that seed germination is highest and seedling mortality lowest during the rainy season (Ceccon et al. 2004). Tamayo-Chim et al. (2012) looked at the effect of drought on seedlings of four legume species and one non-legume over a 2-month period. They found that all species showed detrimental effects of drought (in the form of reduced stem diameter, plant height, leaf area, and biomass), but the

Table 5.1 Select physiological and morphological traits for the most studied tree species of Yucatán

Species	Family	Wood density[1, 2, 3, 4, 5] (gr cm^{-3})	Growth Rings[6]	Midday Water potential[3, 6] (Mpa)	Phenology	$\delta^{18}O$ [7, 8] (‰)	$\delta^{13}C$ [7, 8] (‰)	Root abundance[9] (m)	Sap flow[1] (kg d^{-1})	Specific Leaf area[3, 10] (cm^2 gr^{-1})	annual growth DBH (cm)[4]
Manilkara zapota (L.) Royen	Sapotaceae	0.9	SV		E						
Thouinia paucidentata Radlk.	Sapindaceae	0.9	V		D				14		
Talisia oliviformis (Kunth) Radlk.	Sapindaceae	0.84	V		E	-2.8	-28.7				1
Brosimum alicastrum Sw.	Moraceae	0.82	SV		E	-2.1	-28.6				1
Apoplanesia paniculata Presl	Fabaceae	0.8	SV	-3.7	D				5	119	
Acacia gaumeri Blake	Fabaceae	0.78	V	-3.8	D					131	
Lysiloma latisiliquum (L.) Benth.	Fabaceae	0.78	V		D			< 2.5	16	130	
Caesalpinia gaumeri Greenm.	Fabaceae	0.74	V		D			< 2.5	12		
Caesalpinia yucatanensis Greenm.	Fabaceae	0.69	V		D						
Lonchocarpus xuul Lundell	Fabaceae	0.74	SV		D						
Chlorophora tinctoria (L.) Gaudich. ex Benth.	Moraceae	0.71	SV		E						
Piscidia piscipula (L.) Sarg.	Fabaceae	0.71	V	-4	D	-2.5	-27.4	< 2.5	8	190	3.75
Pithecellobium dulce (Roxb.) Benth.	Fabaceae	0.71	SV		D						
Mimosa bahamensis Benth.	Fabaceae	0.71	V		D						
Karwinskia humboldtiana (Schult.) Zucc.	Rhamnaceae	0.7	SV		D				3		
Malpighia glabra L.	Malpighiaceae	0.7	V		D						
Randia obcordata Watson	Rubiaceae	0.7	V		D				2		
Albizia tomentosa (Micheli) Standl.	Fabaceae	0.69	V		D						
Gymnopodium floribundum Rolfe	Polygonaceae	0.69	V	-3.4	D	-1.5	-26.9	< 2.5	6	140	
Leucaena leucocephala (Lam.) de Wit.	Fabaceae	0.69	V		D					190	3.25
Ehretia tinifolia L.	Boraginacea	0.68	SV		E						2.75
Cordia dodecandra A. DC.	Boraginacea	0.67	V		D	-1	-27.5				3.75
Thouinia paucidentata Radlk.	Sapindaceae	0.67	V		D				14		
Diospyros cuneata Standl.	Ebenaceae	0.64		-2.6	E			< 2.5	2	73	2.75
Guazuma ulmifolia Lam.	Malvaceae	0.63	V		D					150	3
Senna racemosa (Mill.) Irwin & Barneby	Fabaceae	0.62	SV		D				7		3
Guettarda elliptica Sw.	Rubiaceae	0.58	V		D				3		
Trichilia hirta L.	Meliaceae	0.58	V		D						
Vitex gaumeri Greenm.	Lamiaceae	0.56	V		D						
Capparis incana Kunth	Capparaceae	0.54	SV		D						
Cedrela odorata L.	Meliaceae	0.44	V		D		-25.5				
Enterolobium cyclocarpum (Jacq.)	Fabaceae	0.39	SV		D	-2.7	-25.3				
Spondias purpurea L.	Anacardiaceae	0.37	SV		D	-1	-26.5				
Ficus benjamina L.	Moraceae	0.36	I		E						
Ficus cotinifolia Kunth	Moraceae	0.36	I		E	-0.8	-27.2				
Bursera simaruba (L.) Sarg.	Burseraceae	0.31	SV	-0.9	D			< 2.5 and > 5	32	127	4.5
Erythrina standleyana Krukoff	Fabaceae	0.23	I		D						

Species are listed from higher (dark grey) to lower (light grey) wood density. Letters under growth rings indicate level of visibility of growth rings for a given species, with "V" indicating clearly visible and distinguishable growth rings; "SV" indicating growth rings were semi-visible, or present but not easily distinguished; and "I" indicating growth rings were indistinguishable. Leaf phenology: "D" for deciduous and "E" evergreen. The numbers in superscripts indicate the source of the data [1]Reyes-García et al. (2012a), [2]Chan-Dzul (2010), [3]Valdez-Hernández et al. (2010), [4]Palomo-Kumul (2013), [5]Valdez-Hernández (unpublished data), [6]Roig et al. (2005), [7]Estrada-Medina et al. (2013b), [8]Querejeta et al. (2007), [9]Estrada-Medina et al. (2013), [10]Tamayo-Chim et al. (2012), [11]Reyes-García et al. (unpublished data)

extent of the effect on the seedlings and the extent of water limitation that triggered a response differed among species. The legume species (*Albizia lebbeck*, *Leucaena leucocephala*, *Piscidia piscipula*, and *Lysiloma latisiliquum*) displayed higher drought resistance compared to the non-legume species included in the study (*Guazuma ulmifolia*, Malvaceae). A different study, also looked at responses to water stress with deciduous species, including *L. leucocephala*, *P. piscipula*, and *Cordia dodecandra*, but determined the effect of water stress at the time the seedlings showed the first signs of wilt, such that the duration of each treatment differed among species, and no effect on growth (height, root length and biomass) or on water-use efficiency (ratio of carbon gained to water loss, Lambers et al. 1998) was found (Luna-Flores et al. 2012). The main effects of water stress in all three species were related to the water status of the plants, which under stress presented lower water potentials (20 % more negative), low evapotranspiration, low relative water content, and for legumes, lower leaf areas (Luna-Flores et al. 2012). Studies on the effect of drought on seedlings of evergreen species appear scarce and were not found in the literature.

In the Yucatán Peninsula, nutrient and water limitation interact with each other in several ways. For example, in moist soils, seed germination and seedling establishment are limited by nutrient availability: mainly phosphorous (P), potassium (K), nitrogen (N) and manganese (Mn); and in dry soils, certain nutrients (mostly P) may be more limiting than others (des Grades 2006). It has been found that seedlings of deciduous trees such as *Piscidia piscipula*, *Lysiloma latisiliquum*, *Phyllostylon brasiliense* and *Bursera simaruba* have higher growth and survival rates with the addition of P and N (Salinas-Peba et al. 2013; Ceccon et al. 2004). The effect of nutrient addition (N and P) however, may differ depending on forest age, with the addition of N aiding seedling density in younger forests (~10 years), and addition of P decreasing diversity in older forest stands (~60 years, Ceccon et al. 2004).

Re-sprouting is also important to regeneration. Trees capable of basal re-sprouting or of producing suckers are abundant throughout the Yucatán Peninsula. Furthermore, although in the understory of the dry forest areas regeneration through seeds appears to be more prevalent (Ceccon et al. 2004), seedlings regenerated through sprouts appear to survive longer than those regenerated from seed. Given the high incidence of disturbance, this mode of regeneration appears to be a faster and more successful mechanism (in terms of either establishing or maintaining a presence) in the peninsula (Valdez-Hernández et al. 2014).

Wood density can be one of the main indicators and predictors of physiology in tropical trees (Borchert 1994). Valdez-Hernández et al. (2010) found that based on wood density and phenology trees could be placed into three functional groups: deciduous hardwood, deciduous softwood, or leaf-exchanging species. They also found that proximity to an underground water source (*cenote*) could affect species differently depending on the functional group. For example, in deciduous hardwood species (*Acacia gaumeri* and *Apoplanesia paniculata*) the timing of leaf abscission differed among trees growing close (water table at ca. 2 m depth) or far (water table at ca. 10 m depth) from the *cenote*, with individuals retaining their leaves longer at sites near the *cenote*. In contrast, for the leaf- exchanging species (*Gymnopodium floribundum* and *Diospyros cuneata*), distance to the *cenote* did not seem to have a major effect on leaf production and leaf fall, but did influence the production of reproductive structures. Leaf-exchanging species had a higher proportion of flower and fruit production near the *cenote*. For the three functional groups, differences in nutrient availability (higher phosphorous at the site near the *cenote*) may have interacted with water availability in its effect on phenology. Valdez-Hernández et al. (2010) also found that functional groups differed in their water storage capacity and their minimum water potentials. Deciduous hardwood species tolerated lower leaf water potentials (less than −3.0 MPa) compared to the other two groups (deciduous softwood and leaf exchanging species); whereas deciduous softwood species (*Bursera simaruba*) presented high leaf water potentials (0 to −1 MPa), and had the capacity to store water in their stems. This water storage capacity allowed deciduous softwood trees to flower during the dry season (February–April) and produce fruit and seed in the more favorable rainy season. The leaf-exchanging species had leaf water potentials below −2.0 MPa, exhibited a high percent of leaf loss during the dry season (50–70 %) and a fast recovery with rainfall (Valdez-Hernández et al. 2010).

Studies in different parts of northern Yucatán have used isotopic signatures (in the dry season) to identify the source of water used by tree species. These studies have found a diversity of patterns of soil water use that do not appear to conform to patterns of leaf phenology. For example, evergreen species (*Brosimum alicastrum* and *Talisia oliviformis*) and the deciduous species (*Enterolobium cyclocarpum*) appeared to access water stored to depths of within 0.5–3 m of the soil-bedrock profile (Querejeta et al. 2006, 2007). However, for other evergreen species (*Ficus cotinifolia*) and deciduous species (*Cordia dodecandra* and *Spondias purpurea*) water uptake during the dry season appeared to be largely restricted to the upper soil layers (to ~20 cm deep; Querejeta et al. 2007). This shows a distribution of soil water uptake that may decrease interspecific competition. Partitioning of soil water resources has also been found in the savanna ecosystems of Brazil (Jackson et al. 1999) and used to explain maintenance of high species diversity in those ecosystems. Other deciduous species of Yucatán (e.g. *B. simaruba* and *Neomillspaughia emarginata*) were found to distribute their roots in three layers of the soil-bedrock profile (from shallow to deep: *laja*, *sascab* and *coquina*). More than 20 % of their roots reached more than 5 m deep, thus ensuring access to more reliable water sources (Estrada-Medina et al. 2013). The distribution of roots may change with season, as in the dry season the amount of roots in the *sascab* layer (2.5–5.0 m) increases, whereas in the rainy season many roots occur in the *laja* layer (0.3–2.5 m; Estrada-Medina et al. 2013). Also, differential biomass allocation to roots and stems across the peninsula follows the precipitation gradient, with plants showing a higher biomass allocation to stems in areas with higher water availability (Estrada-Medina et al. 2013). However, biomass allocation to roots does not seem to proportionally follow the stem allocation pattern with water gradient, suggesting a greater overall allocation to stem biomass in the seasonally dry tropical forests of Yucatán (Roa-Fuentes et al. 2012).

In addition to leaf loss and the use of different water sources, trees of the Yucatán present a series of other mechanisms to deal with the low water availability and high temperatures of the area. Dominant species, such as those in the legume family, typically have compound leaves with small leaf areas that lose heat quickly through convection; leaves frequently can modify their angle through pulvini. In general, these species present a low ratio of leaf area to total plant biomass (Tamayo-Chim et al. 2012), fast growth, high water use efficiency, and mainly early-seral stage colonization (Hasselquist et al. 2010a; Reyes-García et al. 2012a; Estrada-Medina et al. 2013).

Reyes-García et al. (2012a) studying sap flow in trees, found potential water conservation in legume trees through the development of heartwood in stems greater than 5 cm DBH. In that study, smaller legumes did not present notable heartwood, and volume of water moved was similar to that of other trees of similar diameter. Presence of heartwood in larger trees however, brought about notable differences, with large legumes utilizing less water compared to trees of similar diameter in other species. This apparent development of heartwood could be a potential mechanism that legumes use to add to their success in tropical dry forests.

5.2.2 Tree Growth Rings

The study of growth rings may contribute to understanding the effect environmental stressors have on the physiological response of a given tree, and to reconstruct the environment under which the tree grew and the rings were formed (Roig et al. 2005). In the Yucatán Peninsula, out of 52 tree species characterized, 35 % presented growth rings that could be clearly distinguished. In 52 % of the species growth rings could be identified with some difficulty, and only in 13 % of the species were growth rings not visible (Roig et al. 2005).

The presence of growth rings appears to be related to wood density (Fig. 5.2), with rings more clearly distinguished in trees with higher wood density (wood density > 0.6 g cm^{-3}; $p < 0.05$). For example, the evergreen species *Talisia oliviformis* and *Brosimum alicastrum* that have high wood density (~0.8 g cm^{-3}) present well defined growth rings. In deciduous species such as *Cedrela odorata* and *Bursera simaruba* that have lower wood density (~0.4 g cm^{-3}) growth rings, although identifiable, were not as clearly visible. In the evergreen species *Erythrina standleyana* and *Ficus cotinifolia*, which have still lower wood density (<0.3 g cm^{-3}), growth rings were not visible. So, in addition to seasonality of rainfall, the formation of clearly defined growth rings for tropical tree species may depend on additional factors, many of which are interrelated. Tree wood density has been found to be related to the water status of a plant (Borchert 1994), which is also related to leaf flush and leaf fall. Furthermore, the timing of leaf production and loss will affect whether a tree exhibits continual or seasonal growth, since the presence of leaves in the dry season generates the continuous activity of meristematic tissue (Roig et al. 2005), and hence will influence the presence of growth rings.

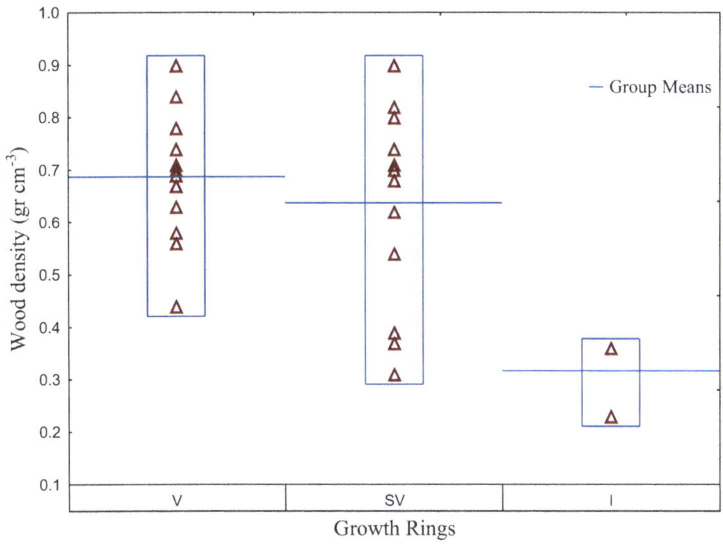

Fig. 5.2 Relationship of growth rings and wood density for 36 tree species found in the Yucatán Peninsula. *Letters* under growth rings indicate level of visibility of growth rings for a given species: "V" indicating clearly visible and distinguishable growth rings, "SV" indicates growth rings were semi-visible or present but not easily distinguished, and "I" indicates that growth rings were indistinguishable. Growth ring data were obtained from Roig et al. (2005) and wood density was obtained from Chán-Dzul (2010), Valdez-Hernández et al. (2010), Palomo-Kumul (2013) and Valdez-Hernández (unpublished data)

5.2.3 Climate Change and Disturbances

Climate change predictions include a potential increase in the frequency and or severity of hurricanes (IPCC 2013). In general, recovery of aboveground forest structure after hurricane disturbance is expected to be relatively fast (Chazdon 2003), but because the Yucatán Peninsula is also subject to frequent fires and other disturbances, the ability of a given site to recover after a hurricane may differ depending on the history of the site (Hasselquist et al. 2010a, b; Vargas et al. 2010). Hasselquist et al. (2010b), studying the response to a large hurricane disturbance in Yucatán at forest sites that differed in time since a fire, found that early and mid-stage seral sites appeared to be more resilient to the hurricane disturbance when compared to the more advanced seral stages. They also found that below-ground processes (fine root density, colonization of arbuscular mycorrhizae (AM) and ectomycorrhizae, and soil nitrogen availability) had the potential to exert strong feedback influences on above ground processes (Hasselquist et al. 2010b). Vargas et al. (2010) also found below ground effects after a hurricane. In that study, the effects after a hurricane differed among thinned and control plots, with both plots exhibiting lower frequency of spore production by AM fungi and increased AM root colonization, which was comparatively greater in the thinned plots than the control plots.

Hurricane frequency and severity can also have an effect on carbon pools both below and aboveground, with large soil carbon discharges reported following hurricane Wilma, where soil CO_2 flux was 18 % higher 1 year after the hurricane (Vargas and Allen 2008). Also, 1 year post-hurricane gross primary productivity was reduced by approximately 15 %, although the pre-hurricane values were reset after 2 years (Vargas 2012). Following a severe disturbance like a hurricane, trees of the region have the capacity to re-mobilize older reserves of carbon for the production of fine roots; an ability that increases with age, size, and volume of the tree (Vargas et al. 2009). In the Yucatán Peninsula, however, the apparent higher biomass allocation to stems together with the higher vulnerability of aboveground biomass to disturbances (Roa-Fuentes et al. 2012) may mean that these forests face greater potential vulnerability to carbon loss.

In addition to the effect of climate change on hurricane frequency and precipitation; changes in the carbon (C), nitrogen (N) and phosphorus (P) cycles are also expected. For tropical forests in general, increases in temperature are expected to stimulate the activity of soil microorganisms (Powers et al. 2009), causing an increased flow of C and N, high CO_2 emissions, and lower net primary productivity (Saxe et al. 2001). Studies that have analyzed the effect of CO_2 and nitrogen fertilization in this region have shown that the ecosystem level response to these treatments can be highly variable. In native palms, under controlled conditions of CO_2 (360 ppm and 750 ppm), CO_2 assimilation, plant growth, and the amount of pigments produced, did not show differences between CO_2 levels (Orellana et al. 2011). Also studies on litter decomposition in *Gymnopodium floribundum* and *Piscidia piscipula* under two scenarios of increased nitrogen deposition found that in moist sites with N enrichment, litter decomposition of *G. floribundum* increased, whereas litter decomposition of *P. piscipula* decreased. Dry sites showed no significant effect of N enrichment on decomposition rates (Bejarano-Castillo 2012).

5.3 CAM Species

Crassulacean acid metabolism (CAM) is an adaptation that some plants have to concentrate carbon dioxide within their cells (Cushman 2001; Dodd et al. 2002; Lüttge 2004). It is common in plants living in environments with long periods without water, but also exists in aquatic plants that inhabit places where CO_2 becomes scarce during the day (Keeley and Rundel 2003). Also, because in non-aquatic species the CO_2 assimilation of CAM occurs at night, when air temperature is lower and relative humidity is higher than during the day, it allows less water loss by transpiration and permits high water use efficiency. In fact, CAM plants have greater water use efficiency than C_3 and C_4 plants under similar conditions (Drennan and Nobel 2000; Winter et al. 2005). Consequently, CAM plants are distinctive of arid and semiarid regions and of forest canopies, as epiphytes, and represent about 6 % of vascular plant species (Winter and Smith 1996).

In the Yucatán Peninsula CAM plants are distributed according to the precipitation gradient which, as mentioned previously, increases from north to south. In the north, with short-statured communities such as dunes and dry deciduous forests, terrestrial CAM plants dominate (Ricalde et al. 2013). In the south, where taller forests exist, CAM plants are predominantly epiphytes or hemiepipytes. In fact, in the northern area, large plantations of a CAM plant, *Agave fourcroydes* (henequen) from which fiber was extracted, constituted the main economic activity of the peninsula for a century (1880s–1970s; García-Quintanilla 1986). These plantations had a large ecological impact on local vegetation, as large areas of forest were destroyed for their creation (González-Iturbe et al. 2002).

This section will focus on CAM plants and the environmental factors that affect their performance. Specifically, their responses to water availability, light, and temperature, will be considered, but also their response to some of the ecological factors that affect their microhabitats such as, deciduousness of host trees, and the distribution of neighbors. Table 5.2 summarizes all physiological studies done for the CAM species of the Yucatán Peninsula.

5.3.1 CAM Terrestrial Species

5.3.1.1 Water Relations

Terrestrial CAM plants have massive succulent stems or leaves that store substantial amounts of water, so negligible changes in photosynthesis occur in these plants during the first days of drought. Specialized storage tissues and the high water use efficiency of the CAM cycle allow continuous CO_2 uptake, since water can be transported from storage tissues to maintain turgor in the photosynthetic tissues (Barcikowski and Nobel 1984; Goldstein et al. 1991; Lüttge 2004; Andrade et al. 2009). Indeed, average relative capacitance, which is the capacity of the tissues to maintain their water potentials when the water content decreases, is more than two times higher for terrestrial CAM plants than that of epiphytic CAM species (Andrade et al. 2009).

In terrestrial Bromeliaceae (a family with many CAM species), plant size, tissue relative capacitance, leaf anatomy, and clonal reproduction all confer drought resistance. For instance, *Bromelia karatas*, which does not have a high organic acid production (González-Salvatierra et al. 2013), is ecologically one of the most important species in dry deciduous forests (Ricalde et al. 2013). Additionally, about half of the plants of the large epiphyte *Aechmea bracteata*, also a bromeliad, live and reproduce on the soil in a dry deciduous forest (Chávez-Sahagún 2014). Yet, the small volume to surface area ratio of terrestrial CAM seedlings makes them too vulnerable to drought and most of them die when the dry season starts (Nobel

Table 5.2 Published physiological studies on CAM terrestrial and epiphytic plants from the Yucatán Peninsula

Species	Family	Physiological responses studied	Reference
Terrestrial CAM			
Agave angustifolia Haw.	Agavaceae	Photosynthesis over a rainfall gradient	Ricalde et al. (2013)
		Growth and water relations	Ricalde-Pérez (2010)
Bromelia karatas L.	Bromeliaceae	Photosynthesis over a rainfall gradient	Ricalde et al. (2013)
		Photoprotective mechanisms	González-Salvatierra et al. (2010)
		Water relations, photosynthesis and leaf morphology	González-Salvatierra et al. (2013)
Acanthocereus tetragonus (L.) Hummelinck	Cactaceae	Photosynthesis over a rainfall gradient	Ricalde et al. (2013)
Hylocereus undatus (Haworth) Britton & Rose		Photosynthesis, growth and establishment	Andrade et al. (2006)
Mammillaria gaumeri (Britton & Rose) Orcutt		Germination and establishment	Cervera et al. (2006)
		Photosynthesis and growth	Cervera et al. (2007)
		Pollinators	Giovanetti et al. (2007)
Nopalea inaperta Schott ex Griffiths		Photosynthesis over a rainfall gradient	Ricalde et al. (2013)
Opuntia dillenii (Ker Gawl.) Haw.		Photosynthesis over a rainfall gradient	Ricalde et al. (2013)
Pereskiopsis scandens Britton & Rose		Photosynthesis over a rainfall gradient	Ricalde et al. (2013)
Pilosocereus gaumeri (Britton & Rose) Backeb.		Photosynthesis over a rainfall gradient	Ricalde et al. (2013)
Selenicereus donkelaarii (Salm-Dyck) Britton & Rose		Photosynthesis over a rainfall gradient	Ricalde et al. (2013)
Stenocereus eichlamii (Britton & Rose) Buxb		Photosynthesis over a rainfall gradient	Ricalde et al. (2013)
Clusia chanekiana Lundell	Clusiaceae	Type of photosynthesis	Vargas-Soto et al. (2009) Vargas-Soto (2010)
Clusia guatemalensis Hemsl.		Type of photosynthesis	Vargas-Soto et al. (2009) Vargas-Soto (2010)

(continued)

Table 5.2 (continued)

Species	Family	Physiological responses studied	Reference
Clusia flava Jacq.		Type of photosynthesis	Vargas-Soto et al. (2009)
		Photosynthesis of seedlings in a common garden	Vargas-Soto (2010)
Clusia rosea Jacq.		Type of photosynthesis	Vargas-Soto et al. (2009)
		Photosynthesis of seedlings in a common garden	Vargas-Soto (2010)
Epiphytic CAM			
Tillandsia balbisiana Schult & Schult F.	Bromeliaceae	Distribution gradients related to seasonal microclimates	Cach-Pérez et al. (2013)
		Light stratification	Chilpa-Galván et al. (2013)
		Seasonal C assimilation (stable isotopic composition)	Goodeet al. (2010)
		Distribution across occupied phorophytes	Zimmerman and Olmsted (1992)
Aechmea bracteata (Sw.) Griseb		Distribution gradients related to seasonal microclimates	Cach-Pérez et al. (2013)
		Photosynthesis over a rainfall gradient	Ricalde et al. (2013)
		Light stratification	Chilpa-Galván et al. (2013)
		Photosynthesis and water relations in two microenvironments	Chávez-Sahagún (2014)
		Light stratification	Zimmerman and Olmsted (1992)
Tillandsia brachycaulos Schltdl.		Plant characteristics and water relations	Andrade (2003)
		Distribution gradients related to seasonal microclimates	Cach-Pérez et al. (2013)
		Light microhabitats, growth and photosynthesis	Cervantes et al. (2005)
		Light stratification	Chilpa-Galván et al. (2013)
		Photoprotective mechanisms and photosynthesis	González-Salvatierra (2009)
		Photoprotective mechanisms	González-Salvatierra et al. (2010)

(continued)

Table 5.2 (continued)

Species	Family	Physiological responses studied	Reference
		Seasonal C assimilation (stable isotopic composition)	Goode et al. (2010)
		Stratification for photoprotective mechanisms and drought tolerance	Graham and Andrade (2004)
Tillandsia dasyliriifolia Baker		Distribution gradients related to seasonal microclimates	Cach-Pérez et al. (2013)
		Seasonal C assimilation (stable isotopic composition)	Goode et al. (2010)
		Photosynthesis over a rainfall gradient	Ricalde et al. (2013)
		Light stratification	Zimmerman and Olmsted (1992)
Tillandsia elongata Kunth var. subimbricata (Baker) L. B. Sm.		Plant characteristics and water relations	Andrade (2003)
		Distribution gradients related to seasonal microclimates	Cach-Pérez et al. (2013)
		Light stratification	Chilpa-Galván et al. (2013)
Tillandsia fasciculata Sw.		Distribution gradients related to seasonal microclimates	Cach-Pérez et al. (2013)
		Light stratification	Chilpa-Galván et al. (2013)
		Seasonal C assimilation (stable isotopiccomposition)	Goode et al. (2010)
Tillandsia juncea (Ruiz and Pav.) Poir.		Distribution gradients related to seasonal microclimates	Cach-Pérez et al. (2013)
Tillandsia polystachia (L.) L.		Distribution gradients related to seasonal microclimates	Cach-Pérez et al. (2013)
Tillandsia pseudobaileyi CS Gardner		Light stratification	Zimmerman and Olmsted (1992)
Tillandsia recurvata (L.) L.		Distribution gradients related to seasonal microclimates	Cach-Pérez et al. (2013)
		Light stratification	Chilpa-Galván et al. (2013)

(continued)

Table 5.2 (continued)

Species	Family	Physiological responses studied	Reference
Tillandsia schiedeana Steud.		Light stratification	Chilpa-Galván et al. (2013)
		Distribution gradients related to seasonal microclimates	Cach-Pérez et al. (2013)
Tillandsia streptophylla Scheidw.		Light stratification	Zimmerman and Olmsted (1992)
Tillandsia usneoides (L.) L.		Distribution gradients related to seasonal microclimates	Cach-Pérez et al. (2013)
Tillandsia utriculata L.		Distribution gradients related to seasonal microclimates	Cach-Pérez et al. (2013)
Tillandsia variabilis Schltdl.		Distribution gradients related to seasonal microclimates	Cach-Pérez et al. (2013)
Tillandsia yucatana Baker		Distribution gradients related to seasonal microclimates	Cach-Pérez et al. (2013)
		Light stratification	Chilpa-Galván et al. (2013)
Brassavola nodosa (L.) Lindl.	Orchidaceae	Light stratification	Zimmerman and Olmsted (1992)
Catasetum integerrimum Hook.		Light stratification	Zimmerman and Olmsted (1992)
Cohniella ascendens (Lindley) Christenson		Drought responses	de la Rosa et al. (2014b)
		Vertical distribution in microclimatic conditions	de la Rosa et al. (2014a)
Cohniella yucatanensis Cetzal and Carnevali		Drought responses	de la Rosa et al. (2014b)
		Vertical distribution in microclimatic conditions	de la Rosa et al. (2014a)
Encyclia alata (Bateman) Schltr.		Climate change	Goode and Allen (2008)
Encyclia belizensis (Reich. f.) Schlechter		Light stratification	Zimmerman and Olmsted (1992)
Encyclia boothiana (Lindl.) Luer		Light stratification	Zimmerman and Olmsted (1992)
Encyclia nematocaulon (BC Rich.) Acuña		Drought responses	de la Rosa et al. (2014b)
		Vertical distribution in microclimatic conditions	de la Rosa et al. (2014a)

(continued)

Table 5.2 (continued)

Species	Family	Physiological responses studied	Reference
Laelia rubescens Lindley		Drought responses	de la Rosa et al. (2014b)
		Vertical distribution in microclimatic conditions	de la Rosa et al. (2014a)
Lophiaris oerstedii (Rchb. f.) A. Jimenez, Carnevali and Dressler		Drought responses	de la Rosa et al. (2014b)
		Vertical distribution in microclimatic conditions	de la Rosa et al. (2014a)
Myrmecophila christinae Carnevali & Gómez-Juárez		Photosynthesis over a rainfall gradient	Ricalde et al. (2013)
Polystachya cerea Lindl.		Light stratification	Zimmerman and Olmsted (1992)
Rhyncholaelia digbyana (Lindl.) Schlechter		Light stratification	Zimmerman and Olmsted (1992)
Schomburgkia tibicinis (Batem.) Batem		Light stratification	Zimmerman and Olmsted (1992)

1988). Consequently, for seedlings of terrestrial CAM plants, there is a higher probability of survival under the canopy of nurse plants (Esparza-Olguín et al. 2002; Flores et al. 2004; Cervera et al. 2006). For instance, seedlings of the rare cactus *Mammilaria gaumeri* growing in a coastal sand dune of northern Yucatán show a two-fold higher volume to surface area ratio when they are growing under the canopy of a local shrub than when they grow receiving 50 % of ambient light (Cervera et al. 2006).

5.3.1.2 Water and CO_2 Uptake

Although CAM photosynthesis can occur during drought, there is a substantial difference in the amount of carbon fixed when compared to the photosynthesis that occurs when water is available. Studies in the Yucatán Peninsula reveal that the greatest tissue acid accumulation (a proxy for CO_2 uptake) of terrestrial CAM plants occurs during the rainy season (Andrade et al. 2006; Cervera et al. 2007; Ricalde et al. 2013; Vargas-Soto 2010; González-Salvatierra et al. 2013; Chávez-Sahagún 2014), when there are lower air temperatures and vapor pressure deficits than during the rest of the year.

Also, because the enzyme that fixes CO_2 at night in CAM plants, phosphoenolpyruvate carboxylase (PEPc), does not discriminate against $^{13}CO_2$ whereas the one that fixes carbon during the day, ribulose bisphosphate carboxylase (RuBisCO), does and favors ^{12}C fixation, investigations of carbon isotopic composition ($\delta^{13}C$) can provide information on the proportion of CO_2 fixed during the night or day (Griffiths 1992). Terrestrial CAM species from northern Yucatán show $\delta^{13}C$ values between −15 and −20 ‰, which indicates strong CAM performance (Ricalde et al. 2013). CAM plants with CO_2 uptake exclusively at night, by PEPc, would be expected to have $\delta^{13}C$ values around −11 ‰ (close to atmospheric values), whereas if all CO_2 is fixed by RuBisCO the values of $\delta^{13}C$ would be approximately −27 ‰ (O'Leary 1988). In northern Yucatán Peninsula, tissue $\delta^{13}C$ values for terrestrial CAM plants of the coastal dunes are 2 ‰ higher than CAM plants from the dry tropical deciduous forests (Ricalde et al. 2013), indicating greater CO_2 fixation through the CAM pathway with decreasing water availability.

One group of plants that shows an interesting array of growth forms and physiological plasticity is the genus *Clusia*, with species exhibiting C_3, C_3/CAM and CAM pathways (Winter and Smith 1996; Winter et al. 2005; Lüttge 2006). In Mexico, there is a preponderance of C_3/CAM *Clusia* species, but in the Yucatán Peninsula the majority of the genus have CAM metabolism (five out of six species, Vargas-Soto et al. 2009). *Clusia rosea* is the only C_3/CAM species found in the peninsula that can change from C_3 to CAM as drought progresses, even at the seedling stage (Vargas-Soto 2010).

5.3.1.3 Plant Responses to Light and Temperature

Depending on their habitat, CAM plants in the Yucatán Peninsula may experience large changes in light environment in different seasons (Graham and Andrade 2004; Cervantes et al. 2005). In dry forests, most of the trees lose their leaves during the dry season, and understory plants may receive up to nine times more light than in the wet season (Graham and Andrade 2004). However, even in the evergreen canopies, changes between seasons, cloud cover, and sunflecks create a highly variable light environment. The physiological adaptations of CAM plants, particularly the ability to tolerate water and light stress, are decisive in their distribution (Graham and Andrade 2004; Reyes-García et al. 2012b). Because physiological responses to these parameters are correlated, it is very difficult to independently analyze the effects of water availability, light, and temperature on CAM plants.

A combination of lack of water and high temperatures can affect growth and reproduction of CAM plants (Andrade et al. 2009). The optimal diurnal/nocturnal temperature for CO_2 uptake of two co-occurring species in the seasonal dry forest of Yucatán, *Agave fourcroydes* and *Hylocereus undatus*, are 30/20 °C (Nobel 1985; Nobel and de la Barrera 2002). Temperatures above or below the optimal, especially nocturnal ones, reduce growth and reproduction in CAM plants (Andrade et al. 2007).

For the commercial hemiepiphytic cactus *H. undatus*, growth is reduced when young plants are exposed or deeply shaded (Andrade et al. 2006). Also, seedlings of the rare cactus *Mammillaria gaumeri* cannot survive under more than 20 % of ambient light, but adults increase growth under 40–80 % of ambient light, indicating that low-magnitude disturbances can be beneficial for populations of this species (Cervera et al. 2006, 2007). However, large disturbances that increase light can be harmful for CAM terrestrial plants. For instance, tissues of

H. undatus and *M. gaumeri* cannot tolerate temperatures above 50 °C (Nobel and de la Barrera 2002; Cervera et al. 2006). Exposure to high temperatures increases respiratory rates and plant tissues must invest photosynthetic products to repair the photosynthetic apparatus from photoinhibitory damage, which leads to lower investment against pathogens and predators (Andrade et al. 2009).

5.3.2 *CAM Epiphytic Species*

5.3.2.1 Water Relations

Epiphytes represent a main group of interest for ecophysiological studies in the Yucatán Peninsula. Because the epiphytes are disconnected from the stored moisture in the ground and must obtain water during precipitation pulses the epiphytic habitat in itself is water limited. Nutrients are also obtained mainly by precipitation, yet water has been reported as the main limiting factor for epiphyte growth (Laube and Zotz 2003).

In the water limited habitats of Yucatán, vascular epiphytes are not as diverse or important in number or in biomass as in the wet forests. Nevertheless, the species show interesting adaptations to deal with the lack of water, excess light, and high temperatures. Studies have focused on the families Bromeliaceae and Orchidaceae, and have found the prevalence of CAM photosynthesis, which is an important water saving mechanism, with only one Bromeliaceae species using C_3 photosynthesis (*Catopsis nutans*, Benzing and Renfrow 1971). This species is found in the wetter, southern region of the peninsula (Cach-Pérez et al. 2013). The isotopic signature of carbon ($\delta^{13}C$) shown in the species of Yucatán denote tight diurnal stomatal closure and the predominance of carbon fixed in the night by the enzyme PEPc (see Sect. 5.3.1.2), as compared to epiphytic CAM species from Trinidad (Griffiths and Smith 1983), distributed over a larger precipitation gradient that may show higher fluctuation in $\delta^{13}C$ values (Fig. 5.3). This low variation in carbon isotope signatures is also evidenced in C_3 trees, which show relatively high $\delta^{13}C$ values, suggesting that RuBisCO is not able to fully discriminate against the heavier isotope, as stomatal closure to conserve water is frequent in the tree species, increasing $\delta^{13}C$ (Hasselquist et al. 2010a; Cach-Pérez 2013; Martin 1994; Fig. 5.3).

The CAM epiphytes exhibit a morphology and physiology modified to tolerate water scarcity and high light incidence (Benzing 1990; Zotz and Hietz 2001; Andrade 2003). Epiphytic bromeliads may store water in succulent tissues or in water-storing tanks, and may absorb water and nutrients through modified leaf trichomes (Reyes-García et al. 2012b). Leaves in bromeliads form a rosette, which, in the case of tank species, has wide leaf bases that form a water impounding reservoir (Fig. 5.4). Thus, the morphology of the leaves in tank species allow a decrease in leaf temperature, which is usually below air temperature in the early morning, promoting dew condensation, contributing to maintain a high water balance (Andrade 2003; Reyes-García et al. 2012b).

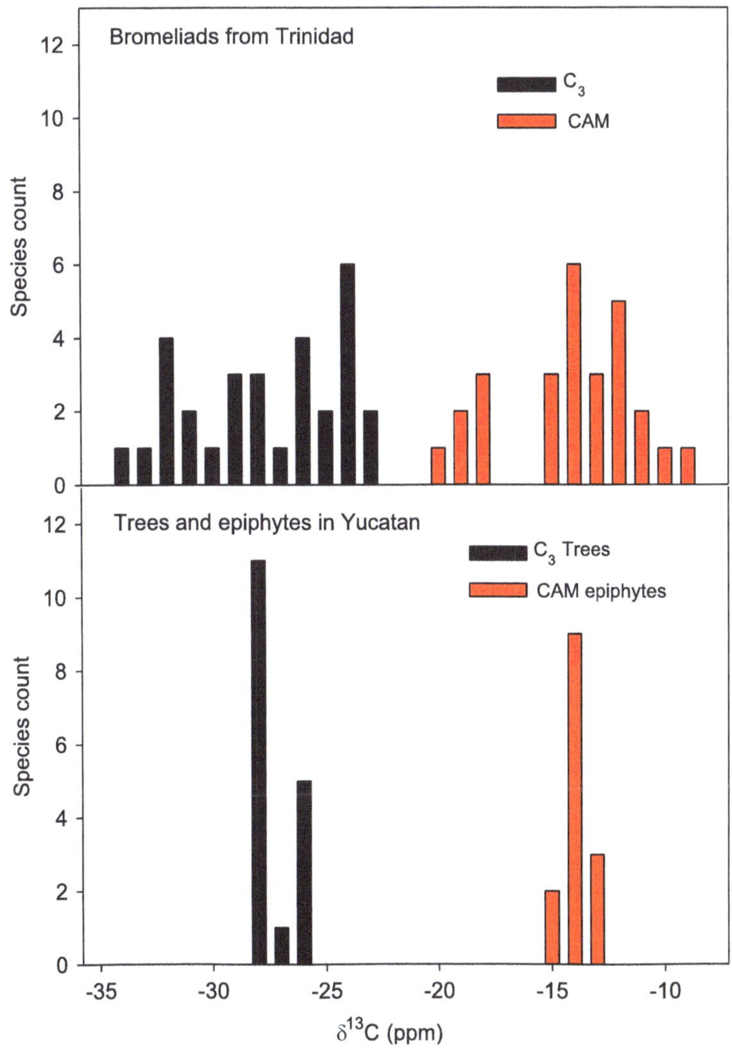

Fig. 5.3 Frequency values of the carbon isotopic signal of Bromeliads from Trinidad (*top panel*) growing in a precipitation gradient of 1000–3000 mm per year (Redrawn from Griffiths and Smith 1983) and C_3 trees and shrubs from Yucatán (*bottom panel*) in a precipitation gradient of 800–1650 mm per year and CAM epiphytic bromeliads in a gradient of 500–1500 mm per year (Hasselquist et al. 2010a, b; Cach-Pérez 2013). No isotopic data are available for *Catopsis nutans*, which is reported to be C_3 (Benzing and Renfrow 1971) and is present in the wetter southern edge of the peninsula (Cach-Pérez et al. 2013)

Alternatively, atmospheric bromeliads lack tanks (Pittendrigh 1948) and may either have succulent leaves or nebulophyte morphology. Nebulophytes are characterized by having very long and thin leaves that have a small boundary layer, allowing for the small fog droplets to be intercepted (Reyes-García et al. 2012b). In the southern region of the Yucatán, where fog formation is frequent, the bromeliad nebulophytes, *Tillandsia festucoides* and *Tillandsia juncea* are common (Cach-Pérez et al. 2013; Ramírez Morillo et al. 2004). The atmospheric *Tillandsia usneoides* shares this environment; this highly modified bromeliad has a tiny succulent stem, but clones of the same individual grow interlinked and can extend for meters and hang as a mesh from trees (Billings 1904), functioning as an effective fog trap.

The different morphologies described in the Bromeliaceae contribute to different interactions with the local microclimate. The more succulent species, which are less in number (representing both tank and atmospheric types), are present in the northern part of the peninsula, while the southern, wetter regions have more diverse epiphytic communities (Cach-Pérez et al. 2013). Life forms of the Orchideaceae have not been defined as well as for the Bromeliaceae, but the morphologies found in the Yucatán suggest a mixture of species that rely on succulence and nebulophytic forms (Fig. 5.4). Orchidaceae species also may show large water storage capacity in their stems or pseudobulbs and have specialized roots with the highly absorptive velamen (de la Rosa et al. 2014b). The lack of water during the prolonged dry seasons in the Yucatán Peninsula, induces partial or total stomatal closure in the evergreen epiphytes (Graham and Andrade 2004). Consequently the

Fig. 5.4 (**a**) *Aechmea bracteata* (bromeliad, tank species), (**b**) *Encyclia nematocaulon* (orchid nebulophyte), (**c**) *Tillandsia juncea* (bromeliad, nebulophyte atmospheric species), (**d**) *Cohniella yucatanensis* (orchid, succulent atmospheric). *Scale bar* shows 5 cm in all images (courtesy of E. de la Rosa, Centro de Investigación Científica de Yucatán, Mexico)

potential of photoinhibition is greatest in epiphytes that grow on more exposed sites or in the upper canopy strata (see Section on photoinhibition 5.3.2.2). The stress induced by reduced CO_2 uptake may be overcome in some CAM species by enhancing CO_2 release from organic acids during the hours of higher light incidence and by using recycled respired CO_2 for carbon fixation (Griffiths et al. 1986).

Thus, precipitation influences epiphytic species diversity, but the density of the individuals is inversely related to vapor pressure deficit (VPD, Cach-Pérez et al. 2013). At low values of VPD, the rate of water loss between rain events is diminished and there is a higher chance of dew condensation (Chilpa-Galván et al. 2013; de la Rosa et al. 2014a); low VPD also stimulates germination in epiphytic bromeliads (Goode and Allen 2009). In the Yucatán Peninsula, the water table depth may be very shallow at some locations, where trees will be able to tap the permanent water source and increase transpiration and tree leaf area, lowering VPD values locally, creating islands of increased epiphyte density (Chilpa-Galván et al. 2013).

5.3.2.2 Plant Responses to Light and Temperature

In general, studies have found that epiphytes show high plasticity in the use of light, but the differential physiological adaptations of each species will influence its distribution within the canopy strata. This is particularly true of the ability to endure water and light stress, which may vary in intensity from the top to the bottom of the canopy (Graham and Andrade 2004; Chilpa-Galván et al. 2013). In the dry forest in Yucatán, the vertical distribution of two epiphytic bromeliads, the atmospheric species *Tillandsia brachycaulos* and the tank species *Tillandsia elongata*, showed that the distribution of the tank species was limited to the top canopy because this environment provided higher precipitation interception and higher dew condensation, compared to the more sheltered lower canopy (Graham and Andrade 2004). Further studies indicated that for *T. brachycaulos* and other atmospheric species that have higher drought tolerance, excess light limits their distribution to the mid to lower canopy strata (Chilpa-Galván et al. 2013; Cervantes et al. 2005). Cervantes et al. (2005) found that individuals of *T. brachycaulos* under full sunlight had lower growth rates and less flower production than those under moderate shade.

During the dry season, concomitant with increased light incidence, as mentioned previously, drought causes stomatal closure preventing the entrance of the CO_2 supply for photosynthesis. Consequently, the potential for photoinhibition is greatest in epiphytes that grow on exposed branches, already receiving high amounts of light, under higher temperatures, and subject to the desiccating action of winds. The interaction of these factors results in a potential increase in photo-oxidative stress and a consequent decrease in the quantum efficiency of photosystem II (ΦPSII; see Fig. 5.5). To prevent long term damage to the photosynthetic apparatus, plants dissipate excess energy as heat (non-photochemical dissipation, NPQ) or re-emit it as electromagnetic radiation in a process known as chlorophyll

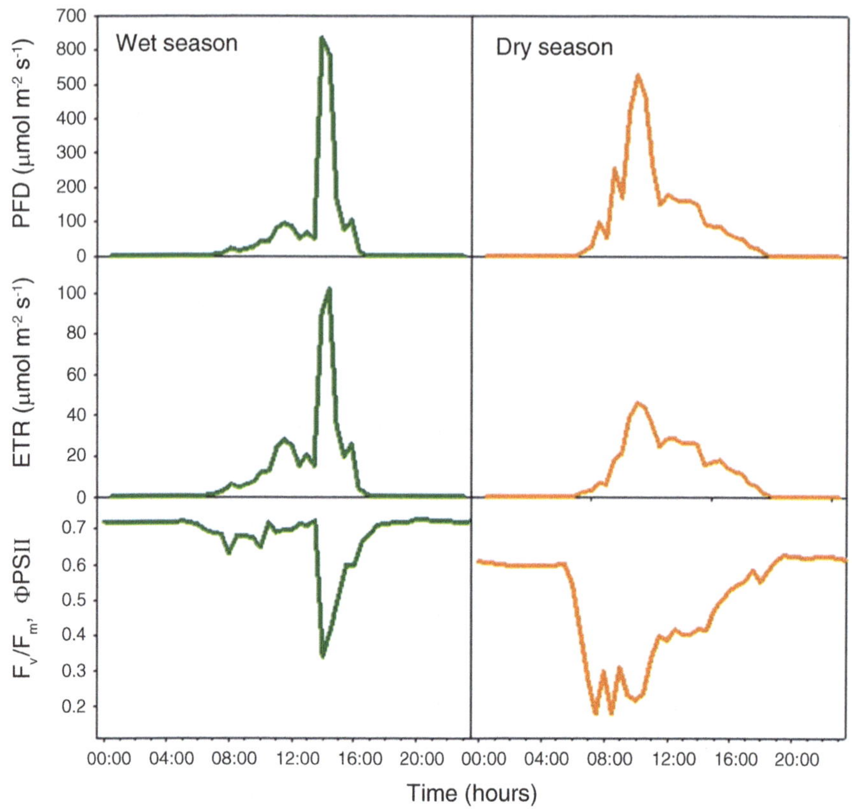

Fig. 5.5 *PFD* Photon flux density, *ETR* electron transport rate and maximum fluorescence (F_v/F_m, under dark conditions) or quantum yield (ΦPSII, in the presence of light) of *Tillandsia utriculata*, an epiphytic bromeliad in a coastal sand dune, during the rainy (*left panels*) and dry (*right panels*) seasons. Measurements were made every 20 m, for 24 h under natural conditions. *T. utriculata* is usually found in the shaded understory, which buffers seasonal changes in incident light. The species undergoes a 6 month dry season (A. Rosado-Calderón and C. Reyes-García, unpublished data)

fluorescence (Maxwell and Johnson 2000). It has been reported that epiphytes showing CAM prevent the production of the harmful reactive oxygen species, controlling photoinhibition and oxidative stress (Niewiadomska and Borland 2008). CAM is advantageous to cope with the desiccating conditions typical of the dry forest (Winter and Smith 1996), because daytime fixation of CO_2 by RuBisCo occurs with closed stomata. This causes high CO_2 concentrations in the cytosol and chloroplasts, favoring RuBisCo's carboxylation activity over its oxygenase activity, providing additional photoprotection through maintenance of electron transport and preventing damage to photosystems (Niewiadomska and Borland 2008).

Usually, excess photons are quickly dissipated as heat and fluorescence (Niyogi 2000), but when the absorption of light energy exceeds the capacity for photosynthesis and the photoprotection mechanisms have been oversaturated, photoinhibition occurs (Takahashi and Murata 2008). Photoinhibition has been defined as a decrease in the photosynthetic efficiency that results from excess light and leads to a partial loss of photochemical capacity and, consequently, diminished growth (Björkman and Demmig 1987; Long et al. 1994; Baker 2008). Therefore, every organism that conducts photosynthesis is potentially vulnerable to injury due to light influences, but the level of susceptibility depends on both environmental factors and physiological limits (Benzing 2000).

In the dry forest, epiphytes, already subjected to stress during the dry season, showed a decrease in photosynthetic efficiency, an increase in antioxidant compounds, and low values of maximum fluorescence (Fig. 5.6), which indicated photoinhibition due to over-excitation of photosystem II (PSII) (Fv/Fm ~0.6; Graham and Andrade 2004; Cervantes et al. 2005; González-Salvatierra et al. 2010). Among the mechanisms of defense against light stress, *T. brachycaulos* increased the production of flavonoids. The production of different types of flavonoids have been reported in a number of species of Bromeliaceae (Saito and Harborne 1983) as a response to high light and other stressful conditions (Benzing 2000). For *T. brachycaulos*, anthocyanins (a type of flavonoid) were located in a single layer under the epidermis on both leaf faces, with higher anthocyanin concentrations in exposed plants during the dry season, confirming their sunscreen role (González-Salvatierra et al. 2010). The strong correlation between incident daily high light and total anthocyanin content suggests that these molecules are involved in photoprotection as a part of a short-term defense to diminish light-derived reactions

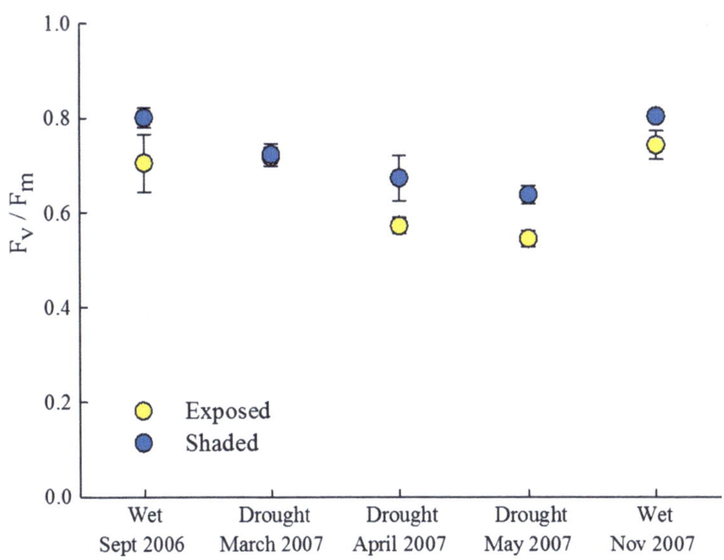

Fig. 5.6 Maximum fluorescence of photosystem II (F_v/F_m) of *T. brachycaulos* under exposed and shaded conditions during sunrise in the dry forest of Dzibilchaltún, Yucatán. Values are mean ± SE, n = 3 (A. Rosado-Calderón and C. Reyes-García, unpublished data)

and prevent the over-energization and over-reduction in the photosynthetic electron transport, protecting the photosynthetic apparatus from the effects of photoxidative stress (Gould et al. 2000; Steyn et al. 2002).

The orchids of the dry forests of Yucatán show similar distribution patterns to those found in bromeliads. In a study with orchids from two dry forests, de la Rosa et al. (2014a) show that species are more frequent in the mid-canopy stratum, and are rarely found in the lower stratum. Light measurements indicate that the lower stratum of the wettest forest received low light during the wet season (only ~2 % of incident total light), which may be too low to maintain photosynthetic rates in orchids (Ter Steege and Cornelissen 1989). Also, de la Rosa et al. (2014a) found that even though vertical canopy strata show environmental differences, the orchids had higher physiological variation between seasons than between strata. At the beginning of the dry season, the maximum quantum efficiency (F_v/F_m) of the orchids was close to the theoretical optimum (0.8–0.83; Björkman and Demmig 1987), indicating that the plants were not under environmental stress (Horton et al. 1996; Demmig-Adams et al. 2012). However, by the end of the dry season, when environmental conditions such as light, VPD and leaf temperature increased, most species show a decrease in photosynthetic activity, with values of F_v/F_m between 0.4 and 0.7 indicating moderate photoinhibition (de la Rosa et al. 2014a).

The high temperatures exhibited in the Yucatán Peninsula are another stress factor for the native epiphytic species. CAM photosynthesis should alleviate part of the stress, as the high CO_2 concentrations within the mesophyll abolish the temperature dependent process of photorespiration (Winter and Smith 1996). Yet, the metabolic sensitivity to temperature has not been directly studied. An example from the coastal sand dune scrub of the northern region of the peninsula shows the photoinhibition of the epiphyte *Tillandsia utriculata* in the dry season (Fig. 5.5; Rosado-Calderón and Reyes unpublished data). In this case, the epiphyte receives similar light in the dry and wet seasons, but photoinhibition results from the combination of water stress, as the relative water content has diminished from 80 % in the wet season to 40 % in the dry season, and from temperature stress, as maximum temperatures escalate from 23 °C in the wet season to 38.7 °C in the dry season. This results in a diminished day (ΦPSII) and night (F_v/F_m) quantum yield of PSII (Fig. 5.5) that lowers the electron transport rates during the day. Further studies are needed to separate the effects that temperature and water can have on the photosynthetic apparatus.

Studies have dealt with leaf heat dissipation, which is an important issue in CAM plants that cannot transpire to cool off during the day. There is a tradeoff between having large volumes of water storage tissue to survive during drought and reducing leaf area, which allows higher heat dissipation through convection by having a small boundary layer (de la Rosa 2014a). High leaf temperatures and excess light are avoided by the epiphytes of the Yucatán by limiting their distribution mostly to the more shaded mid-canopy stratum (Cach-Pérez et al. 2014). Conversely, in sites with highly reflective karstic soils, the lower canopy stratum is subjected to intense heat with low leaf heat dissipation that has been shown to induce mortality in epiphytes (Cervantes et al. 2005; Mondragón et al. 2004).

5.4 Concluding Remarks and Gaps in Knowledge

Despite the increase in research on the ecology and physiology of vascular plants of the Yucatán, many gaps in knowledge continue to exist. As mentioned in Sect. 5.1, physiological adaptations have been studied in only about 2 % of the total vascular plants of the Yucatán, and even these have not been completely characterized, since studies focus on specific variables (Tables 5.1 and 5.2). This suggests a large gap in information on the basic ecophysiology of many of the species of the area. Among the species studied, most of the information obtained has been for adults and gaps in information exist for seedlings (mainly on their physiology and factors that affect their survivorship). Although studies on below ground processes have started to increase, due in part to the difficulty of such studies, large gaps of information continue to exist regarding below ground processes (e.g. interactions between nutrients and microbes, the effects of soil microbial communities on the mobilization of nutrients or on nutrient sequestration, and potential relationships between levels of mycorrhizal colonization and successional stage; Hasselquist et al. 2010b). But also relationships between levels of mycorrhizal colonization and species (Ceccon et al. 2004), and species type (e.g. pioneer, versus late succession species; Ceccon et al. 2003) also merit further examination. Additional studies are needed to look at the synergistic effects of above and below ground processes, including the relationship of the effects of mycorrhizal colonization with succession and disturbance.

To follow on findings by Reyes-García et al. (2012a), whereby larger legumes presented more heartwood, developmental studies in legumes are needed in order to elucidate if there is an active process whereby drier conditions lead to more heartwood development in larger plants. Also, general ecological surveys, perhaps by collecting wood samples, would be needed to determine potential differences in the extent of heartwood in legumes across precipitation gradients. Active responses of plants to drought, potentially through cell signaling in laying down heartwood but also in growth, total canopy leaf area and osmotic response to drought are also needed in order to elucidate the extent to which responses to drought are dynamic or set (phylogenetic).

Since the main limiting factor in Yucatán is water, studies have focused on characterizing the water relations of tree species. However it is also necessary to study carbon assimilation directly and determine photosynthetic rates of species under different conditions. Due to the technical difficulties involved, the study of photosynthesis in trees has been relegated. However, novel techniques such as eddy covariance are being implemented in the Yucatán Peninsula to study the exchange of carbon, water and energy between the atmosphere and the forest.

The coastal dune scrubland on the north of the peninsula is a narrow area of endemism with several species adapted to arid environments (Espadas et al. 2003). Unfortunately, at present, this area is under considerable pressure due to human activities and several populations of the rare cactus *Mammillaria gaumeri* have been lost (Ferrer et al. 2011). Despite the fact that the dry tropical deciduous forests have been subject to natural disturbances and centuries of human use, CAM terrestrial plants in these forests still possess ecological importance (González-Iturbe et al. 2002; Ricalde et al. 2013). For conservation purposes, additional physiological studies, including all life stages along with demographic studies, should be applied to CAM terrestrial species of these communities to determine sites for the reintroduction of threatened species (de la Barrera and Andrade 2005; Reyes-García and Andrade 2009).

The genus *Clusia* deserves special attention. There is evidence that more than six species occur on the peninsula, but they have not been described (Vargas-Soto 2010). Under climate change scenarios for the region, *Clusia* species could become ecologically important, and should be included in reforestation programs, so more physiological and ecological studies on these species are needed.

CAM plants are good models for biophysical studies. Succulence does not allow leaves and stems to move under normal wind conditions, and energy balance and water relations can be modeled and then more easily predicted (Niklas and Hammond 2013). Also, for photosynthesis, an environmental productivity index has been tested for many CAM species, mainly crops (Nobel 1988, 1996; Andrade et al. 2009). This index can help predict plant photosynthesis and growth under different environmental scenarios, which can allow us to model plant productivity responses to global climate change.

Even though the epiphytes are among the best studied groups of plants in the Yucatán, some topics have not been addressed, such as limitation by nutrients, and how this can change across the different environments found in the peninsula. Differential growth rates, germination rates and seed production among species are other relevant topics that lack study. It is difficult to predict the response of the epiphytic species to climate change. Temperatures in the peninsula are expected to increase with global warming (Orellana et al. 2009), but the temperature limits of the different species have not been tested, so there is little information on how this will affect them. Predictions are uncertain on whether the peninsula will get drier or wetter, but changes in seasonal weather patterns have already been observed, prolonging the dry season, or the wet season in different years. This affects perennial epiphytes by modifying the whole microenvironment as trees lose or conserve their leaves for longer periods. These changes in leaf cover have been shown to affect both productivity, and induce stress, but there has been no attempt to integrate how these changes will affect annual survival rate. If the favorable, wet growing season is too short, the epiphyte may not have enough reservoirs to survive a prolonged drought, when photosynthesis drops and photoinhibition is induced (Reyes-García and Griffiths 2009). In contrast, some of the epiphytic species have high drought tolerance conveyed by traits that may be unfavorable under wetter conditions. For example, stomata are blocked by the water retained on the surface on species with very high trichome density, and these species may have local extinctions under a wetter climate. Furthermore, epiphytes have been seen to be very sensitive to changes in climate (Cach-Pérez et al. 2014) and should be among the first life form to respond to climate change and land use change.

Environmental instability as a result of the effects of climate change on the Yucatán Peninsula generates the following questions: How will plant communities under conditions of climate change be distributed? Will climate change modify the floristic composition of plant associations? Which species will dominate and what are the strategies that will allow them a better adaptation to environmental change? How are C_4 plants reacting in the peninsula? What are the physiological strategies of invasive species? How can we link the plant physiology with conservation strategies? These questions can be answered by modeling plant physiological parameters and environmental variables. However, to achieve an understanding on how plant communities respond to climate change the number of species studied must be increased. With this, well-characterized functional groups may be identified allowing for the potential modeling of C_3 and CAM plants.

Acknowledgements We thank EA Graham, PS Nobel, R Orellana, C Cervera, MF Ricalde, E de la Barrera, E Rengifo, R Barceló, L Simá, M Mandujano, R Us-Santamaría, G Vargas-Soto, K Winter, M Giovanetti, L Santiago, M Cach-Pérez, O Hernández-González, E Chávez-Sahagún, J Palomo-Kumul, T McElroy, A Rosado-Calderón, NC Chilpa-Galván, M. Tamayo-Chim, D Mondragón, E de la Rosa-Manzano, S. Cervantes and undergraduate students from Kennesaw State University (Georgia, USA), for their contribution to our studies. We are grateful for funding from CONAFOR-CONACYT 9765, Fondo Mixto CONACYT-gobierno del estado de Yucatán YUC-2003-C02-042, SEP-CONACYT 36931, UC Mexus-CONACYT, SEP-CONACYT 48344/24588, CONABIO GU002, SEP-CONACYT 177842, SEP-CONACYT 80181, SEMARNAT-CONACYT 107916 and NSF RUI 0516387.

References

Andrade JL. Dew deposition on epiphytic bromeliad leaves: an important event in a Mexican tropical dry deciduous forest. J Trop Ecol. 2003;19:479–88.

Andrade JL, Rengifo E, Ricalde MF, Simá JL, Cervera JC, Vargas-Soto G. Microambientes de luz, crecimiento y fotosíntesis de la pitahaya (*Hylocereus undatus*) en un agrosistema de Yucatán, México. Agrociencia. 2006;40:687–97.

Andrade JL, de la Barrera E, Reyes-García C, Ricalde MF, Vargas-Soto G, Cervera JC. El metabolismo ácido de las crasuláceas: diversidad, fisiología ambiental y productividad. Bol Soc Bot Mex. 2007;81:37–51.

Andrade JL, Cervera JC, Graham EA. Microenvironments, water relations and productivity of CAM plants. In: de la Barrera E, Smith WK, editors. Perspectives in biophysical plant ecophysiology: a tribute to Park S. Nobel. Mexico: Universidad Nacional Autónoma de México; 2009. p. 95–120.

Baker NR. Chlorophyll fluorescence: a probe of photosynthesis in vivo. Annu Rev Plant Biol. 2008;59:89–113.

Barcikowski W, Nobel PS. Water relations of cacti during desiccation: distribution of water in tissues. Bot Gaz. 1984;145:110–5.

Bejarano-Castillo M. Dinámica del C y del N en el suelo de bosques tropicales estacionalmente secos ante los escenarios de incremento en la deposición atmosférica. PhD Dissertation. México DF: Universidad Nacional Autónoma de México; 2012.

Benzing DH. Vascular epiphytes. General biology and related biota. Nueva York: Cambridge University Press; 1990.

Benzing DH. Bromeliaceae: a profile of an adaptative radiation. Cambridge: Cambridge University Press; 2000.
Benzing DH, Renfrow A. Significance of the patterns of CO_2 exchange to the ecology and phylogeny of the Tillandsioideae (Bromeliaceae). Bull Torrey Bot Club. 1971;98:322–7.
Billings FH. A study of *Tillandsia usneoides*. Bot Gaz. 1904;38:99–121.
Björkman O, Demmig B. Photon yield of O_2 evolution and chlorophyll fluorescence characteristics at 77 K among vascular plants of diverse origins. Planta. 1987;170:489–504.
Boose ER, Foster D, Barker Plotkin A, Hall B. Geographical and historical variation in hurricanes across the Yucatán Peninsula. In: Gomez-Pompa A, Allen M, Fedick SL, Jimenez-Osornio JJ, editors. The Lowland Maya: three millennia at the human–wildland interface. Binghamton, NY: Haworth Press; 2003. p. 495–516.
Borchert R. Soil and stem water storage determine phenology and distribution of tropical dry forest trees. Ecology. 1994;75(5):1437–49.
Cabrera E, Sousa M, Téllez O. Imágenes de la Flora Quintanarroense. México, DF: CIQRO-UNAM; 1982. p. 219.
Cach-Pérez MJ. Bromeliáceas epifitas de la Península de Yucatán como indicadoras de los posibles efectos del cambio climático regional. PhD Dissertation. Mérida, México: Centro de Investigación Científica de Yucatán; 2013.
Cach-Pérez MJ, Andrade JL, Chilpa-Galván N, Tamayo-Chim M, Orellana R, Reyes-García C. Climatic and structural factors influencing epiphytic bromeliad community assemblage along a gradient of water-limited environments in the Yucatán Peninsula, Mexico. Trop Conserv Sci. 2013;6:283–302.
Cach-Pérez M, Andrade JL, Reyes-García C. La susceptibilidad de las bromeliáceas epífitas al cambio climático. Bot Sci. 2014;92:157–68.
Ceccon E, Olmsted I, Vázquez-Yanes C, Campo-Alves J. Vegetation and soil properties in two tropical dry forests of differing regeneration status in Yucatán. Agrociencia. 2002;36:621–31.
Ceccon E, Huante P, Campo J. Effects of nitrogen and phosphorus fertilization on the survival and recruitment of seedlings of dominant tree species in two abandoned tropical dry forests in Yucatán, Mexico. Forest Ecol Manage. 2003;182:387–402.
Ceccon E, Sánchez S, Campo J. Tree seedling dynamics in two abandoned tropical dry forest of differing successional status in Yucatán, Mexico: a field experiment with N and P fertilization. Plant Ecol. 2004;170:277–85.
Cervantes S, Graham E, Andrade JL. Light microhabitats, growth and photosynthesis of an epiphytic bromeliad in a tropical dry forest. Plant Ecol. 2005;179:107–18.
Cervera JC, Andrade JL, Simá JL, Graham EA. Microhabitats, germination, and establishment for *Mammillaria gaumeri* (Cactaceae), a rare species from Yucatán. Int J Plant Sci. 2006;167:311–9.
Cervera JC, Andrade JL, Graham EA, Durán R, Jackson PC, Simá JL. Photosynthesis and optimal light microhabitats for a rare cactus, *Mammillaria gaumeri*, in two tropical ecosystems. Biotropica. 2007;39:620–7.
Chán-Dzul A. Diversidad florística y funcional a través de una cronosecuencia de la selva mediana subperennifolia en la zona de influencia de la Reserva de la Biosfera Calakmul, Campeche, México. MSc. Dissertation. Turrialba, Costa Rica: Centro Agronómico Tropical de Investigación y Enseñanza; 2010.
Chávez-Sahagún E. Microhabitats y estacionalidad en *Aechmea bracteata* (Sw.) Griseb (Bromeliaceae) en la selva baja de Dzibilchaltún, Yucatán. Master thesis. Mérida, México: Centro de Investigación Científica de Yucatán, AC; 2014.
Chazdon RL. Tropical forest recovery: legacies of human impact and natural disturbances. Perspect Plant Ecol Evol Syst. 2003;6:51–71.
Chilpa-Galván N, Tamayo-Chim M, Andrade JL, Reyes-García C. Water table depth may influence the asymmetric arrangement of epiphytic bromeliads in a tropical dry forest. Plant Ecol. 2013;214:1037–48.

Cushman JC. Crassulacean acid metabolism. A plastic photosynthetic adaptation to arid environments. Plant Physiol. 2001;127:1439–48.
de la Barrera E, Andrade JL. Challenges to plant megadiversity: how environmental physiology can help? New Phytol. 2005;167:5–8.
de la Rosa E, Andrade JL, Zotz G, Reyes-Garcia C. Epiphytic orchids in tropical dry forests of Yucatán, Mexico—Species occurrence, abundance and correlations with host tree characteristics and environmental conditions. Flora. 2014a;209:100–9.
de la Rosa E, Andrade JL, Zotz G, Reyes-García C. Respuestas fisiológicas a la sequía de cinco especies de orquídeas epífitas en dos selvas secas de la Península de Yucatán. Bot Sci. 2014b;92:607–16.
Demmig-Adams B, Cohu CM, Muller O, Adams III WW. Modulation of photosynthetic energy conversion efficiency in nature: from seconds to seasons. Photosynth Res. 2012;113:75–88.
des Grades E. Limiting factors for the establishment of *Cordia dodecandra* ADC and *Bixa orellana* Lon semi-arid calcareous soils in Yucatán, Mexico. PhD Dissertation. Herford, Germany: Center for Development Research (Zef) Department of Ecology and Natural Resource Management, Rheinischen Friedrich-Wilhelms-Universität zu Bonn; 2006.
Dodd AN, Borland AM, Haslam RP, Griffiths H, Maxwell K. Crassulacean acid metabolism: plastic, fantastic. J Exp Bot. 2002;53:569–80.
Drennan PM, Nobel PS. Responses of CAM species to increasing atmospheric CO_2 concentrations. Plant Cell Environ. 2000;23:767–81.
Duno R, Ramírez Morillo IM, Tapia-Muñoz JL, Carnevali G. Plantas vasculares de Quintana Roo. In: Pozo de la Tijera C, editor. Riqueza biológica del estado de Quintana Roo: un análisis para su conservación (tomo 2). México, D.F: ECOSUR-CONABIO-Gobierno del Estado de Quintana Roo, Programa de Pequeñas Donaciones (PNUD); 2011. p. 607.
Espadas C, Durán R, Argáez J. Phytogeographic analysis of taxa endemic to the Yucatán peninsula using geographic information systems, the domain heuristic method and parsimony analysis of endemicity. Divers Distrib. 2003;9:313–30.
Esparza-Olguín L, Valverde T, Vilchis-Anaya E. Demographic analysis of a rare columnar cactus (*Neobuxbaumia macrocephala*) in the Tehuacán Valley, Mexico. Conserv Biol. 2002;103:349–59.
Espejel I. A phytogeographical analysis of coastal vegetation in the Yucatán Peninsula. J Biogeogra. 1987;14:499–519.
Estrada-Medina H, Graham RC, Allen MF, Jiménez-Osornio JJ, Robles-Casolco S. The importance of limestone bedrock and dissolution karst features on tree root distribution in northern Yucatán, México. Plant Soil. 2013;362:37–50.
Ferrer M, Durán R, Méndez M, Dorantes A, Dzib G. Dinámica poblacional de genets y ramets de *Mammillaria gaumeri* cactácea endémica de Yucatán. Bol Soc Bot Mex. 2011;89:83–105.
Flores Guido JS, Duran Garcia RD, Ortiz Diaz JJ. Comunidades vegetales terrestres. In: Durán R, Méndez M, editors. Biodiversidad y desarrollo humano en Yucatán. México: CICY, PPD-FMAM, CONABIO, SEDUMA; 2010. p. 125–9.
Flores J, Briones O, Flores A, Sánchez-Colón S. Effect of predation and solar exposure on the emergence and survival of desert seedlings of contrasting life-forms. J Arid Environ. 2004;58:1–18.
García-Quintanilla A. Los tiempos en Yucatán. Editorial Claves Latinoamericanas. Mexico City, Mexico; 1986. p. 166.
Giovanetti M, Cervera JC, Andrade JL. Pollinators of an endemic and endangered species *Mammillaria gaumeri* (Cactaceae) in its natural habitat (coastal dune) and in a botanical garden. Madroño. 2007;54:286–92.
Goldstein G, Andrade JL, Nobel PS. Differences in water relations parameters for the chlorenchyma and the parenchyma of *Opuntia ficus-indica* under wet versus dry conditions. Aust J Plant Physiol. 1991;18:95–107.

González-Iturbe JA, Olmsted I, Tun-Dzul F. Tropical dry forest recovery after long term Henequen (sisal, *Agave fourcroydes* Lem.) plantation in northern Yucatán, Mexico. Forest Ecol Manage. 2002;167:67–82.

González-Salvatierra C. Antioxidantes y fotoprotección en dos especies con metabolismo ácido de las crasuláceas en una selva baja de Yucatán. PhD Dissertation. Mérida, México: Centro de Investigación Científica de Yucatán; 2009. p. 177.

González-Salvatierra C, Andrade JL, Escalante-Erosa F, García-Sosa K, Peña-Rodríguez LM. Antioxidant content in two CAM bromeliad species as a response to seasonal light changes in a tropical dry deciduous forest. J Plant Physiol. 2010;167:792–9.

González-Salvatierra C, Andrade JL, Orellana R, Peña-Rodríguez LM, Reyes-García C. Microambientes de luz y morfología y fisiología foliar de *Bromelia karatas* (Bromeliaceae) en una selva baja caducifolia de Yucatán, México. Bot Sci. 2013;91:75–84.

Goode LK, Allen FA. The impacts of hurricane Wilma on the epiphytes of El Edén Ecological Reserve, Quintana Roo, México. J Torrey Bot Soc. 2008;135(3):377–87.

Goode LK, Allen FA. Seed germination conditions and implications for establishment of an epiphyte, *Aechmea bracteata* (Bromeliaceae). Plant Ecol. 2009;204:179–88.

Goode LK, Erhardt EB, Santiago LS, Allen MF. Carbon stable isotopic composition of soluble sugars in *Tillandsia* epiphytes varies in response to shifts in habitat. Oecologia. 2010;163:583–90.

Gould KS, Markham KR, Smith RH, Goris JJ. Functional role of anthocyanins in the leaves of *Quintinia serrata* A. Cunn. J Exp Bot. 2000;51:1107–15.

Graham E, Andrade JL. Drought tolerance associated with vertical stratification of two co-occurring epiphytic bromeliads in a tropical dry forest. Am J Botany. 2004;91:699–706.

Griffiths H. Carbon isotope discrimination and the integration of carbon assimilation pathways in terrestrial CAM plants. Plant Cell Environ. 1992;15:1051–62.

Griffiths H, Smith JAC. Photosynthetic pathways in the Bromeliaceae of Trinidad: relations between life-forms, habitat preference and the occurrence of CAM. Oecologia. 1983;60:176–84.

Griffiths H, Lüttge U, Stimmel KH, Crook CE, Griffiths NM, Smith JAC. Comparative ecophysiology of CAM and C_3 bromeliads. III. Environmental influences on CO_2 assimilation and transpiration. Plant Cell Environ. 1986;9:385–93.

Hasselquist NJ, Allen MF, Santiago LS. Water relations of evergreen and drought-deciduous trees along a seasonally dry tropical forest chronosequence. Oecologia. 2010a;164:881–90.

Hasselquist NJ, Santiago LS, Allen MF. Belowground nitrogen dynamics in relation to hurricane damage along a tropical dry forest chronosequence. Biogeochemistry. 2010b;98:89–100.

Horton P, Ruban A, Walters R. Regulation of light harvesting in green plants. Annu Rev Plant Biol. 1996;47:655–84.

Ibarra-Manríquez G, Villaseñor JL, Durán-García R. Riqueza de especies y endemismo del componente arbóreo de la Península de Yucatán, México. Bol Soc Bot Mex. 1995;57:49–77.

IPCC. Climate change 2013: The physical science basis. Contribution of working group I to the fifth assessment report of the intergovernmental. 2013. http://www.ipcc.ch/report/ar5/wg1/. Accessed 20 Jun 2014.

Jackson PC, Meinzer FC, Bustamante M, Goldstein G, Franco A, Rundel PW, Caldas L, Igler E, Causin F. Partitioning of soil water among tree species in a Brazilian Cerrado ecosystem. Tree Physiol. 1999;19:717–24.

Keeley JE, Rundel PW. Evolution of CAM and C_4 carbon-concentrating mechanisms. Int J Plant Sci. 2003;164(3 Suppl):S55–77.

Lambers H, Chapin FS, Pons TL. Plant physiological ecology. New York, NY: Springer; 1998.

Laube S, Zotz G. Which abiotic factors limit vegetative growth in a vascular epiphyte? Funct Ecol. 2003;17:598–604.

Long SP, Humphries S, Falkowski P. Photoinhibition of photosynthesis in nature. Annu Rev Plant Physiol Plant Mol Biol. 1994;45:633–62.

Luna-Flores W, Estrada-Medina H, Jiménez-Osornio JJ, Pinzón-López LL. Efecto del estrés hídrico sobre el crecimiento y eficiencia del uso del agua en plántulas de tres especies arbóreas caducifolias. Terra Latinoamericana. 2012;30(4):343–53.

Lüttge U. Ecophysiology of crassulacean acid metabolism (CAM). Ann Bot. 2004;93:629–52.

Lüttge U. Photosynthetic flexibility and ecophysiological plasticity: questions and lessons from *Clusia*, the only CAM tree in the neotropics. New Phytol. 2006;171:7–25.

Martin EG. Physiological ecology of the Bromeliaceae. Bot Rev. 1994;60:1–82.

Maxwell K, Johnson G. Chlorophyll fluorescence—a practical guide. J Exp Bot. 2000;51:659–68.

Mondragón D, Durán R, Ramírez I, Valverde T. Temporal variation in the demography of the clonal epiphyte *Tillandsia brachycaulos* (Bromeliaceae) in the Yucatán Peninsula, Mexico. J Trop Ecol. 2004;20:189–200.

Niewiadomska E, Borland AM. Crassulacean acid metabolism: a cause or consequence of oxidative stress in planta? In: Lüttge U, Beyschlag W, Murata J, editors. Progress in botany 69. Berlin: Springer; 2008. p. 247–66.

Niklas KJ, Hammond T. Biophysical effects on plant competition and coexistence. Funct Ecol. 2013;27:854–64.

Niyogi KK. Safety valves for photosynthesis. Curr Op Plant Biol. 2000;3:455–60.

Nobel PS. PAR, water and temperature limitations on the productivity of cultivated *Agave fourcroydes* (henequen). J Appl Ecol. 1985;22:157–73.

Nobel PS. Environmental biology of Agaves and Cacti. New York, NY: Cambridge University Press; 1988.

Nobel PS. Responses of some North American CAM plants to freezing temperatures and doubled CO_2 concentrations: implications of global climate change for extending cultivation. J Arid Environ. 1996;34(2):187–96.

Nobel PS, de la Barrera E. High temperatures and net CO_2 uptake, growth, and stem damage for the hemiepiphytic cactus *Hylocereus undatus*. Biotropica. 2002;34:225–31

O'Leary MH. Carbon isotopes in photosynthesis. Fractionation techniques may reveal new aspects carbon dynamics in plants. BioScience. 1988;38:328–36.

Orellana R, Espadas C, Conde C, Gay C. Atlas. Escenarios de cambio climático en la Península de Yucatán. Centro de Investigación Científica de Yucatán, A.C., Universidad Nacional Autónoma de México, Consejo Nacional de Ciencia y Tecnología, Gobierno de Yucatán, Mérida; 2009. p. 111.

Orellana R, Carrillo L, Espadas C. Las arecáceas de la península de Yucatán ante el cambio climático: aproximación ecofisiológica y de distribución. In: Sánchez-Rojas G, Ballesteros-Barrera C, Pavón NP, editors. Cambio climático aproximaciones para el estudio de su efecto sobre la biodiversidad. México: Universidad Autónoma del Estado de Hidalgo; 2011. p. 79–114.

Palomo-Kumul J. Diferencias estacionales en la densidad de madera y el contenido relativo de agua en 16 especies arbóreas tropicales. Dissertation. Quintana Roo, México: Instituto Tecnológico de Chetumal; 2013.

Pittendrigh CS. The bromeliad-anopheles-malaria complex in Trinidad. I-The bromeliad flora. Evolution. 1948;2:58–89.

Powers JS, Montgomery RA, Adair EC, Brearley FQ, DeWalt SJ, Castanho CT, Chave J, Deinert E, Ganzhorn JU, Gilbert M, González-Iturbe JA, Bunyavejchewin S, Grau HR, Harms KE, Hiremath A, Iriarte-Vivar S, Manzane E, De Oliveira AA, Poorter L, Ramanamanjato JB, Salk C, Varel A, Weiblen GD, Lerdau MT. Decomposition in tropical forests: a pan-tropical study of the effects of litter type, litter placement and mesofaunal exclusion across a precipitation gradient. J Ecol. 2009;97:801–11.

Querejeta JI, Estrada-Medina H, Allen MF, Jiménez-Osornio JJ, Ruenes R. Utilization of bedrock water by *Brosimum alicastrum* tres growing on shallow soil a top limestone in a dry tropical climate. Plant Soil. 2006;287:187–97.

Querejeta JI, Estrada-Medina H, Allen MF, Jiménez-Osornio JJ. Water source partitioning among trees growing on shallow karst soils in a seasonally dry tropical climate. Oecologia. 2007;152:26–36.
Ramírez Morillo IM, Carnevali G, Chi F. Guía ilustrada de las Bromeliaceae de la porción mexicana de la Península de Yucatán. México: Centro de Investigación Científica de Yucatán; 2004.
Reyes-García C, Andrade JL. Crassulacean acid metabolism under global climate change. New Phytol. 2009;181:754–7.
Reyes-García C, Griffiths H. Strategies for survival of perennial bromeliads in seasonally dry forests. In: de la Barrera E, Smith WK, editors. Perspectives in biophysical plant ecophysiology: a tribute to Park S. Nobel. Mexico: Universidad Nacional Autónoma de México; 2009. p. 121–51.
Reyes-García C, Andrade JL, Simá JL, Us-Santamaría R, Jackson PC. Sapwood to heartwood ratio affects whole-tree water use in dry forest legume and non-legume trees. Trees. 2012a;26 (4):1317–30.
Reyes-García C, Mejia-Chang M, Griffiths H. High but not dry: diverse epiphytic bromeliad adaptations to exposure within a seasonally dry tropical forest community. New Phytol. 2012b;193:745–54.
Ricalde MF, Andrade JL, Durán R, Dupuy JM, Simá JL, Us-Santamaría R, Santiago L. Environmental regulation of carbon isotope composition and crassulacean acid metabolism in three plant communities along a water availability gradient. Oecologia. 2013;164:871–80.
Ricalde-Pérez MF. Fisiología ecológica de plantas CAM en dos hábitats de la penínusula de Yucatán: matorral de duna costera y selva baja caducifolia. PhD Dissertation. Mérida, México: Centro de Investigación Científica de Yucatán, A. C.; 2010.
Roa-Fuentes L, Campo J, Parra-Tabla V. Plant biomass allocation across a precipitation gradient: an approach to seasonally dry tropical forest at Yucatán, Mexico. Ecosystems. 2012;15 (8):1234–44.
Roig FA, Jimenez-Osornio JJ, Villanueva-Diaz J, Luckmand B, Tiessene H, Medina A, Noellemeyerf EJ. Anatomy of growth rings at the Yucatán Peninsula. Dendrochronologia. 2005;22:187–93.
Saito N, Harborne JB. A cyaniding glycoside giving scarlet coloration in plants of the Bromeliaceae. Phytochemistry. 1983;22:1735–40.
Salinas-Peba L, Parra-Tabla V, Munguia-Rosas M, Campo J. Survival and growth of dominant tree seedlings in a tropical dry forests of Yucatán: site and fertilization effects. J Plant Ecol. 2013. doi:10.1093/jpe/rtt055.
Sánchez SO, Islebe G. Tropical forest communities in southeastern Mexico. Plant Ecol. 2002;158:183–200.
Saxe H, Cannell MGR, Johnsen Ø, Ryan MG, Vourlitis G. Tree and forest functioning in response to global warming. New Phytol. 2001;149(3):369–99.
Steyn WJ, Wand SJE, Holcroft DM, Jacobs G. Anthocyanins in vegetative tissues: a proposed unified function in photoprotection. New Phytol. 2002;155:349–61.
Takahashi S, Murata N. How do environmental stresses accelerate photoinhibition? Trends Plant Sci. 2008;13:178–82.
Tamayo-Chim M, Reyes-García C, Orellana R. A Combination of forage species with different responses to drought can increase year-round productivity in seasonally dry silvopastoral systems. Agrofor Syst. 2012;84(2):287–97.
Ter Steege H, Cornelissen JHC. Distribution and ecology of vascular epiphytes in lowland rain forest of Guayana. Biotropica. 1989;21:331–9.
Torres W, Méndez M, Dorantes A, Durán R. Estructura, composición y diversidad del matorral de duna costera en el litoral yucateco. Bol Soc Bot Méx. 2010;86:37–51.
Valdez-Hernández M, Andrade JL, Jackson PC, Rebolledo-Vieyra M. Phenology of five tree species of a tropical dry forest in Yucatán, Mexico: effects of environmental and physiological factors. Plant Soil. 2010;329:155–71.

Valdez-Hernández M, Sánchez O, Islebe GA, Snook LK, Negreros-Castillo P. Recovery and early succession after experimental disturbance in a seasonally dry tropical forest in Mexico. Forest Ecol Manage. 2014;334:331–43.

Vargas R. How a hurricane disturbance influences extreme CO_2 fluxes and variance in a tropical forest. Environ Res Lett. 2012;7:035704. doi:10.1088/17489326/7/3/035704.

Vargas R, Allen MF. Diel patterns of soil respiration in a tropical forest after hurricane Wilma. J Geophys Res. 2008;113, G03021. doi:10.1029/2007JG000620.

Vargas R, Trumbore SE, Allen MF. Evidence of old carbon used to grow new fine roots in a tropical forest. New Phytol. 2009;182(3):710–8.

Vargas R, Hasselquist N, Allen EB, Allen MF. Effects of a hurricane disturbance on aboveground forest structure, arbuscular mycorrhizae and belowground carbon in a restored tropical forest. Ecosystems. 2010;13:118–28.

Vargas-Soto G. Regulación ambiental de la fotosíntesis de algunas especies de *Clusia* L. (Clusiaceae) en México. PhD Dissertation. Mérida, México: Centro de Investigación Científica de Yucatán, AC; 2010.

Vargas-Soto G, Andrade JL, Winter K. Carbon isotope composition and mode of photosynthesis in *Clusia* species from Mexico. Photosynthetica. 2009;47:33–40.

White DA, Hood CS. Vegetation patterns and environmental gradients in tropical dry forests of the northern Yucatán Peninsula. J Veg Sci. 2004;15:151–60.

Winter K, Smith JAC. An introduction to crassulacean acid metabolism: biochemical principles and biological diversity. In: Winter K, Smith JAC, editors. Crassulacean acid metabolism. Biochemistry, ecophysiology and evolution. Berlin: Springer; 1996.

Winter K, Aranda J, Holtum JAM. Carbon isotope composition and water-use efficiency in plants with Crassulacean acid metabolism. Funct Plant Biol. 2005;32:381–8.

Zaldivar-Jimenez A, Herrera-Silveira JA, Teutli-Hernandez C, Comin FA, Andrade JL, Coronado-Molina C, Perez-Ceballos R. Conceptual framework for mangrove restoration in the Yucatán Peninsula. Ecol Res. 2010;28:333–4.

Zimmerman KJ, Olmsted CI. Host tree utilization by vascular epiphytes in a seasonally inundated forest (Tintal) in Mexico. Biotropica. 1992;24:402–7.

Zotz G, Hietz P. The physiological ecology of vascular epiphytes: current knowledge, open questions. J Exp Bot. 2001;52(364):2067–78.

Chapter 6
Bee–Plant Interactions: Competition and Phenology of Flowers Visited by Bees

Rogel Villanueva-Gutiérrez, David W. Roubik, and Luciana Porter-Bolland

Abstract We present results of the flowering phenology of most plant species visited by European and Africanized honey bee (*Apis mellifera ligustica* and *A. mellifera scutellata,* respectively) in the Yucatán Peninsula. Colonies from both bee types visited the largest number of plant species at the end of the wet season (September and October) and the beginning of the dry season (November). A calendar is presented to indicate the community phenology of the floral resources of *Apis mellifera*. Comparisons were made in order to assess potential competition between both honey bee groups and between honey bees and native bees in relation to their food resources. Trees were also a constant pollen resource for *Apis mellifera* and *Melipona beecheii*, a native stingless bee. Solitary bees and *M. beecheii* bee seemed to change their floral resource use, both show '**resource partitioning**' to avoid competition. For example, two important plant families, Anacardiaceae and Euphorbiaceae, were lost to competing honey bees, but compensated for by greater use of Fabaceae, Rubiaceae, and Sapotaceae among solitary bees. Invasive generalist pollinators may, however, cause specialized competitors to fail, especially in less biodiverse environments. Deforestation, hurricanes, and fires are three factors that affect the habitat and food resources for bee colonies. Within agricultural areas, having large areas of natural vegetation, corridors, or strips of vegetation between the crop fields is important to favor adequate diversity of natural pollinators for pollination of crop plants.

R. Villanueva-Gutiérrez (✉)
Conservación de la Biodiversidad, El Colegio de la Frontera Sur, Apdo. Postal 424, Av. Centenario km 5.5, CP 77014 Chetumal, Quintana Roo, Mexico
e-mail: rvillanu@ecosur.mx

D.W. Roubik
Smithsonian Tropical Research Institute, Balboa, Republic of Panama

Smithsonian Tropical Research Institute, Unit 9100, DPO, AA 34002, USA
e-mail: roubikd@si.edu

L. Porter-Bolland
Instituto de Ecología, A.C., Red de Ecología Funcional, Carretera Antigua a Coatepec 351, CP 91070 Xalapa, Veracruz, Mexico
e-mail: luciana.porter@inecol.mx

G.A. Islebe et al. (eds.), *Biodiversity and Conservation of the Yucatán Peninsula*,
DOI 10.1007/978-3-319-06529-8_6

Keywords Solitary bees • *Melipona beecheii* • *Apis mellifera* • Flowering phenology • Melissopalinology • Competition • Nectar sources • Pollen sources • Foraging behavior

6.1 Introduction

Plants offer bees the resources they utilize to grow and reproduce, and plants benefit in turn from bee pollination, which enhances or facilitates sexual reproduction and genetic recombination. The interactions between any given plant and one or more of its pollinators may be intricate, but most obey certain general rules or performance standards. Obviously, in the Yucatán Peninsula there is no evolutionary history between the honey bee *Apis mellifera* and any native plant. However, many of their interrelationships allow one to investigate the behavioral and ecological adjustments of bees to the appearance and abundance of their resources. Here we explore some of those interactions in depth by including the most significant pollinating organism on earth, which has now become a separate biological entity in all of tropical America: the Africanized variety of *Apis mellifera*, here termed AHB.

From an economic point of view, pollination represents an important ecological service for wild and cultivated plants (Klein et al. 2007). Up to 90 % of terrestrial vascular plants potentially benefit from animal pollination (Ashworth et al. 2009; Ollerton et al. 2011). Insects in general, but particularly bees, represent one of the most important groups for pollination. An estimated 30 % of all crop species and as much as 75 % of major crop species, need or benefit from bee pollination (Roubik 1995; Meléndez-Ramírez et al. 2002; Klein et al. 2007). Managed bees are also important since not only can they be directed for enhancing pollination in agricultural systems, but also because of the economic importance of bee products.

To understand the interactions between plants and bees, information on flowering phenologies is of great importance. Phenology refers to the timing of recurring biological events in the animal and plant world, and its study pertains to causes of their timing with regard to biotic and abiotic factors. In relation to plants, phenology is the study of the periodicity mainly of leafing, flowering and fruiting of individual species, while at the community level, flowering phenology trends are also often considered (Lieth 1974; Roubik et al. 2005; Kuswaha et al. 2013).

From the bee's perspective, the phenology of plants determines resource availability in time and space. From the human perspective, knowing such information is important not only for bee management but for understanding how bee communities face current challenges brought about by environmental change. Changes that have been identified as threats to bee diversity and pollination services relate to effects of habitat destruction, climate change, biological invasions, biocides or chemical agents used to control other organisms (Burkle and Alarcón 2011; Roubik 2014), and the impact these products have on bee management (Villanueva-Gutiérrez et al. 2014). More information is urgently needed in order to establish the relationship between the availability of resources to bees and their interactions with the aforementioned environmental factors.

The Yucatán Peninsula presents an interesting setting for studying flowering phenology of plants visited by bees, including the effects of challenges for pollinators. Bee management in the area has been an economic activity of historical importance, although exotic honey bees were used frequently only after the 1950s (Labougle-Rentería and Zozaya-Rubio 1986; Calkins 1975). This may be seen as the result of a combination of a rich flora available for bees and the Mayan tradition of beekeeping (Rico-Gray et al. 1991). Although until recently beekeeping was conducted with native bees, today the use of the Old World and now primarily African descendants of that primarily African *Apis* (Africanized honeybee hybrid) species, frequently mislabeled as "the European honey bee", is managed throughout the Yucatán Peninsula.

The African honey bee *Apis mellifera scutellata* was introduced in Brazil in 1956 for honey production improvement. It quickly gave rise to the AHB, which is still much like its African progenitors in behavior and ecology. The replacement of European *Apis mellifera ligustica* with an African subspecies, their hybridization, and the current predominance of Africanized bees in most Neotropical habitats has been extensively reviewed in many journals and books, and will not be further reviewed here.

The arrival of Africanized honey bees in the Yucatán Peninsula was reported in September 1987 (Barrios-Delgado et al. 1990). The Africanization process took around 5 years in the area. This represents a longer period, by 2 years, than was reported in other places, probably because of the large number of European honey bee colonies that were present in the region. When there are very few Africanized honey bees compared to numbers of the European honeybees in an area, as occurred on the island of Puerto Rico (Galindo-Cardona et al. 2013), the Africanization process results in fewer behavioral or ecological changes among wild or naturalized honey bees. A key factor in the Neotropics is that there were almost no feral or naturalized honey bees in tropical America, before AHB arrival (Roubik 1989).

Here we present information on flowering phenology of plants visited by bees. The information is drawn from several studies conducted in the Yucatán Peninsula regarding flower visits by European and Africanized honey bees. One objective of these studies was to assess the impact of Africanization, both on European honey bees and native bees. The studies took place when the Africanized bees were just arriving to the Yucatán Peninsula (1988–1991). Comparisons were made in order to assess competition between both honey bee groups in relation to their food resources (Villanueva-Gutiérrez 1994, 1999, 2002; Villanueva-Gutiérrez and Roubik 2004).

6.2 Flowering Phenology Studies

A handful of studies have been conducted in the southeast of Mexico in relation to bee resources and flowering phenology. Villanueva-Gutiérrez (1984) related the flowering phenology of a low tropical deciduous forest in Veracruz with the flowering periods in which European honey bees visited the flora. Gentry (1995) focused on how different dry forest plant communities differ from each other. He found that differences in the severity of the dry season as well as rainfall patterns cause differing leafing patterns. Borchert et al. (2004) also found a relation between flowering periodicity and soil humidity, comparing flowering times of similar species in different areas. Porter-Bolland (2003) determined the important melliferous and polliniferous species in relation to their flowering phenology and beekeeping management in La Montaña, north of the Calakmul Biosphere Reserve in the state of Campeche, Mexico. Parra-Tabla and Vargas (2004) found that, as in other food deceit orchids (Roubik 2014), natural selection in *Myrmecophila christinae* favors individuals that flower early or late in relation to the population's flowering peak. However, results also suggest that a fluctuating regime of selective events act on flowering time of this species. Salinas-Peba and Parra-Tabla (2007) described the reproductive phenology, breeding system and pollination of *Manilkara zapota* in two contrasting environments: medium-height, subdeciduous forest, and homegardens or "*solares*" in a Mayan community in the state of Yucatán. Significant differences were found between environments both in the temporal distribution of flower and mature fruit production, as well as in the proportion of mature fruits.

The data were obtained by registering the pollen species identified in pollen load samples of honey bees. We also made a calendar to indicate the community phenology of the floral resources during our two-year study. It is part of a long-term study on bees to determine the most important honey and pollen sources for European and Africanized honey bees in the Sian Ka'an Biosphere Reserve (Villanueva-Gutiérrez 1994, 1999, 2002; Roubik and Villanueva-Gutiérrez 2009; Villanueva-Gutiérrez and Roubik 2004). Here we also discuss the competition that took place between the two *A. mellifera* subspecies, as well as between Africanized honeybees and native bees.

6.3 A Case Study

6.3.1 Study Area

Field work began in June 1989 and was completed in May 1991. Two sites were chosen to conduct fieldwork in the area of the Sian Ka'an Biosphere Reserve (SKBR). The first one, Palmas (site 1), is located bordering the reserve (33 km south of Felipe Carrillo Puerto) while the second one, Santa Teresa (site 2), is located within the reserve (13 km northeast of Felipe Carrillo Puerto).

Vegetation was medium stature tropical forest (selva mediana subperennifolia), low stature forest (selva baja subcaducifolia) and low inundated tropical forest (selva baja inundable) (according to the classification of Miranda and Hernández-X 1963), and included patches of secondary vegetation. The area surrounding Palmas were more disturbed than those in the immediate vicinity of St. Teresa. This was because the first site was near a road and there was more agricultural activity. According to the classification of Koeppen (1936), the type of climate that exists in the study area is Aw, which is defined as hot subhumid, with a mean annual temperature over 22 °C and annual precipitation between 700 and 1500 mm.

The area foraged often by honey bees has a mean range of 1.7 km, with most foraging occurring within 6 km of the nest or hive (Levin and Glowska-Konopacka 1963; Winston 1987). In some cases, European and Africanized honey bees have been observed foraging at a range of 10 km from their colonies (Visscher and Seeley 1982; Vergara 1983; Roubik 1989). The mean range of 8 km that usually applies to Africanized honey bees implies that each colony could have access to at least 20,000 ha, or 200 km^2.

6.3.2 Pollen Analysis

At both study sites, Palmas and St. Teresa, 15 hives containing colonies headed by mated European honey bee queens, and 15 hives containing Africanized colonies derived from Africanized wild colonies [identified morphometrically by the method of Daly and Balling (1978)] were selected. The Africanized colonies were collected from the study areas. Modified Ontario Agricultural College (O.A.C.) pollen traps (Smith and Adie 1963; Waller 1980) were placed at the base of the hives to sample the pollen loads that the honey bees carry on their hind legs. Pollen load samples were obtained from these hives for determination and quantification of food sources for these two bee types.

Pollen loads were collected for 48 h each month during a period of 2 years. A total of 204 pollen load samples were obtained in both sites (102 from European honey bees and 102 from Africanized honey bees). The pollen grains from each sample were homogenized using a Sonicator cell disrupter (O'Rourke and Buchmann 1991), then acetolyzed using the Erdtman technique (1943) and mounted on slides with glycerin jelly. During the 2 years, we registered the flowering period of the flora and collected plant specimens. A pollen atlas (Palacios Chávez et al. 1991) was consulted for the identification of pollen grains from the honey and pollen load samples. In order to have a complete calendar of the identified pollen resources used by honeybees, we also used data obtained from honey samples (Villanueva-Gutiérrez 1994). In this way, sample composition could be analyzed in terms of species richness, pollen percentage frequency and mean percentage frequency per total number of samples.

6.3.3 Species Visited by Honey Bees and Their Timing

A total of 168 different pollen species were identified, belonging to 41 different plant families. Those that contributed the largest number of pollen species were Fabaceae, Asteraceaee, Boraginaceae, Convolvulaceae, Euphorbiaceae, Sapindaceae, Poaceae, Myrtaceae, Sapotaceae and Tiliaceae. Fabaceae and Asteraceae families are very important in the bee diet; both families were represented in the pollen load samples, with 12 % and 11 % of the total number of pollen species respectively. The most frequent pollen species in the European bee samples for years 1 and 2 were *Cecropia peltata*, *Metopium brownei*, *Lonchocarpus* sp. 2, *Viguiera dentata*, *Eragrostis* sp. 1, *Panicum* sp. 1, *Bursera simaruba*, *Trema micrantha*, *Eupatorium albicaule*, *Eugenia* sp. 1 and *Pluchea* sp. 1. For the Africanized honey bee samples the most frequent pollen species during both years were *Cecropia peltata*, *Metopium brownei*, *Lonchorpus* sp. 2, *Eragrostis* sp. 1, *Eupatorium albicaule*, *Viguiera dentata*, *Mimosa bahamensis*, *Rinchospora microcarpa*, *Eugenia* sp. 1, *Panicum* sp. 1, *Eupatorium albicaule* and *Bidens* sp. 1 (see also Villanueva-Gutiérrez 1999).

The pattern in which pollen species were identified in pollen load samples lets us understand their availability through time. We separated the plant species identified according to their habit: (1) trees, (2) shrubs, (3) herbs, (4) woody vines and (5) trees or shrubs. Group (5) refers to taxa (family, genus or species level) that have been described as either trees or bushes. Arranging the data according to the total number of pollen species found monthly in the pollen load samples, two graphs were made showing the annual flowering phenology of the community of plants used by honey bees both by total and according to the species' habit (Figs. 6.1 and 6.2).

A table was made in the form of a calendar, indicating the periods in which the different pollen species were collected by European and Africanized bees (Appendix). Some species provide pollen to the honey bees for short periods of 1 month or less, and others for longer periods. Species that are present in sampled pollen loads for long periods of time include *Buxus bartletti*, *Cordia* sp. 4, *Melanthera aspera*, *Mimosa bahamensis*, *Pluchea* sp. 1, *Psidium sartorianum* (9 months), *Eupatorium* sp. 2, *Pouteria* sp. 1 (10 months), *Bursera simaruba*, *Metopium brownei*, *Parthenium hysterophorus*, *Trema micrantha*, *Thouinia canescens* (11 months), and *Cecropia peltata* and *Eugenia* sp. 1 (12 months).

The distribution of flowering periods according to species habit reveals a small fluctuation in the number of trees species visited by both honey bees (combined data) during the year. The largest number of tree species was visited in September (18) and November (20), and the smallest in January (11). Shrubs had the largest number of species visited in January (10), July (9), September (10) and November (13), and the smallest number in May (1). Herbs' contribution to pollen had a large peaked sharply in October (25 species) and dropped in June (4). Woody vine species were not much visited by the honey bees. The largest number of species was found in November (8), while from January to June and in October, bees visited

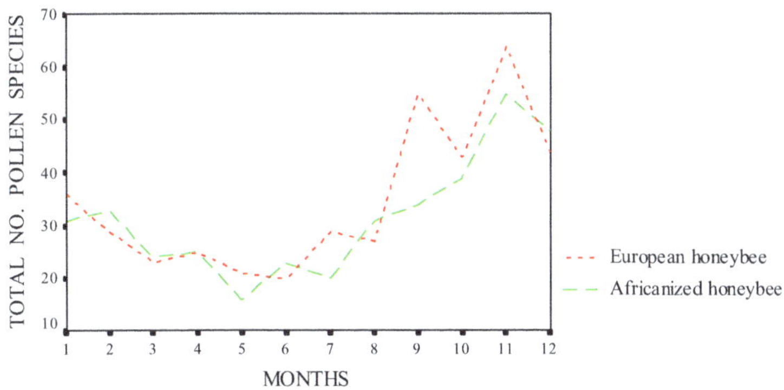

Fig. 6.1 Total number of pollen species found in European and Africanized honeybee pollen load samples (January to December)

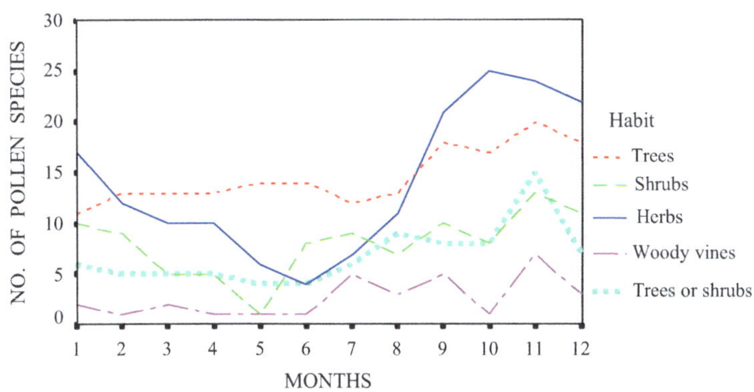

Fig. 6.2 Distribution of the pollen species according to their habit: trees, shrubs, herbs, woody vines and trees or shrubs (January to December)

only one or two species. The fifth group of plants, which includes either trees or shrubs, has the largest number of species visits in November (15) and the smaller ones in May and June (3 in both months).

Colonies of both bee types visited the largest number of plant species at the end of the wet season (September and October) and beginning of the dry season (November). The species visited included herbs, shrubs and trees. Considering only the number of pollen species per colony from both bee types, the largest number of plants visited by honey bees ranged from 10 to 31 species. The smallest number of plant species (per colony) visited by both bee types was in May and June, at the beginning of the wet season, and the species were mainly trees and shrubs. Colonies visited from 3 to 11 species per colony at that time.

Of the 168 total plant species visited by both subspecies of *A. mellifera* in this study, European bees visited 84 % and Africanized bees visited 64 %. Perhaps surprisingly, only 48 % of the plants visited by bees were common to both honey bee types. Only 10 species (6 %) of those shared by both bee types were used intensively (Villanueva-Gutiérrez 1999; Villanueva-Gutiérrez and Roubik 2004).

6.4 Phenology of Plants Visited for Pollen Collection

The pattern followed by the total number of plant species that were visited by both European and Africanized honey bees was very similar to the one followed by the mean number of species visited per colony. European and Africanized bees visited the largest number of species at the end of the wet season and early dry season (September to January), showing a small drop on the number of species visited during October. The same pattern has been observed during this period by Ceballos-Martínez (1987) in Sian Ka'an Reserve in areas of similar vegetation types.

According to Villanueva-Gutiérrez (1999, 2002) the most frequent pollen species in both European and Africanized bee pollen load samples were *Cecropia peltata*, *Metopium brownei*, *Lonchocarpus* sp. 2, *Viguiera dentata*, *Bursera simaruba*, *Eragrostis* sp. 1, *Eupatorium albicaule*, *Panicum* sp. 1 and *Eugenia* sp. 1. These represent 71 % of all the mean percentage frequencies in the samples. Although using more plant species, the European bees specialize much more heavily on a few plants than do Africanized honey bees (Villanueva-Gutiérrez and Roubik 2004). European bees in the Yucatán Peninsula may have responded to exotic Africanized honey bee competition in two ways, by shifting flower choice and by increasing resource specialization. Interestingly, when competition is most intense, solitary native bees with no food reserves or colonies, seem specialize more on fewer resource species—the same as European honey bees, when faced with competition from feral Africanized honey bees (Roubik and Villanueva-Gutiérrez 2009; Roubik 2009).

As with the honey samples, competition might also occur if one or more of the most frequent pollen species become increasingly scarce, unless honey bees can switch to other abundant pollen sources (Villanueva-Gutiérrez and Roubik 2004; Roubik 2009). Competition often increases during part of a long drought period or when there is very high precipitation.

6.5 Nectar Competition Between European and Africanized Honey Bees

The major period of honey production at the sites in this study began in January and ended in May. According to our results, the most important nectar sources for bees during January and February are herbs, while trees comprise their main nectar source from March to May—the driest months of the dry season. Pollen of *Gymnopodium floribundum*, *Viguiera dentata*, *Metopium brownei*, *Bursera simaruba*, *Lonchocarpus* spp. *Piscidia piscipula*, and *Samyda yucatanensis* were the most common in honey samples (Villanueva-Gutiérrez 1994).

In flowering phenology studies of plants visited by honey bees at La Montaña area in Hopelchen, Campeche (Porter-Bolland 2003; Porter-Bolland et al. 2009), honey production began a little later, in mid February, and also ended in May (at the beginning of the rainy season). The difference may be that the forested area of La Montaña honey bees mostly rely on arboreal resources, rather than herbs.

Porter-Bolland (2010) identified plant species visited by honey bee according to beekeeper observations. She identified 146 species, belonging to 101 genera and representing 35 families. The latter is slightly less than the 168 species (from 41 different families) identified in the study presented in this chapter using a more precise method (the identification of plant sources through pollen loads on bees). The most important families identified by Porter-Bolland also included Fabaceae, Euphorbiaceae, Sapindaceae, and Sapotaceae (mostly trees), and Asteraceae and Convolvulaceae (mostly herbs and vines), as in this study.

The number of tree species visited by honey bees was constant during the year except for two small peaks in September and November. Trees represent a constant resource of nectar and pollen, even during periods in which there is not much nectar flow or pollen collection (Fig. 6.1 and Appendix). Trees were also found to be a constant pollen resource for *Apis mellifera* in a study in low deciduous forest in Veracruz, Mexico (Villanueva-Gutiérrez 1984). Porter-Bolland (2003) also found trees to be of great importance as bee resources in the study conducted in La Montaña area cited above, in an area covered by medium- and low-stature tropical forest located not far from Sian Ka'an, but inland of the peninsula in the state of Campeche.

Trees represent the most important pollen and nectar sources during the latter half of the dry period of the year (March to May); herbs are the most important nectar and pollen sources during the end of the wet period (October) and the first half of the dry period (November to February). The total number of plant species visited by the honey bees represented around 20 % of all the Angiospermae present at the Sian Ka'an area study sites (Villanueva-Gutiérrez 1994, 1999, 2002).

Between eight and ten species of shrubs were visited during the year, except at the end of the dry season (May) when sharp decline occurred. The group of species categorized as either trees or shrubs in habit presents a pattern similar to that of trees. Woody vines were not much visited by bees, except during July, September and November, when bees visited most of the Bignoniaceous and Sapindaceae species.

For this study, the invasive Africanized honey bee was compared side by side with re-introduced European *Apis mellifera ligustica* in post-invasion experiments, including adequate bee colony density and replication, and to make comparisons to pre-invasion foraging ecology. We compared which of the plant species were visited by European and Africanized honey bees (the species they shared), and which ones were visited only by either of them. The volumes of pollen introduced to the colonies were also considered and compared to give a better measure of the real diet being used by both honey bee subspecies.

Almost half of all flowering species in the study areas were visited by both Africanized and European honey bee types. That is, only 44 % of the plants visited by honey bees were shared by both groups, indicating a differential utilization of available resources.

Pollen from flowering species that do not produce any nectar were found in the honey samples. This is the case with *Cecropia peltata* (one of the most abundant species in the European and Africanized honey samples during the wet and the dry seasons). Also, 11 Poaceae, 6 Cyperaceae, 2 *Mimosa*, 2 *Malpighia* and 1 Piperaceae were found (Villanueva-Gutiérrez 1994). The pollen of these nectarless plants was probably collected by the foraging bees and introduced to the hives where some of it became trapped in the honey. Many of the activities undertaken by honey bees within the colony (dancing, food transmission, food handling, cleaning and grooming) probably contribute to the dispersal "contamination" of honey by pollen.

6.6 Bee Competition for Pollen and Nectar Sources

6.6.1 Pollen Volumes Collected by European and Africanized Honey Bees

We also calculated the volume of pollen collected by European and Africanized honey bees (Villanueva-Gutiérrez and Roubik 2004), so we could provide detailed quantitative comparisons of the volume and kind of pollen used by the two honeybee subspecies, and also corroborate our findings with the flower species used by apiary European honey bees at the same sites, 2 years before the African honeybee invasion began. Although major resource species were similar, the richest diet was that of the European bees, while *quantitatively*, African bees were more generalized and European bees more specialized. Morisita-Horn similarities were 0.76 for pollen use by volume sp^{-1}, and 0.55 for taxon-specific intensity of use. European bees were evidently displaced from their previous floral resources, to the extent that their principal diet item was novel based on pre-invasion records, as was a maximum monthly diet breadth recorded for both bees. We suggest resident European bees that compete with invasive honeybees may increase both diet breadth and quantitative specialization.

Frequency differences may also be influenced by factors such as the amount of pollen and nectar produced by each species, the availability of pollen and nectar from each plant species on days when pollen loads and honey were sampled, and the distance of those resources from the hives. These factors, together with the color, odor, and morphology of flowers, determine the preference of honeybees for particular nectar and pollen sources. It can be observed that there is a high reliance on only a few species: the first five, for example, comprise almost 50 % of all the mean percentage frequencies (Villanueva-Gutiérrez 1994, 1999; Villanueva-Gutiérrez and Roubik 2004). These polliniferous species represent an important resource for beekeeping, considering that the Yucatán Peninsula is one of the most important honey production regions of the world, with over 10 hives per square kilometer), due to the large number of colonies managed mostly by the Mayan people (Paxton et al. 1991; Echazarreta et al. 1997).

6.6.2 Ecological Studies on Solitary Bees in the Yucatán Peninsula

A population dynamics study was conducted with solitary bees, Megachilidae and *Centris*, in the Sian Ka'an Biosphere Reserve from 1988 to 2005. Wooden traps were used to collect these bees and the pollen they store within their nest for feeding their brood (Roubik and Villanueva-Gutiérrez 2009). We found that drought and heavy rain or hurricane damage had severe effects on bee populations. Honeybee competition, and ostensibly pollination of native plants, caused changes in local pollination ecology. Natural disasters made a large negative impact on native bee populations, but sustained presence of Africanized honey bees were not affected.

Shifts in floral hosts by native bees were common. Two important plant families, Anacardiaceae and Euphorbiaceae, were lost to competing honeybees, but compensated for by greater use of Fabaceae, Rubiaceae, and Sapotaceae among native bees (Roubik and Villanueva-Gutiérrez 2009). Invasive generalist pollinators may, however, cause specialized competitors to fail, especially in less biodiverse environments.

6.6.3 Conservation and Food Resources of the Stingless Bee **Melipona beecheii**

Melipona bees visit the flora of mature vegetation or low disturbed vegetation. The plants that have been reported as important nectar or pollen sources for this bee, (some based on beekeeper reports, are *Gymnopodium floribundum* ('dzidzilche'), *Piscidia piscipula* ('jabín'), *Lysiloma latisiliquum* ('tzalám'), *Eugenia buxifolia* ('pichi che)', *Eugenia axilaris* ('ich juju'), *Guazuma ulmifolia* '(pixoy'), *Swartzia cubensis* ('k'ataloox'), *Lonchocarpus longystilus* ('balché'), *Coccoloba cozumelensis* ('boob'), *Vitex gaumeri* ('ya axnik'), *Solanum lanceifolium* ('silclimuch'), *Senna* sp and *Solanum* sp. (Villanueva-Gutiérrez et al. 2005a; Parra-Romero 2015). This bee also seems to switch their food sources after the arrival of Africanized honey bee in order to avoid competition (Villanueva-Gutiérrez 1994; Villanueva-Gutiérrez and Colli Ucán 2005; Parra-Romero 2015). In the Yucatán Peninsula there is a low number of *Melipona* bee populations in the wild (Cairns et al. 2005) but a number of colonies are kept in hives of some type. Deforestation, hurricanes, and fires are three factors that affect food resources for bee colonies. Many *Melipona* beekeepers report that they think their bees are dying from lack of food. They have witnessed, over time, that colonies produce less honey and that the population is diminishing in size to the point where the *Melipona* cannot defend against their natural enemies (Villanueva-Gutiérrez et al. 2005b).

6.7 Strategies for the Conservation of the Forest and Pollinators in the Yucatán Peninsula

Because of certain competitive advantages shown by Africanized honey bees, bee management programs should be prioritized. Bee management should be encouraged with a strong component of genetic improvement to lower Africanization rates and control feral populations. Bee management with native bees should also be promoted since the activity has been decreasing in most of the Yucatán Peninsula.

On the other hand, forest protection is important for the conservation of bee floral resources, as well as the regulation of agricultural activities. For this, legislation to protect forested areas from complete destruction, especially near villages and cultivated areas should be made more effective, including the encouragement of different conservation strategies, such as both Natural Protected Areas and community conservation initiatives. Other strategies can be the development of projects at the local level for reforestation and restoration. Within agricultural areas, having large natural vegetation areas, corridors, or strips of vegetation between the crop fields favor adequate diversity of natural pollinators for the pollination of crop plants. It is also important to allow the growth of herbs, shrubs, and trees that will be important sources of food for stingless bees. At the same time, planting native trees around meliponaries will provide important sources of food for stingless bees when they are located in backyards of the homes of local residents.

Mechanized agriculture is becoming a common practice in the contemporary Yucatán Peninsula, so the control of herbicides and pesticides is important. It has been demonstrated that honey bees do visit the flowers of genetically modified (GM) soybean, and there is a high risk of contamination not only of the honeys produced in the Yucatán Peninsula, but also the honeys produced throughout Mexico (Villanueva-Gutiérrez et al. 2014). The aforementioned strategies are vital for the conservation of pollinators of many native plants and crops. Native stingless bee colonies as well as many other foragers are vital to sustain plant communities.

6.8 Acknowledgements

We want to thank to Wilberto Colli-Ucán, Victor Ceballos-Martínez, and Margarito Tuz-Novelo for help in the fieldwork and Rodrigo Villanueva-Ceballos for his help in making the graphs.

Appendix

Calendar of Plants Visited for Honey and Pollen Collection by Honey Bees (Sampling Period of 2 Years)

Angiospermae	Jan	Feb	Mar	Apr	May	Jun	Jul	Aug	Sep	Oct	Nov	Dec
Acanthaceae												
Bravaisia tubiflora											+	
Justicia campechiana	+		+			+	+	+	+	+	+	+
Amaranthaceae												
Amaranthus sp. 1	+											
Iresine celosia					+							
Anacardiaceae												
Metopium brownei	+	+	+	+	+	+	+		+	+	+	+
Spondias sp. 1						+	+					
Annonaceae												
Annona sp. 1										+		
Apocynaceae												
Mandevilla sp. 1					+							
Araliaceae												
Dendropanax arboreus									+	+	+	
Arecaceae												
Chamaedorea sp. 1								+	+		+	
Cocos nucifera	+	+							+	+	+	+
Thrinax radiata				+								
Arecaceae 1		+	+					+	+			
Arecaceae 2		+						+				
Asteraceae												
Baccharis heterophylla	+											

(continued)

Angiospermae	Jan	Feb	Mar	Apr	May	Jun	Jul	Aug	Sep	Oct	Nov	Dec
Bidens squarrosa	+	+					+					
Bidens sp. 1	+							+	+	+	+	+
Eupatorium albicaule	+	+	+						+	+	+	+
Eupatorium campechense	+											
Eupatorium sp. 1								+	+	+	+	
Eupatorium sp. 2	+	+	+	+		+	+	+	+	+	+	
Melanthera aspera		+		+	+	+		+	+	+	+	+
Melanthera nivea								+	+		+	
Mikania sp. 1			+									
Parthenium hysterophorus	+	+	+	+	+	+	+		+	+	+	+
Pluchea sp. 1	+		+	+			+	+	+	+	+	+
Vernonia cinerea	+	+										
Viguiera dentata	+	+	+	+							+	
Asteraceae 1	+											
Asteraceae 3				+								
Asteraceae 4			+									
Asteraceae 5										+		
Bignoniaceae												
Amphilophium paniculatum		+					+		+	+	+	
Arrabidaea floribunda									+		+	
Cydista sp. 1								+			+	
Cydista sp. 2								+	+			
Bignoniaceae 1										+	+	
Bombacaceae												
Ceiba aesculifolia									+	+	+	
Ceiba pentandra					+							
Pseudobombax ellipticum	+		+	+	+		+			+	+	+
Boraginaceae												
Cordia alliodora						+						
Cordia curassavica								+	+			
Cordia dodecandra	+											+
Cordia gerascanthus											+	

(continued)

Angiospermae	Jan	Feb	Mar	Apr	May	Jun	Jul	Aug	Sep	Oct	Nov	Dec
Cordia sp. 1							+				+	
Cordia sp. 3					+			+	+			
Cordia sp. 4	+	+	+	+		+		+	+	+	+	
Cordia sp. 5											+	
Heliotropium sp. 1		+		+								
Heliotropium sp. 2											+	
Burseraceae												
Bursera simaruba	+	+	+	+	+	+		+	+	+	+	+
Bursera sp. 1	+		+	+	+	+	+		+		+	
Buxaceae												
Buxus bartlettii	+	+		+		+	+	+		+	+	+
Cecropiaceae												
Cecropia peltata	+	+	+	+	+	+	+	+	+	+	+	+
Combretaceae												
Bucida spinosa					+							
Conocarpus erecta							+					
Convolvulaceae												
Evolvulus alsinoides										+	+	+
Ipomoea sp. 1											+	
Ipomoea sp. 2	+										+	+
Ipomoea sp. 3	+										+	
Ipomoea sp. 4	+											
Jacquemontia nodiflora											+	
Jacquemontia aff. *pentantha*	+											
Merremia cisoides									+			+
Merremia tuberosa	+	+										
Operculina pinnatifida											+	+
Cucurbitaceae												
Cayaponia sp. 1								+	+	+	+	
Cucurbita maxima									+			
Cyperaceae												
Rhynchospora microcarpa		+					+	+	+	+	+	+
Cyperaceae 1									+	+	+	

(continued)

Angiospermae	Jan	Feb	Mar	Apr	May	Jun	Jul	Aug	Sep	Oct	Nov	Dec
Cyperaceae 3										+		
Euphorbiaceae												
Croton campechianus										+		
Croton chichenensis							+				+	+
Croton sp. 1	+	+					+				+	+
Croton sp. 2				+					+			
Euphorbia sp. 1						+	+					
Euphorbia sp. 2							+	+	+		+	+
Euphorbia sp. 3								+				
Euphorbiaceae 1							+					
Fabaceae												
Bauhinia divaricata											+	
Bauhinia herrerae											+	+
Caesalpinia gaumeri		+	+		+							
Caesalpinia sp. 1		+									+	
Caesalpinia sp. 3								+				
Gliricidia sepium						+						
Leucaena leucocephala	+	+	+	+					+	+	+	+
Lonchocarpus rugosus	+							+	+	+	+	
Lonchocarpus sp. 1			+	+								
Lonchocarpus sp. 2		+	+	+	+	+					+	
Lysiloma latisiliquum					+	+						
Mimosa bahamensis	+	+	+	+	+				+	+	+	+
Mimosa pudica	+	+		+					+	+	+	+
Nissolia fruticosa			+	+						+	+	
Piscidia piscipula					+	+						
Pithecellobium albicans							+					
Pithecellobium dulce								+				
Stylosanthes humilis												+

(continued)

Angiospermae	Jan	Feb	Mar	Apr	May	Jun	Jul	Aug	Sep	Oct	Nov	Dec
Fabaceae 1						+						
Fabaceae 2											+	
Flacourtiaceae												
Casearia nitida						+						
Laetia thamnia							+					
Samyda yucatanensis	+		+	+	+				+		+	+
Labiatae												
Hyptis pectinata									+			
Hyptis sp. 1												+
Leonotis nepetaefolia												+
Ocimum micranthum					+							
Labiatae 1									+			
Lauraceae												
Nectandra coriacea										+		
Loranthaceae												
Psittacanthus mayanus										+		
Malpighiaceae												
Malpighia emarginata										+	+	
Malpighia glabra										+		
Malvaceae												
Abutilon permolle	+	+							+	+	+	+
Hampea trilobata									+			
Hibiscus sp. 1								+	+	+	+	
Hibiscus sp. 2								+	+	+	+	+
Malvaceae 2									+			+
Moraceae												
Chlorophora tinctoria	+	+	+							+	+	+
Moraceae 1								+	+			
Myrtaceae												
Calyptranthes millspaughii	+											
Eugenia mayana										+	+	
Eugenia sp. 1	+	+	+	+	+	+	+	+	+	+	+	+
Eugenia sp. 2										+	+	
Eugenia sp. 3							+	+			+	

(continued)

Angiospermae	Jan	Feb	Mar	Apr	May	Jun	Jul	Aug	Sep	Oct	Nov	Dec
Myrcianthes fragans		+						+	+		+	+
Psidium sartorianum			+	+	+	+	+		+	+	+	+
Poaceae												
Chloris sp. 1								+	+	+	+	
Eragrostis sp. 1		+	+	+								
Panicum sp. 1	+	+							+	+	+	+
Panicum sp. 2										+	+	
Panicum sp. 3												+
Paspalum sp. 1								+	+	+	+	
Zea mays									+	+		
Polygonaceae												
Coccoloba sp. 1							+	+	+	+	+	
Coccoloba sp. 2							+	+			+	+
Gymnopodium floribundum							+	+	+	+	+	+
Ranunculaceae												
Clematis dioica			+	+	+	+						
Rhamnaceae												
Colubrina arborescens											+	
Rubiaceae												
Asemnanthe pubescens							+	+				
Chiococca alba							+					+
Exostema caribaeum								+				
Guettarda combsii				+								
Rutaceae												
Esenbeckia berlandieri								+				+
Sapindaceae												
Exothea diphylla												+
Paullinia sp. 1	+		+		+		+	+	+		+	+
Serjania racemosa	+											
Serjania yucatanensis		+	+	+						+	+	+
Serjania sp. 2							+					
Serjania sp. 4											+	
Thouinia canesceras	+	+	+	+		+	+	+	+	+	+	+
Sapindaceae 1								+				

(continued)

Angiospermae	Jan	Feb	Mar	Apr	May	Jun	Jul	Aug	Sep	Oct	Nov	Dec
Sapotaceae												
Bumelia sp. 1											+	
Pouteria mammosa												+
Pouteria sp. 1	+	+	+	+	+	+			+	+	+	+
Pouteria sp. 2			+	+	+				+	+	+	
Sapotaceae 1	+						+				+	
Sapotaceae 2												+
Sapotaceae 3										+		
Scrophulariaceae												
Capraria biflora									+			
Smilacaceae												
Smilax spinosa		+	+									
Tiliaceae												
Corchorus sp. 1												+
Heliocarpus donnell-smithii												+
Heliocarpus sp. 1	+										+	
Heliocarpus sp. 2												+
Luehea speciosa												+
Triumfetta dumetorum	+	+										
Ulmaceae												
Trema micrantha	+	+	+	+		+	+	+	+	+	+	+
Vervenaceae												
Verbenaceae 1								+				

References

Ashworth L, Quesada M, Casas A, Aguilar R, Oyama K. Pollinator-dependent food production in Mexico. Biol Conserv. 2009;142:1050–7.

Barrios-Delgado C, Pérez-Dominguez E, Sánchez-Navarro LM. Efectos y repercusiones de la abeja africana en la Península de Yucatán. In: Proceedings of the IV American Seminar on beekeeping, Mazatlán, Sinaloa. Secretaría de Agricultura y Recursos Hidráulicos and Gobierno del estado de Sinaloa, Mexico; 1990, p. 15–33.

Borchert R, Meyer SA, Felger RS, Porter-Bolland L. Environmental control of flowering periodicity in Costa Rican and Mexican tropical dry forests. Glob Ecol Biogeogr. 2004;13:409–25.

Burkle LA, Alarcón R. The future of plant—pollinator diversity: understanding interaction networks across time, space, and global change. Am J Bot. 2011;98(3):1–11.

Cairns CE, Villanueva-Gutiérrez R, Koptur S, Bray DB. Bee populations, forest disturbance, and Africanization in Mexico. Biotropica. 2005;37(4):686–92.

Calkins CF. The introduction of *A. mellifera* to the Yucatán Peninsula of Mexico. Gleanings in Bee Culture. 1975;103:202–3.
Ceballos-Martínez VM. Evaluación del polen recolectado por la abeja *Apis mellifera ligustica* Spinola en la Resrva de la Biósfera de Sian Ka'an, Quintana Roo, México. Bch thesis. University of Veracruz; Córdoba, Ver., Mexico; 1987, p. 24.
Daly HV, Balling SS. Identification of Africanized honeybees in the Western Hemisphere by discriminant analysis. J Kansas Entomol Soc. 1978;5:857–69.
Echazarreta CM, Quezada-Euan JJG, Medina LM, Pasteur KL. Beekeeping in the Yucatán Peninsula: development and current status. Bee World. 1997;78(3):115–27.
Erdtman G. An introduction to pollen analysis. Waltham, MA: Chronica Botánica Co.;1943. 239p.
Galindo-Cardona A, Acevedo-Gonzalez JP, Rivera-Marchard B, Giray T. Genetic structure of the gentle Africanized honey bee population (gAHB) in Puerto Rico. BMC Genet. 2013;14:65.
Gentry AH. Diversity and floristic composition of neotropical dry forests. In: Bullock SH, Mooney HA, Medina F, editors. Seasonally dry tropical forests. Cambridge: Cambridge University Press; 1995.
Klein AM, Vaissiere BE, Cane JH, Steffan-Dewenter I, Cunningham SA, Kremen C, Tscharntke T. Importance of pollinators in changing landscapes for world crops. Proc R Soc B Biol Sci. 2007;274(1608):303–13.
Koeppen W. *Das gepographische system der klimate in handbuch der klimatologie.* Band I, Teil C; Berlin, Germany; 1936. p. 44.
Kuswaha CP, Tripathi SK, Tripathi BD, Singh KP. Patterns of tree phenological diversity in dry tropics. Acta Ecol Sin. 2013;31:179–85.
Labougle-Rentería JM, Zozaya- Rubio JA. La apicultura en México. Ciencia y Desarrollo. 1986;69:17–35.
Levin MD, Glowska-Konopacka S. Responses of foraging honeybees in alfalfa to increasing competition from other colonies. J Apic Res. 1963;2(1):33–42.
Lieth H. Phenology and seasonality modeling. Ecological studies, vol. 8. New York: Springer; 1974.
Meléndez-Ramírez V, Magaña-Rueda S, Parra-Tabla V, Ayala R, Navarro J. Diversity of native bee visitors of cucurbit crops (Cucurbitaceae) in Yucatán México. J Insect Conserv. 2002;6 (135–147):2002.
Miranda F, Hernández-X J. Los tipos de vegetación de México y su clasificación. Bol Soc Bot Méx. 1963;28:29–179.
Ollerton J, Winfree R, Tarrant S. How many flowering plants are pollinated by animals? Oikos. 2011;120(3):321–6.
O'Rourke MK, Buchmann SL. Standardized analytical techniques for bee-collected pollen. Entomol Soc Am. 1991;20:507–13.
Palacios Chávez R, Ludlow-Wiechers B, Villanueva-Gutiérrez R. Flora palinológica de la Reserva de la Biosfera de Sian Ka'an, Quintana Roo, México. Mexico: Centro de Investigaciones de Quintana Roo;1991. 321 pp.
Parra-Romero HI. Recursos poliníferos de importancia para la abeja *Melipona beecheii* durante el período húmedoen el campus de ECOSUR, Chetumal. Bch. thesis. Instituto Tecnológico de Chetumal. Chetumal, Quintana Roo, Mexico; 2015.
Parra-Tabla V, Vargas CF. Phenology and phenotypic natural selection on the flowering time of a deceit-pollinated tropical orchid, *Myrmecophila christinae*. Ann Bot. 2004;94:243–50.
Paxton RJ, Echazarreta-González C, García M. Africanized honey bees in Yucatán, Mexico: a detailed survey. Am Bee J. 1991;131:646–8.
Porter-Bolland L. La apicultura y el paisaje maya. Estudio sobre la fenología de floración de las especies melíferas y su relación con el ciclo apícola en La Montaña, Campeche, México. JSTOR. University of California Press, Mexican Studies-Estudios mexicanos. 2003;19 (2):303–30.

Porter-Bolland L. Estudio de caso: flora melífera de Campeche. En: Villalobos-Zapata, GJ, Mendoza Vega J. (Coord.), 2010. La Biodiversidad de Campeche: Estudio de Estado. Comisión Nacional para el Conocimiento y Uso de la Biodiversidad (CONABIO), Gobierno del Estado de Campeche, Universidad Auónoma de Campeche, El Colegio de la Frontera Sur. Mexico. Pg; 2010.

Porter-Bolland L, Medina Abreo ME, Montoy-Koh P, Martín-Ek G, May Pacheco G. Flora Melífera de la Montaña, Campeche: su importancia para la apicultura y la vida diaria. Comisión Nacional para el Conocimiento y Uso de la Biodiversidad e Instituto de Ecología, A. C., Xalapa, Ver. México; 2009, p. 322 ISBN: 978-607-7579-11-3.

Rico-Gray V, Chemas A, Mardujano S. Uses of tropical deciduous forest species by the Yucatecan Maya. Agr Syst. 1991;14:149–61.

Roubik DW. Ecology and natural history of tropical bees. Cambridge: Cambridge University Press; 1989.

Roubik DW. Pollination of cultivated plants in the tropics. Agricultural Services Bulletin 1118. Rome: FAO; 1995.

Roubik DW. Ecological impact on native bees by the invasive Africanized honey bee. Acta Biol Colombiana. 2009;14(2):114–24.

Roubik DW, Villanueva-Gutiérrez R. Invasive Africanized honey bee impact on native solitary bees: a pollen resource and trap nest analysis. Biol J Linn Soc. 2009;98:152–60.

Roubik D W, Sakai S, Hamid AA, editors. Pollination ecology and the rain forest. Ecological studies 174. Sarawak Studies. New York: Springer; 2005.

Roubik DW, editor. Pollinator safety in agriculture. Rome: FAO and UNEP; 2014.

Roubik DW, Bernhardt P, Meier S. Orchids and Neotropical tropical pollinators, since Darwin's time. In: Darwin's orchids: then and now. Chicago: University of Chicago Press; 2014. p. 229–61.

Salinas-Peba L, Parra-Tabla V. Phenology and pollination of *Manilkara zapota* in forest and homegardens. Forest Ecol Manage. 2007;248:136–42.

Smith MV, Adie A. A new design in pollen traps. Can Bee J. 1963; 74: 4,5,8.

Vergara C. Rango de vuelo y cuantificación de los recursos colectados por abejas africanizadas en un bosque tropical de Panamá. Unpublished data report. Balboa, Panama: Smithsonian Tropical Research Institute; 1983.

Villanueva-Gutiérrez R. Plantas de importancia apícola en el ejido de Plan del Río, Veracruz, México. Biotica. 1984;9(3):279–340.

Villanueva-Gutiérrez R. Nectar sources of European and Africanized honeybees (*Apis mellifera* L.) in the Yucatán Peninsula, Mexico. J Apic Res. 1994;33(1):44–58.

Villanueva-Gutiérrez R. Pollen resources used by European and Africanized honey bees in the Yucatán Peninsula, Mexico. J Apic Res. 1999;38(1–2):105–11.

Villanueva-Gutiérrez R. Polliniferous plants and foraging strategies of *Apis mellifera* in the Yucatán Peninsula, Mexico. Rev Biol Trop. 2002;50(3/4):1035–44.

Villanueva-Gutiérrez R, Roubik DW. Why are African honey bees and not European bees invasive? Pollen diet diversity in community experiments. Apidologie. 2004;35:481–91.

Villanueva-Gutiérrez R, Colli Ucán W. Plantas nectaríferas visitadas por la abeja maya Xunancab (*Melipona beecheii*) en Quintana Roo. Proceedings of the XIX American Seminario of Beekeeping, Campeche, Campeche. SAGARPA, Unión Nacional de Apicultores and Gobierno del Estado de Campeche; 2005.

Villanueva-Gutiérrez R, Buchamann S, Donovan AJ, Roubik DW. Crianza y manejo de la abeja xunancab en la Península de Yucatán. Tucson Arizona: ECOSUR-The BEE Works; 2005a.

Villanueva-Gutiérrez R, Roubik WD, Colli-Ucán W. Extinction of *Melipona beecheii* and traditional beekeeping in the Yucatán Peninsula. Bee World. 2005b;86(2):35–41

Villanueva-Gutiérrez R, Echazarreta González C, Roubik DW, Moguel-Ordoñez Y. Transgenic soybean pollen (*Glycine max* L.) in honey from the Yucatán Peninsula, Mexico. Scientific Rep. 2014; 4: 4002.

Visscher PK, Seeley TD. Foraging strategy of honey bee colonies in a temperate deciduous forest. Ecology. 1982;63:1790–801.

Waller GD. A modification of the O.A.C. pollen trap. Am Bee J. 1980;120:119–21.

Winston M. The biology of the honey bee. Cambridge, MA: Harvard University Press;1987. 281p.

Chapter 7
Natural and Human Induced Disturbance in Vegetation

Odilón Sánchez-Sánchez, Gerald Alexander Islebe, Pablo Jesús Ramírez-Barajas, and Nuria Torrescano-Valle

Abstract The present chapter analyses natural and human disturbance which impacted the vegetation of the Yucatán Peninsula. The relevance of the slash-and-burn system is outlined, a system used for many centuries, and the relevance of fire in this practice. Slash-and-burn and fires provide information about the present structure and composition of the tropical forests of the Yucatán Peninsula. Several seral stages are identified, which agree well with seral stages recognized by modern Maya farmers. In relation to natural disturbance, we analyze the influence of hurricanes and their immediate and general effects on the vegetation and related fauna. Hurricanes and droughts are relevant agents of disturbance, as hurricanes cause considerable structural damage to forests and contribute to the accumulation of large quantities of dry biomass. This biomass can act as fuel for large forest fires. We discuss the natural dynamics that have characterized the vegetation of the Yucatán Peninsula through time, showing that it is a resilient ecosystem.

Keywords Composition change • Fire • Hurricanes • Maya farmers • Milpa • Perturbation • Succession • Structural damage • Tropical forest • Wildlife

7.1 Introduction

The vegetation of the Yucatán Peninsula has been subject to the pressures of natural and human-mediated disturbance for a long time. Paleoecological evidence shows that human influences have been a key factor for the last 3500 years (Islebe et al. 1996; Leyden et al. 1998), when Maya culture began to emerge. The

O. Sánchez-Sánchez (✉)
Centro de Investigaciones Tropicales de la, Universidad Veracruzana, Ex Hacienda Lucas Martín, Privada de Araucarias s/n, Col. Periodistas, 91019 Xalapa, Veracruz, Mexico
e-mail: odsanchez@uv.mx

G.A. Islebe • P.J. Ramírez-Barajas • N. Torrescano-Valle
Departamento Conservación de la Biodiversidad, El Colegio de la Frontera Sur Unidad Chetumal, Avenida Centenario km. 5.5, Apartado Postal 424, CP 77014, Chetumal, Quintana Roo, Mexico
e-mail: gislebe@ecosur.mx; pjbajaras@gmail.com; ntorresca@ecosur.mx

G.A. Islebe et al. (eds.), *Biodiversity and Conservation of the Yucatán Peninsula*,
DOI 10.1007/978-3-319-06529-8_7

demographic peak in the late Classic period (300–900 AD) forced the Maya to develop advanced agroforestry systems. To fulfill the needs of a burgeoning population, agricultural areas increased during the Maya Classic period (Flannery 1982; Siemens 1978; Gómez-Pompa et al. 1987). The vegetation provided additional uses (Barrera et al. 1976; Folan et al. 1979). Lambert and Arnason (1982) suggested that the present floristic composition of tropical forests in the Yucatán Peninsula is tied to agroforestry practices of the ancient Maya, considering the abundance of arboreal species such as breadnut or Maya nut (*Brosimum alicastrum* Sw.), guaya (*Talisia oliviformis* [H.B.K.] Radlk) and sapodilla (*Manilkara zapota* [L.] P. Royen), among other species. The presence of many pyramids and other structures that are covered by vegetation in the tropical forests of the Maya realm is evidence that forests established in former agricultural areas and ceremonial centers after 900 AD. Population reduction continued until the Spanish conquest and establishment of Mexico as a country (de Landa 1566/1982; Hammond 1982). However, over the last 170 years, the vegetation of the Yucatán has been placed under enormous pressure of a different nature and intensity. With the steady increase in population, shifting cultivation increased (Read and Lawrence 2003; Dalle et al. 2011), based on an ancient system that has been used for millennia until the present.

From 1850 onwards, large areas of forests were cleared to plant *henequén* (*Agave fourcroydes* Lem.) for the production of rope fiber and a traditional Mexico alcoholic beverage, *Licor del henequén*. For more than a century, henequen plantations were the main agricultural activity and source of revenue for the State of Yucatán. Between 1890 and 1945, exploitation of sapodilla latex for *chicle* production was an important economic activity, mainly in the States of Campeche and Quintana Roo. This was a sustainable activity, which helped to maintain the forest in good condition for conservation purposes. When demand for chicle declined, timber extraction became the main forest activity, causing the near-extinction of *caoba* or Honduran mahogany (*Swietenia macrophylla* King) and Spanish cedar (*Cedrela odorata* L.). More recently and since the early 1970s, extensive animal husbandry and sugar cane production has expanded across the Yucatán Peninsula, converting large forested areas into crop or pasture lands. The Yucatán Peninsula can be further characterized as a hurricane-sensitive region, given that it is influenced by two major hurricane areas, i.e., the eastern Caribbean and the Atlantic region, the latter being an area under regular hurricane influence. During the period between 1871 and 2013, about 50 hurricanes hit the eastern coast, while 45 hurricanes passed by the Yucatán Peninsula (Jáuregui et al. 1980; NOAA 2015; Fig. 7.1). These metereological phenomena are the main agents of natural disturbance in the Yucatán Peninsula, and experienced an increase in number and intensity since the 1980s, which is driven by a changing climate due to increased greenhouse gas concentrations in Earth's atmosphere.

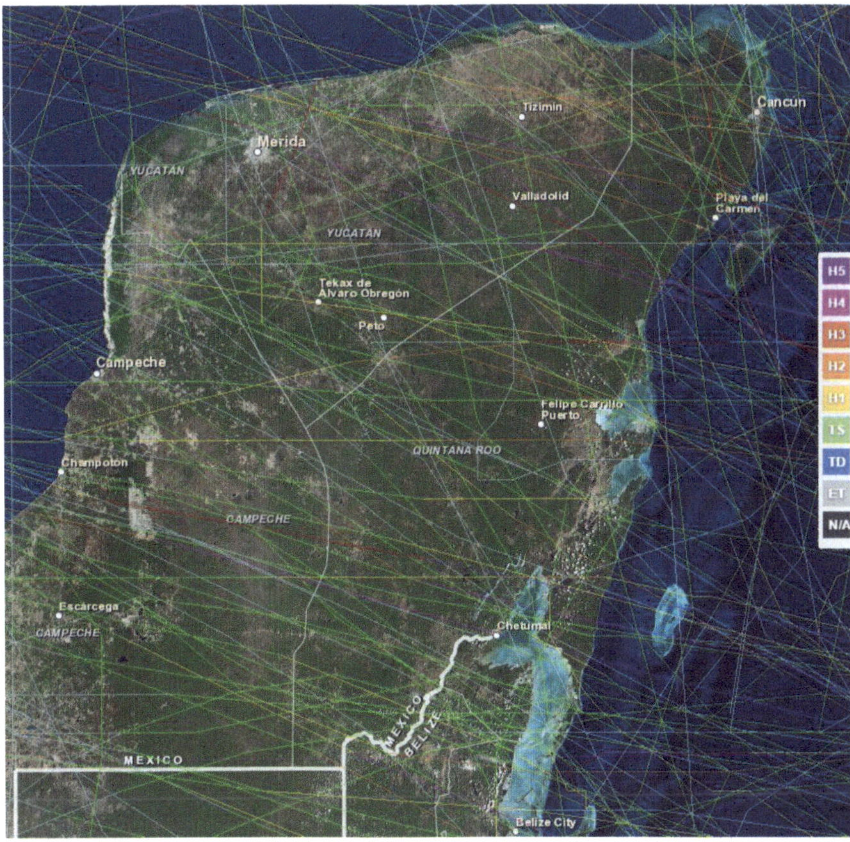

Fig. 7.1 Hurricane incidence between 1843 and 2013, based on NOAA database. http://www.coast.noaa.gov/hurricanes/ *H5* hurricane 5 on Saffir-Simpson scale, *H4* hurricane 4 on Saffir-Simpson scale, *H3* hurricane 3 on Saffir-Simpson scale, *H2* hurricane 2 on Saffir-Simpson scale, *H1* hurricane 1 on Saffir-Simpson scale, *TS* tropical storm, *TD* tropical depression

7.2 Importance of Fire as Disturbance Agent in Vegetation

The application of fire in the Yucatán Peninsula continues to be the common practice of Maya farmers, who use the slash-and-burn system to manage their *milpas* (fields of corn, bean and squash, among other crops; Toledo et al. 2008). Detailed knowledge of milpa management and *cosmovision*, understanding the world, is provided by Atran (1993), and by Ford and Nigh (2009). In local communities, burning is performed by organized groups, which establish *guardarrayas* (fire control strips) to prevent fires from extending to nearby areas. Fire control strips are important, given that prescribed fires occasionally burn out of control and can potentially cause large-scale forest fires. Thus, it is not surprising that most forest fires are detected during months with low levels of precipitation

(April, May and June; 87 % of all fires) After hurricanes strike the forest, large forest fires can develop due to the large quantities of fallen leaves, branches and trees that accumulate on the ground, and which act as a natural fuel. These fires can cover hundreds or thousands of hectares of forest (López-Portillo et al. 1990).

In comparing areas that were affected by fires during hurricane impacts versus areas where fires occurred without hurricanes from 1993 to 2004, 123 reported fires damaged 97,560 ha of forest, compared to 8132 ha that were damaged by fire alone (CONAFOR 2005). It should be mentioned only two hurricanes that were reported in 1995 (Roxanne and Opal) affected *ca.* 700,000 ha of forest. Due to effective fire prevention at that time, only 2200 ha were damaged (less area than the average area for the previous hurricane-free period). In a second case (1975), Hurricane Carmen (1974) caused forest fires, which damaged 97,655 ha, while forest fires that were incurred by the effects of Hurricane Gilbert (1989) damaged more than 135,000 ha (López-Portillo et al. 1990). In most cases, the causes of fires were that they were not controlled during slash-and-burn operations. Tropical forests of the Yucatán Peninsula are ecosystems that are sensitive to the effects of fire, but some species may respond positively to the presence of fire. Alternatively, fire is a disturbance that plays a role in favoring certain species and habitats, such as Honduran mahogany (Snook and Negreros-Castillo 2004) and associated species. Although fires are one of the main agents that have shaped the present tropical forest communities of the Yucatán Peninsula (Sánchez-Sánchez and Islebe 2002), their impacts and dynamics have not been well documented. In an early study by Pérez (1980) regarding the climate and forest fires in Quintana Roo, he discussed the ecological and economic aspects of fire, mentioning that its notable effects included reducing forest diversity and changing ecological conditions of the areas where fires develop. Damage of woody products, which could be commercially valuable, range from simple burns to the destruction of adult trees and the additional damage that is caused to trees by insects and diseases as a consequence of their exposure to fire. Some researchers, such as Gómez-Pompa et al. (2003) and Dickinson et al. (2001), mention that fire can irreversibly damage younger trees and juveniles, causing the disappearance of forests over the long-term. The damage that fires cause to soil, upon which tree species depend for their growth (Urquiza-Haas et al. 2007; Vargas et al. 2008), reduces vegetation cover temporarily. Soil is exposed to the action of wind and water, thereby accelerating processes of soil erosion and degradation (Castellanos et al. 2001). Microclimate is also influenced, an effect that might be difficult to detect in early stages; if fires are more frequent, changes become more noticeable (Allen and Rincon 2003).

In 1989, a year after Hurricane Gilbert, a long dry period caused several fires that affected large areas of northern Quintana Roo. About 130,000 ha of tropical forest burned. In analyzing the effects of burning on tropical forest, López-Portillo et al. (1990) conducted preliminary assessments of the 1989 fires. These authors report that the fires exerted differential effects on the vegetation, creating a mosaic of varying degrees of disturbance. Areas of disturbance were classified as: (1) superficial damage, where only the herb and shrub layer was affected; (2) intermediate damage, where some trees were burned along with the herbs and shrubs; and

(3) complete damage, where intense fires affected all vegetation layers, including the canopy. In the assessment, 28,200 ha were recorded as suffering complete damage, 33,500 ha experienced intermediate damage, and 46,300 ha had superficial damage. Concerning the regeneration of plant species, a comparison between sampling sites (hurricane-impacted sites vs hurricane- and fire-impacted sites), the following results were obtained: at the hurricane sites, only 46 % of all recorded species grew from seeds, while the rest were classified as resprouters; and 12 % of the species recovered from seeds and sprouts. Frequency per square meter was 48.6 %. In the sites that were affected by both hurricanes and fire, only resprouting was observed. Cover estimate in the herb layer was 14.8 %, which was 70 % less than the exclusively hurricane-damaged sites. The authors noted that burned tree and palm species presented 86 % of the species found in the fire-damaged areas. Survival was calculated at 33.5 %, including those individuals that were capable of resprouting from the base of the trunk. The tallest trees suffered the highest mortality, but there were survivors in all diameter classes. In the burned areas, the number of individuals was reduced by nearly 50 % compared to the hurricane damaged area; however, the range of life-forms (trees, shrubs, and vines) remained constant. Most species corresponded to the tree layer, and of these, a substantial fraction (25 %) were canopy species common to less disturbed forests of Quintana Roo. Considering these results and the known recovery of forests after slash-and-burn practices, we could consider that forests have developed a high degree of resilience. However, the occurrence of repeated forest fires in the burned areas of 1989 and the shortening of fallow periods have caused an interruption in the natural successional process and favored conditions for the wide dispersal of bracken fern (*Pteridium aquilinum* [L.] Kuhn), which is common in disturbed tropical forest areas (Schneider 2006; Schneider and Fernando 2010). At present, areas invaded by this fern present one of the main risks of forest fires. Detailing the role of fire in forest structure and dynamics of the Yucatán Peninsula is crucial, particularly forest fires that occur over large areas, as these cause the greatest damage on both ecological and economic scales. Also, large forest fires threaten ecosystem services and continuity of forests in the region. Due to climate change, an increase in fires is expected in the future, as forests will likely be more sensitive to drought events (Valdez-Hernández et al. 2014). Hardesty et al. (2005) pointed out that alteration of the fire regime in these ecosystems is the main threat and that such changes will affect conservation of key species, and rare and threatened species.

7.3 Slash-and-Burn Systems, and Maya Nomenclature of Different Seral Stages of Tropical Forest

In the Yucatán Peninsula, the main economic activity of Maya farmers is milpa-based agriculture, an ancient practice which is based on shifting cultivation and traditional techniques that have been known for the last 4000 years (Ucan-Ek

Table 7.1 Maya nomenclature on seral stages

Maya terminology	Milpa type
Sak'aab kool (kool: step away)	Recently abandoned milpa, early succession
Hubche' (hub: tangle; che': stick, tree)	Early succession, 1–5 years post-disturbance
K'ambal hubche' (k'ambal: mature)	Succession, 5–10 years post-disturbance
Ka'anal hubche' (Ka'anal: tall)	Succession, 10–15 years post-disturbance
Kelemche' (kelem: strength)	Succession, 15–30 years post-disturbance
Ka'anal k'aax (k'aax: tropical forest)	Old-growth forest, >30-years-old

et al. 1984; Hammond 1982). This system consists of felling tropical forest in areas of 0.5–3 ha or more (Sánchez-Sánchez and Islebe 2002). Trees that have specific uses for forestry, fodder, religious purposes, shade or fruit crops are frequently excluded from the felling process. This use is part of a strategy considering the holistic use of a various species (Toledo et al. 2008). Gómez-Pompa (1987) suggested the possibility that Maya nut, sapodilla, *zapote prieto* or black sapote (*Dyospiros digyna* [J.F. Gmel.] Perrier), mamey sapote (*Pouteria sapota* [Jacq.] H.E. Moore and Stearn), cainito (*Chrysophyllum cainito* L.), pawpaw or soursop (*Annona* sp.), and *pimenta* or allspice (*Pimenta dioica* [L.] Merr.) were remnants of ancient silvicultural practices of the Maya culture, which could explain their presence and dominance.

Local farmers give specific names to various stages of milpa-based agriculture (Table 7.1).

First-year milpa is called *milpa* (*ch'akbe'en*) and offers the greatest yield. The second year of sowing is called *cañada* (*sak'aab*) and results in lower yields. After the second yield, the field is abandoned to natural succession; vegetation regeneration depends upon certain factors, which are determined by previous management practices. After nearly 30 years, vegetation recovers and the process of slash-and-burn is repeated with the sowing of new *milpas*. The fallow period is known as *barbecho* and is a factor critical for the functioning of the system. Dalle et al. (2011) noted the danger of degradation if the fallow period is shortened to just 4 or 5 years. Gómez-Pompa (1987) concluded that in isolated conditions and at low population densities, these agroforestry systems function well. Maya nomenclature, which defines the seral stages of this system, have been described for the Yucatán Peninsula by Flores and Ucan (1983) and Gómez-Pompa (1987) and is consistence with field observations.

A study by Sánchez-Sánchez et al. (2007) analyzed a successional sequence in northern Quintana Roo that compared tree species in different seral stages (Table 7.1), including *Ka'anal k'aax* (mature forest >70-years-old), *K'ambal hubche'* (5-year-old slash-and-burn), *Ka'anal hubche'* (15-year-old slash-and-burn), and *Kelemche'* (20-year-old slash-and-burn), together with plots subjected to fires (2-year-old, 13-year-old). They found that mature tropical forest had the highest number of species (72), while the lowest number of species (40) was recorded in the 13-year-old fire-successional stage plot. In this study, the distance between the sampling plot and surrounding vegetation was ≤50 m for the 13-year-old successional stage. Floristic richness is lower where fire has been the main cause of disturbance, while floristic richness after 5 years is relatively high in the slash-and-burn system (Valdez-Hernández et al. 2014). Sánchez-Sánchez et al. (2007) and Valdez-Hernández et al. (2014) have noted that differences in numbers of recorded species in the milpa were related to management practices implemented by local farmers. This management included opening the forest canopy, while protecting useful arboreal species, and making fire breaks to protect vegetation from fires. These actions favor mainly vegetative regeneration, the establishment of seeds from remaining trees, and recruitment of propagules from nearby areas. Sánchez-Sánchez et al. (2007) recorded 62 species in a 20-year-old plot, which was just ten species less than the number recorded in mature forest >60-years-old.

In the study described above, the regeneration of slash-and-burn could be classified as a form of auto-succession, where rapid re-establishment of the floristic composition is achieved by two effective mechanisms of regeneration: resprouting by remaining vegetative structures that are resistant to fire, and recruitment of germinants from the soil seed bank that can survive fire, such as like *Sabal yapa* C. Wright ex Bec, thatch palm (*Thrinax radiata* Lodd. ex Schult. and Schult. f.) and Mexican silver palm (*Coccothrinax readii* H.J. Quero). The latter three palm species are characteristic of northern Quintana Roo (Quero 1992). Seeds are also brought in by dispersal nearby areas. Thus, two types of dispersal operate in these forests, *i.e.*, resprouting and seed dispersal, observations that consistent with those of similar forest types (López-Portillo et al. 1990; Valdez-Hernández et al. 2014). With respect to dominant plant families, the Fabaceae is the most important family, followed by the Myrtaceae, Rubiaceae, Euphorbiaceae and Sapotaceae (Ceccon et al. 2002; Sánchez-Sánchez and Islebe 2002; Snook and Negreros-Castillo 2004; Vandecar et al. 2011; Dupuy et al. 2012; Valdez-Hernández et al. 2014). Many species in these families are heliophytes, hence it is not unusual to find them as components of mature forest, including false tamarind (*Lysiloma latisiliquum* [L.] Benth.), *mata ratón* (*Gliricidia sepium* [Jacq.] Kunth ex Walp.), glassywood or *ronrón* (*Astronium graveolens* Jacq.), Spanish cedar (*C. odorata*), *pochote* (*Ceiba aesculifolia* [Kunth] Britten & Baker F.), and caimitillo (*Chrysophyllum mexicanum* Brandegee ex Standl). Of all of the described seral stages, farmers prefer *Ka'anal k'aax* (old-growth forest >30-years-old) in which to apply slash-and-burn. However, mature forest stands are hard to find, so farmers will apply slash-and-burn to younger seral stages, or even to mature *hubche'ob*. The total available area of *ejidos* is limited, while continual population growth is observed. The traditional strategy of management appears to have reached some limits, which could cause a severe ecological imbalance (Dalle et al. 2011). This response is frequently observed in the central parts of Quintana Roo, where a large proportion of Maya farmers live.

The *Hampea trilobata-Metopium brownei-Bursera simaruba* plant community (Sánchez-Sánchez and Islebe 2000), which is distributed in the east-central region of the peninsula between Tulum and Carrillo Puerto, has abundant *che che'en* or black poisonwood (*M. brownei* Roxb.) and *S. yapa*, which are both fire resistant, together with gumbo-limbo or *chaca* (*Bursera simaruba* [L.] Sarg.) and tzitzilché (*Gymnopodium floribundum* Rolfe), which are abundant species in secondary vegetation. This community has low canopy heights of 12–15 m, suggesting that it may be in early successional stages of 8–10 years. The resulting lack of advanced successional areas prompts farmers to use these early successional forests for milpa with consequent lower yields (Dalle et al. 2011).

7.4 Importance of Hurricanes as Factors of Disturbance in Vegetation of the Yucatán Peninsula

Hurricanes are considered to be among the most dangerous and destructive natural phenomena, due to the strength of their winds (up to 300 km/h), heavy rainfall and heavy flooding. The flat relief of the Yucatán Peninsula increases its vulnerability to hurricanes, as it favors the inland advance of these storms. There is no substantial variation in elevation or presence of mountain barriers that would slow or impede wind flow (Jáuregui et al. 1980; Wilson 1980; Whigham et al. 1991; Sánchez-Sánchez and Islebe 1999, 2000; Boose et al. 2003; Sánchez-Sánchez et al. 2007). Over the past 60 years, the most destructive hurricanes in the region were Janet and Hilda (1955), Behula (1967) and Carmen (1974). One of the most destructive hurricanes of the 1980s was Gilbert, with wind speeds of more than 280 km/h and maximum gusts of 350 km/h. Minimum pressure in the eye of the hurricane was 886 millibars (88.6 kpa, lowest value ever recorded); the storm was catalogued as a category five hurricane following the Saffir-Simpson scale (Jáuregui et al. 1980). After Gilbert, the peninsula was struck more frequently by strong hurricanes. Between 1990 and 2005, nine major hurricanes caused severe damage; 2005 was an important year as three hurricanes that made landfall, Emily, Stan and Wilma (Saffir-Simpson scale 4), caused the most destruction (Sánchez-Sánchez et al. 2007; Bonilla-Moheno 2012). In 2006, no hurricane was recorded, but Hurricane Dean entered southern Quintana Roo as a category five storm in 2007 (Islebe et al. 2009). The destructive effects that hurricanes have on vegetation in the wider Caribbean area have been detailed in a series of studies by Webb (1958), Craighead and Gilbert (1962), Dittus (1985), Brokaw and Grear (1991), Walker (1991), Imbert and Portecop (2008) and Zimmerman et al. (2014). Few detailed analyses of hurricanes and their biological effects or evaluations of biodiversity exist (Tanner et al. 1991; Zimmerman et al. 1996; Vandecar et al. 2011). The most detailed analysis is still that of Whitmore (1989), who conducted research in the Solomon Islands. A special volume on hurricane effects on vegetation (Lodge and McDowell 1991) emphasized that hurricanes are disturbances that are capable of changing processes in an ecosystem on both a short- and a long-term basis, and though the periodicity of hurricane impacts seems low (years or decades), those processes are shorter than the life cycle of the canopy or ecosystem. Lodge and McDowell (1991), therefore, considered that cyclones or hurricanes have profound effects on population dynamics, soil development and nutrient cycling over relevant temporal and evolutionary scales across most of the Caribbean region. In the case of the Yucatán Peninsula, the first description of a hurricane was provided by the Roman Catholic Bishop of Yucatán, Diego de Landa, who wrote in 1566:

> a wind came . . . it became stronger, converting into hurricane . . . and this wind caused the fall of all trees, . . . and the earth remained without trees, and the present trees seem to have been planted at the same time, as watching them from higher ground, they all look they have been cut by the same scissor (de Landa 1566/1982).

Detailed effects of hurricanes on the Yucatán Peninsula have not been evaluated until recently by Rodríguez et al. (1989), Sánchez-Sánchez and Herrera (1990), Olmsted et al. (1990), Whigham et al. (1991), Sánchez-Sánchez and Islebe (1999, 2002), Boose et al. (2003), Sánchez-Sánchez et al. (2007) and Bonilla-Moheno (2012). The first five studies refer to the effects of Hurricane Gilbert on the vegetation of northern Quintana Roo. Sánchez-Sánchez et al. (2007) and Bonilla-Moheno (2012) refer to damage caused by Hurricanes Wilma and Emily. Most of the studies describe the major kinds of damage incurred to the vegetation, such as falling trees, defoliation, and accumulation of plant biomass on the ground. Sánchez-Sánchez and Islebe (1999) provide a detailed description of the damage that the tropical forest suffered following a category five hurricane. These authors compare post-hurricane data with previous structural and compositional data of Sánchez-Sánchez (1987). The tree and shrub layer experienced 100 % defoliation by the hurricane, while the herb layer was completely covered with leaves and twigs. The quantity of fallen leaves was about 4.5 Mg/ha dry mass. Whigham et al. (1991) report a much higher value one month after Hurricane Gilbert, *viz.*, 1500 g/m^2 (15 Mg/ha). These litterfall data, when combined with adverse climatic conditions that are incurred by drought and the use of slash-and-burn, suggests a debris load that more than sufficient to fuel large-scale fires (a total of 135,000 ha; López-Portillo et al. 1990). With respect structural damage, mean canopy height (between 3 and 8 m) height decreased by 6 %, while the upper canopy decreased by 9 % (between 8.1 and 16 m). The reduced canopy height of tropical forests is clearly related to hurricane activity (Sánchez-Sánchez and Islebe 1999; Islebe et al. 2009). Given that hurricanes are recurring events, they periodically prune the canopy, which incurs a higher energetic cost to trees, explaining the delay in growth.

Whigham et al. (1991) report a tree mortality of 12.6 % (182 out of 1447) for a tropical forest in northern Quintana Roo 2 years after hurricane Gilbert. A 33 % decrease in the number of individuals and a 12 % decrease in basal area were observed. Of all of the fallen trees, 96 % belonged to small diameter classes (class 1 = 3–10 cm). This contrasts with the findings of Walker (1991), who worked in El Valle, Puerto Rico, and reported for that the tallest and larger diameter classes were uprooted or their trunks were split. Trees with large diameters that are located in tropical forest near the Caribbean coast are well rooted and, therefore, more resistant to strong winds. Sánchez-Sánchez and Islebe (1999) report that even though most trees of diameter class > 20 cm remained standing, most accrued damaged to their branches, 100 % exhibited damage to secondary branches, and 55 % received damage to their main branches. Damage at branches additionally caused the release of vines and epiphytes. Damage also varied along the main stem: 50 %, middle parts of the trunk; was 15 %, tree base; and 35 % at root level. The understory lost 51 % of all individuals and cover was decreased by 70 %. Diversity of species and the floristic composition did not change substantially, given that five months after disturbance, only four new secondary species were recorded in the understory. A study of 40 species of secondary vegetation of 15 years of age, after being damaged by Hurricane Wilma in 2005, showed that foliar recovery was extremely rapid (Sánchez-Sánchez et al. 2007). Foliar recovery of the different seral stages took only two months and seven days. Species in the Fabaceae showed the greatest capacity for response, probably due to their deciduous character, making those species more competitive. Greater knowledge is required regarding the response of plant communities and individual species to catastrophic events, such as strategies of recovery, time to recovery, species mortality, adaptations developed by species to resist, and relevance to restoration and aspects of forestry.

7.5 Hurricanes and Habitat Affectation of Vertebrates

The effect of disturbance on habitat and its consequences on the fauna has been poorly studied, despite the close relationship of habitat, organisms that occupy the habitat and their changing abundances (Ramírez-Barajas et al. 2012a). In tropical forests, a large number of tree species are present that produce fruits and attract frugivores as a strategy for seed dispersal (Fleming et al. 1987; Chapman et al. 1994; Milton 2008). Habitat quality can be measured in terms of the availability of fruits or tree species that are of potential use to frugivores. The deterioration of the habitat can be used as base line to analyze the relationship between ecosystem health and faunal abundance. In the Yucatán Peninsula, there is a long history of hurricanes, which have caused damage of variable strength, but few studies have evaluated the effect of these disturbances on biodiversity and ecosystems. In 2007, Hurricane Dean made landfall as a category five storm according to the Saffir-Simpson scale. To evaluate the effect of this hurricane on the vegetation, Ramírez-Barajas et al. (2012b) sampled a total 20,000 m^2 in plots that were located along a gradient from high to low damage. The results of their study indicated that 40–56 % of all trees at a distance of 40 km from the hurricane center were damaged, which was progressively reduced to 1 % damage at 120 km from the hurricane center. The diversity of tree species was assessed and the number of species quantified based on traditional ecological knowledge of hunters and farmers who older were 18 years and old; 65 % of all tree species were reported as sources of food (leaves, flowers, seeds, fruits) for vertebrate herbivores and omnivores. Nine large vertebrates were sampled and their relative abundance was estimated by means of an abundance index of tracks along the same gradient of damage (Ramírez-Barajas et al. 2012a, b). Percentage vegetation damage was negatively correlated with faunal abundance, particularly with herbivores ($F_{1,18} = 24.72$; $P < 0.0001$; Pearson $r = -0.76$). The abundances of frugivores such as the Central American agouti *Dasyprocta punctate* and the paca *Cuniculus paca* ($F_{1,18} = 6.7$; $P < 0.01$; Pearson $r = -0.52$;), together with frugivore browsers such as the white-tailed deer *Odocoileus virginianus* and brocket deer *Mazama* sp. ($F_{1,18} = 8.5$; $P = 0.009$; Pearson $r = -0.57$), exhibited significant decreases with increasing damage to the vegetation. Vertebrates that depended most upon the vegetation as a food source were the most affected, while abundances of species with omnivorous feeding habits did not significantly respond to damage, despite a weak negative correlation for some species, *e.g.*, great curassow *Crax rubra* and ocellated turkey *Agriocharis ocellata* (Pearson $r = -0.19$) and for some, a positive correlation for white-nosed coati *Nasua narica* and collared peccary *Tayassu tajacu* (Pearson $r = 0.25$), and nine-banded armadillo *Dasyprocta punctata* (Pearson $r = 0.26$). In the case of carnivorous vertebrates that were present in the area (i.e., jaguar *Panthera onca*; puma *Puma concolor*; ocelot *Leopardus pardalis*; margay *Leopardus wiedii*; and hog-nosed skunk *Conepatus semistriatus*), Hernández-Díaz et al. (2012) report a weak relationship between abundance and the degree vegetation damage. Although those species are associated with large forest-covered areas (Emmons 1988), they can be found in a wide variety of habitats (Emmons 1988; Villa-Meza et al. 2002; Trolle and Kery 2003; Dillon and Kelly 2007). These species are considered opportunistic carnivores, as their diets include more than 20 prey species (Konecny 1989; Farrell et al. 2000; Villa-Meza et al. 2002). Unlike the other species that have been previously discussed, it was expected that carnivore abundances would not respond negatively to changes in habitat that were caused by hurricanes. Changes reflect other aspects of biology and ecology, including adjustments to diet according to the availability or behavior of prey species and the availability of shelter and habitat use (Hernández-Díaz et al. 2012).

7.6 Conservation Outlook

Land use change is part of the natural history of the Yucatán Peninsula, whether it be at millennial scales (Carrillo-Bastos et al. 2010) or with respect to recent land use changes (Turner et al. 2001). The insufficient time that is given for fallow land to recover is one of the main challenges, which lies beyond understanding of ecological knowledge of tropical forests of the Yucatán Peninsula, and requires holistic conservation approaches. Conservation strategies should involve fauna, as a key element to habitat functioning. Conversion to secondary forests changes carbon sequestration by more than 30 % and nutrient cycles are out of balance (Eaton and Lawrence 2009). So, effective conservation measures must consider resilience of forests and alternative land use practices to guarantee future biodiversity.

References

Allen MF, Rincon E. The changing global environment and the lowland Maya: past patterns and current dynamics. In: Gomez-Pompa A, Allen MF, Fedick SL, Jimenez-Osornio JJ, editors. The lowland maya area: three millennia at the human-wildland interface. Binghamton, NY: Haworth Press; 2003. p. 13–30.

Atran S. Itza Maya tropical agro-forestry. Curr Anthropol. 1993;34(5):633–700.

Barrera AM, Barrera A, López F. Nomenclatura Etnobotánica Maya. México: Colección científica INAH; 1976.

Bonilla-Moheno M. Damage and forest recovery structure and composition after two subsequent hurricanes in the Yucatán Peninsula. Caribbean J Caribbean Sci. 2012;46:240–8.

Boose E, Foster DR, Plotkin AB, Hall A. Geographical and historical variation in hurricanes across the Yucatán Peninsula. In: Gomez-Pompa A, Allen MF, Fedick SL, Jimenez-Osornio JJ, editors. The lowland Maya area: three millennia at the human-wildland interface. Binghamton, NY: Haworth Press; 2003. p. 495–516.

Brokaw NVL, Grear JS. Forest structure before and after Hurricane Hugo at three elevations in the Luquillo mountains, Puerto Rico. Biotropica. 1991;23:386–92.

Carrillo-Bastos A, Islebe GA, Torrescano Valle N. Holocene vegetation and climate history of central Quintana Roo, Yucatán Península, Mexico. Rev Palaeobot Palynol. 2010;160:189–96.

Castellanos J, Jaramillo VJ, Sanford RL, Kauffman JB. Slash-and-burn effects on fine root biomass and productivity in a tropical dry forest ecosystem in Mexico. Forest Ecol Manage. 2001;148:41–50.

Ceccon E, Olmsted I, Vazquez-Yanes C, Campo J. Vegetation and soil properties in two tropical dry forests of differing regeneration status in Yucatán. Agrociencia. 2002;36:21–34.

Chapman CA, Wrangham R, Chapman LJ. Indices of habitat-wide fruit abundance in tropical forest. Biotropica. 1994;26:160–71.

CONAFOR. Información estadística de incendios en áreas forestales por Entidad Federativa; 2005.

Craighead FC, Gilbert VC. The effects of hurricane Donna on vegetation of Florida. Q J Florida Acad Sci. 1962;25:1–28.

Dalle SP, Pulido MT, de Blois S. Balancing shifting cultivation and forest conservation: lessons from a "sustainable landscape" in southeastern Mexico. Ecol Appl. 2011;21:1557–72.

De Landa D. Relación de las cosas de Yucatán; 1566.

Dickinson MB, Hermann SM, Whigham DF. Low rates of background canopy-gap disturbance in a seasonally dry forest in the Yucatán Peninsula with a history of fires and hurricanes. J Trop Ecol. 2001;17:895–902. doi:10.1017/ S0266467401001663.

Dillon A, Kelly MJ. Ocelot *Leopardus pardalis* in Belize: the impact of trap spacing and distance moved on density estimates. Oryx. 2007;41:469–77.

Dittus WPJ. The influence of cyclones on the tropical evergreen forest of Sri Lanka. Biotropica. 1985;17:1–14.

Dupuy JM, Hernández-Stefanoni JL, Hernández-Juárez RA, Tetlela-Rangel E, López-Martínez JO, Leyequien-Abarca E, Tun-Dzul F, May-Pat F. Patterns and correlates of tropical dry forest structure and composition in a highly replicated chronosequence in Yucatán, Mexico. Biotropica. 2012;44:151–62.

Eaton JM, Lawrence D. Loss of carbon sequestration potential after several decades of shifting cultivation in the southern Yucatán. Forest Ecol Manage. 2009;258:949–58.

Emmons LH. A field study of ocelots in Peru. Revue d'Écologie Terra Vie. 1988;43:133–57.

Farrell LE, Roman J, Sunquist ME. Dietary separation of sympatric carnivores identified by molecular analysis of scats. Mol Ecol. 2000;9:1583–90.

Flannery K. Maya subsistence: studies in memory of Dennis E. New York Puleston: Academic; 1982.

Fleming TH, Breitwisch R, Whitesides GH. Patterns of tropical vertebrate frugivore diversity. Ann Rev Ecol Syst. 1987;18:91–109.

Flores SJ, Ucan EE. Nombres usados por los mayas para designar a la vegetación. Xalapa, Veracruz, México: Instituto Nacional de Investigaciones sobre Recursos Bióticos; 1983.

Folan W, Fletcher LA, Kintz ER. Fruit, fiber, bark and resin: social organization of a Maya urban center. Science. 1979;204:697–701.

Ford A, Nigh R. Origins of the Maya forest garden: Maya resource management. J Ethnobiol. 2009;29:213–36.

Gómez-Pompa A. On Maya silviculture. Mexican Stud. 1987;3:1–17.

Gómez-Pompa A, Flores S, Sosa V. The "Pet Kot": A Man-Made tropical forest of the Maya. Interciencia. 1987;12:10–5.

Gómez-Pompa A, Allen MF, Fedick SL, Jimenez-Osornio JJ. The lowland Maya area: three millennia at the human- wildland interface. Binghamton, NY: Haworth Press; 2003.

Hammond N. Ancient Maya civilization. New Jersey: Rutgers University Press; 1982.

Hardesty J, Myers RL, Fulks W. Fire, ecosystems, and people: a preliminaty assessment of fire as a global conservation issue. The George Wright Forum. 2005;22:78–87.

Hernández-Díaz M, Ramírez-Barajas PJ, Chávez C, Schmook B, Calmé S. Presencia y abundancia relativa de carnívoros en una selva dañada por el huracán Dean (2007). Rev Mex Biodivers. 2012;83:790–801.

Imbert D, Portecop J. Hurricane disturbance and forest resilience: assessing structural vs. functional changes in a Caribbean dry forest. Forest Ecol Manage. 2008;255:3494–501.

Islebe GA, Hooghiemstra H, Brenner M, Hodell DA, Curtis JA. A Holocene vegetation history from lowland Guatemala. The Holocene. 1996;6:265–71.

Islebe GA, Torrescano-Valle N, Valdez-Hernández M, Tuz-Novelo M, Weissenberger H. Efectos del impacto del huracán Dean en la vegetación del sureste de Quintana Roo México. Foresta Veracruzana. 2009;11:1–6.

Jáuregui E, Vidal F, Cruz J. Los ciclones y tormentas tropicales en Quintana Roo; 1980.

Konecny MJ. Movement patterns and food habits of four sympatric carnivore species in Belize, Central America. In: Redford KH, Eisenberg JF, editors. Advances in neotropical mammalogy. Gainesville: Sandhill Crane; 1989. p. 243–64.

Lambert JDH, Arnason JT. Ramón and Maya Ruins: an ecological, not an economic relation. Science. 1982;216:298–9.

Leyden BW, Brenner M, Dahlin B. Cultural and climatic history of Cobá, a lowland Maya city in Quintana Roo, Mexico. Q Res. 1998;49:111–22.

Lodge DJ, McDowell DW. Summary of ecosystem-level effects of Caribbean hurricanes. Biotropica. 1991;23:373–8.

López-Portillo J, Keyes M, González A, Cabrera F, Sánchez O. Los incendios de Quintana Roo ¿Catástrofe ecológica o evento periódico? Ciencia y Desarrollo. 1990;16:43–57.

Milton K. Macronutrient patterns of 19 species of Panamanian fruits from Barro Colorado Island. Neotrop Primates. 2008;15:1–7.

NOAA. 2015. http://coast.noaa.gov/hurricanes/. Accessed 20 Mar 2015.

Olmsted I, Palma J, Pérez del Valle L, Castillo J, Moreno Y. Efecto de las perturbaciones huracán e incendio sobre la estructura y composición de la selva mediana subperennifolia en el norte de Quintana Roo. Oaxtepec, Morelos, Mexico: Congreso Mexicano de Botánica Programas y resúmenes; 1990.

Pérez G. El clima y los incendios forestales en Quintana Roo. In: Memorias del simposio Quintana Roo: Problemática y perspectivas. México: CIQRO Instituto de Geografía de la UNAM; 1980. p. 65–80.

Quero HJ. Las palmas silvestres de la Península de Yucatán. Publicaciones Especiales No. 10. México, DF: Instituto de Biología, UNAM; 1992. 63 p.

Ramírez-Barajas PJ, Islebe GA, Calmé S. Impact of Hurricane Dean (2007) on game species of the Selva Maya. Biotropica. 2012a;44:402–11.

Ramírez-Barajas PJ, Islebe GA, Torrescano-Valle N. Perturbación post-huracán Dean en el hábitat y la abundancia relativa de vertebrados mayores en la Selva Maya, Quintana Roo, México. Rev Mex Biodivers. 2012b;83:1194–207.

Read L, Lawrence D. Recovery of biomass following shifting cultivation in dry tropical forests of the Yucatán. Ecol Appl. 2003;13:85–97.
Rodríguez C, Vera GC, Carrillo A, Chavelas J, Escoto P, Parraguirre L. Evaluación de daños en el área afectada por el hurácan Gilberto y el incendio ocurrido en el presente año. Instituto Nacional de Investigaciones Forestales y Agropecuarias. Mexico: Inédito; 1989.
Sánchez-Sánchez O. Estructura y composición de la selva mediana subperennifolia presente en el Jardín Botánico del CIQRO, Puerto Morelos, Quintana Roo. BSc thesis, Universidad Veracruzana, Mexico; 1987.
Sánchez-Sánchez O, Herrera P. Cambio structural de la selva mediana subperennifolia del Jardín Botánico del CIQRO, después del paso del Huracán Gilberto. Oaxtepec, Morelos, México: XI Congreso Mexicano de Botánica Programa y resúmenes; 1990.
Sánchez-Sánchez O, Islebe GA. Hurricane Gilbert and structural changes in a tropical forest in south-eastern Mexico. Glob Ecol Biogeogr. 1999;8:29–38.
Sánchez-Sánchez O, Islebe GA. La selva del noroeste del estado de Quintana Roo: distribución y dominancia de especies arbóreas. Foresta Veracruzana. 2000;2:11–7.
Sánchez-Sánchez O, Islebe GA. Tropical forest communities of southeastern Mexico. Plant Ecol. 2002;158:183–200.
Sánchez-Sánchez O, Islebe GA, Valdéz-Hernández M. Flora arbórea y caracterización de gremios ecológicos en distintos estados sucesionales de la selva mediana de Quintana Roo. Foresta Veracruzana. 2007;9:17–26.
Schneider LC. The effect of land management practices on bracken fern invasion in the region of Calakmul, Mexico. J Lat Am Geogr. 2006;5:91–107.
Schneider LC, Fernando DN. An untidy cover: Invasion of bracken fern in the shifting cultivation systems of Southern Yucatán, Mexico. Biotropica. 2010;42:41–8.
Siemens A. Karst and the Pre-Hispanic Maya in the Southern Lowlands. In: Harrison PD, Turner II BL, editors. Pre-Hispanic Maya agriculture. Albuquerque: University of New Mexico Press; 1978. p. 117–44.
Snook LK, Negreros-Castillo P. Regenerating mahogany (Swietenia macrophylla King) on clearings in Mexico's Maya forest: the effects of clearing method and cleaning on seedling survival and growth. Forest Ecol Manage. 2004;189:143–60.
Tanner EVJ, Kapos V, Healey JR. Hurricane effects on forest ecosystems in the Caribbean. Biotropica. 1991;23:513–21.
Toledo VM, Barrera-Bassols N, García-Frapolli E, Alarcón-Chaires P. Uso múltiple y biodiversidad entre los mayas yucatecos (México). Interciencia. 2008;33:345–52.
Trolle M, Kery M. Estimation of ocelot density in the Pantanal using capture-recapture analysis of camera trapping data. J Mammal. 2003;84:607–14.
Turner II BL, Villar SC, Foster D, Geoghegan J, Keys E, Klepeis P, Lawrence D, Mendoza PM, Manson S, Ogneva-Himmelberger Y, Plotkin AB, Perez Salicrup D, Chowdhury RR, Savitsky B, Schneider L, Schmook B, Vance C. Deforestation in the Southern Yucatán Peninsular Region: an integrative approach. Forest Ecol Manage. 2001;154:353–70.
Ucan-Ek E, Narváez M, Puch A, Chan C. El cultivo del maíz en el ejido de Mucel, Pixoy, Valladolid, Yucatán. SEP. Mexico D.F.: Dirección General de Culturas Populares; 1984.
Urquiza-Haas T, Dolman PM, Peres CA. Regional scale variation in forest structure and biomass in the Yucatán Peninsula, Mexico: Effects of forest disturbance. Forest Ecol Manage. 2007;247:80–90.
Valdez-Hernández M, Sánchez-Sánchez O, Islebe GA, Snook L, Negreros-Castillo P. Recovery and early succession after experimental disturbance in a seasonally dry tropical forest in Mexico. Forest Ecol Manage. 2014;334:331–43.
Vandecar KL, Lawrence D, Richards D, Schneider L, Rogan J, Schmook B, Wilbur H. High mortality for rare species following hurricane disturbance in the Southern Yucatán. Biotropica. 2011;43:676–84. doi:10.1111/j.1744-7429.2011.00756.x.

Vargas R, Allen MF, Allen EB. Biomass and carbon accumulation in a fire chronosequence of a seasonally dry tropical forest. Glob Chang Biol. 2008;14:109–24. doi:10.1111/j.1365-2486.2007.01462.x.
Villa-Meza A, Meyer EM, González CAL. Ocelot (*Leopardus pardalis*) food habits in a tropical deciduous forest of Jalisco, México. Am Midl Nat. 2002;148:146–54.
Walker LR. Tree damage and recovery from Hurricane Hugo in Luquillo Experimental forest, Puerto Rico. Biotropica. 1991;23:379–85.
Webb LJ. Cyclones as an ecological factor in tropical lowland rain forest, North Queensland. Aust J Bot. 1958;6:220–8.
Whigham DF, Olmsted I, Cano EC, Harmon ME. The impact of Hurricane Gilbert on trees, litterfall, and woody debris in a dry tropical forest in the northeastern Yucatán Peninsula. Biotropica. 1991;23:434–41.
Whitmore TC. Changes over twenty-one years in the Kolombangara rain forests. J Ecol. 1989;77:469–83.
Wilson EM. Physical geography of the Yucatán Peninsula. In: Moseley EH, Terry ED, editors. Yucatán: a world apart. Birmingham, Alabama: University of Alabama Press; 1980. p. 5–40.
Zimmerman JK, Willig MR, Walker LR, Silver WL. Introduction: disturbance and Caribbean ecosystems. Biotropica. 1996;28:414–23.
Zimmerman JK, James AH, Shiels AB, Bithorn J, Carmona SM, Brokaw NVL. Seven-year responses of trees to experimental hurricane effects in a tropical rainforest, Puerto Rico. Forest Ecol Manage. 2014;332:64–74.

Chapter 8
Conservation and Use

Juan Manuel Dupuy Rada, Rafael Durán García, Gerardo García-Contreras, José Arellano Morín, Efraim Acosta Lugo, Martha Elena Méndez González, and María Andrade Hernández

Abstract The Yucatán Peninsula has a great biological, ecological and cultural diversity, marked historically by the Mayan civilization and biologically by its high endemism, and affinity with nearby regions. It is botanically well known with an estimated total richness of over 2300 vascular plant species, but there is uncertainty about the number of rare species (195 rare woody species are estimated to occur in its forests). Further research on the population dynamics, distribution and ecology of rare and endemic species of the peninsula is needed. Its biodiversity also provides various ecosystem services, such as provisioning, including about 680 species of medicinal plants, 145 edible species and 130 timber-species; and regulation of air and water quality (terrestrial ecosystems of the peninsula protect and help regulate one of the largest karstic aquifers worldwide), and climate. The Yucatán Peninsula has the second-largest forested area in Latin America after the Amazon basin, and harbors the largest natural reservoir of carbon in Mexico (3554 Pg). However, this biodiversity and the environmental services it provides are threatened by human activities linked to development policies focused on short-term economic benefits, leading to forest fires and land use change for agriculture, tourism, roads or urban development. Thus, by 2010, the primary vegetation occupied only 22 % of the region. Fortunately, protected areas cover 21.6 % of the total land area and, along with other initiatives (Mesoamerican Biological Corridor-Mexico), and sustainable productive activities (beekeeping, alternative tourism), have conserved large tracts of forests, along with the biodiversity and environmental services they provide.

J.M.D. Rada (✉) • R.D. García • M.E.M. González
Centro de Investigación Científica de Yucatán, A. C. Unidad de Recursos Naturales, Calle 43 No. 130, Colonia Chuburná de Hidalgo, CP 97200 Mérida, Yucatán, Mexico
e-mail: jmdupuy@cicy.mx; duran@cicy.mx; mar@cicy.mx

G. García-Contreras • E.A. Lugo • M.A. Hernández
Pronatura Península de Yucatán, A. C., Mérida, Yucatán, Mexico
e-mail: mandrade@pronatura-ppy.org.mx

J.A. Morín
Secretaría de Desarrollo Urbano y Medio Ambiente del Gobierno del Estado de Yucatán, Mérida, Yucatán, Mexico

G.A. Islebe et al. (eds.), *Biodiversity and Conservation of the Yucatán Peninsula*,
DOI 10.1007/978-3-319-06529-8_8

Keywords Development policies • Environmental services • Forest fires • Land degradation • Land-use change • Mesoamerican Biological Corridor • Plant biodiversity • Protected areas • Sustainable use

8.1 Introduction

From a geographical and geomorphological perspective, the Yucatán Peninsula includes the Mexican states of Campeche, Quintana Roo and Yucatán, as well as northern Belize and Guatemala (Ferrusquía-Villafranca 1993). It is a region of high biodiversity, but also high ecological and cultural diversity, marked historically by the social and cultural development of Maya culture (Durán and Méndez 2010; Ramírez-Carrillo 2010). The Mexican part of the peninsula is considered one of the country's biotic provinces (Barrera 1962; Rzedowski 1978; Morrone 2002); considering its geographic position, its flora and fauna exhibit a strong relationship with neighboring areas like Central America, the Caribbean Basin and southeast Mexico. It also presents a considerable number of endemic species (Estrada-Loera 1991; Durán et al. 1998; Ibarra-Manríquez et al. 2002).

The region presents tropical vegetation formed by several types of forest, which historically occupied large tracts, like the evergreen, semi-evergreen and subdeciduous forest, as well as deciduous forests (Miranda 1958; Flores and Espejel 1994; Olmsted et al. 1999). In fact, the area occupied by forests in the Yucatán Peninsula forms the one of the largest forest tracts in Latin America, second only to the Amazon basin. In addition to forests, a variety of different vegetation types are present on the peninsula, such as coastal dune vegetation and mangroves in the coastal areas, low flooded forests, palm vegetation, flooded grasslands and hydrophytic communities found inside the forest matrix (see Chap. 3; Espejel 1982; Trejo-Torres et al. 1993; Olmsted et al. 1999; Zaldivar et al. 2010).

In tandem with its high biodiversity, the peninsula has a long history of natural resource use by its indigenous inhabitants, the lowland Maya, who recognized the alimentary, medicinal, forestry, and ritual value of its native flora and fauna (see Chap. 3; Espejel 1982; Trejo-Torres et al. 1993; Olmsted et al. 1999; Zaldivar et al. 2010). However, in recent decades ecosystems have been severely affected by a suite of different human activities. The development of agricultural and husbandry practices has transformed and destroyed large extensions of forests and caused the drying of wetlands (Hernández-Barrios 2002; Andrade 2010). The region plays host to one of the largest centers of tourism in the country, which combined with chaotic urban development and an increasingly extensive road network, has caused changes to numerous terrestrial ecosystems. Due to its geographical position, this region in Mexico also experiences extreme weather events, like hurricanes, which regularly impact the peninsula, causing natural disturbances, the negative effects of which are compounded by consequent flooding and forest fires (Jáuregui et al. 1980). To deal with these processes of degradation and the concomitant pressures on the biodiversity of the Yucatán Peninsula, multiple conservation initiatives were launched, including the establishment of natural protected areas, the conservation of species in situ and ex situ in botanical gardens, and the development of sustainable productive activities which use natural resources while helping to conserve communities and ecosystems.

8.2 Plant Diversity

The vegetation of the Yucatán Peninsula is strongly associated with the edaphic conditions of the region and its karstic substrate. It is composed of a mosaic of plant communities, in which high and low forests dominate, frequently interspersed with flooded communities and water bodies (lakes and sinkholes). In the region different types of plant associations abound, from forests, mangroves, and coastal shrub vegetation, to *petenes* (forest islands surrounded by salt water marshes, that receive discharges from the acquifer), *sabanas* (savannahs), and different associations of aquatic plants (see Chap. 3; Durán 1987; Campos and Durán 1991; Flores and Espejel 1994; Olmsted et al. 1999).

In the most southern part of the region, some areas of evergreen high forest can be found, alternating with medium statured forest. Evergreen high forests are more structurally complex communities, and more diverse in their floristic composition. Medium statured forest, semi-evergreen and subdeciduous forest cover a greater proportion of the region, blanketing most of Campeche and Quintana Roo and the south of the State of Yucatán. Although diverse in their composition and structure, medium statured forests are less diverse than high forests, especially in regards to epiphytes. Low deciduous and subdeciduous forest also covers a large portion of the northwestern part of the peninsula, large parts of the state of Yucatán, a small portion of Campeche and some areas along the coast of Quintana Roo. Low forests are communities with a high diversity of trees. Due to their rocky, shallow soils, their understories are floristically and structurally simpler than other forests.

A characteristic vegetation type of the area is its low flooded forests, which develop in low-lying, poorly drained depressions that accumulate precipitation for several months. This period of flooding alternates with a long period of drought causing water stress to the vegetation. These communities are formed by the few species of trees and shrubs capable of resisting stressful hydrological conditions (Olmsted and Durán 1986). Epiphytes are present, while understory species are nearly absent.

As described earlier, the forest matrix has additional vegetation types that occupy smaller areas like *petenes*, palm vegetation, flooded grasslands, and aquatic vegetation, among others (Olmsted et al. 1999). Along the coast, dune communities are present and mangroves grow in flooded areas, bordering a wide area of wetlands with different species of hydrophytes in salt, brackish and fresh water (Espejel 1982; Trejo-Torres et al. 1993; Zaldivar et al. 2010). This diversity of communities gives the peninsula a high environmental heterogeneity and floristic diversity.

8.2.1 Plant Richness

The floristic composition of the Yucatán Peninsula is one of the best-explored in Mexico. Botanical studies go back to the nineteenth century with the explorations of Arthur C. Schott, Charles Frederick Millspaugh and George F. Gaumer. In the first half of the twentieth century, works by Paul Standley and Cyrus Lundell continued the botanical exploration (Standley 1936). From 1980 onwards, this research was strengthened by the floristic studies by Mexican institutions like the National Institute for the Study of Biotic Resources (Instituto Nacional de Investigación sobre Recursos Bióticos, INIREB), the Biological Institute of the Autonomous University of México (Instituto de Biología de la UNAM, IBUNAM), the Quintana Roo Center for Research (Centro de Investigaciones de Quintana Roo, CIQRO), and the Yucatán Center for Scientific Research (Centro de Investigación Científica de Yucatán, CICY). Later, other institutions, like the universities of Yucatán, Campeche, and Quintana Roo and El Colegio de la Frontera Sur (ECOSUR) have continued to advance the state of botanical research.

Several botanical publications describe the floristic richness of the region (Millspaugh 1895, 1896; Standley 1930; Lundell 1934; Sousa and Cabrera 1983; Sosa et al. 1985; Durán et al. 2000; Arellano et al. 2003; Gutiérrez-Báez 2003; Carnevali et al. 2010a), as well as its phytogeographical relationships (Estrada-Loera 1991; Durán et al. 1998; Ibarra-Manríquez et al. 2002; Espadas-Manrique et al. 2003). The most recent inventory shows that the flora encompasses more than 2300 species, with 956 genera and 161 botanical families (Carnevali et al. 2010a).

8.2.2 Endangered Species

Among the plant species that deserve special attention and should be considered priority for conservation, are three groups of species: first, those that are endemic to the Yucatán Peninsula. Second, species that are considered rare, either because they are not common in communities where they grow, or because they have a reduced spatial distribution, or highly specialized habitat requirements. Third, species that are considered threatened because their populations have been reduced by human activities, especially those that lead to degradation and loss of ecosystems.

At present only 40 native species of the Yucatán Peninsula are legally protected (SEMARNAT 2010). Just seven of these are endemic to the peninsula, despite the documented existence of 198 endemic species (Carnevali et al. 2010b). Some of the protected species have a wide distribution, as is the case of the four species of mangroves and some aquatic herbs. Their protection status is based on their importance to high value ecosystems, which are under threat in the entire country. Other species that are widely distributed, like cedar (*Cedrela odorata*), guayacán (*Guaiacum sanctum*) and kulinché (*Astronium graveolens*), are severely threatened due to timber extraction.

An unknown number of species, as many as several hundred, could be catalogued as rare, but to date no clear evaluation of their status is available (Durán and Trejo 2010). Tetetla et al. (2012) estimate that 195 woody species are rare in the Yucatán Peninsula, considering only species present in forest communities, but leaving out species of coastal, mangrove and dune communities, as well as all herbaceous species. Therefore the actual number of rare species could be much higher.

8.2.3 Useful Flora

Maya communities which have inhabited the Yucatán Peninsula for several centuries have a long tradition of natural resource use, based on their cosmovision and a deep knowledge of their environment. Ethnobotanical studies have documented the diversity of species of plants and animals which indigenous communities use to fulfill their physical, cultural and spiritual needs, as well as local uses of biodiversity (Durán and Méndez 2010; Villalobos-Zapata 2010; Pozo et al. 2011).

Databases of useful species developed by CICY (Durán et al. 2012) show that more than 130 timber species are used in the region, including species used for the construction of rural homes and the manufacture of furniture and tools. The same researchers recorded 145 food species, 88 of them native and 57 introduced. In addition, the authors reported 680 plant species used for medicinal purposes, mainly for gastrointestinal diseases, skin and respiratory problems (Méndez and Durán 1997). Ninety-seven species were recorded as ornamentals, mainly trees, with some palms and shrubs, and 36 melliferous species were noted, 28 of which were native to the peninsula. For use in handicrafts, 27 species (25 native) were documented, while 24 native species were reportedly used for firewood. Twelve fodder species were identified, and five species were recorded as dyestuffs. It is noteworthy that, of the useful species documented, 64 are used in an industrial manner, for their wood, fiber, fruits, seeds, or pigments and substances (Durán et al. 2012).

8.2.4 Environmental Services of Plant Biodiversity

The documented uses in the previous section form part of the environmental services provided by biodiversity, or ecosystem services, that is, all the benefits human society receive from ecosystems (MA 2005). Particularly, the plant uses enlisted in the previous section can be considered as goods, or provision services, which are the most directly for our benefit, as they are part of the sustenance of human life. However, humans obtain from ecosystems many more benefits that are less tangible and easy to recognize, which can be classified into three other types of services (Table 8.1; CONABIO 2006; MA 2005). Regulatory services encompass complex processes by which ecosystems regulate or modulate the conditions of the

Table 8.1 Environmental services of Yucatan ecosystems

Provision	Regulatory	Cultural	Support
Food	Biodiversity	Recreational (tourism)	Biodiversity
Medicines	Pollination	Environmental education	Soil protection
Genetic heritage	Air quality	Science and research	Permanence of human communities
Scenic landscape	Soil erosion	Uses and traditions (spiritual values)	Connectivity
Water quantity	Biogeochemical cycles	Appropriation of territories	Primary productivity
Organic soil layer	Hydrological cycles	Feeling of well being	Biogeochemical cycles
Forest products (timber & non timber)	Climate		
	Vector regulation		
	Impact of natural disasters		

environment, including regulation of climate, soil erosion, flooding and droughts, air and water quality and disease vectors, among others. Cultural services depend strongly on collective perceptions and the sociocultural context, and include recreational benefits (e.g. tourism), educational, aesthetic and spiritual or religious benefits, as well as societies' knowledge, perceptions and classifications of their natural environment (Balvanera et al. 2009; De Groot et al. 2005). Finally, support services are all those basic ecological processes which guarantee the adequate functioning of ecosystems and the flux of other services, and include primary productivity, the soil genesis, biogeochemical cycles of nutrients and protection of biodiversity, among others (Table 8.1).

Biodiversity modulates the supply of nearly all ecosystem services, like primary productivity (which in turn supplies food and other goods), soil development and maintenance, and nutrient cycling (affecting nutrient availability, soil fertility and agricultural production). Biodiversity also alters the hydrological cycle (and therefore the quantity, quality and seasonality of water), and plays a role in climate regulation and regulation of pollinators, pests and disease vectors, affecting human health and the primary sector (Balvanera et al. 2009; Díaz et al. 2005; Hooper et al. 2005).

Over 7 million ha of the Yucatán Peninsula's forests are still well-conserved (vegetation of more than 40 years of age, Table 8.2), including continuous patches of more than half a million ha (Garcia-Contreras 2014). This mosaic of forests fulfills an important role in the functioning of one of the world's largest karstic aquifers (Bauer-Gottwein et al. 2011). It also hosts the largest natural carbon stock in the country (de Jong et al. 2007), but is under strong pressure from deforestation and degradation by human activities (García-Contreras 2014; see Sect. 8.3).

Table 8.2 Area and density of above-ground carbon stocks in the Yucatán Peninsula

	Area		Stored carbon	
Cover type	Thousands of ha	%	Mt	%
Agro.-Urban	4223.6	31.3	691.5	19.5
Forest < 40 years	2100.5	15.6	477.9	13.4
Forest > 40 years	7181.5	53.2	2384.7	67.1
Total	13,505.7		3554.0	

Source: Land use change 1974–2014, Pronatura Península de Yucatán. 2014 and map of carbon density in Woody biomass of forests in Mexico, Alianza MREDD+ 2013

Therefore, in this region, the Early Action Area for REDD+ (Reduction of Emissions from Deforestation and forest Degradation and enhancement of forest carbon stock) Puuc-Chenes was established, covering an area of 1,381,924 ha (Alianza México REDD+ 2015). Besides, the peninsula contains three intensive carbon monitoring sites as part of the strengthening REDD+ and south-south cooperation project of the National Forest Commission-United Nations Development Program (Proyecto Fortalecimiento REDD+ y Cooperación Sur-Sur 2015). These sites contribute to the parametrization of models of carbon dynamics at different scales, and developing a system of measuring, reporting and verification (MRV) of anthropogenic emissions of greenhouse gases due to deforestation and forest degradation, and carbon absorption due to regeneration and forest management.

The largest reservoirs of carbon in the peninsula are located in high forests and medium statured forests in the south, as this area hosts the largest trees (in diameter and height), many with high commercial value (generally species with high wood-density and therefore high carbon content), compared to other vegetation types on the peninsula (Pennington and Sarukhán 2005). In each vegetation type, the largest reservoirs are found in older successional stages of abandoned agricultural areas, where vegetation is most developed and more biomass has accumulated (Vargas et al. 2008; Hernández-Stefanoni et al. 2011; Dai et al. 2014). At the same time, these areas have high agricultural potential, and are therefore susceptible to land-use changes (Dupuy et al. 2012).

According to data from Alianza México REDD+ (2013), the total above-ground carbon reservoir of the Yucatán Peninsula is 3554 Pg (10^{15} g). As shown in Table 8.2, forests older than 40 years occupy 53 % of the peninsula and contain 67 % of its above-ground carbon stocks. The largest reservoir of above-ground carbon is in the state of Campeche (43.6 % of all), followed by Quintana Roo (35.2 %) and last, Yucatán (21.2 %). The differences in carbon reservoirs among the three states are due to differences in territorial extent and distribution of vegetation types in each state, as high forest and medium forest are mainly located in Campeche and Quintana Roo, as well as to differences in deforestation rates in each state (see Sect. 8.3).

Despite the relevance of the environmental services provided by the Yucatán Peninsula's biodiversity, their quantity, quality and spatial and temporal distribution are strongly affected by human development, which has been achieved without adequate planning or consideration of regional sustainability.

8.3 Threats and Present Condition of the Vegetation

Almost 80 % of the vegetation of the Yucatán Peninsula has been deforested or degraded, a trend driven by several factors, including land use change transforming large tracts of forest to agricultural land or to urban and tourism development. Additionally, forest fires, development of road infrastructure and services, the high extraction rates of species of economic relevance, and use of products for domestic use have also contributed to deforestation and land degradation (García-Contreras 2014).

Historically, but particularly during the last four decades, a host of social and economic events have driven the conversion of natural ecosystems into productive systems, without a due process of planning or establishment of limits to the growth of each activity (Ramírez-Carrillo 2010). For example, driven by federal settlement incentives, the population of the peninsula has increased enormously the last 40 years, from just over one million in 1970 to over four million in 2010 (Table 8.3), with the consequent growth of cities and increased social and productive pressures on land.

During those years, population growth has concentrated in a few cities in the region. In Yucatán more than 42 % of the population is concentrated in Merida, while the city of Campeche hosts 31 % of the population of Campeche State. In Quintana Roo nearly half of the population lives in the municipalities of Benito Juárez, Solidaridad and Othon P. Blanco, due to the tourism development of Cancun, the Riviera Maya and, to a lesser extent, Chetumal.

The main threats to the vegetation of the Yucatán Peninsula and its biodiversity and environmental services are deforestation and land degradation (Fig. 8.1). Loss of forest cover, caused mainly by land-use change for agriculture and urban development, and by fires, has a particularly high impact. Activities driving these land-use changes are road-building, urbanization, commercial agriculture, extensive livestock raising, quarrying, and to a lesser degree the traditional milpa and intensive animal husbandry. Forest fires, which also diminish forest cover, are mainly caused by agricultural practices, followed by hunting, open garbage dumps, and tourism. The primary driver of forest degradation is extraction of timber and non-timber products, followed by contamination by rural waste, and open garbage dumps (Fig. 8.1, García-Contreras 2014).

Table 8.3 Population increase in the three states of the Yucatán Peninsula from 1970 to 2010

Year	Yucatán	Campeche	Quintana Roo	Total
1970	758,355	251,556	88,150	1,098,061
1980	1,063,733	420,553	225,985	1,710,271
1990	1,362,940	535,185	493,277	2,391,402
2000	1,658,210	690,689	874,963	3,223,862
2010	1,955,577	822,441	1,325,578	4,103,596

Source: INEGI

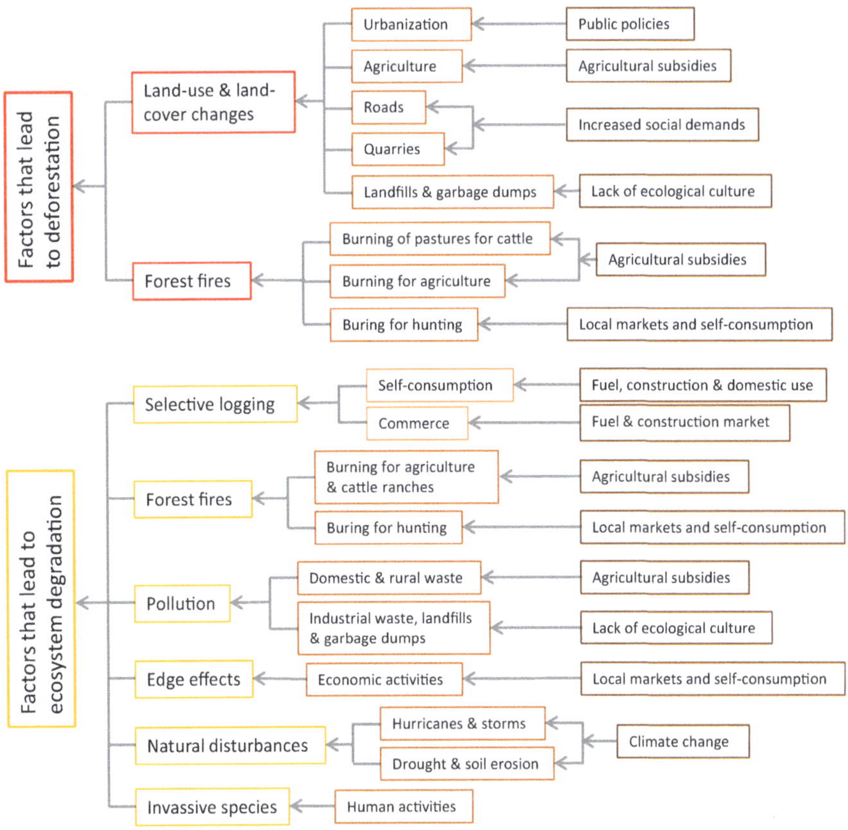

Fig. 8.1 Primary threats to ecosystems of the Yucatán Peninsula and their causes. *Source*: García-Contreras (2014)

8.3.1 Land Use Change

Land use conversion causes loss of forest cover and hence damages biodiversity and environmental services. For example, with deforestation, a large proportion of an area's stored carbon is released to the atmosphere as CO_2 and other greenhouse gases, thereby not only reducing the mitigation potential of tropical forests, but also further contributing to climate change (Pan et al. 2011).

The Yucatán Peninsula has an extension of 13,781,229 ha, of which 90.43 % was historically covered by low, medium and high forests (dry and humid), while the remaining areas (9.57 %) were wetlands. In the year 2010, only 22 % was still covered by primary vegetation, 59.4 % by secondary vegetation (herbaceous, shrub and arboreal) and 18.1 % by urban and agricultural areas (Table 8.4). The most affected vegetation types were deciduous and subdeciduous medium forests: only 1.8 % of their original cover remains. Low subdeciduous and evergreen forests have also been hit hard, retaining just 7.5 % of their original cover, while medium and

Table 8.4 Historical (potential) cover and recorded cover in 2010 in the Yucatán Peninsula (area in thousands of ha)

Vegetation type	Potential vegetation	Veg. 2010	Change (%)
Agriculture	0	2284.4	16.6
Urban	0	148.2	1.1
Water bodies	111.0	71.5	−35.6
Without vegetation	0	57.4	0.4
Mangroves	492.2	443.9	−9.8
Secondary mangrove vegetation	0	16.2	3.3
Palm vegetation	9.1	6.4	−29.2
Hydrophylic vegetation	705.5	604.0	−14.4
Coastal dunes	20.4	10.9	−46.6
Low deciduous and subdeciduous forest	1316.9	450.0	−65.8
Secondary vegetation of low deciduous and subdeciduous forest	0	938.8	71.3
Low semi evergreen and evergreen forest	1134.7	84.9	−92.5
Secondary vegetation of low semi evergreen and evergreen vegetation	0	16.5	1.5
Medium stature deciduous and subdeciduous forest	4031.7	72.3	−98.2
Secondary vegetation of medium deciduous and subdeciduous forest	0	3593.9	89.1
Medium and high semi-evergreen and evergreen forest	5959.7	1360.4	−77.2
Secondary vegetation of medium and high semi evergreen and evergreen forest	0	3621.6	60.8

Source: Pronatura Península de Yucatán, 2015

high semi-evergreen and evergreen forests still cover 22.8 % of their original extent. The least affected ecosystems are mangroves with 90.2 % of their original cover (Table 8.4).

The oldest record of forest cover is from 1974, with subsequent datasets available for 1994, 2000, 2007 and 2014. These data permit analysis of changes in cover over the last 40 years. Table 8.5 shows how forest cover changed in the region, as the area occupied by humans increased.

Forest cover decreased from 89.2 % in 1974 to 79.3 % in 2014, a net loss of about 10 %. However, in the different periods analyzed, variations in forest cover can be observed, due to deforestation, abandonment and natural regeneration. The period 1994–2000 is the time frame that exhibits the highest net deforestation (11.1 %), relative to the previous period (1974–1994). These reductions are due to areas converted for agriculture. The period 2000–2007 is the only one showing a net recovery of forest cover, an increase of nearly 7.5 % relative to the previous period. In the last period evaluated, a slight net loss of forest cover (1.8 %) is observed (Table 8.5, Fig. 8.2).

The spatial distribution of these land use changes was not uniform (Fig. 8.2). Historically, the state of Yucatán has experienced the highest deforestation rates,

Table 8.5 Changes in land cover of the Yucatán Peninsula (area in thousands of ha)

	1974	1994	2000	2007	2014
Vegetation	12,294.1	12,091.5	10,288.7	11,295.0	10,976.1
Human-occupied	1487.1	1689.7	3492.5	2486.3	2805.1

Source: Pronatura Península de Yucatán, 2015

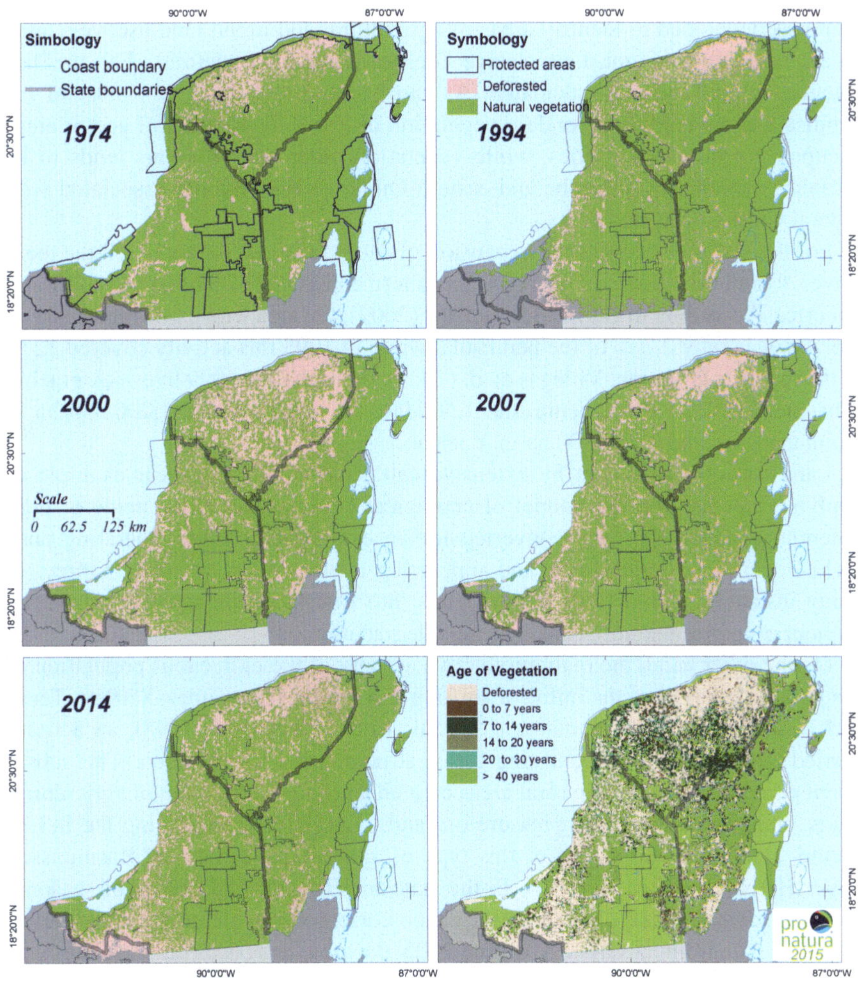

Fig. 8.2 Land use change of the last 40 years, *marking areas* with and without natural vegetation. Based on the first five maps (1974, 1994, 2000, 2007 and 2014) areas were identified which remained used for agriculture and urban areas, areas with vegetation and areas with continuous changes (deforestation-afforestation). The age for the last category was estimated. *Source*: Pronatura Yucatán Península, 2015

especially in the northeast. In the state of Campeche a huge loss of forest cover is also observed, mainly in the region of Escarcega and Candelaria (southern Campeche), due to extensive livestock production. Quintana Roo is the state with lowest deforestation rates, although in 2014 some large deforested areas (mainly grasslands for cattle) can be identified, for example in the La Union area, and along the Highway Jose Maria Morelos-Chetumal.

Based on the land cover data in the years 1974, 1994, 2000, 2007 and 2014, a map was developed to identify areas of agricultural and urban land use, vegetated areas, and areas of constant change (deforestation-afforestation). For the last category, stand age was calculated. As can be observed (Fig. 8.2), most of the peninsula is covered by secondary vegetation of less than 40 years of age or areas occupied by human activities, while vegetation older than 40 years tends to be located in southern Campeche and eastern Quintana Roo, mainly associated with federal natural protected areas.

For the last four decades the expansion of cattle husbandry constituted the main driver of land use change in the region, a pattern consistent with other areas in Latin America. According to Hernández-Barrios (2002), the area with cattle husbandry in the 70's was only 2.7 % of the peninsula, while in 1995 this activity covered 12 % of the region's land area. Villegas et al. (2001) report that, in 1999 livestock grazing subsumed 28.3 % of the peninsula: 1,500,000 ha in Campeche, 1,200,000 ha in Quintana Roo, and 1,200,000 ha in Yucatán.

Land use change driven by extensive cattle husbandry is a clear example of applying an inappropriate model of ecological exploitation. Ecosystems of high potential productivity were converted for an economic activity generating poor yields, occupying large areas, and supporting a mainly low-wage workforce. In many instances, this activity also extends into forests where cattle graze freely, without proper regulation, causing forest degradation.

On the other hand, the main productive activity of the indigenous population of the peninsula has been the milpa or shifting cultivation (Hernández-X 1959; Terán and Rasmunssen 1994; Hernández-X et al. 1995; Terán et al. 1998), an activity carried out extensively, modifying forest structure and composition. The subsequent population increase in rural areas created a need for expansion of agricultural lands, and placed increased pressure on land, considerably shortening the fallow period, a fundamental aspect of this type of agriculture (Terán and Rasmunssen 1994; Hernández-X et al. 1995). In the state of Yucatán, 150,000 ha of seasonal corn are sown annually, 80,000 ha of which correspond to recently cleared lands. Based on data from SAGARPA, during 2005, the area for agricultural practices was 201,000 ha in Yucatán, 38,000 ha in Quintana Roo and 42,000 ha in Campeche.

8.3.2 *Forest Fires*

The Yucatán Peninsula leads Mexico in forest fire damage. For example, 42 % of mature forest areas that burned in Mexico from 2003 to 2014 were located in the

Table 8.6 Forest area (in thousands of ha) damaged by fires from 2003 to 2014

State	Damaged area (thousands of ha)	% of total fire-affected area
Campeche	40.1	14.6
Quintana Roo	51.7	18.9
Yucatán	23.5	8.6
Rest of Mexico	158.5	57.9

Source: Centro Nacional de Control de Incendios Forestales (CENCIF) de la Comisión Nacional Forestal (2015)

peninsula, mainly in Quintana Roo and Campeche (Table 8.6; CONAFOR 2015). Forest fires reduce the mitigation capacity of ecosystems and directly trigger the release of greenhouse gases, risking human health, reducing vegetation cover, biodiversity and environmental services.

Use of fire in agriculture is prevalent throughout the region. Most fires stem from agricultural practices. The second cause of fires is roads, mainly via land-clearing for their construction, or started by motorists with discarded cigarettes or glass bottles (Secretaría de Medio Ambiente y Recursos Naturales 2015). The damage to vegetation produced by hurricanes like Isidor in 2002, Emily and Wilma in 2005, and Dean in 2009 cause a considerable increase of fuel material, both standing and fallen dead biomass, which increases the frequency, strength and extent of area affected by fires in the following dry seasons.

Apart from the incidence and track of the hurricanes that affect the peninsula, the probability of forest fires is distributed heterogeneously, depending on fuel load and human impact. Based on heat points, which reflect sources of ignition and areas with higher fuel loads and more human impact, a map of fire risk was developed (Fig. 8.3). It can be observed that agricultural areas of southern Campeche, central Yucatán and northern Quintana Roo are high risk and should give high priority to fire prevention mechanisms, including regulation of use of fire as a tool for agricultural practices and implementation of a fire warning system during the dry season.

Considering the main threats to the vegetation of the Yucatán Peninsula and its biodiversity, we developed a map of human impact (Fig. 8.4). Most of the area of the Yucatán Peninsula is subject to medium to high human impact, associated with activities that lead to ecosystem destruction, fragmentation and degradation. Those activities are generally linked to government-promoted development policies, which favor exploitation focused on short-term economic benefits, without management to protect natural resources (Leff 1990). Those policies have characterized the exploitation of the natural resources of the peninsula, causing damage to forests and displacing autochthonous productive systems. Fortunately, as can be observed in Figs. 8.2, 8.3, and 8.4, the establishment of protected areas has helped to counter the effects of these development policies to some degree.

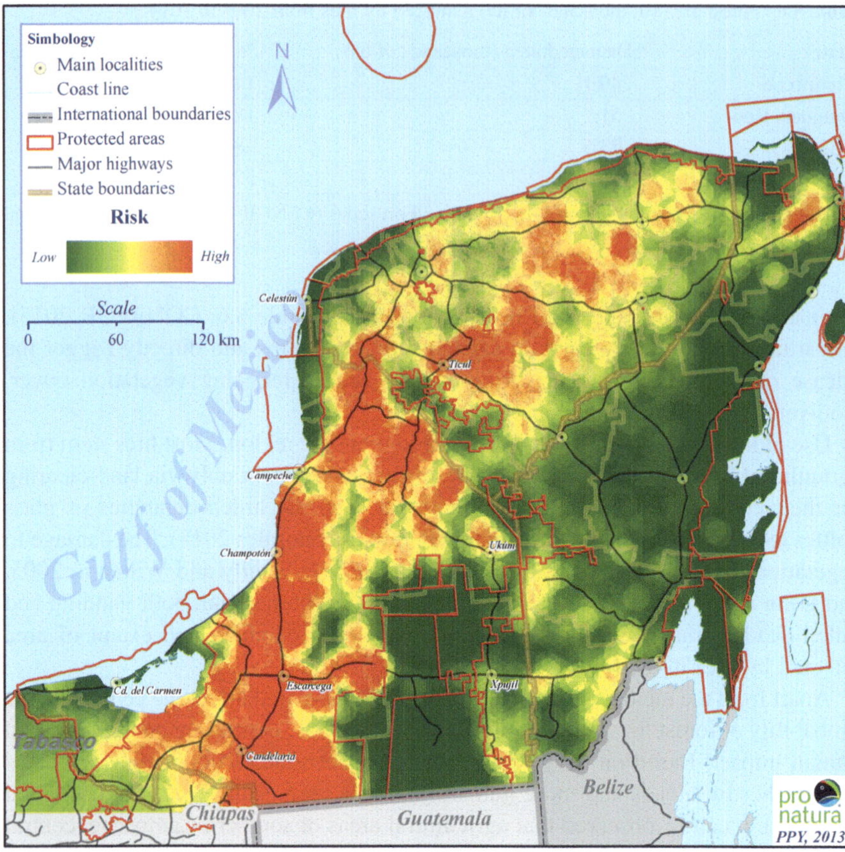

Fig. 8.3 Fire risk of the Yucatán Peninsula, based on heat points and areas of mayor human impact and biomass fuel loads. *Source*: Pronatura Península de Yucatán, 2013

8.4 Strategies for Conservation and Use of Plant Diversity

8.4.1 Conservation

In Mexico, Protected Natural Areas (ANP, by their Spanish acronym) represent the government's main instrument for conservation of biological diversity. Besides working to protect and conserve landscapes, ecosystems and species, ANPs have served as a focal point for the development of activities promoting sustainable use of biodiversity and involving participation of local communities (Benítez Torres and Villalobos-Zapata 2010; Ruiz Barranco and Arellano 2010; Prezas 2011; Secretaría de Ecología y Medio Ambiente 2014; Secretaría de Medio Ambiente y Aprovechamiento Sustentable del Gobierno de Campeche 2014; CONANP 2015; Secretaría de Desarrollo Urbano y Medio Ambiente 2015).

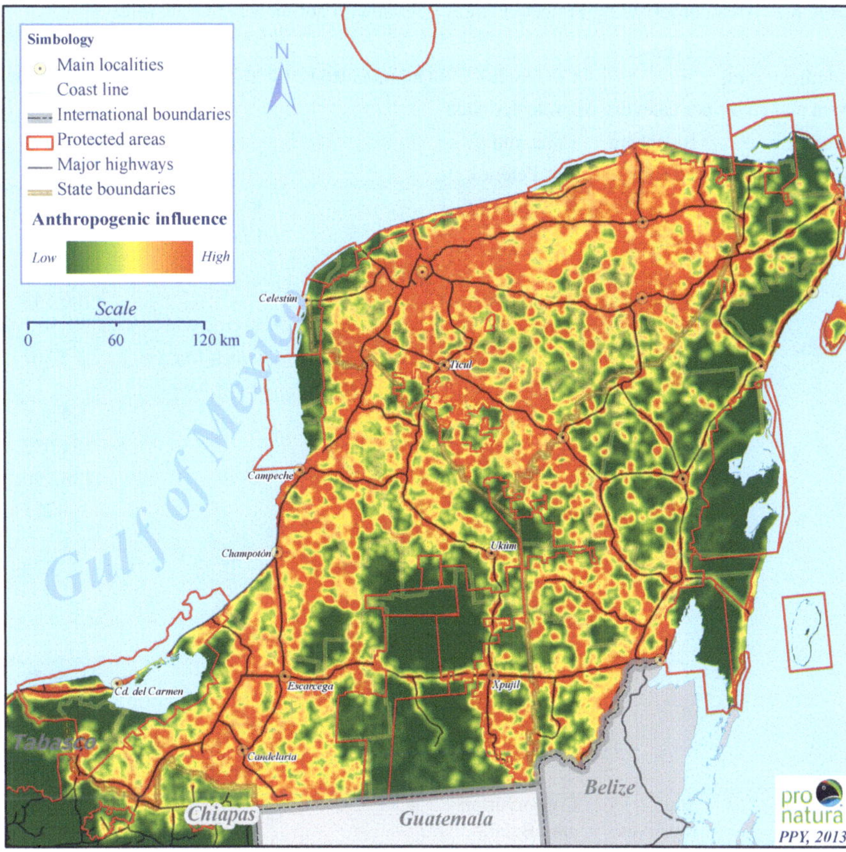

Fig. 8.4 Human impact on ecosystems of the Yucatán Peninsula. *Source*: Pronatura Península de Yucatán, 2012

The Yucatán Peninsula has 39 protected areas covering a total land area of 2,977,752 ha, which represents 21.6 % of the entire peninsula (Table 8.7). The analysis of land use changes on the peninsula in the last four decades (Table 8.5, Fig. 8.2), the risk of forest fires (Fig. 8.3) and anthropogenic influence (Fig. 8.4), indicates that protected natural areas in the region have contributed significantly to the conservation of large tracts of forests and other tropical ecosystems, along with their biodiversity and the environmental services they provide, such as sequestration of atmospheric CO_2.

Although all types of vegetation in the region are represented in these protected areas, studies to determine if the area currently protected in the Yucatán Peninsula is sufficient to sustain ecological processes or maintain viable populations of native species indicate that current networks of protected areas do not adequately cover all the eco-regions or priority sites within the area devoted to conservation

Table 8.7 Area conserved by Protected Areas (the data do not include marine reserves)

Administration	Type of area	Campeche	Quintana Roo	Yucatán	Area (ha)
Area and cover per category of protected area					
Federal	Biosphere Reserver	Calakmul			723,185
		Los Petenes			282,858
		Ría Celestún		Ría Celestún	81,482
				Ría Lagartos	60,348
			Sian Ka'an		528,148
	Areas of protection for flora and fauna	Laguna de Términos			706,148
			Otoch Ma'ax Yetel Kooh	Otoch Ma'ax Yetel Kooh	5367
			Uaymil		89,118
			Yum Balam		154,052
			Bala'an Ka'ax		128,390
			Nichupte Mangroves		4257
			North portion and Eastern coast of Cozumel		37,829
			Contoy Island		5126
			Tulum		664
			Xcalak reef		17,949
				Dzibilchaltun	539
State	State Reserves	Conservation area Balam-Kin			110,990
		Conservation area Balam-Kú			409,200
			Kabah Urban Park		41
			Manatí Lagoon refuge for Flora and Fauna		203
			State refuge for Flora and Fauna Lagoon System Chacmochuch		1915
			Refuge for Flora and Fauna Colombia Lagoon		1114
			State Reserve Lagoon System Chichankanab		11,610

(continued)

Table 8.7 (continued)

Administration	Type of area	Campeche	Quintana Roo	Yucatán	Area (ha)
Area and cover per category of protected area					
			Bacalar Lagoon Park		5367
			State Reserve Cozumel Forests and marshes		19,846
				El Palmar State Reserve	50,177
				Dzilam State Reserve	69,039
				State Park Yalahau Lagoon	5683
				San Juan Bautista Tabi and Sacnicté Anex Natural Proteceted Area	1356
				Kabah State Park	947
				Mangroves and Marches of the Northern Coast of Yucatan State Reserve	54,777
				Ring of Cenotes Geohydrological Reserve	219,208
				Puuc Biocultural State Reserve	135,849
Town	Town Reserves	"Salto Grande" Urban Park			1570
		Laguna Ik			28,820
			Ombligo verde		4064
				Cuxtal Reserve	10,757
	Total	8	18	13	2,977,752

Source: National Commission of Protected Natural Areas (CONANP), 2015

(CONABIO-CONAP-TNC-PRONATURA-FCF, UAN L 2007; CONABIO-PNUD 2009; Durán and Pacheco 2010; Koleff and Urquiza-Haas 2011). In addition to increasing the number and extent of areas under protection, the authors recommend strengthening the network of federal, state and municipal protected areas, by using a full range of conservation tools, including management units for the conservation of wildlife (UMA), programs for sustainable forest management and forest certification, and voluntary conservation areas (VCA), as well as the program of payment for environmental services (PES). In addition, the Mesoamerican Biological Corridor-Mexico seeks to maintain connectivity in some areas, particularly in the areas of influence of protected areas in the region, allowing free movement of wildlife, as well as encouraging conservation of flora.

The Yucatán Peninsula also has other strategies for in situ and ex situ conservation, namely the establishment of botanical gardens such as that of Yucatán Center of Scientific Research (CICY) which houses about 700 species (Centro de Investigación Científica de Yucatán 2014), the botanical garden of the College of the Southern Border (ECOSUR) at Puerto Morelos, which is the country's largest, with 65 ha of tropical evergreen forest, preserving several collections and 170 species (El Colegio de la Frontera Sur 2015), and the Autonomous University of Carmen which protects 25 ha of mangrove and also houses several collections (Universidad Autónoma del Carmen 2014). In addition, the Mayab Medicinal Gardens Network houses about 250 species of medicinal plants; and the CICY germplasm bank contains over 1200 collections of 242 species and several collections of live plants (Hernández 2015).

8.4.2 Use and Sustainable Use

The use of forest resources in the region has not always been sustainable, and has gone through different stages. In the middle of the last century, a 25-year forestry-project was launched for timber production in northeast Yucatán. Unfortunately this experience resulted in the devastation of forest resources, as it focused on the extraction of the highest quality individual trees from the forest, leaving poor quality trees behind at the end of the process (Galletti 1994; Merino 2004). Later, during the 1980s, the so called "Plan Piloto" (pilot plan) was promoted in the states of Quintana Roo and Campeche, which tried to encourage the participation of communities and ejidos in management and forestry (Argüelles 1999; Merino 2004; Santos et al. 2005). A major achievement of Plan Piloto was that *campesinos* were given and assumed responsibility for the care of forest, which in turn has allowed large forested areas to persist without major changes in land use. Although in recent years there has been an increased in the volume of tropical timber species utilized in the Yucatán Peninsula, this increase is considered very limited in scope, since total timber production in the region represents less than 2 % of national production. Furthermore, the territorial extent of forest areas under active management is still very limited, especially in the State of Yucatán.

At present, other productive activities in the region have also contributed to biodiversity conservation through sustainable use of natural resources. Examples include the beekeeping programs, sustainable agriculture, and alternative tourism and conservation initiatives supported, through the Small Grants Program of the Fund for Global Environment in the Yucatán Peninsula, among others (Small

Grants Program 2008). This program has supported numerous beekeeping projects which together kept an area of approximately 472,000 ha of forest under productive management, while forestry projects cover an area of approximately 100,000 ha of forests under management (Durán et al. 2013).

However, to achieve conservation and sustainable use of biodiversity, improvements in effectiveness and efficiency of public policies are needed. Similarly, the region lacks tools designed for management of natural resources that contribute to an equilibrium among social, economic and environmental benefits of ecosystems in the peninsula, and improved knowledge about biodiversity and environmental goods and services provided to society.

8.5 Conclusions and Prospects

The Yucatán Peninsula is a region of great biological, ecological and cultural diversity, and historically has been marked by the development of the Mayan people and biologically by its affinity with the neighboring regions of Central America, the Caribbean Basin and Southeast Mexico, as well as by having high endemism. This region contains several types of forests that constitute the second-largest forested area in Latin America after the Amazon basin, and other coastal and hydrophilic vegetation, resulting in high environmental heterogeneity and floristic diversity.

The Yucatán Peninsula is one of the best-explored regions of Mexico with an estimated total number of vascular plant species surpassing 2300 species. However, there is still uncertainty concerning the precise number of rare species (there are an estimated 195 rare woody species in the forests of the peninsula alone), so given their vulnerability to extinction, botanical studies focusing on this species group are urgently needed. In addition, a review and update of legal protection of the most vulnerable species is needed. Currently, only 40 species are legally protected and of these only seven are endemic, though conservative estimates suggest the occurrence of 198 endemic species in the region. So, clearly further research on population dynamics, distribution and ecological requirements of rare and endemic plant species of the peninsula is required.

In addition to the intrinsic value of the plant diversity of the Yucatán Peninsula, it is important to highlight its utility value, since vegetation is the basis of a wide variety of environmental and ecosystem services such as provisioning, including about 680 medicinal species, 145 edible species and 130 timber-species; cultural, including tourism; and regulation of air and water quality (terrestrial ecosystems of the peninsula protect and help regulate one of the largest karstic aquifers worldwide), floods, droughts, and climate. Concerning climate, the peninsula is home of the largest natural reservoir of carbon in Mexico, estimated at 3554 Pg (10^{15} g) of above-ground carbon, which is concentrated in semi-deciduous forests towards the south of the peninsula and in the vegetation of late successional stages.

However, the biodiversity of the Yucatán Peninsula and the environmental services that it provides have been threatened by various human activities linked to development policies that have focused on short-term economic benefits, without proper planning. The main threats to biodiversity in the region are land use change that has transformed large tracts of forests into agricultural land and tourist developments, roads, or urban areas; and the occurrence of forest fires. As a result of these processes, by 2010, primary vegetation occupied only 22 % of the territory of the peninsula, while agricultural, urban areas and roads occupied 18.1 % and secondary vegetation (mostly under 40 years old) occupied 59.4 %. The Yucatán Peninsula has also been heavily impacted by forest fires, accounting for 42 % of the national extent of damage to mature forests in Mexico during the period 2003–2014.

Fortunately, the state has also developed conservation initiatives to address the processes of deterioration and pressures on biodiversity of the peninsula, mainly through the establishment of protected areas covering 21.6 % of the land area of the region. The analysis of changes in land use over the past four decades, anthropogenic influence and risk of forest fires, indicate that conserved areas have contributed significantly to the conservation of large tracts of forests and other tropical ecosystems. Conserved areas also contribute to biodiversity and the environmental services it provides, such as sequestration of atmospheric CO_2. In this sense, the importance of the region for climate change mitigation through REDD+ has recently been recognized, particularly through the establishment of a REDD+ Early Action Area of 1,381,924 ha, and three sites of intensive forest carbon monitoring. The Yucatán Peninsula is therefore a pioneer in the implementation of this international strategy, which aims to preserve and increase the ability of ecosystems to capture and store atmospheric CO_2. It also proposes to contribute to training and governance for rural residents and to create alternative sources of income to promote more sustainable development and lower emissions of greenhouse gases.

It is also important to note other conservation efforts, such as the Mesoamerican Biological Corridor-Mexico. This corridor seeks to maintain connectivity in the areas of influence of protected areas in the region, allowing free movement of wildlife and promoting the conservation of flora and fauna. It further supports the management of wildlife conservation units (UMA), voluntary conservation areas and program of payment for environmental services (PES). All these initiatives endeavor to ensure the permanence and function of ecosystems, and contribute to maintaining viable native populations throughout their ranges. In addition to this, the development of sustainable productive activities such as beekeeping, sustainable agriculture and alternative tourism, supported by initiatives such as the Small Grants Program of the Fund for Global Environment, have allowed for sustainable use of natural resources in the peninsula. However, improvements are still needed in the implementation and enforcement of effective public policies and in developing tools for conservation and use of natural resources of the Yucatán Peninsula. Ideally, these policies will serve to harmonize social, economic and environmental benefits from ecosystems, as well as to improve the knowledge and sustainable use of biodiversity from the region.

References

Alianza México REDD+. Densidad de carbono en la biomasa leñosa aérea de los bosques y selvas en México. Woods Hole Research Center, The Nature Conservancy, CONAFOR, Proyecto México Noruega, CONABIO; 2013.

Alianza México REED+. Áreas de Acción Temprana. 2015. http://www.alianza-mredd.org/areas-de-accion-temprana/region-puuc-y-los-chenes-en-Yucatán-y-campeche#.VRwr0fmUeCk. Accessed 31 Mar 2015.

Andrade Hernández M. Transformación de los ecosistemas naturales por actividades antropogénicas. In: Durán R, Méndez M, editors. Biodiversidad y desarrollo humano en Yucatán. CICY, PPD-FMAM, CONABIO, SEDUMA; 2010. p. 316–9.

Arellano-Rodríguez JA, Flores Guido JS, Tun-Garrido J, Cruz-Bojórquez MM. Nomenclatura, forma de vida, uso, manejo y distribución de las especies vegetales de la península de Yucatán. Etnoflora Yucatánense, Fascículo 20, UADY, Mérida; 2003.

Argüelles SA. Diagnóstico de las poblaciones de caoba en México. Mexico: Chetumal; 1999.

Balvanera P, Cotler H, et al. Estado y tendencias de los servicios ecosistémicos. In: Capital Natural de México, vol. II: Estado de conservación y tendencias de cambio. México: CONABIO; 2009. p. 185–245.

Barrera A. La Península de Yucatán como Provincia Biótica. Revista de la Sociedad Méxicana de Historia Natural. 1962;23:71–105.

Bauer-Gottwein P, et al. Review: the Yucatán Peninsula karst aquifer. Mex Hydrogeol J. 2011;19:507–24.

Benítez Torres JA, Villalobos-Zapata GJ. Sitios prioritarios para la conservación. In: Villalobos-Zapata GJ, Mendoza-Vega J (coord) La Biodiversidad en Campeche: Estudio de Estado. México: CONABIO, Gobierno del Estado de Campeche, Universidad Autónoma de Campeche, El Colegio de la Frontera Sur; 2010. p. 588–606.

Campos G, Durán R. La vegetación de la Península de Yucatán. In: Apuntes del Curso-Taller para maestros "El jardín botánico como herramienta didáctica". Mérida: Centro de Investigación Científica de Yucatán; 1991. p. 23–35.

Carnevali G, Tapia JL, Duno R, Ramírez I. Flora Ilustrada de la Península de Yucatán: Listado Florístico. Mérida: CICY; 2010a.

Carnevali G, Duno R, Ramírez I, Tapia JL. Diversidad de la Flora. In: Durán R, Méndez M, editors. Biodiversidad y desarrollo humano en Yucatán. CICY, PPD-FMAM, CONABIO, SEDUMA. 2010b. p. 420–3.

Centro de Investigación Científica de Yucatán, AC. Jardín Botánico Regional-Xíitbal neek. 2014. http://www.cicy.mx/Documentos/CICY/Servicios/Jardin_Botanico/2013/Lineamientos_del_Jardin_Botanico_Regional.pdf. Accessed 31 Mar 2015.

CONABIO. Capital natural y bienestar social. Comisión Nacional para el Conocimiento y Uso de la Biodiversidad, México; 2006.

CONABIO-CONANP-TNC-PRONATURA-FCF, UAN L. Análisis de vacíos y omisiones en conservación de la biodiversidad terrestre de México: espacios y especies. México: CONABIO, CONANP, TNC-Programa México, Pronatura AC, Facultad de Ciencias Forestales, UNAL; 2007.

CONABIO-PNUD. México: capacidades para la conservación y el uso sustentable de la biodiversidad. México: CONABIO, PNUD; 2009.

CONAFOR. Incendios. 2015. http://www.conafor.gob.mx/web/temas-forestales/incendios/. Accessed 31 Mar 2015.

CONANP. Áreas Protegidas Decretadas. 2015. http://www.conanp.gob.mx/que_hacemos/. Accessed 2 Mar 2015.

De Jong B, Jiménez Ferrer G, Soto-Pinto L. Agroforestería y secuestro de carbono. In: Jiménez G, Nahed J, Soto L, editors. Agroforestería pecuaria en Chiapas, México. San Cristóbal de las Casas: ECOSUR; 2007. p. 27–31.

Dai Z, Birdsey RA, Johnson K, Dupuy JM, Hernández-Stefanoni JL, Richardson K. Modeling carbon stocks in a secondary tropical dry forest in the Yucatán Peninsula, Mexico. Water Air Soil Pollut. 2014;225:1925.

De Groot R, Ramakrishnan PS, van de Berg A, et al. Cultural and amenity services. In: Hassan R, Scholes R, Ash N, editors. Ecosystems and human well-being: current state and trends, Findings of the condition and trends working group of the millennium ecosystem assessment, vol. 1. Washington, DC: Island Press; 2005. p. 455–76.

Díaz S, Tilman D, Fargione J, et al. Biodiversity regulation of ecosystem services. In: Hassan R, Scholes R, Ash N, editors. Ecosystems and human well-being: current state and trends, Findings of the condition and trends working group of the millennium ecosystem assessment, vol. 1. Washington, DC: Island Press; 2005. p. 297–329.

Dupuy Rada JM, Hernández-Stefanoni JL, Hernández-Juárez R, et al. Efectos del cambio de uso del suelo en la biomasa y diversidad de plantas leñosas en un paisaje de bosque tropical seco en Yucatán. Investigación Ambiental Ciencia y Política Pública. 2012;4:130–40.

Durán R. Descripción y análisis de la estructura y composición de la vegetación de los petenes del noroeste de Campeche, México. Biótica. 1987;12:181–98.

Durán R, Trejo-Torres JC, Ibarra-Manríquez G. Endemic phytotaxa of the Peninsula of Yucatán. Harv Pap Bot. 1998;3:263–314.

Durán R, Campos G, Trejo-Torres JC, et al. Listado Florístico de la Península de Yucatán. Mérida: CICY-PNUD; 2000.

Durán R, Méndez M. Biodiversidad y desarrollo humano en Yucatán. Mérida: CICY, PPD-FMAM, CONABIO, SEDUMA; 2010.

Durán R, Pacheco LR. Papel de las Áreas Naturales Protegidas en la conservación de la biodiversidad. In: Durán R, Méndez M, editors. Biodiversidad y desarrollo humano en Yucatán. Mérida: CICY, PPD-FMAM, CONABIO, SEDUMA; 2010. p. 420–3.

Durán R, Chi-May F, Méndez M. Base de Datos de Especies Útiles de la Península de Yucatán. Mérida: CICY; 2012.

Durán R, Trejo-Torres JC. Plantas vasculares prioritarias para la conservación. In: Durán R, y Méndez M, editors. Biodiversidad y desarrollo humano en Yucatán. Mérida: CICY, PPD-FMAM, CONABIO, SEDUMA; 2010. p. 194–6.

Durán R, Catilla M, McManus MP. Experiencias exitosas en materia de Biodiversidad y Cambio Climático del Programa de Pequeñas Donaciones, PPD-FMAM, PNUD; 2013.

El Colegio de la Frontera Sur. Ya'ax'che, Jardín Botánico Dr. Alfredo Barrera Marín. 2015. http://www.ecosur.mx/jb/YaaxChe/home.html. Accessed 31 Mar 2015.

Espadas C, Durán R, Argáez J. Phytogeographic analysis of taxa endemic to the Yucatán Peninsula using geographic information systems, the DOMAIN heuristic method and parsimony analysis of endemicity. Divers Distrib. 2003;9:313–30.

Espejel I. Sand dunes ecology in the Peninsula of Yucatán. México/Upsala: Institute of Ecological Botany/Uppsala University; 1982.

Estrada-Loera E. Phytogeographic relationships of the Yucatán Peninsula. J Biogeogr. 1991;18:687–97.

Ferrusquía-Villafranca I. Geology of México: a synopsis. In: Ramamoorthy TP, Bye R, Lot A, Fa J, editors. Biology diversity of Mexico: origins and distribution. Oxford: Oxford University Press; 1993. p. 3–108.

Flores S, Espejel I. Tipos de vegetación de la Península de Yucatán. Mérida: UADY, Sostenibilidad Maya; 1994.

Galletti H. Actividades forestales y su desarrollo histórico. Estudio integral de la frontera México-Belice. Análisis socioeconómico. Tomo I. CIQRO, Chetumal; 1994. p. 131–98.

García-Contreras G. Modelo geoespacial para el mapeo de las áreas con potencial para la restauración productiva del paisaje rural en la península de Yucatán. PPY, UICN, CPIC. 2014.

Gutiérrez-Báez C. Listado florístico actualizado del estado de Campeche, México. Campeche: Universidad Autónoma de Campeche; 2003.

Hernández A. Llenan el 'Arca de Noé' yucateca con semillas locales. Milenio Novedades, Miércoles 25 de febrero de 2015; 2015.
Hernández-Barrios JC. Análisis del impacto del crecimiento de la ganadería bovina sobre el medio ambiente de la Península de Yucatán. México: Facultad de Ciencias, UNAM; 2002.
Hernández-Stefanoni JL, Dupuy JM, Tun-Dzul F, May-Pat F. Influence of landscape structure and stand age on species density and biomass of a tropical dry forest across spatial scales. Landsc Ecol. 2011;26:355–70.
Hernández-X E. La agricultura en la Península de Yucatán. In: Beltrán E, editor. Los Recursos Naturales del Sureste y su Aprovechamiento Vol III. México: IMERNAR; 1959.
Hernández-X E, Bello E, Levy S. La milpa en Yucatán I. Colegio de Posgraduados, Montecillo; 1995.
Hooper DU, Chapin FS, Ewell JJ, et al. Effects of biodiversity on ecosystem functioning: a consensus of current knowledge. Ecol Monogr. 2005;75:3–35.
Ibarra-Manríquez G, Villaseñor JL, Durán R, Meave J. Biogeographical analysis of the tree flora of the Yucatán Peninsula. J Biogeogr. 2002;29:17–29.
Jáuregui E, Vidal J, Cruz F. Los ciclones y tormentas tropicales en Quintana Roo durante el periodo 1871-1978. In: Quintana Roo: Problemática y Perspectivas. Simposio. Cancún: CIQRO-Instituto de Geografía UNAM; 1980. p. 47–63.
Koleff P, Urquiza-Haas T (coords). Planeación para la conservación de la biodiversidad terrestre en México: retos en un país megadiverso. México: CONABIO–CONANP; 2011.
Leff E. Medio ambiente y desarrollo en México. México: CIIH-UNAM/Miguel Ángel Porrúa; 1990.
Lundell CL. Preliminary sketch of the phytogeography of the Yucatán Peninsula. Contrib Am Archaeol. 1934;12:257–321.
Méndez M, Durán R. Diagnóstico del conocimiento etnobotánico actual de las plantas medicinales de la Península de Yucatán. Bol Soc Bot México. 1997;60:15–24.
Merino L. Las políticas forestales de conservación en México y en Quintana Roo. In: Armijo N, Llorens C, editors. Uso, conservación y cambio en los bosques de Quintana Roo. Chetumal: Universidad de Quintana Roo-CONACYT; 2004. p. 15–42.
Millennium Ecosystem Assessment (MA). Ecosystems and human well-being: a framework for assessment. Washington, DC: Island Press; 2005.
Millspaugh CF. Contribution to the flora of Yucatán. Publ. Field Columbian Museum. Bot Ser. 1895;1:1–56.
Millspaugh CF. Contribution II to the coastal and plain flora of Yucatán. Publ. Field Museum. Bot Ser. 1896;1:279–339.
Miranda F. Estudios acerca de la vegetación. In: Beltrán E, editor. Los Recursos Naturales del Sureste y su Aprovechamiento. México: IMERNAR; 1958.
Morrone JJ. Presentación sintética de un nuevo esquema biogeográfico de América Latina y El Caribe. Sociedad Entomológica Aragonesa. Monografías Tercer Milenio. 2002;2:267–75.
Olmsted I, Durán R. Aspectos ecológicos de la selva baja inundable en la reserva de la Biósfera Sian Ka'an, en Quintana Roo. Biótica. 1986;11:151–79.
Olmsted I, González-Iturbe JA, Granados-Castellanos J, Durán R, Tun-Dzul F. Vegetación de la Península de Yucatán. In: García de Fuentes A, Córdoba J, Chico P, editors. Atlas de Procesos Territoriales de Yucatán, Mérida; 1999. p. 183–94.
Pan Y, Birdsey RA, Fang J, et al. A large and persistent carbon sink in the world's forests. Science. 2011;333:988–93.
Pennington TD, Sarukhán J. Arboles tropicales de México. Manual para la identificación de las principales especies. Colección Ediciones Científicas Universitarias. 3ª ed. México: UNAM, FCE; 2005.
Pozo C et al. Áreas potenciales para la conservación. In: Pozo C, Armijo Canto N, Calmé S, editors. Riqueza Biológica de Quintana Roo. Un análisis para su conservación, Tomo I. ECOSUR, CONABIO, Gobierno del Estado de Quintana Roo y Programa de Pequeñas Donaciones (PPD), México DF; 2011. p. 310–4.

Prezas B. Áreas naturales protegidas en Quintana Roo. In: Pozo C, Armijo Canto N, Calmé S, editors. Riqueza Biológica de Quintana Roo. Un análisis para su conservación, Tomo I. ECOSUR, CONABIO, Gobierno del Estado de Quintana Roo y Programa de Pequeñas Donaciones (PPD), México DF; 2011. p. 300–9.

Proyecto Fortalecimiento REDD+ y Cooperación Sur-Sur. Áreas de Trabajo. 2015. http://www.mrv.mx/index.php/es/mrv-m/areas-de-trabajo/mexsmic/78-pmn.html. Accessed 31 Mar 2015.

Ramírez Carrillo LA. Las relaciones peligrosas: sociedad, naturaleza y construcción de la modernidad. In: Durán R, y Méndez M, editors. Biodiversidad y desarrollo humano en Yucatán. CICY, PPD-FMAM, CONABIO, SEDUMA; 2010. p. 29–34.

Ruiz Barranco H, Arellano J. Áreas Naturales Protegidas. In: Durán R, Méndez M, editors. Biodiversidad y desarrollo humano en Yucatán. CICY, PPD-FMAM, CONABIO, SEDUMA; 2010. p. 414–9.

Rzedowski J. La vegetación de México. México: LIMUSA; 1978.

Santos JV, Mas PK, López C, Snook LK. El manejo forestal y la caoba en los ejidos de la Zona Maya, México. Desarrollo histórico, condiciones actuales y perspectivas. Revista Recursos Naturales y Ambiente. 2005;44:27–36.

Secretaría de Desarrollo Urbano y Medio Ambiente. Sistema de Áreas Naturales Protegidas del Estado de Yucatán. 2015. http://www.seduma.Yucatán.gob.mx/areas-naturales/sanpy.php. Accessed 5 Mar 2015.

Secretaría de Ecología y Medio Ambiente. Áreas Naturales Protegidas de Quintana Roo. 2014. http://sema.qroo.gob.mx/descargas_web/DESCARGAS//01.2/01anp/anpquintanaroo.pdf. Accessed 6 Mar 2015.

Secretaría de Medio Ambiente y Aprovechamiento Sustentable del Gobierno de Campeche. Áreas Naturales Protegidas. 2014. http://www.smaas.campeche.gob.mx/anp/. Accessed 4 Mar 2015.

Secretaría de Medio Ambiente y Recursos Naturales. Incendios. 2015. www.conafor.gob.mx/web/temas-forestales/incendios. Accessed 26 Feb 2015.

SEMARNAT. Norma Oficial Mexicana, Protección ambiental-Especies nativas de México de flora y fauna silvestres-Categorías de riesgo y especificaciones para su inclusión, exclusión o cambio-Lista de especies en riesgo, DOF, diciembre 2010.

Small Grants Program. Annual Country Program Report Mexico 2007, Mérida. 2008.

Sosa V, Flores JS, Rico-Gray V, Lira R, Ortíz JJ. Etnoflora Yucatánense. Lista florística y sinonimia maya. Xalapa: INIREB; 1985.

Sousa M, Cabrera E. Listados florísticos de México II. Flora de Quintana Roo. México: Instituto de Biología, UNAM; 1983.

Standley PC. Flora of Yucatán. Publ. Field Columbian Museum. Bot Ser. 1930;3:157–492.

Standley PC. Las relaciones geográficas de la Flora Mexicana. Anales del Instituto de Biología, Universidad Nacional Autónoma de México, 7; 1936. p. 9–16.

Terán S, Rasmunssen C. La milpa de los Mayas: La agricultura de los mayas prehispánicos y actuales en el noreste de Yucatán. Mérida: DANIDA; 1994.

Terán S, Rasmunssen C, May Cauiche O. Las Plantas de la Milpa entre los Mayas. Mérida: Fundación Tum Ben Kin; 1998.

Tetetla Rangel E, Durán García R, Hernández Stefanoni JL, Dupuy JM. Distribución espacial de la riqueza de especies leñosas raras de la Península de Yucatán y su relación con las áreas naturales protegidas. Trop Conserv Sci. 2012;5:320–39.

Trejo-Torres JC, Durán R, Olmsted I. Manglares de la Península de Yucatán. In: Salazar-Vallejo G, editor. Biodiversidad Marina y Costera de México. México: CONABIO-CIQRO; 1993. p. 660–72.

Universidad Autónoma del Carmen. Jardín Botánico Universitario. 2014. http://www.unacar.mx/comunicacion_social_unacar/unacar/noticia.php?id=114. Accessed 31 Mar 2015.

Vargas R, Allen MF, Allen EB. Biomass and carbon accumulation in a fire chronosequence of a seasonally dry tropical forest. Glob Chang Biol. 2008;14:109–24.

Villalobos-Zapata GJ. Áreas Naturales Protegidas en Campeche. In: Villalobos-Zapata GJ, Mendoza Vega J (coords) La Biodiversidad en Campeche: Estudio de Estado. México: CONABIO,

Gobierno del Estado de Campeche, Universidad Autónoma de Campeche, ECOSUR; 2010. p. 580–7.
Villegas G, Bolaños A, Olguín L. La Ganadería en México. México, DF: Instituto de Geografía-PlazaValdés; 2001.
Zaldivar A, Herrera-Silveira J, Teutli C, Hernández R, Caamal J. Manglares. In: Durán R, Méndez M, editors. Biodiversidad y desarrollo humano en Yucatán. CICY, PPD-FMAM, CONABIO, SEDUMA; 2010. p. 138–9.

Part III
Fauna

Chapter 9
Diversity and Eco-geographical Distribution of Insects

Jorge L. León-Cortés, Ubaldo Caballero, and Marisol E. Almaraz-Almaraz

Abstract This chapter presents an integrative analysis on diversity levels, distributional patterns, and the implications of habitat preferences and geographic distributions of insect groups in the Yucatán Peninsula. Over 3000 insect species have been documented in the Yucatán Peninsula. Yet the number of species recorded may represent no more than 20–30 % of the total number of species. Taxonomic orders that recorded the highest number of species were: diurnal Lepidoptera (N = 510 species), Hymenoptera (602), and Coleoptera (478). Differences in insect species numbers for the three peninsular states were largely due to differences in sampling efforts. Maximum entropy models that were applied to a range of insect species showed that most suitable areas were fairly well distributed towards central and southern areas of the Yucatán Peninsula, with some spatial distributions matching "El Petén" or arid/dry Yucatán, while some others showed rather 'idiosyncratic' distributions. The habitat and geographic categories analysis of 221 butterfly species revealed a declining proportion of species that were restricted to primary habitats with increasing geographic range ($\chi^2 = 11.23$, df = 2, p = 0.004), and an increasing proportion of widespread butterfly species that make use of modified habitats ($\chi^2 = 40.7$, df = 2, $p < 0.001$). The present status of butterfly diversity (and perhaps many other insects) in the Yucatán Peninsula revealed important changes in species composition (i.e. the prevalence of a large fraction of weedy species). It is possible that dramatic habitat changes over much of the peninsula throughout the course of its history, together with the current accelerated impact of habitat modification might have precipitated changes in species composition and diversity.

Keywords Species rarefaction curves • Spatial distribution model • Intra-peninsular regionalization • Habitat use • Geographic distribution • Species composition • Insect conservation

---

J.L. León-Cortés (✉) • U. Caballero • M.E. Almaraz-Almaraz
Departamento de Conservación de la Biodiversidad, El Colegio de la Frontera Sur Unidad San Cristóbal de las Casas, Carretera Panamericana y Periférico Sur s/n, Barrio de María Auxiliadora, AP 63, CP 29000 San Cristóbal de las Casas, Chiapas, Mexico
e-mail: jleon@ecosur.mx; ucaballerou@gmail.com; egipsol_2105@hotmail.com

G.A. Islebe et al. (eds.), *Biodiversity and Conservation of the Yucatán Peninsula*,
DOI 10.1007/978-3-319-06529-8_9

9.1 Introduction

Insects include >50 % of all known species and are essential components of all terrestrial and freshwater aquatic ecosystems except those of the polar regions (Collins and Thomas 1991). Insects intervene in many biological processes: nutrient recycling, seed dispersal, and pollination, which contribute in an important way to maintaining the diversity and functioning of the majority of terrestrial ecosystems (Erwin 1982; Didham et al. 1996).

Yet this major group has been largely ignored in diversity and conservation assessments at the regional scale; instead most programs involve preventing whole-scale ecosystem destruction, combating pollution, or taking deliberate measures to conserve vertebrates or plants (León-Cortés et al. 2004; New 2008). Unfortunately, measures that are successful in conserving vertebrates or plants do not always guarantee that insect populations will be adequately maintained (New et al. 1995). Unlike most vertebrate or plant groups in the Neotropics, an important number of insect groups occur at disproportionate levels of species richness, or have specific habitat requirements (Lewis et al. 1998). These characteristics have made tropical insects a key group for investigations into the effects of small-scale disturbance (Spitzer et al. 1997), selective logging, forest fragmentation (Daily and Ehrlich 1995), and spatial population dynamics (León-Cortés et al. 2004; Almaraz-Almaraz et al. 2013).

To our knowledge, no previous report has attempted an integrative analysis on the levels of diversity and distributional patterns, nor to discuss the implications of habitat preferences and geographic distributions of major insect groups in the Yucatán Peninsula. This chapter aims to describe and analyze the patterns of insect species diversity, regional spatial distribution, and the relationship between species habitat preferences and geographic ranges of insect groups on the Yucatán Peninsula.

Based on the information from entomological collections in Mexico, electronic databases, entomological literature, and our own surveys from various locations in the region, we documented levels of diversity for major insect groups in the peninsula. We ask: What is the documented (and expected) level of insect diversity in the Yucatán Peninsula and the associated Mexican states? Second, based on distributional records for key insect groups in the Yucatán Peninsula (e.g., butterflies, sphinx moths, hymenopterans), we explored spatial distribution patterns and compared them with previously proposed regionalization schemes (e.g., Ramírez-Barahona et al. 2009): Do insect groups show comparable distributional patterns as related to other vertebrate or plant groups? And thirdly, we wanted to determine whether current species composition and distributional patterns in adequately-recorded taxa in the peninsula (e.g., butterflies) relate to aspects of habitat use and geographic distribution. Can current species composition tell us something about past and present habitat disturbance in the Yucatán Peninsula? Are geographically widespread species more common in the Yucatán Peninsula?

9.1.1 Entomo-Diversity Setting

Despite the interest in aspects of Mexican "entomo-diversity" in recent years, we still are far from producing a complete picture that document detailed levels of diversity and distribution of insects in the Yucatán Peninsula. The data that have been produced on the taxonomy and biogeography of major insect groups, constitute a significant advance on the current entomological knowledge (e.g. Pozo et al., 2003; Delfín-González and Chay-Hernández 2010), although there are considerable gaps for highly diverse insect groups, namely Coleoptera, Hymenoptera, and Lepidoptera.

Knowledge of insects in the Yucatán Peninsula is patchy and varies greatly among taxonomic groups. In some insect groups it is likely that more than 80 % of the species have already been registered and nominated (e.g., butterflies), while in others perhaps less than 5–10 % of the species have been recorded (e.g., weevils). Moreover, even when a great number of local checklists for some orders and families of insects have been reported (Maya-Martínez et al. 2005; Pozo et al. 2003; Delfín-González and Chay-Hernández 2010), few compilations of regional or state-level checklists for taxonomic orders are available (e.g. Delfín-González et al., 2010; Ruíz-Cancino et al., 2014).

First, to provide an update on the number of insect species so far cataloged in the Yucatán Peninsula, and for each peninsular state, we gathered insect records from public electronic databases (i.e. National Commission for Biodiversity, CONABIO, Pozo et al., 2005; The Global Biodiversity Information Facility, GBIF), records from scientific public collections and their curators (i.e. Museum of Zoology, Faculty of Science, UNAM; National Insect Collection of the Institute of Biology, UNAM; University of Kansas Entomology Collection; The McGuire Centers for Lepidoptera Research and Insect Conservation at the University of Florida; the Canadian National Collection of Insects, Arachnids and Nematodes of Agriculture and Agri-Food Canada; the Natural History Museum (London), Entomological Collection of El Colegio de la Frontera Sur, ECOSC-E), relevant entomological literature from the Yucatán Peninsula (e.g. Maya-Martínez et al., 2005; Pozo et al., 2003; Delfín-González and Chay-Hernández 2010; Morón et al., 2014; Ruíz-Cancino et al., 2014), as well as entomological records arising from field work (e.g. León-Cortés et al., 2003).

Second, we used data from a well-studied insect group (butterflies) to generate rarefaction curves for incidence values of species check-lists and to examine levels of diversity completeness (see Gotelli and Colwell 2001; Colwell et al., 2012), particularly for: (i) butterfly species richness across the entire Yucatán Peninsula; (ii) butterfly species richness at each peninsular state: Campeche, Quintana Roo and Yucatán (for comparison rarefaction curves were extrapolated to a maximum number of incidences—the case for Campeche and Quintana Roo-or to double the number of incidences—the case for Yucatán state); and (iii) species richness for six representative families within the butterfly super-families Papilionoidea and Hesperioidea. For this last set of calculations (iii) rarefaction curves were extrapolated to double the number of incidences—except for the case of Nymphalidae. All rarefaction curves were generated using EstimateS (Colwell 2013).

The completeness of butterfly inventory richness (observed vs. predicted values) was also examined using the incidence-based coverage estimator (ICE). ICE is a non-parametric estimator which reduces potential biases produced by under-sampling for estimated values of species richness (Magurran 2004).

In total, 21 taxonomic insect orders have been documented over the entire Yucatán Peninsula (Table 9.1). Taxonomic orders that recorded the highest number of species were: diurnal Lepidoptera (510 species included in 53,184 records), Hymenoptera (602 species and 25,371 records), and Coleoptera (478 species and 3741 records). Taxonomic orders such as Isoptera, Embioptera and Protura (Entognatha) have been poorly recorded.

Hymenoptera, Coleoptera, and diurnal Lepidoptera are the most diverse insect taxonomic orders for Yucatán State (N = 602, N = 478, and N = 425 species, respectively). In contrast, diurnal Lepidoptera recorded the highest species diversity in Quintana Roo (N = 450) and Campeche (N = 458; Table 9.1).

Insect diversity values broken down for major taxonomic orders (i.e. Lepidoptera, Coleoptera, Hymenoptera, and Diptera) at each peninsular state are shown in Tables 9.2, 9.3, 9.4, 9.5. For butterflies, Hesperiidae (N = 145 in Campeche state) and Nymphalidae correspond to the best documented groups for the three peninsular states (i.e. 155, 131, and 74 species, for Campeche, Quintana Roo, and Yucatán states, respectively, Table 9.2). For beetles, Quintana Roo and Yucatán states recorded the highest number of species (i.e. 78 and 74 species in the Cerambycidae family, respectively), whereas Campeche registered an important number of Scolytidae species (N = 68; Table 9.3). Members of Culicidae family within Diptera order recorded the highest diversity levels in Quintana Roo (N = 69 species), followed by Campeche (N = 62), and Yucatán, (N = 53 species; Table 9.4). For Hymenoptera, a disproportionate number of parasitoid species has been recorded in Yucatán state, particularly within families Ichneumonidae and Braconidae (N = 230 and N = 110 species, respectively), as well as for Apoidea family (N = 140 species; Table 9.5).

Butterfly species rarefaction curves are shown in Fig. 9.1. For the entire Peninsula, the accumulation curve for butterfly species failed to reach an asymptote (Fig. 9.1a). Visual examination of the curve shapes suggests that the differences in numbers of species that were accumulated among the three peninsular states were largely due to differences in sampling effort. Asymptotic accumulation patterns for Campeche and Quintana Roo suggest that few additional species would be recorded by further sampling and, therefore, their regional species checklists are nearly complete. Additional effort will be required to increase the number of butterfly species in Yucatán State (Fig. 9.1b).

Species rarefaction curves for selected butterfly families in the Yucatán Peninsula and for each State are shown in Fig. 9.2. The accumulation pattern for Nymphalidae and Hesperiidae was fairly asymptotic across the entire peninsula and for each state: an important proportion of Nymphalidae and Hesperiidae members (two highly diverse butterfly groups), has been recorded. Further sampling effort could produce a few additional species, mainly in Yucatán state.

Table 9.1 Recorded number of insect species per taxonomic order in the Yucatán Peninsula

Order	Number of species			Source[a]
	Campeche	Quintana Roo	Yucatán	
Coleoptera	301	298	478	6, 12, 25
Collembola*	15	70	29	22
Dermaptera	0	0	1	6
Diplura*	0	0	2	6, 24
Diptera	125	309	482	6, 7, 11, 25
Embioptera	0	1	0	6
Hemiptera	19	20	17	25
Hymenoptera	83	257	602	1, 2, 5, 6, 8, 9, 15, 16, 19, 25
Isoptera	5	1	4	6
Lepidoptera (d)	458	450	425	10, 13, 25
Lepidoptera (n)	149	247	152	4, 14, 25
Mecoptera	0	1	1	18
Neuroptera	2	8	17	6
Odonata	68	76	62	21
Orthoptera	14	18	27	6
Plecoptera	0	0	0	6
Protura*	0	1	0	23
Psocoptera	96	35	43	20
Siphonaptera	3	1	2	17
Thysanoptera	5	2	2	6
Thysanura	0	1	1	6
Trichoptera	1	0	0	3
Total	1344	1796	2347	

d = diurnal, n = nocturnal

[a]1 = Ruíz-Cancino et al. (2002), 2 = Reyes et al. (2009), 3 = Chamorro and Holzenthal (2010), 4 = Hernández-Baz (2010), 5 = Delfín-González and Chay-Hernández (2010), 6 = Delfín-González et al. (2010), 7 = Manrique and González-Moreno (2010), 8 = Meléndez et al. (2010), 9 = Meneses et al. (2010), 10 = Rodríguez et al. (2010), 11 = Ibáñez-Bernal (2011), 12 = Morón (2011), 13 = Pozo (2011), 14 = Prado et al. (2011a, b), 15 = Vásquez-Bolaños (2011), 16 = González-Moreno and Bordera (2012), 17 = Acosta-Gutiérrez (2014), 18 = Contreras-Ramos et al. (2014), 19 = Coronado-Blanco and Zaldívar-Riverón (2014), 20 = García-Aldrete (2014), 21 = González-Soriano and Novelo-Gutiérrez (2014), 22 = Palacios-Vargas (2014), 23 = Palacios-Vargas and Figueroa (2014), 24 = Palacios Vargas and García-Gómez (2014), 25 = The Global Biodiversity Information Facility, GBIF (2015), 26 = Hernández-Baz et al. (2013), 27 = Romero (2013), 28 = Bousquet (1996), 29 = Bousquet (2012), 30 = Shpeley and Ball (2000), 31 = Noguera (2014), 32 = Ordóñez-Reséndiz et al. (2014), 34 = Zaragoza-Caballero and Pérez-Hernández (2014), 35 = Reyes-Castillo (2002), 36 = Reyes-Novelo and Morón (2005), 37 = Equihua-Martínez and Burgos (2002), 38 = Navarrete-Heredia and Newton (2014), 39 = Cifuentes-Ruíz and Zaragoza-Caballero (2014), 40 = Ortega et al. (2010), 41 = González-Moreno et al. (2011).

*These taxa are hexapods like insects.

The species rarefaction curves for Pieridae and Papilionidae for the Yucatán Peninsula and for each state reached fairly complete asymptotes (except by Yucatán state, where additional sampling effort might produce new species records; Fig. 9.2).

Table 9.2 Number of species of Lepidoptera in the Yucatán Peninsula

Family	Number of species		
	Campeche	Quintana Roo	Yucatán
Arctiidae	7	108	58
Crambidae	1	12	1
Hesperiidae	145	98	34
Lycaenidae	60	45	8
Noctuidae	3	9	1
Notodontidae	0	1	2
Nymphalidae	155	131	74
Oecophoridae	1	0	0
Papilionidae	24	24	12
Pieridae	27	26	26
Riodinidae	47	33	8
Saturniidae	6	11	0
Sphingidae	84	63	78
Total	560	561	302

Table 9.3 Number of species of Coleoptera in the Yucatán Peninsula

Family	Number of species		
	Campeche	Quintana Roo	Yucatán
Bruchidae	59	44	51
Buprestidae	2	1	6
Carabidae	14	8	27
Cerambycidae	12	78	74
Chrysomelidae	39	20	73
Ciidae	1	1	1
Curculionidae	3	2	2
Elateroidea	14	18	42
Lycidae	2	3	0
Leiodidae	2	1	2
Melolonthidae	17	41	64
Passalidae	8	8	6
Platypodidae	7	2	1
Scarabaeidae	19	40	32
Scolytidae	68	11	15
Staphylinidae	32	10	37
Tenebrionidae	2	8	43
Trogidae	0	2	2
Total	301	298	478

However, further sampling effort will be required to increase the number of species for Lycaenidae and Riodinidae groups across the entire Yucatán Peninsula.

Finally, the estimated total number of butterfly species indicated a reasonable level of completeness (ICE = 84, 82, and 70 %, for Campeche, Quintana Roo, and Yucatán States, respectively), with relatively low levels of species uniqueness

Table 9.4 Number of species of Diptera in the Yucatán Peninsula

Family	Number of species		
	Campeche	Quintana Roo	Yucatán
Anastrepha	9	8	14
Asilidae	0	0	24
Bombyliidae	0	0	28
Calliphoridae	1	1	7
Ceratopoginidae	7	8	57
Chironomidae	7	0	1
Choloropidae	0	2	3
Culicidae	62	69	53
Dolichopodidae	0	0	22
Muscidae	0	0	5
Mycetophilidae	0	0	17
Psychodidae	16	23	30
Sarcophagidae	3	1	18
Stratiomyiidae	0	0	25
Syrphidae	0	0	32
Tabanidae	19	25	28
Tachinidae	0	0	33
Tephritidae	0	0	18
Other families[a]	1	172	67
Total	125	309	482

[a]Number species contained in 23 families

Table 9.5 Number of species of Hymenoptera in the Yucatán Peninsula

Family	Number of species		
	Campeche	Quintana Roo	Yucatán
Apoidea	32	90	140
Aphelinidae	1	4	6
Braconidae	10	62	110
Chalcididae	0	0	20
Encyrtidae	0	4	8
Formicidae	35	88	88
Ichneumonidae	5	9	230
Total	83	257	602

among samples (on average, 33 %, 12 %, and 38 %, for Campeche, Quintana Roo, and Yucatán States, respectively, Table 9.6).

9.1.2 Insect Distribution Patterns

Intra-peninsular regionalization, as proposed by Lundell (1934; based on climate, physiography and plants), Goldman and Moore (1946; based on birds, mammals

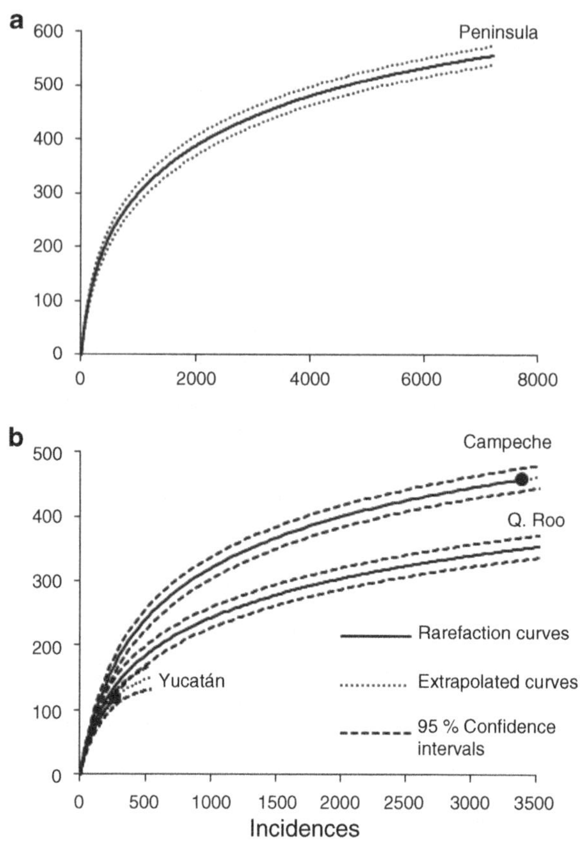

Fig. 9.1 Species incidence rarefaction curves with 95 % confidence intervals (*dashed lines*) for the butterfly fauna of (**a**) the entire Yucatán Peninsula, and (**b**) each peninsular state (for Yucatán state only partial incidence data are shown, see Table 9.6)

and plants) and Lee (1980; based on snakes), recognized a different number of regional sub-divisions (see Ibarra-Manríquez et al. 2002). The simplest bio-geographical regionalization of the Yucatán Peninsula distinguished the NW portion from the remaining area, or what could be referred to as arid/dry Yucatán and El Petén (Ramírez-Barahona et al. 2009). Geographic and environmental factors (e.g. vegetation and soil type) have been used as criteria to distinguish "limits" between these two areas, but clearly some other factors, such as the ecological features of species, call for a more cautious interpretation of regionalization schemes in the peninsula. Here we examine representative insect species distributions in the Yucatán Peninsula, aiming to distinguish broad distributional patterns from data-sets of key insect groups (i.e. butterflies, sphinx-moths, and hymenopterans). The assemblage included a range of species mobilities, spatial dynamics and habitat association.

Based on known distribution data sets combined with spatial distributional modeling, we produce species distribution maps from incomplete presence-only data-sets, using maximum entropy modeling ('MaxEnt'; Phillips et al., 2004, 2006). MaxEnt uses a machine-learning algorithm to estimate the target probability

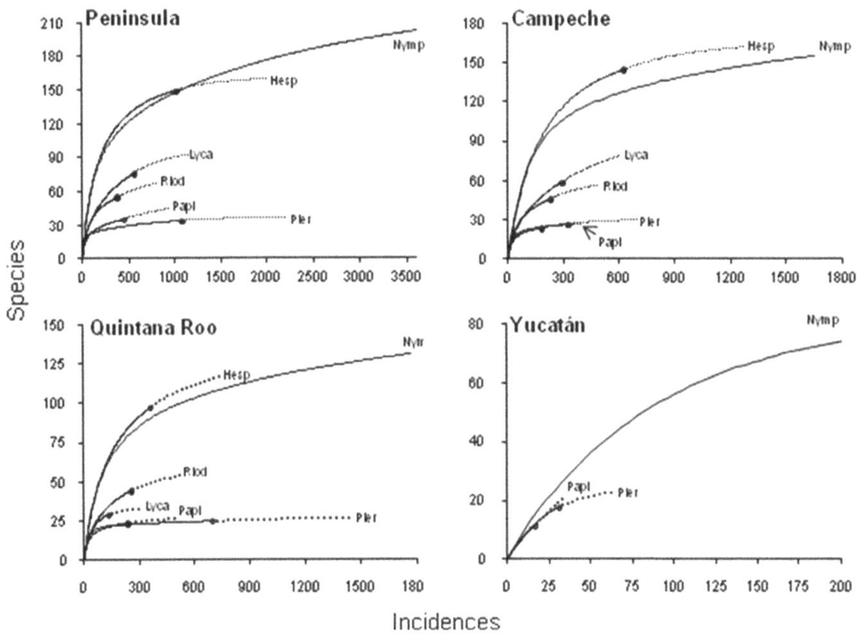

Fig. 9.2 Species incidence rarefaction curves with 95 % confidence intervals (*dashed lines*) for Nymphalidae (Nymp), Hesperiidae (Hesp), Lycaenidae (Lyca), Riodinidae (Riod), Papilionidae (Papi), and Pieridae (Pier) butterflies, for the entire Yucatán Peninsula (Peninsula), and for the peninsular states Campeche and Quintana Roo (for Yucatán state only partial incidence data are shown, see Table 9.6)

distribution of maximum entropy (Phillips et al., 2006). Species distribution raster maps represent distributional areas where each pixel represents a maximum entropy value in the range 0–100. As these values approach zero, the resulting probability of finding a species decreases.

To develop a distribution model that is calibrated in one area and re-projected into a larger area (that is not overly fit to our data), we followed Elith et al. (2010) to establish parameter settings, i.e. using only the linear, quadratic, hinge features, plus regularization multiplier = 2.0, (our best estimate; Elith et al., 2010), which resulted in response curves and maps balanced over and under fitting of the models. In particular, the projections of our insect distributions mapped fairly far into areas that were not where our training data were, and hence our mapping predictions fell in the same basic areas with not a whole lot of difference in generalization.

Species lists for two insect groups (i.e. butterflies, and sphinx-moths) that have been relatively well sampled throughout the Yucatán Peninsula were compiled from the previously mentioned entomological sources (see "Entomo-diversity setting"). We selected these groups because they have a well-known taxonomy and have been widely sampled in the Yucatán Peninsula over the past 20–30 years. In addition, we included some representatives of Hymenoptera (a highly diverse insect group) to compare with the general patterns observed for the two Lepidoptera groups.

Table 9.6 The observed (Species) and estimated total number of butterfly species (ICE), and relative uniqueness values among samples (see text), for Campeche, Quintana Roo, Yucatan, and for the entire (Overall) Yucatan Peninsula

						Regional	Species richness[a]						
State	Individuals	Species[a]	Incidences	Locations	ICE[b] (%)	uniques[c] (%)	Hesperiidae	Lycaenidae	Nymphalidae	Papilionidae	Pieridae	Riodinidae	Uranidae
Campeche	35,259	467	3411	94	548 (84)	151 (33)	145	60	155	24	27	56	0
Quintana Roo	18,806	357	3523	170	432 (82)	44 (12)	98	45	131	26	26	30	1
Yucatán[d]	361	347	268	44	–	45 (38)	82	62	110	20	28	45	0
Overall	54,426	582	7202	308	655 (85)	240 (43)	149	87	203	36	34	72	1

[a]The number of species recorded by Llorente-Bousquets et al. (2006, 2014), Martínez-Noble et al. (2015) and this work
[b]Estimated species richness (% recorded)
[c]Uniques species by state (% recorded)
[d]A small sample size for incidence data prevented us to obtain ICE values

The Hymenoptera is one of the four great orders of insects, with over 300,000 species as a conservative world total. Therefore it is difficult to make any general observations about so many species without citing many exceptions, but our purpose here was to contrast observed distributional patterns in the Yucatán Peninsula. All taxonomic insect records were validated by us. Geo-referenced records were verified using Google earth ® 2015-*digital globe*. In total, over 79,000 distribution records of 111 butterfly species, 68 species of sphinx-moths, and 20 species of hymenopterans were compiled. For simplicity's sake and generality, only representative spatial distributions for each insect group were shown.

Butterflies are well-known insects, and detailed information about their ecology and life histories has led to substantial progress being made regarding their distributions and conservation status, particularly in Nearctic and Palearctic environments. Previous observations indicate that important population processes, e.g. oviposition rates, time of adult feeding, and mate location, appear to be restricted by environmental variations (Boggs and Murphy 1997). Also, members of this group have been touted as potential surrogates for habitat conservation due to their responsiveness to environmental change (Molina-Martínez et al. 2013).

Sphinx moths (sphingids or hawkmoths) are characterized as medium to large-sized moths, with strong, narrow wings and streamlined, conical bodies (Haber 1983). Most species are nocturnal and come readily at artificial light sources, but some are diurnal or crepuscular. A few species have very broad distributions (such as *Agrius cingulata*, which is found throughout the New World, from Canada to Argentina), but most have much more restricted ranges (León-Cortés 2000; León-Cortés et al. 2006). The distribution of sphinx moths is closely related to that of their host plants, and in a given locality, larvae feed on a single genus or several closely related genera within a single family (León-Cortés 2000; Dan Janzen pers. comm.) However, across its range, a sphinx moth species may be recorded on numerous plants and hence the degree to which the local specialization observed by some authors (e.g. Janzen 1984, 1986) that is generally true remains to be studied.

MaxEnt models that were applied to a range of butterfly species (species using unmodified or modified habitats, and exhibiting different geographic ranges in the Neotropics, see Table 9.8 in Appendix), are shown in Fig. 9.3a–c. Most suitable areas for relatively specialized butterflies (as in the case of *Fountainea eurypyle* and *Prepona pylene*; Fig. 9.3) were restricted to the southeast portion of the Yucatán Peninsula. However, species associated to unmodified but that make use of modified habitats—the case for *Pyrrhogyra otalis*, *Historis odius*, *Agraulis vanillae* and *Anthanassa frisia*-exhibited fairly suitable areas distributed towards the central and southern areas of the Yucatán Peninsula.

Figure 9.3d–f shows selected sphinx-moth distribution maps for the Yucatán Peninsula. In particular, Central-American species (e.g., *Manduca wellingi*) showed suitable areas distributed towards eastern parts of the Yucatán Peninsula. South-American species reaching into tropical Mexico (e.g., *Madoryx oiclus*) showed suitable areas through most of the peninsula. Some widespread and common sphinx moths (e.g., *Eumorpha satellitia*, *Protambulyx strigilis*), have potential suitable areas distributed towards the north and eastern coastal plain in the Yucatán

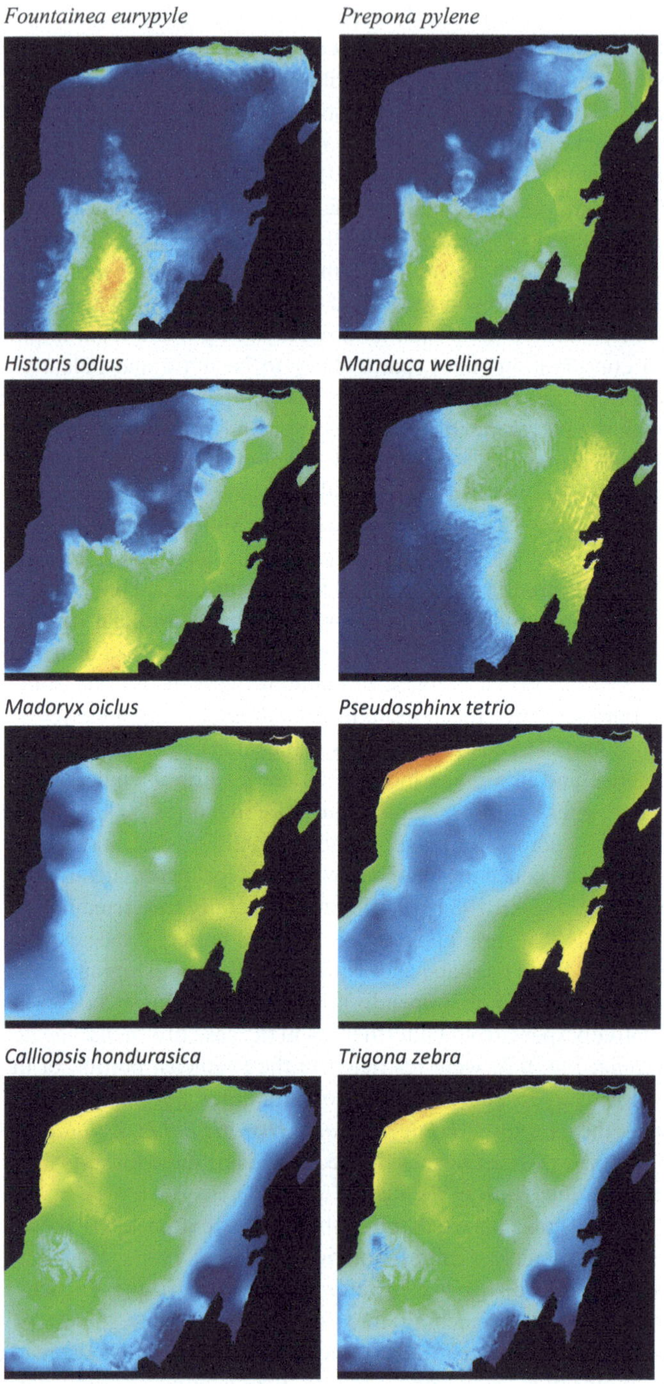

Fig. 9.3 Potential distribution maps for butterfly, sphinx moth, and hymenopteran species in the Yucatán Peninsula (see text). The image uses *colors* to indicate predicted probability that conditions are suitable, with *red* indicating high probability of suitable conditions for the species, *green*

Peninsula. Interestingly, other widespread sphinx-moths (e.g., *Pseudosphinx tetrio*) exhibited potential distributions somewhat restricted to the coastal plains in Yucatán Peninsula. Although the host plants of widespread species such as *Cocytius duponchel*, actually occur in the Yucatán Peninsula (Dan Janzen, personal communication), these sphinx moths have been poorly recorded there, possibly because the vegetation had suffered dramatic destruction during the Maya period (Schreiber 1978), but also possibly due to collecting bias (common and widespread species are often under-recorded).

In contrast to butterflies and sphinx-moths, the majority of distribution maps for hymenopterans suggested that suitable areas are widely distributed across most of the Yucatán Peninsula region (Fig. 9.3g, h). Some hymenopterans also showed highly suitable areas towards northern parts of the peninsula—Hymenoptera groups have been intensively collected towards Yucatán State (e.g., see the case for *Calliopsis hondurasica* and *Trigona zebra*), but otherwise these maps merely confirm that a highly diverse and widely distributed insect group in the Neotropics—such as the Hymenoptera-would be expected to be fairly well distributed in the Yucatán Peninsula region.

Based on present analyses, the matching of insect distributions to a given intra-peninsular regionalization proposed by others (e.g. Lundell 1934; Goldman and Moore 1946; Lee, 1980) it is not a straightforward matter. Although spatial distributions for some insect representatives roughly match to "El Petén" or arid/dry Yucatán (see specialized butterfly species distributions matching "El Petén" region: *Fountainea eurypyle* and *Prepona pylene* (Fig. 9.3). Some other distributions could simply be regarded as 'idiosyncratic' (see sphinx-moth distribution maps, Fig. 9.3d–f). Whatever the precise biological explanation, widespread sphinx-moth species in the Yucatán Peninsula generally have wide distributions at larger (national) scales (Spearman Rank Correlation $= 0.71$, $p < 0.001$, $N = 107$).

In general, climate can constrain the ranges of species at large scales (i.e. from global to regional), while habitat-related variables seem to operate at landscape and local scales (Pearson and Dawson 2003; Thuiller et al. 2004). The maximum entropy analysis helped us in recovering potential distributional patterns in spite of distributional data absence of many of the taxa studied. Some of these distribution maps clearly demonstrate that, in addition to bio-geographical history, a variety of ecological factors at different temporal and spatial scales might operate and shape an insect's regional distribution.

We must recognize that modeling the distribution of an insect could be challenging because many bionomic processes can affect species distributions, but their effects are often complex and do not have significant effects at scales larger than local sites (Hortal et al. 2010). In fact, suitable distribution areas can be delimited

Fig. 9.3 (continued) indicating conditions typical of those where the species is found and *lighter shades* of *blue* indicating low predicted probability of suitable conditions (see Phillips et al. 2006)

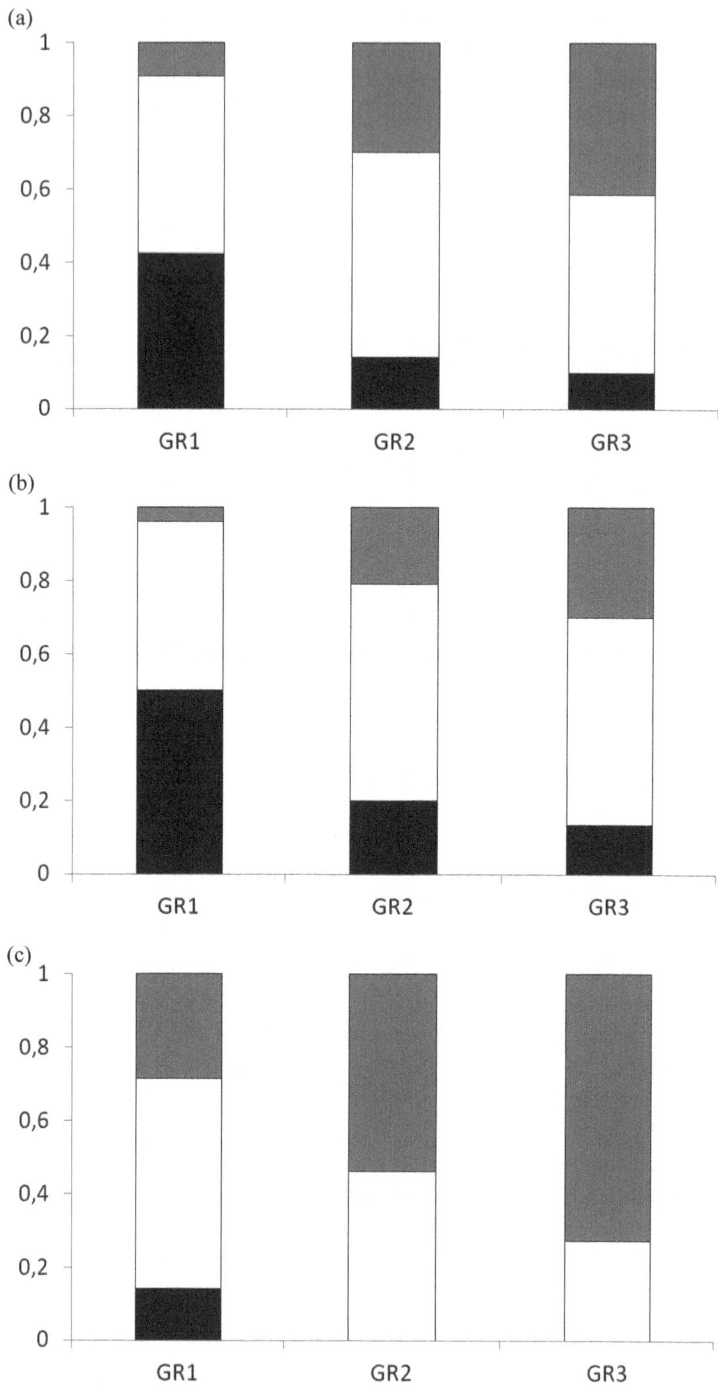

Fig. 9.4 The proportion of butterfly species in the Yucatán Peninsula using unmodified habitats (*solid*), unmodified but making some use of modified habitats (*white*), and modified habitats (*grey*), with increasing geographic ranges (G1, G2, G3, see text). (**a**) All butterfly species

from bio-geographical considerations (Soberón 2010), but also from considering a more analytical framework on the selective use of 'absence data' to describe the absence of species due to bio-geographical effects, but also from either lack of suitable environment or simply lack of recording effort (Lobo et al. 2010). In both approaches, accounting for species absences from suitable areas is crucial to obtain a realistic description of the distribution of species (Hortal et al. 2010).

9.1.3 Habitat Use and Geographic Distribution

Historical human and natural destructive forces have been influential in greatly modifying the forest cover along with the associated fauna in the Yucatán Peninsula. Since the area of primary forest and other semi-natural habitats have suffered dramatic changes in much of the peninsula over the course of its history, it is likely that this has also precipitated changes in animal species composition and diversity (León-Cortés et al. 2003).

We analyzed the habitat and geographic distribution data for Nymphalidae, Papilionidae, and Pieridae butterflies, from field observations and from the information given in the species accounts of De Vries (1987), Tyler et al., (1994), Opler and Wright (1999), Scott (1997), León-Cortés et al., (2003), Llorente-Bousquets et al., (1997, 2006), Luis-Martínez et al., (2003, 2010), and Vargas-Fernández et al., (2008). The present analysis deals with 221 butterfly species in the Nymphalidae (73 %), Papilionidae (14 %), and Pieridae (13 %). The butterfly groups included a range of species dispersal ability, spatial dynamics and habitat specialization.

A habitat use designation was assigned to each species. Butterfly species were classified as occupying unmodified vegetation, when no mention was made that butterflies occupy secondary or agricultural habitats (1); as occupying unmodified vegetation but making use of modified habitats, but in most cases the latter are used rarely by these species, and usually only when modified and unmodified habitats are adjacent (2); and as occupying modified vegetation, when mention was made that they occupy secondary, agricultural, or other human-modified habitats (3) (Thomas 1991; León-Cortés et al. 2003).

Although the habitat and geographic range categories were somewhat arbitrary, we were interested in the general pattern, rather than the habitat-use patterns of particular species. We also wanted to examine a possible overall association between habitat use and geographic range within each family. We calculated the proportion of species in each geographic range category that was restricted to unmodified or modified habitats. Thus, the whole sample had up to three values. Values were

Fig. 9.4 (continued) (Nymphalidae, Papilionidae, and Pieridae); (**b**) Nymphalidae; (**c**) collated data for Papilionidae and Pieridae families

examined according to a declining or increasing proportion of species that were restricted to primary or modified habitats with increasing geographic range.

The geographic range categories mentioned were defined as follows:

1. Butterfly species restricted to southeast Mexico and Central America, but not further south of Panama.
2. Species distributed from North America to Central America, and/or southern Mexico to northern South America.
3. Species showing widespread distributions in the Neotropics (reaching Brazil or Bolivia).

The habitat and geographic categories assigned to each of the 221 species included in the analysis are shown in Appendix (Table 9.8). In total, 39 species (17 %) were placed in the unmodified habitats category '1', 118 species (53 %) in habitat category '2' and 64 species (28 %) in the modified habitat category '3', indicating a significant larger proportion of butterfly species associated to modified habitat categories ($\chi^2 = 44.26$, df = 2, $p < 0.001$).

We found a significant association between habitat use and geographic range for 221 butterfly species of three taxonomic families over the entire Yucatán Peninsula (Table 9.7, rows 1 and 3, columns 1 and 3; $\chi^2 = 15.08$, df = 1, $p < 0.005$). Species with narrow geographic ranges were usually grouped in unmodified habitats, whereas widespread species were much more likely to be classified as making at least some use of human-modified habitats.

In addition, we quantified a significant declining proportion of species restricted to primary habitats with increasing geographic range (collating data for the three butterfly families: $\chi^2 = 11.23$, df = 2, $p = 0.004$), as well as a significant increasing proportion of widespread butterfly species that make some use of modified habitats with increasing geographic range ($\chi^2 = 40.7$, df = 2, $p < 0.001$; Fig. 9.4). Trends analyzing habitat use and geographic range for Nymphalidae, and the collated data for Papilionidae and Pieridae, remained comparable.

Our analyses on a large butterfly family (Nymphalidae) resulted in a mixture of relatively widespread and common species that can exploit disturbed forest habitat, and a species group that is adapted to undisturbed primary forests (Fig. 9.4b). Geographically restricted or endemic butterfly species are often regarded as biotope specialists, although they can be highly abundant species at sites where they occur.

Table 9.7 Association between habitat use and geographic ranges for Nymphalidae, Papilionidae and Pieridae butterflies in the Yucatán Peninsula.[a] For habitat use and geographic range descriptions see text

	Geographic range		
Habitat use	1	2	3
Unmodified habitat	14	21	4
Unmodified/modified habitat	16	82	20
Modified habitat	3	44	17

$\chi^2 = 15.08$, df = 1, $p < 0.005$

[a]Habitat use and geographic range are significantly related (columns 1 vs. 3)

Some of these relatively "specialized" butterflies eventually occupy disturbed habitats: *Eueides isabella eva, Hamadryas februa ferentina, H. guatemalena guatemalena, H. amphinome mexicana, Memphis moruus, Pyrrhogyra neaerea hypsenor, Taygetis virgilia, Temenis laothoe hondurensis*. Most of these 'specialists' occur on naturally successional habitats and so are able to occupy human-modified areas that provide the same successional conditions. If successional habitat is widespread, as would seem to be case in many Yucatán Peninsula landscapes, then species would be distributed across a large range; conversely, if the habitat is rare, then the species would be narrowly distributed (León-Cortés et al. 2003).

Nearly 28 % of the species reported here are considered to be as butterflies associated with disturbed habitats (mostly within families Papilionidae and Pieridae, see Fig. 9.4a, c). Some of these butterflies are migratory habitat generalists with non-specialized host plant requirements and wide distributions throughout the Neotropics, and could rapidly invade disturbed rainforest patches in the Yucatán Peninsula, e.g. *Heraclides thoas, Ascia monuste, Phoebis sennae, P. argante*. An increasing proportion of secondary vegetation throughout the Yucatán Peninsula has resulted in a decreasing proportion of endemic plant species (which have been removed by the modification of primary forest), and widespread species will be left inhabiting secondary vegetation.

Although our conclusions are preliminary, the present status of butterfly diversity in the Yucatán Peninsula revealed important changes in species composition (i.e. the prevalence of a large fraction of weedy species throughout). It is possible that dramatic habitat changes over much of the peninsula throughout the course of its history, along with the current accelerated impact of habitat modification might have precipitated changes in animal species composition and diversity. In line with this result, it would be worth about the present status of other insect groups in the peninsula and the factors which are affecting changes in community structures.

9.1.4 Conservation Perspectives

An important number of recent surveys have improved the level of recording of insects throughout the Yucatán Peninsula (e.g. Pozo et al., 2003; Maya-Martínez et al., 2005; Delfín-González and Chay-Hernández 2010; Prado et al., 2011a). Over 3000 insect species have already been documented in the Yucatán Peninsula, which comprises a substantial percentage of the known insect fauna of Mexico and the world. However, the number of species so far recorded might represent no more than 20–30 % of the likely total number of species in the Yucatán Peninsula.

Certain taxonomic families of Lepidoptera (i.e. Papilionidae, Pieridae, Nymphalidae, Hesperiidae, Sphingidae, Saturniidae), Coleoptera (Melolonthidae, Scarabaeidae, Passalidae, and Scolytidae), Hymenoptera (i.e. Apoidea), and Diptera (i.e. Culicidae and Tabanidae), have been relatively well collected in Yucatán Peninsula.

In contrast, other Lepidoptera groups (Tortricidae, Noctuidae, Geometridae and Pyralidae), Coleoptera (Curculionidae—possibly the most diverse insect family-Staphylinidae, Chrysomalidae), Hymenoptera (Ichneumonidae) have been little studied because they are extremely diverse or because their taxonomic status of each of these groups has not been adequately resolved (but see Prado et al., 2011b). A more comprehensive picture of regional insect diversity could be achieved with further records from these groups.

There are regions of the Yucatán Peninsula that deserve special attention, not only because they might support rare or area-restricted insects, but because the regions themselves are poorly known, e.g., southern Campeche state, central and northern parts of the peninsula have not achieved a similar level of insect sampling. Nevertheless we must admit that for many areas and for an important number of insect groups in the Yucatán Peninsula, the need for distributional information far outstrips the resources available to entomologists. Recording of insect groups in the Yucatán Peninsula has privileged national biosphere reserves and parks, but knowledge that an insect is present in a given area, provides little information on the likely number of individuals and/or populations that occur there or in adjacent areas. Such distribution records might be difficult to interpret because presence may mean one sighting in many recording hours, or many sightings from many recorders over many visits. Similarly, absence of a species may represent true absence or a lack of adequate assessment effort in that location. It is a matter of some urgency that entomologists and conservation managers in the Yucatán Peninsula should be able to generate atlases and distribution maps from insect groups to measure rates of decline, identify areas of endemism and confirm "hotspots" of diversity (Groombridge 1992; Thomas and Abery 1995; León-Cortés et al., 1999; Gaston et al., 2000).

In contrast, threatened insects in the Yucatán Peninsula are not currently included on the Mexican official list of species of concern (except by *Danaus plexippus plexippus*; NOM-059-SEMARNAT-2010). However, species like *Megasoma elephas* (Coleoptera), *Pseudostigma accedens* (Odonata), deserve particular attention as their populations are threatened by habitat fragmentation and disturbance, and by illegal animal trafficking. Endemic butterfly species and subspecies in the Yucatán Peninsula, namely *Battus philenor acauda; B. loadamas copanae; Troilides torquatus mazai; Phyciodes milita mexicanus; Morpho achilles montezuma; Taygetis weymeri; Epargyreus deleoni, Codatractus Yucatánus, Hamadryas julitta* (Luis-Martínez et al. 2000; Warren 2000; Prado et al. 2011b), do merit individual attention, given that some of these species are of unusual evolutionary or ecological importance (Vane-Wright et al., 1991).

Ecological aspects of insects in the Yucatán Peninsula are far less well known. Although we could be able to make some generalizations from local studies or surveys outside of this particular region (León-Cortés et al., 2004; Almaraz-Almaraz et al., 2013; Molina-Martínez et al., 2013), there is lack of knowledge in many aspects on major groups of insects e.g. host plants (see however Essens and Hernández-Stefanoni 2013; Essens et al., 2014), behavior, population structure, population control, distributions and ranges, and population changes in space and time.

Overall, insects have been recognized as efficient long-distance plant pollinators (Janzen 1986). By recognizing insects as targets for conservation, co-existing and co-dependent organisms may also be safeguarded (host plants, natural enemies), while preserving inherent ecological services (e.g., maintenance of food webs). Many South and Central American insect species reach their distribution limits in the Yucatán Peninsula, where extensive habitat transformation has taken place. About 80–90 % of the forests of southern Mexico have been logged into farmland, cattle pasture land and secondary vegetation, and logging intensity has accelerated in the last 20 years. It would be extremely unlikely that marginal or isolated populations—despite their potential mobility—can replace themselves over the long term.

Acknowledgments We were supported by CONACyT (CB-2008-01: 102875 to JLL-C), and by El Colegio de la Frontera Sur. The authors wish to thank Manuel Ramos-Martínez and Manuel Girón-Intzin for helping organize insect 'massive' data-bases. We thank Alexandra D. Syphard for advice on spatial distribution map analyses, and to Gerald Islebe and WFJ Parsons for comments and suggestions on the manuscript.

Appendix

Table 9.8 Habitat use (Habitat) and geographic range (Range) for butterfly species and sub-species in the families Nymphalidae, Papilionidae and Pieridae from the Yucatan Peninsula

Family	Genus	Species	Subspecies	Habitat	Range
Nymphalidae	*Actinote*	*guatemalena*	*guerrerensis*	2	1
Nymphalidae	*Adelpha*	*serpa*	*celerio*	2	2
Nymphalidae	*Adelpha*	*naxia*	*epiphicla*	2	3
Nymphalidae	*Adelpha*	*iphicleola*	*iphicleola*	2	2
Nymphalidae	*Adelpha*	*barnesia*	*leucas*	2	2
Nymphalidae	*Adelpha*	*phylaca*	*phylaca*	2	3
Nymphalidae	*Adelpha*	*nea*	*sentia*	2	1
Nymphalidae	*Adelpha*	*basiloides*		2	2
Nymphalidae	*Adelpha*	*escalantei*		2	1
Nymphalidae	*Adelpha*	*fessonia*	*fessonia*	2	2
Nymphalidae	*Agraulis*	*vanillae*	*incarnata*	4	2
Nymphalidae	*Anaea*	*troglodyta*	*aidea*	2	2
Nymphalidae	*Anartia*	*fatima*	*fatima*	4	2
Nymphalidae	*Anartia*	*jatrophae*	*luteipicta*	3	2
Nymphalidae	*Anthanassa*	*ardys*	*ardys*	2	2
Nymphalidae	*Anthanassa*	*dracaena*	*phlegias*	2	2
Nymphalidae	*Anthanassa*	*ptolyca*	*ptolyca*	3	2
Nymphalidae	*Anthanassa*	*frisia*	*tulcis*	4	2
Nymphalidae	*Anthanassa*	*argentea*		2	2
Nymphalidae	*Anthanassa*	*ptolyca*	*amator*	2	2
Nymphalidae	*Antirrhea*	*lindigii*	*casta*	2	1
Nymphalidae	*Archaeoprepona*	*demophon*	*centralis*	2	2
Nymphalidae	*Archaeoprepona*	*demophoon*	*gulina*	2	3
Nymphalidae	*Archaeoprepona*	*demophoon*	*mexicana*	1	1
Nymphalidae	*Archaeoprepona*	*demophon*	*occidentalis*	1	2
Nymphalidae	*Archaeoprepona*	*meander*	*phoebus*	1	2
Nymphalidae	*Asterocampa*	*idyja*	*argus*	2	2
Nymphalidae	*Biblis*	*hyperia*	*aganisa*	3	3
Nymphalidae	*Caligo*	*telamonius*	*memnon*	3	2
Nymphalidae	*Caligo*	*eurilochus*	*sulanus*	2	2
Nymphalidae	*Caligo*	*uranus*		2	2
Nymphalidae	*Callicore*	*texa*	*grijalva*	2	1
Nymphalidae	*Callicore*	*texa*	*titania*	2	2
Nymphalidae	*Castilia*	*ofella*	*ofella*	3	2
Nymphalidae	*Castilia*	*griseobasalis*		3	1
Nymphalidae	*Castilia*	*myia*		3	2
Nymphalidae	*Catonephele*	*numilia*	*esite*	1	2
Nymphalidae	*Catonephele*	*numilia*	*immaculata*	1	1

(continued)

Table 9.8 (continued)

Family	Genus	Species	Subspecies	Habitat	Range
Nymphalidae	*Catonephele*	*mexicana*		2	2
Nymphalidae	*Cepheuptychia*	*glaucina*		1	3
Nymphalidae	*Chlosyne*	*erodyle*	*erodyle*	2	1
Nymphalidae	*Chlosyne*	*gaudialis*	*gaudialis*	2	2
Nymphalidae	*Chlosyne*	*hippodrome*	*hippodrome*	3	3
Nymphalidae	*Chlosyne*	*janais*	*janais*	3	2
Nymphalidae	*Chlosyne*	*lacinia*	*lacinia*	3	2
Nymphalidae	*Chlosyne*	*marina*	*marina*	2	2
Nymphalidae	*Chlosyne*	*theona*	*theona*	3	2
Nymphalidae	*Cissia*	*confusa*		1	3
Nymphalidae	*Cissia*	*labe*		2	2
Nymphalidae	*Cissia*	*pompilia*		2	2
Nymphalidae	*Cissia*	*pseudoconfusa*		2	2
Nymphalidae	*Cissia*	*similis*		3	2
Nymphalidae	*Cissia*	*terrestris*		2	3
Nymphalidae	*Cissia*	*themis*		2	1
Nymphalidae	*Cissia*	*undina*		1	1
Nymphalidae	*Colobura*	*dirce*	*dirce*	2	3
Nymphalidae	*Consul*	*electra*	*electra*	2	2
Nymphalidae	*Cynthia*	*cardui*		4	4
Nymphalidae	*Danaus*	*eresimus*	*montezuma*	3	3
Nymphalidae	*Danaus*	*plexippus*	*plexippus*	2	4
Nymphalidae	*Danaus*	*gilippus*	*thersippus*	3	2
Nymphalidae	*Dione*	*juno*	*huascuma*	3	2
Nymphalidae	*Doxocopa*	*laure*	*laure*	2	3
Nymphalidae	*Doxocopa*	*pavon*	*theodora*	2	2
Nymphalidae	*Dryadula*	*phaetusa*		3	3
Nymphalidae	*Dryas*	*iulia*	*moderata*	2	2
Nymphalidae	*Dynamine*	*postverta*	*mexicana*	3	3
Nymphalidae	*Dynamine*	*artemisia*		1	1
Nymphalidae	*Dynamine*	*ate*		2	1
Nymphalidae	*Dynamine*	*dyonis*		1	2
Nymphalidae	*Dynamine*	*theseus*		1	2
Nymphalidae	*Eresia*	*clara*	*clara*	1	3
Nymphalidae	*Eryphanis*	*aesacus*	*aesacus*	1	2
Nymphalidae	*Eueides*	*isabella*	*eva*	2	2
Nymphalidae	*Eueides*	*aliphera*	*gracilis*	3	2
Nymphalidae	*Eunica*	*alcmena*	*alcmena*	2	2
Nymphalidae	*Eunica*	*tatila*	*tatila*	1	2
Nymphalidae	*Eunica*	*monima*		1	2
Nymphalidae	*Euptoieta*	*claudia*	*daunius*	2	2
Nymphalidae	*Euptoieta*	*hegesia*	*meridiania*	3	3

(continued)

Table 9.8 (continued)

Family	Genus	Species	Subspecies	Habitat	Range
Nymphalidae	*Fountainea*	*eurypyle*	*confusa*	2	2
Nymphalidae	*Fountainea*	*eurypyle*	*glanzi*	1	2
Nymphalidae	*Fountainea*	*glycerium*	*yucatanum*	2	1
Nymphalidae	*Fountainea*	*halice*	*maya*	2	1
Nymphalidae	*Hamadryas*	*feronia*	*farinulenta*	2	3
Nymphalidae	*Hamadryas*	*februa*	*ferentina*	3	4
Nymphalidae	*Hamadryas*	*glauconome*	*glauconome*	1	2
Nymphalidae	*Hamadryas*	*guatemalena*	*guatemalena*	2	2
Nymphalidae	*Hamadryas*	*guatemalena*	*marmarice*	2	2
Nymphalidae	*Hamadryas*	*amphinome*	*mazai*	2	2
Nymphalidae	*Hamadryas*	*amphinome*	*mexicana*	2	2
Nymphalidae	*Hamadryas*	*julitta*		1	1
Nymphalidae	*Heliconius*	*erato*	*petiverana*	3	3
Nymphalidae	*Heliconius*	*ismenius*	*telchinia*	1	2
Nymphalidae	*Heliconius*	*charithonia*	*vazquezae*	3	2
Nymphalidae	*Hermeuptychia*	*hermes*		2	3
Nymphalidae	*Historis*	*acheronta*	*acheronta*	2	3
Nymphalidae	*Historis*	*odius*	*dious*	2	3
Nymphalidae	*Ithomia*	*patilla*	*patilla*	3	2
Nymphalidae	*Junonia*	*genoveva*	*nigrosuffusa*	3	2
Nymphalidae	*Junonia*	*evarete*	*ssp. nov.*	2	3
Nymphalidae	*Junonia*	*coenia*		2	2
Nymphalidae	*Libytheana*	*carinenta*	*mexicana*	2	2
Nymphalidae	*Lycorea*	*halia*	*atergatis*	1	2
Nymphalidae	*Magneuptychia*	*libye*		1	2
Nymphalidae	*Manataria*	*maculata*		1	2
Nymphalidae	*Marpesia*	*chiron*	*marius*	2	3
Nymphalidae	*Marpesia*	*petreus*		3	2
Nymphalidae	*Mechanitis*	*polymnia*	*lycidice*	2	2
Nymphalidae	*Mechanitis*	*menapis*		2	2
Nymphalidae	*Memphis*	*philumena*	*xenica*	2	2
Nymphalidae	*Memphis*	*pithyusa*	*pithyusa*	2	3
Nymphalidae	*Memphis*	*artacaena*		2	2
Nymphalidae	*Memphis*	*forreri*		2	2
Nymphalidae	*Memphis*	*hedemanni*		2	2
Nymphalidae	*Memphis*	*herbacea*		2	2
Nymphalidae	*Memphis*	*moruus*		2	2
Nymphalidae	*Memphis*	*oenomais*		3	2
Nymphalidae	*Mestra*	*dorcas*	*amymone*	3	2
Nymphalidae	*Microtia*	*elva*	*elva*	2	2
Nymphalidae	*Morpho*	*helenor*	*montezuma*	2	2
Nymphalidae	*Myscelia*	*cyaniris*	*alvaradia*	2	2

(continued)

Table 9.8 (continued)

Family	Genus	Species	Subspecies	Habitat	Range
Nymphalidae	*Myscelia*	*cyananthe*	*cyananthe*	2	2
Nymphalidae	*Myscelia*	*cyaniris*	*cyaniris*	2	2
Nymphalidae	*Myscelia*	*ethusa*	*chiapensis*	2	1
Nymphalidae	*Myscelia*	*ethusa*	*ethusa*	2	2
Nymphalidae	*Nessaea*	*aglaura*	*aglaura*	1	2
Nymphalidae	*Nica*	*flavilla*	*bachiana*	1	1
Nymphalidae	*Nica*	*flavilla*	*ssp*	1	1
Nymphalidae	*Oleria*	*paula*		2	2
Nymphalidae	*Opsiphanes*	*cassina*	*fabricii*	2	2
Nymphalidae	*Opsiphanes*	*quiteria*	*quirinus*	1	2
Nymphalidae	*Opsiphanes*	*cassiae*	*ssp*	1	1
Nymphalidae	*Opsiphanes*	*tamarindi*	*tamarindi*	2	2
Nymphalidae	*Pareuptychia*	*metaleuca*	*metaleuca*	1	2
Nymphalidae	*Pareuptychia*	*ocirrhoe*		1	1
Nymphalidae	*Philaethria*	*diatonica*		1	1
Nymphalidae	*Phyciodes*	*phaon*	*phaon*	2	2
Nymphalidae	*Phyciodes*	*tharos*	*tharos*	1	2
Nymphalidae	*Prepona*	*laertes*	*octavia*	1	2
Nymphalidae	*Prepona*	*pylene*	*philetas*	1	1
Nymphalidae	*Pseudodebis*	*zimri*		2	2
Nymphalidae	*Pteronymia*	*cotytto*	*cotytto*	1	2
Nymphalidae	*Pyrrhogyra*	*neaerea*	*hypsenor*	2	2
Nymphalidae	*Pyrrhogyra*	*otolais*	*otolais*	1	2
Nymphalidae	*Siderone*	*galanthis*	*ssp. nov.*	2	1
Nymphalidae	*Siproeta*	*stelenes*	*biplagiata*	2	3
Nymphalidae	*Siproeta*	*epaphus*	*epaphus*	2	2
Nymphalidae	*Smyrna*	*blomfildia*	*datis*	2	2
Nymphalidae	*Splendeuptychia*	*kendalli*		1	1
Nymphalidae	*Taygetis*	*mermeria*	*excavata*	2	2
Nymphalidae	*Taygetis*	*leuctra*		2	2
Nymphalidae	*Taygetis*	*thamyra*		2	3
Nymphalidae	*Taygetis*	*virgilia*		1	3
Nymphalidae	*Taygetis*	*weymeri*		1	1
Nymphalidae	*Tegosa*	*guatemalena*		2	2
Nymphalidae	*Temenis*	*laothoe*	*hondurensis*	2	2
Nymphalidae	*Thessalia*	*theona*	*theona*	3	2
Nymphalidae	*Tigridia*	*acesta*		2	3
Nymphalidae	*Yphthimoides*	*renata*		2	2
Nymphalidae	*Zaretis*	*callidryas*		2	2
Nymphalidae	*Zaretis*	*ellops*		2	2
Papilionidae	*Battus*	*philenor*	*acauda*	2	1
Papilionidae	*Battus*	*laodamas*	*copanae*	2	2

(continued)

Table 9.8 (continued)

Family	Genus	Species	Subspecies	Habitat	Range
Papilionidae	*Battus*	*laodamas*	*iopas*	2	2
Papilionidae	*Battus*	*philenor*	*philenor*	2	2
Papilionidae	*Battus*	*polydamas*	*polydamas*	4	4
Papilionidae	*Heraclides*	*thoas*	*autocles*	3	2
Papilionidae	*Heraclides*	*androgeus*	*epidaurus*	2	2
Papilionidae	*Heraclides*	*anchisiades*	*idaeus*	3	2
Papilionidae	*Heraclides*	*ornythion*	*ornythion*	2	2
Papilionidae	*Heraclides*	*astyalus*	*pallas*	3	2
Papilionidae	*Heraclides*	*rogeri*	*rogeri*	1	1
Papilionidae	*Heraclides*	*torquatus*	*tolus*	2	2
Papilionidae	*Heraclides*	*cresphontes*		3	2
Papilionidae	*Mimoides*	*ilus*	*branchus*	3	2
Papilionidae	*Mimoides*	*ilus*	*occiduus*	2	2
Papilionidae	*Mimoides*	*phaon*	*phaon*	2	2
Papilionidae	*Parides*	*iphidamas*	*iphidamas*	4	1
Papilionidae	*Parides*	*photinus*	*photinus*	3	2
Papilionidae	*Parides*	*erithalion*	*polyzelus*	2	2
Papilionidae	*Parides*	*erithalion*	*trichopus*	2	2
Papilionidae	*Parides*	*sesostris*	*zestos*	4	2
Papilionidae	*Parides*	*montezuma*		2	2
Papilionidae	*Priamides*	*pharnaces*		4	2
Papilionidae	*Protesilaus*	*macrosilaus*	*penthesilaus*	2	1
Papilionidae	*Protographium*	*epidaus*	*epidaus*	3	2
Papilionidae	*Protographium*	*epidaus*	*fenochionis*	3	1
Papilionidae	*Protographium*	*agesilaus*	*neosilaus*	2	2
Papilionidae	*Protographium*	*philolaus*	*philolaus*	3	2
Papilionidae	*Pterourus*	*menatius*		2	2
Papilionidae	*Pterourus*	*menatius*	*victorinus*	2	2
Papilionidae	*Troilides*	*torquatus*	*tolus*	2	2
Pieridae	*Abaeis*	*nicippe*		3	2
Pieridae	*Anteos*	*clorinde*		4	3
Pieridae	*Anteos*	*maerula*		3	3
Pieridae	*Aphrissa*	*statira*	*statira*	3	3
Pieridae	*Ascia*	*monuste*	*monuste*	4	3
Pieridae	*Charonias*	*theano*	*nigrescens*	2	1
Pieridae	*Colias*	*philodice*	*philodice*	3	2
Pieridae	*Dismorphia*	*theucharila*	*fortunata*	2	2
Pieridae	*Eurema*	*arbela*	*boisduvaliana*	3	2
Pieridae	*Eurema*	*albula*	*celata*	2	2
Pieridae	*Eurema*	*daira*	*eugenia*	3	2
Pieridae	*Eurema*	*mexicana*	*mexicana*	4	2
Pieridae	*Ganyra*	*josephina*	*josepha*	3	2

(continued)

Table 9.8 (continued)

Family	Genus	Species	Subspecies	Habitat	Range
Pieridae	*Glutophrissa*	*drusilla*	*tenuis*	2	3
Pieridae	*Kricogonia*	*lyside*		2	2
Pieridae	*Nathalis*	*iole*		3	2
Pieridae	*Phoebis*	*agarithe*	*agarithe*	2	4
Pieridae	*Phoebis*	*sennae*	*marcellina*	3	4
Pieridae	*Phoebis*	*philea*	*philea*	4	3
Pieridae	*Phoebis*	*argante*		3	2
Pieridae	*Phoebis*	*neocypris*	*virgo*	3	3
Pieridae	*Pieriballia*	*viardi*	*viardi*	2	2
Pieridae	*Pyrisitia*	*lisa*	*centralis*	4	2
Pieridae	*Pyrisitia*	*nise*	*nelphe*	3	2
Pieridae	*Pyrisitia*	*dina*	*westwoodi*	3	2
Pieridae	*Pyrisitia*	*proterpia*		3	2
Pieridae	*Rhabdodryas*	*trite*		2	1
Pieridae	*Zerene*	*cesonia*	*cesonia*	2	3

For habitat use, geographic range descriptions and data sources see text. Taxonomic authorities followed Llorente-Bousquets et al. (2006)

References

Acosta-Gutiérrez R. Biodiversidad de Siphonaptera en México. Revista Mexicana de Biodiversidad. 2014;85:S345–52.

Almaraz-Almaraz ME, León-Cortés JL, Molina-Martínez A. The population ecology and conservation of Pterourus esperanza (Insecta: Lepidoptera): an ancestral swallowtail butterfly in the northern Sierra of Oaxaca, Mexico. Ann Entomol Soc Am. 2013;106(6):753–60.

Boggs CL, Murphy DD. Community composition in mountain ecosystems: climatic determinants of montane butterfly distributions. Glob Ecol Biogeogr Lett. 1997;6:39–48.

Bousquet Y. Taxonomic revision of Nearctic, Mexican, and West Indian Oodini (Coleoptera: Carabidae). Can Entomol. 1996;128:443–537.

Bousquet Y. Catalogue of Geadephaga (Coleoptera, Adephaga) of America, North of Mexico. ZooKeys. 2012;245:1–1722.

Chamorro ML, Holzenthal R. Taxonomy and phylogeny of New World Polyplectropus Ulmer, 1905 (Trichoptera: Psychomyioidea: Polycentropodidae) with the description of 39 new species. Zootaxa. 2010;2582:1–252.

Cifuentes-Ruíz P, Zaragoza-Caballero S. Biodiversidad de Tenebrionidae (Insecta: Coleoptera) en México. Revista Mexicana de Biodiversidad. 2014;85:S325–31.

Collins NM, Thomas JA. The conservation of insects and their habitats. London: Academic Press; 1991.

Colwell RK, Chao A, Gotelli NJ, Lin SY, Mao CX, Chazdon RL, Longino JT. Models and estimators linking individual-based and sample-based rarefaction, extrapolation and comparison of assemblages. J Plant Ecol. 2012;5:3–21.

Colwell RK. *Estimates*: statistical estimation of species richness and shared species from samples (Software and User's guide), Version 9.0.0 Persistent. 2013. http://purl.oclc.org/estimates

Contreras-Ramos A, Sarmiento-Cordero M, Penny N. Biodiversidad de Mecoptera en México. Revista Mexicana de Biodiversidad. 2014;85:S339–44.

Coronado-Blanco J, Zaldívar-Riverón A. Biodiversidad de Braconidae (Hymenoptera: Ichneumonoidea) en México. Revista Mexicana de Biodiversidad. 2014;85:S372–8.

Daily GC, Ehrlich PR. Preservation of biodiversity in small rainforest patches: rapid evaluations using butterfly trapping. Biodivers Conserv. 1995;4:35–55.

Delfín-González H, Chay-Hernández D. Riqueza de Hymenoptera. In: Duran R, Méndez M (comps) Biodiversidad y desarrollo humano en Yucatán. Yucatán, México: CYCY, PPD-FMAM, SEDUMA; 2010. p. 239–40.

Delfín-González H, Meléndez V, Manrique P, Reyes-Novelo E, Chay D. Insectos. In: Duran R, Méndez M (comps) Biodiversidad y desarrollo humano en Yucatán. Yucatán, México: CYCY, PPD-FMAM, SEDUMA; 2010. p. 226–8.

De Vries JP. The butterflies of Costa Rica and their natural history. Vol. I. Papilionidae, Pieridae and Nymphalidae. Princeton, NJ: Princeton Academic Press; 1987.

Didham RK, Ghazoul J, Stork NE, Davis AJ. Insects in fragmented forests: a functional approach. Trends Ecol Evol. 1996;11(6):255–60.

Elith J, Kearney M, Phillips S. The art of modelling range-shifting species. Methods Ecol Evol. 2010;1(4):330–42.

Equihua-Martínez A, Burgos S. Scolytidae. In: Llorente B, Morrone JJ (comps) Biodiversidad, Taxonomía y Biogeografía de Artrópodos de México: Hacia una síntesis de su conocimiento. CONABIOIBUNAM, vol. III. México, DF; 2002. p. 539–57.

Erwin TL. Tropical forests: their richness in Coleoptera and other arthropod species. Coleopt Bull. 1982;36:74–5.

Essens T, Hernández-Stefanoni JL. Mapping Lepidoptera and plant alpha-diversity across a heterogeneous tropical dry forest using field and remotely sensed data with spatial interpolation. J Insect Conserv. 2013;17(4):725–36.

Essens T, Leyequién E, Pozo C, Vester HF, Hernández-Arana HA. Effects of climate and forest age on plant and caterpillar diversity in the Yucatán, Mexico. J Trop Ecol. 2014;30(5):419–34.

García-Aldrete A. Biodiversidad de Psocoptera (Insecta: Psocodea) en México. Revista Mexicana de Biodiversidad. 2014;85:S252–6.

Gaston KJ, Blackburn TM, Greenwood JJ, Gregory RD, Quinn RM, Lawton JH. Abundance—occupancy relationships. J Appl Ecol. 2000;37:39–59.

Goldman EA, Moore RT. Biotic provinces of Mexico. J Mammal. 1946;26(4):347–60.

González-Moreno A, Bordera S. The Ichneumonidae (Hymenoptera: Ichneumonoidea) of Ría Lagartos Biosphere Reserve, Yucatán, México. Zootaxa. 2012;3230:1–51.

González-Moreno A, Marcos-García MA, Manrique-Saide P. Registros nuevos de especies de sírfidos (Diptera: Syrphidae) para Yucatán, México. Revista Mexicana de Biodiversidad. 2011; 82:301–3.

González-Soriano E, Novelo-Gutiérrez R. Biodiversidad de Odonata en México. Revista Mexicana de Biodiversidad. 2014;85:S343–251.

Gotelli NJ, Colwell RK. Quantifying biodiversity: procedures and pitfalls in the measurement and comparison of species richness. Ecol Lett. 2001;4:379–91.

Groombridge B. Global-biodiversity: status of the earths living resources. A report compiled by the world conservation monitoring centre. London: Chapman & Hall; 1992.

Haber WA. Check list of Sphingidae. In: Janzen D, editor. Costa Rican natural history. Chicago, IL: Chicago University Press; 1983. p. 645–50.

Hernández-Baz F. Arctiidae: palomillas nocturnas. In: Duran R, Méndez M (comps) Biodiversidad y desarrollo humano en Yucatán. Yucatán, México: CYCY, PPD-FMAM, SEDUMA; 2010. p. 245–6.

Hernández-Baz F, Morón MA, Chamé E, González JM. Ctenuchina and Euchromiina (Lepidoptera: Erebidae: Arctiinae) of three biological research stations of Mexico's ECOSUR. J Lepid Soc. 2013;67:145–55.

Hortal J, Roura-Pascual N, Sanders N, Rahbek C. Understanding (insect) species distributions across spatial scales. Ecography. 2010;33:51–3.

Ibáñez-Bernal S. Moscas, mosquitos y jejenes. In: Pozo C. (comp) Riqueza biológica de Quintana Roo. Un análisis para su conservación. Tomo 2, Capítulo 4, Invertebrados terrestres, El Colegio de la Frontera Sur, Conabio, Gobierno del Estado de Quintana Roo y Programa de Pequeñas Donaciones. México, D. F.; 2011. p. 202–4.

Ibarra-Manríquez G, Villaseñor JL, Durán R, Meave J. Biogeographical analysis of the tree flora of the Yucatán Peninsula. J Biogeogr. 2002;29:17–29.

Janzen DH. Two ways to be a tropical big moth: Santa Rosa saturniids and sphingids. Oxf Surv Evol Biol. 1984;1:85–140.

Janzen DH. Biogeography of an unexceptional place: what determines the saturniid and sphingid moth fauna of Santa Rosa national park, Costa Rica, and what does it mean to conservation biology? Brenesia. 1986;25(26):51–87.

Lee JC. An ecogeographic analysis of the herpetofauna of the Yucatán Peninsula, Miscellaneous Publication Museum of Natural History, vol. 67. Lawrence, KS: University of Kansas; 1980. p. 1–79.

León-Cortés JL. Sphingoidea (Lepidoptera). In: Llorente Bousquets J, González E, Papavero N (comps) Biodiversidad, taxonomía y biogeografía de artrópodos de México: Hacia una síntesis de su conocimiento, vol. II. Facultad de Ciencias, UNAM, Mexico, D.F.; 2000. p. 483–500.

León-Cortés JL, Cowley M, Thomas CD. Detecting decline in a formerly widespread species: how common is the common blue butterfly Polyommatus Icarus? Ecography. 1999;22:643–50.

León-Cortés JL, Jones RW, Gómez-Nucamendi OL. A preliminary assessment of the butterfly fauna of El Edén ecological reserve: species richness and habitat preferences. In: Gómez-Pompa A, Allen MF, Fedick SL, Jiménez-Osornio JJ (comps) The lowland Maya area: three millennia at the human-wildland interface. New York, NY: The Haworth Press; 2003. p. 261–76.

León-Cortés JL, Pérez-Espinoza F, Marín L, Molina-Martínez A. Complex habitat requirements and conservation needs of the only extant Baroniinae swallowtail butterfly. Anim Conserv. 2004;7(3):241–50.

León-Cortés JL, Kitching IJ, Molina-Martínez A. Distributional patterns of the family Sphingidae (Lepidoptera). In: Morrone JJ, Llorente Bousquets J (comps.) Componentes Bióticos Principales de la Entomofauna Mexicana. Las Prensas de Ciencias. México, D.F.: Universidad Nacional Autónoma de México; 2006. p. 629–47.

Lewis OT, Wilson RJ, Harper MC. Endemic butterflies on Grande Comore: habitat preferences and conservation priorities. Biol Conserv. 1998;85:113–21.

Llorente-Bousquets J, Oñate LO, Luis-Martínez AM, Vargas-Fernández I. Papilionidae y Pieridae de México: distribución Geográfica e Ilustración. Facultad de Ciencias, UNAM.-CONABIO. México, DF; 1997.

Llorente-Bousquets J, Luis-Martínez AM, Vargas-Fernández I. Apéndice General de Papilionoidea: Lista Sistemática, Distribución Estatal y Provincias Biogeográficas. In: Morrone JJ, Llorente-Bousquets J (comps) Componentes Bióticos Principales de la Entomofauna Mexicana. México, DF: Las Prensas de Ciencias, UNAM; 2006. p. 945–1009.

Llorente-Bousquets J, Vargas I, Luis A, Trujano M, Hernández B, Warren AD. Biodiversidad de Lepidoptera en México. Revista Mexicana de Biodiversidad. 2014;85:353–71.

Lobo JM, Jiménez-Valverde A, Hortal J. The uncertain nature of absences and their importance in species distribution modelling. Ecography. 2010;33:103–14.

Luis-Martínez A, Llorente-Bousquets J, Vargas IF, Gutiérrez AL. Síntesis preliminar del conocimiento de los Papilionoidea (Lepidoptera: Insecta) de México. In: Martin F, Morrone JJ, Melic A (comps.) Monografías Tercer Milenio (Boletín Sociedad Entomológica Aragonesa). 2000. p. 275–85.

Luis-Martínez AM, Llorente-Bousquets J, Vargas-Fernández I. Nymphalidae de México I (Danainae, Apaturinae, Biblidinae y Heliconiinae): distribución Geográfica e Ilustración. México, DF: UNAM.-CONABIO; 2003.

Luis-Martínez AM, Llorente-Bousquets J, Vargas-Fernández I, Pozo C. Nymphalidae de México III (Nymphalinae): distribución geográfica e ilustración. México, D.F.: UNAM, ECOSUR; 2010.

Lundell C. Preliminary sketch of the phytogeography of the Yucatán Peninsula. In: Contributions to American archaeology, vol. II, No. 12., Publ. No. 436. Washington, DC: Carnegie Institute of Washington; 1934. p. 253–321.

Magurran AE. Measuring biological diversity. Oxford: Blackwell Publishing; 2004.

Manrique P, González-Moreno A. Moscas y mosquitos. In: Duran R, Méndez M (comps) Biodiversidad y desarrollo humano en Yucatán. Yucatán, México: CYCY, PPD-FMAM, SEDUMA; 2010. p. 229–30.

Martínez-Noble JI, Meléndez-Ramírez V, Delfín-González H, Pozo C. Mariposas de la selva mediana subcaducifolia de Tzucacab, con nuevos registros para Yucatán, México. Revista Mexicana de Biodiversidad. 2015;86:348–57.

Maya-Martínez A, Pozo C, May-Uc E. Las mariposas (Rhopalocera: Papilionidae, Pieridae y Nymphalidae) de la selva alta subperenifolia de la región de Calakmul, México, con nuevos registros. Folia Entomol Mex. 2005;44(2):123–43.

Meléndez RV, Reyes-Novelo E, Parra V, Quezada J, Meneses C. Diversidad de abejas silvestres. In: Duran R, Méndez M (comps) Biodiversidad y desarrollo humano en Yucatán. CYCY, PPD-FMAM, SEDUMA; 2010. p. 226–8.

Meneses CL, Meléndez RV, Parra TV, Navarro AJ. Bee diversity in fragmented landscape of the Mexican neotropic. J Insect Conserv. 2010;14:323–34.

Molina-Martínez A, León-Cortés JL, Regan HM. Climatic and geometric constraints as driving factors of butterfly species richness along a Neotropical elevational gradient. J Insect Conserv. 2013;17:1169–80.

Morón MA. Escarabajos. In: Pozo C. (comp) Riqueza biológica de Quintana Roo. Un análisis para su conservación. Tomo 2, Capítulo 4, Invertebrados terrestres, El Colegio de la Frontera Sur, Conabio, Gobierno del Estado de Quintana Roo y Programa de Pequeñas Donaciones. México, DF; 2011. p. 182–5.

Morón MA, Nogueira G, Rojas-Gómez V, Arce-Pérez R. Biodiversidad de Melolonthidae (Coleoptera) en México. Revista Mexicana de Biodiversidad. 2014;85:S298–302.

Navarrete-Heredia JL, Newton AF. Biodiversidad de Staphylinidae (Insecta: Coleoptera) en México. Revista Mexicana de Biodiversidad. 2014;85:S332–8.

New TR, Pyle RM, Thomas JA, Thomas CD, Hammond PC. Butterfly conservation management. Annu Rev Entomol. 1995;40:57–83.

New TR. Insect species conservation. Cambridge: Cambridge University Press; 2008.

Noguera F. Biodiversidad de Cerambycidae (Coleoptera) en México. Revista Mexicana de Biodiversidad. 2014;85:S290–7.

NORMA Oficial Mexicana NOM-059-SEMARNAT-2010. Protección ambiental-Especies nativas de México de flora y fauna silvestres-Categorías de riesgo y especificaciones para su inclusión, exclusión o cambio-Lista de especies en riesgo;2010.

Opler AP, Wright AM. A field guide to Western butterflies. New York, NY: Houghton Mifflin; 1999.

Ordóñez-Reséndiz M, López-Pérez S, Rodríguez-Mirón G. Biodiversidad de Chrysomelidae (Coleoptera) en México. Revista Mexicana de Biodiversidad. 2014;85:S271–8.

Ortega A, Mis P, Elizondo A, Harbach R, Siller Q, Fernández I. The mosquitoes of Quintana Roo State, Mexico (Diptera: Culicidae). Acta Zoológica Mexicana (ns). 2010;26:33–46.

Palacios-Vargas J. Biodiversidad de Collembola (Hexapoda: Entognatha) en México. Revista Mexicana de Biodiversidad. 2014;85:S220–31.

Palacios-Vargas J, Figueroa D. Biodiversidad de Protura (Hexapoda: Entognatha) en México. Revista Mexicana de Biodiversidad. 2014;85:S232–5.

Palacios-Vargas J, García-Gómez A. Biodiversidad de Diplura (Hexapoda: Entognatha) en México. Revista Mexicana de Biodiversidad. 2014;85:S236–42.

Pearson RG, Dawson TP. Predicting the impacts of climate change on the distribution of species: are bioclimate envelope models useful? Glob Ecol Biogeogr. 2003;12(5):361–71.

Phillips SJ, Dudik M, Schapire RE. A maximum entropy approach to species distribution modeling. In: Proceedings of the 21st International conference on machine learning. New York, NY: ACM Press; 2004. p. 655–62.

Phillips SJ, Anderson RP, Schapire RE. Maximum entropy modeling of species geographic distributions. Ecol Model. 2006;190(3):231–59.

Pozo C, Luis-Martinez A, Uc S, Salas N, Maya A. Butterflies (Papilionoidea and Hesperioidea) of Calakmul, Campeche, Mexico. Southwest Nat. 2003;48:505–25.

Pozo, C. Salas, N, Maya-Martínez, A. y B. Prado. Uso y monitoreo de los recursos naturales en el Corredor Biológico Mesoamericano (áreas focales Xpujil-Zoh Laguna y Carrillo Puerto). Subproyecto Mariposas. En: Pozo, C y S. Calmé. El Colegio de la Frontera Sur. Unidad Chetumal. *Bases de datos* SNIB2010-CONABIO proyectos No. BJ002, Y006, Q049_Mariposas, J112_Mariposas y P028-Mariposas. México, D.F. 2005.

Pozo C. Mariposas. In: Pozo C. (comp) Riqueza biológica de Quintana Roo. Un análisis para su conservación. Tomo 2, Capítulo 4, Invertebrados terrestres, El Colegio de la Frontera Sur, Conabio, Gobierno del Estado de Quintana Roo y Programa de Pequeñas Donaciones, México, DF; 2011. p. 186–91.

Prado B, Domínguez E, Pozo C. Palomillas. In: Pozo C. (comp) Riqueza biológica de Quintana Roo. Un análisis para su conservación. Tomo 2, Capítulo 4, Invertebrados terrestres, El Colegio de la Frontera Sur, Conabio, Gobierno del Estado de Quintana Roo y Programa de Pequeñas Donaciones. México, DF; 2011a. p. 192–5.

Prado B, Pozo C, Valdez-Moreno M, Hebert PDN. Beyond the colours: discovering hidden diversity in the Nymphalidae of the Yucatán Peninsula in Mexico through DNA barcoding. PLoS One. 2011b;6(11):1–11.

Ramírez-Barahona S, Torres-Miranda A, Palacios-Ríos M, Luna-Vega I. Historical biogeography of the Yucatán Peninsula, Mexico: a perspective from ferns (Monilophyta) and lycopods (Lycophyta). Biol J Linn Soc. 2009;98:775–86.

Reyes NE, Meléndez RV, Ayala BR, Delfín GH. Bee faunas (Hymenoptera: Apoidea) of six natural protected areas in Yucatán, Mexico. Entomol News. 2009;120:530–44.

Reyes-Castillo P. Passalidae. In: Llorente B, Morrone JJ (comps) Biodiversidad, Taxonomía y Biogeografía de Artrópodos de México: Hacia una síntesis de su conocimiento. CONABIOIBUNAM, Vol. III, México, DF; 2002. p. 467–83.

Reyes-Novelo ER, Morón MA. Fauna de Coleoptera Melolonthidae y Passalidae de Tzucacab y Conkal, Yucatán, México. Acta Zoológica Mexicana (ns). 2005;21:15–49.

Rodríguez F, Villalobos M, Pozo C. Lepidópteros diurnos. In: Duran R, Méndez M (comps) Biodiversidad y desarrollo humano en Yucatán. Yucatán, México: CYCY, PPD-FMAM, SEDUMA; 2010. p. 242–4.

Romero G. Estudio faunístico de la familia Bruchidae (Insecta: Coleoptera) del estado de Morelos, México. Doctorate Dissertation, Colegio de Postgraduados, Montecillo, Texcoco, Edo de México; 2013.

Ruíz-Cancino E, Coronado-Blanco JM, Martínez-Ramírez J. Contribución al conocimiento de Ichneumonidae (Hymenoptera) del Sureste de México. Serie Publicaciones Científicas CIDAFF-UAT. Cd. Victoria, Tamaulipas; 2002.

Ruíz-Cancino E, Rafaelevich-Kasparyan D, González-Moreno A, Khalaim AI, Coronado-Blanco JM. Biodiversidad de Ichneumonidae (Hymenoptera) en México. Revista Mexicana de Biodiversidad. 2014;85:385–91.

Schreiber H. Dispersal centres of Sphingidae (Lepidoptera) in the Neotropical region. The Hague: Dr. W. Junk B.V. Publishers; 1978. 195pp.

Scott JA. The butterflies of North America: a natural history and field guide. Stanford, CA: Stanford University Press; 1997.

Shpeley D, Ball GE. A taxonomic review of the subtribe Pericalina (Coleoptera: Carabidae: Lebiini) in the Western Hemisphere, with descriptions of new species and notes about classification and zoogeography. Insecta Mundi. 2000;14:1–185.

Soberón JM. Niche and area of distribution modeling: a population ecology perspective. Ecography. 2010;33:159–67.
Spitzer K, Jaroša J, Havelka J, Lepš J. Effect of small-scale disturbance on butterfly communities of an Indochinese montane rainforest. Biol Conserv. 1997;80:9–15.
Thomas CD. Habitat use and geographic ranges of butterflies from the wet lowlands of Costa Rica. Biol Conserv. 1991;55:269–81.
Thomas CD, Abery JCG. Estimating rates of butterfly decline from distribution maps: the effect of scale. Biol Conserv. 1995;73:59–65.
Thuiller W, Brotons L, Araújo MB, Lavorel S. Effects of restricting environmental range of data to project current and future species distributions. Ecography. 2004;27(2):165–72.
Tyler H, Brown KS, Wilson K. Swallowtail butterflies of the Americas: a study in biological dynamics, ecological diversity, biosystematics and conservation. Gainesville, FL: Scientific Publishers; 1994.
Vane-Wright RI, Humphries CJ, Williams PH. What to protect?—Systematics and the agony of choice. Biol Conserv. 1991;55(3):235–54.
Vargas-Fernández I, Llorente-Bousquets J, Luis-Martínez AM, Pozo C. Nymphalidae de México II (Libytheinae, Ithomiinae, Morphinae y Charaxinae): distribución geográfica e ilustración. México, D.F.: UNAM.-CONABIO; 2008.
Warren AD. Hesperioidea (Lepidoptera). In: Llorente J, González E, Papavero N (comps.) Biodiversidad, taxonomía y biogeografía de artrópodos de México: hacia una síntesis de su conocimiento, vol. 2. México, D. F.: Universidad Nacional Autónoma de México; 2000. p. 535–80.
Vásquez-Bolaños M. Lista de especies de hormigas (Hymenoptera: Formicidae) para México. Dugesiana. 2011;18:95–133.
Zaragoza-Caballero S, Pérez-Hernández C. Biodiversidad de cantaroideos (Coleoptera: Elateroidea [Cantharidae, Lampyridae, Lycidae, Phengodidae, Telegeusidae]) en México. Revista Mexicana de Biodiversidad. 2014;85:S279–89.

Web Sites Consulted

The Global Biodiversity Information Facility, GBIF. 2015. http://www.gbif.org/NatureServe, http://www.natureserve.org/es
Norenburg J, Ritgers R. Encyclopedia of life global access to knowledge about life on Earth. 2015. http://eol.org/
Tree of life. Lepidoptera and some other life forms. 2015. http://ftp.funet.fi/index/Tree_of_life/intro.html
Warren AD, Davis KJ, Stangeland EM, Pelham JP, Grishin NV. Illustrated lists of American butterflies. 2013. http://www.butterfliesofamerica.com/

Chapter 10
Large Terrestrial Mammals

Rafael Reyna-Hurtado, Georgina O'Farrill, Cuauhtémoc Chávez, Juan Carlos Serio-Silva, and Guillermo Castillo-Vela

Abstract The Yucatán Peninsula contains some of the largest tracts of tropical forest in Mexico. These forests host six species of ungulates, including the largest and last survivor of the Neotropical megafauna, the Central American Tapir; one of the rarest ungulate species in Mexico, the White-lipped Peccary; and one endemic species of deer, the Gray Brocket. The Yucatán Peninsula is also home to another peccary species, two more deer species, five felid species, including the jaguar and the puma, and three species of primates. Most of these species face serious conservation threats, as their habitat is increasingly fragmented and because they are among the preferred targets of subsistence hunters. Some of these species require large areas of habitat in good conservation status to fulfill their basic needs for survival. Several research projects undertaken in the past years, and some currently being carried out, have addressed a lack of basic ecological information in this region. Among the ungulates, ecological research has focused on tapir, white-lipped peccary and the three deer species. For felids, scientific attention has been focused on the two largest species, the jaguar and puma; and all three primate species have received scientific attention recently, although more studies have focused on the black howler monkey. This chapter is an attempt to summarize what is currently known about these, the largest mammal species of the Yucatán Peninsula, and to point out gaps in the existing information. Such information is

R. Reyna-Hurtado (✉) • G. Castillo-Vela
Departamento Conservación de la Biodiversidad, El Colegio de la Frontera Sur, Unidad Campeche, CP 24500 Campeche, Mexico
e-mail: rreyna@ecosur.mx; gcastillo@ecosur.mx

G. O'Farrill
University of Toronto, 25 Harbord Street, Toronto, ON, Canada, M5S 3G5
e-mail: georgina.ofarrill@gmail.com

C. Chávez
Departamento de Ciencias Ambientales Unidad Lerma, Universidad Autónoma Metropolitana, CP 52006 Edo de México, Mexico
e-mail: j.chavez@correo.ler.uam.mx

J.C. Serio-Silva
Red de Biología y Conservación de Vertebrados, Instituto Nacional de Ecología, A.C., CP 91070 Xalapa, Veracruz, Mexico
e-mail: juan.serio@inecol.mx

G.A. Islebe et al. (eds.), *Biodiversity and Conservation of the Yucatán Peninsula*,
DOI 10.1007/978-3-319-06529-8_10

absolutely necessary to design conservation and management plans for these highly interesting and endangered species.

Keywords Endangered species • Hunting • Tropical ungulates • Large cats • Mexican primates • Maya forest • Yucatán Peninsula • Coastal ecosystems • Wildlife conservation • Wildlife research

10.1 Introduction

The Yucatán Peninsula is an extraordinary region in Mexico that embraces a vast diversity of habitats and wildlife, mixed with rich cultural traditions and an historical legacy as the center of ancient and modern Mayan civilizations. This area has a rich geological history and has witnessed several wildlife colonization and extinction events. Sixty-five millions years ago an asteroid impacted the northern portion of the peninsula, which caused the Chikchulub crater and a mass extinction that affected the whole planet's biota (Hildebrand et al. 1991). This event changed the composition of biotic communities around the world and is sometimes theorized as the cause of the extinction of the dinosaurs (Vazquez-Dominguez and Arita 2010). The peninsula has undergone several climatic events in its history, changing its profile and biota; the advance and retreat of ice sheets caused sea level to rise and decrease several times. The Yucatán Peninsula and Central America have served as a bridge between communities of North and South American biota (Vazquez-Dominguez and Arita 2010). The wildlife community in this region is therefore a product of these geological and climatic events and is composed of elements of the Neoarctic and Neotropical realms, creating a unique wildlife community (Arroyo-Cabrales and Álvarez 2003).

Evidence of human habitation of the Yucatán Peninsula goes back 7500–9000 years. Since then, humans have always been present in the region, with periods where more than 50,000 people lived in pre-Hispanic cities such as Tikal or Calakmul. According to Vazquez-Dominguez and Arita (2010), the area currently hosts a population similar in size to when Spanish first arrived in 1519. In the Yucatán Peninsula, the community of large vertebrates (the term usually refers to birds and mammals that weigh more than 10 kg) that Spaniards found when they first arrived more than 500 years ago still persists.

This chapter examines the mammals among those large vertebrates, which comprise a wide variety of species, a number of them endangered. Six species of ungulates inhabit the Yucatán Peninsula; one is the last representative of the Perissodactyla order in Central America, the Central American Tapir (*Tapirus bairdii*) and the largest of all Neotropical terrestrial mammals. The Artiodactyla order is represented with two species of peccaries, the white-lipped peccary (*Tayassu pecari*) and the collared peccary (*Pecari tajacu*) and three species of deer, the white-tailed deer (*Odocoileus virginianus*), the grey brocket deer (*Mazama pandora*), and the Central American red brocket deer (*Mazama temama*).

These ungulate species are primary target species among subsistence hunters, and they are a source of high-quality animal protein for many rural families in the area (Jorgenson 1995; Escamilla et al. 2000; Weber 2000; Reyna-Hurtado and Tanner 2007; Urquiza-Haas et al. 2009; Ramírez-Barajas et al. 2012). Five species of felids inhabit the peninsula forests. The jaguar (*Panthera onca*) is the largest of all, followed by the puma (*Puma concolor*), the ocelot (*Leopardus pardalis*), the jaguarondi (*Herpailurus yagouaroundi*) and finally the margay (*Leopardus wiedii*). Three species of primates are also found in this area: the spider monkey (*Ateles geoffroyi*), the black howler monkey (*Alouatta pigra*) and the mantled howler monkey (*Alouatta palliata*). Felids and primates are also target species among sport hunters and illegal pet traders, respectively. Almost all these species need large areas of habitat to maintain viable populations and some of them require large ranges, as individuals cross long distances to fulfill their basic needs (Reyna-Hurtado et al. 2009; Ceballos et al. 2002; O'Farrill et al. 2011; Sarabia-Hernandez and Reyna-Hurtado unpub. data). These species represent the largest Neotropical mammals, and all of them face various threats for their survival.

In this chapter we explore current knowledge on 10 of these 14 species. We also explore their conservation status, and suggest some future directions and research priorities for protecting and conserving these amazing species in the Yucatán Peninsula. Tapirs, jaguars, pumas, deer, and herds of peccaries still roam the forest understory while spider monkeys and groups of howler monkeys find their way through the forest canopy. Currently, the region fosters some of the largest remaining tracts of well-conserved tropical forest in Mexico, and its value for the protection of large Neotropical wildlife is becoming more evident as other areas of the country lose their original forest cover. In addition, for species that require large ranges to maintain viable populations, such as the jaguar, tapir, or white-lipped peccary, the Yucatán Peninsula represents a space of hope for their conservation (Reyna-Hurtado et al. 2009; Naranjo 2009; Ceballos et al. 2002).

10.2 Ungulates

*10.2.1 Central American or Baird's Tapir (***Tapirus bairdii***)*

The common names of this species in the Yucatán Peninsula include *danta*, *anteburro*, or *tzimin* in Mayan. The Central American tapir is the only representative of the order Perissodactyla in Mexico. Tapirs are phylogenetically related to horses and rhinoceros and belong to the family Tapiridae. Baird's tapir is a massive rotund animal that weighs up to 300 kg, reaches up to 1.3 m in length and has a protuberant nose that give them their signature profile (Emmons and Feer 1997).

The distribution of Baird's tapir in the Yucatán Peninsula includes the Biosphere Reserves of Calakmul and Sian Ka'an in the south and east, as well as the communal areas adjoining them in the States of Campeche and Quintana Roo.

There are reports of tapirs in Los Petenes Biosphere Reserve north of Campeche (Naranjo et al. 2015), in the Chenes region of the central areas of Campeche (Zuñiga-Fuentes 2009; GCV pers. observ.), and in coastal ecosystems of northern Yucatán (pers. comm., fishermen from Sisal, Yum Balam, Yucatán). While isolated populations may exist in these ecosystems, these reports have not been confirmed. Baird's tapirs are classified as endangered according to the Mexican list of species at risk (NOM-059, Semarnat 2010), and by the IUCN Red List (http://www.iucnredlist.org).

Baird's tapirs prefer well-conserved forests with available water (Fig. 10.1a), but also use low flooded forest while avoiding dry deciduous forest; they also utilize secondary/perturbed forest to some degree if they are not being hunted (Reyna-Hurtado and Tanner 2005). In fact, in a study using tracks as indicator of relative abundance, tapirs were three times more abundant in communal forests (0.42 tracks/km) contiguous to the Calakmul Biosphere Reserve than within the interior of the protected area (0.03 tracks/km) (Reyna-Hurtado and Tanner 2007). This research showed that communal forests in Southern Yucatán Peninsula may present better habitat quality than some parts of the Biosphere Reserve. The higher abundance of water outside the Calakmul protected area (O'Farrill et al. 2014) and the fact that tapirs are rarely the prey of subsistence hunters in this region might explain why the population persists and even grows more abundant than the one in the nearby protected area (Reyna-Hurtado and Tanner 2007). In the Calakmul Region, some arguments offered against killing tapirs include their excessive weight, which increases the work of butchering the animals and the chance of wasting meat; the unpalatable taste of the meat; and general respect for an animal that is rarely seen by hunters (RRH, GO, S. Calmé unpubl. data). Likewise, in Quintana Roo, Jorgenson (1995) found that local residents did not hunt tapir because they are difficult to kill and to transport.

Tapirs depend on water all year-round, repeatedly visiting water bodies such as *aguadas* (water reservoirs and sinkholes), lakes, streams and rivers. Since 2008 a monitoring effort focused on *aguadas*, using camera traps within Calakmul Biosphere Reserve, has shown that tapir is one of the most frequent species photographed (Reyna-Hurtado et al. 2010), with an abundance of 37.6 individuals/1000 trap-nights (Pérez-Cortez et al. 2012). A recent study suggests that tapirs can also persist in dry areas such as the north of the Calakmul Biosphere Reserve when there are water sources available (Carrillo et al. 2015). Though tapirs are mainly nocturnal in the peninsula (Pérez-Cortez et al. 2012), they will visit *aguadas* during the day in the hottest months of the year.

Tapirs have important roles in the ecosystem. Through seed dispersal, seed predation and herbivory, tapirs shape the structure of the vegetation in their habitat (O'Farrill et al. 2013). In particular, tapirs are the only animals that can disperse large seeds (those once dispersed by extinct megafauna), giving them a unique function in the ecosystem (Hansen and Galetti 2009; O'Farrill et al. 2013). In the Yucatán Peninsula, Baird's tapirs are one of the few known species that can successfully disperse large seeds of tree species such as zapote (*Manilkara zapota*; O'Farrill et al. 2006). In an in-depth study of the relationship between tapirs and

Fig. 10.1 (**a**) Central American tapir (*Tapirus bairdii*) Photo: Rafael Reyna-Hurtado. (**b**) White-lipped peccaries (*Tayassu pecari*). Photo: Rafael Reyna-Hurtado. (**c**) Gray brocket deer (*Mazama pandora*). Photo: Pablo J. Ramírez-Barajas

zapote trees, O'Farrill et al. (2012) found that seeds that pass through the digestive tract of tapir are able to germinate at a similar rate to seeds that did not pass through the tapir's digestive system, therefore facilitating long-distance dispersal of the zapote tree. Despite the presence of what we believe are stable populations of Baird's tapir in the Yucatán Peninsula, the long-distance movement patterns of this species may be under threat, given current rates of habitat fragmentation due to increased human activities (road infrastructure and agriculture), and changes in water resource availability due to climate change (O'Farrill et al. 2013). The disruption of the movement patterns of this species can have detrimental and cascading effects on other species, and on the functional role of this important long-distance disperser of large seeds, putting this species at risk of becoming functionally extinct (McConkey and O'Farrill 2015).

The Maya forest that comprises the forest of the Greater Calakmul Region and the forest in the Peten area in Guatemala may harbor the largest tapir population in its whole distribution range (Naranjo 2009). Although this population may be the largest, several threats to it exist. For example, in the forest surrounding the Calakmul Biosphere Reserve, there has been an increase in tapir conflicts with local residents due to tapirs' crop raiding (M. Sanvicente, RRH, S. Calmé pers. obs.). Future studies aimed at identifying the causes of the increase in conflicts should allow us to suggest whether they are caused by further encroachment of croplands on tapir habitat, by an increasing tapir population, or by a more aggregated distribution of the tapir population. Identifying the cause of this increase in conflicts may help prevent further killing of tapirs in the future. In addition, the isolation of protected areas from communal forest caused by the road network and the subsequent deforestation belt that always accompanies it also presents a major threat. This fragmentation can prevent tapirs from moving between protected areas and the surrounding forests, leading to genetic isolation, which offers a poor scenario for the future of the species. Finally, the central area of the Yucatán Peninsula is predicted to become drier if climate change intensifies, with a predicted reduction of up to 48 % in annual rainfall for the end of the century (Magrin et al. 2007). The effect of this reduction in rainfall may have drastic consequences on the availability of water resources for tapirs (O'Farrill et al. 2014).

Despite an increase in tapir research in the Yucatán Peninsula in recent years, we believe that such research still needs to address the following areas:

1. Movement ecology. This will provide information on areas most frequented by tapirs and therefore areas that need to be conserved to maintain a stable and connected population. In addition, information on how far tapirs are dispersing seeds from parental trees will provide information on their unique long-distance seed dispersal role. Research on tapir's dispersal capacities and connectivity among populations must be given priority due to current habitat encroachment.
2. Population estimates. We need to know the status of the population under different human-induced and ecological conditions. Population estimates

(relative abundance, density, e.g.) are very important information for shaping management and conservation plans and initiatives.

3. Distribution. Research must be carried out in the northern and central areas of the Campeche and Yucatán States with emphasis on coastal ecosystems to confirm the species' presence and to estimate its population status, if possible. Confirmation of tapir presence in areas that are defined as stable tapir habitat is important for understanding the limits of this species' distribution.

*10.2.2 White-Lipped Peccary (*Tayassu pecari*)*

The white-lipped peccary is known in the Yucatán Peninsula as *senso*, *jabalín*, *hauilla* and, in Mayan, *kitam* (Reyna-Hurtado et al. 2014). The white-lipped peccary belongs to the order Artiodactyla and is one of the two peccary species of the Tayassuidae family (known as the New World pigs) that exist in Mexico. White-lipped peccaries are pig-like animals with long black hairs over their entire body except for a white area under the cheeks and the lips that give them its name. At more than a meter in length, 60 cm in height and weighing up to 40 kg, the white-lipped peccary is larger than its congener, the collared peccary, and has longer legs (Emmons and Feer 1997). The white-lipped peccary is a social animal that lives in groups of up to 300 individuals with anecdotal reports of 700 or even 1000 individuals in a single group (Fragoso 2004; R. Bodmer pers. comm.; Mayer and Wetzel 1987).

White-lipped peccary distribution in the Yucatán Peninsula is generally restricted to the large protected areas such as the Calakmul and Sian Ka'an Biosphere Reserves, and the large communal forests that surround these protected areas (Reyna-Hurtado et al. unpubl. data). However, in the last years, three additional populations have been documented, all of them in coastal ecosystems of the Yucatán Peninsula. Two populations have been found in the surroundings of the Laguna de Términos protected area in the Campeche State (Hidalgo-Mihart and Contreras-Moreno 2012), and one in the Dzilam de Bravo State Reserve in the north of Yucatán State (C. Alcerreca pers. comm.). The White-lipped Peccary is classified as endangered by the Mexican list of species at risk (NOM-059, Semarnat 2010), and as Vulnerable by the IUCN Red list (http://www.iucnredlist.org).

White-lipped peccary is among the favorite game species for both subsistence and sport hunters in the area, and it is estimated that its historical distribution range in Mexico has been reduced by 84 % in the last 40 years (Altrichter et al. 2012). The species' complete elimination from some areas outside the Calakmul reserve has been documented, with only small numbers surviving in some communal forests (Reyna-Hurtado 2009). White-lipped peccaries are easily hunted in the dry season at the few remaining water bodies in forests, *aguadas* or savannas, and it has been documented that subsistence hunting combined with sport hunting can eradicate a whole group in a single dry season (Reyna-Hurtado et al. 2010; E. Quijano Hernández pers. comm.).

White-lipped peccaries are specialist animals that prefer humid tall and well-conserved forest with available water (Sowls 1997). In the Yucatán Peninsula the white-lipped peccary disproportionately uses medium semi-perennial forest and low-flooded forest, while avoiding dry semi-deciduous forest (Reyna-Hurtado and Tanner 2005; Reyna-Hurtado et al. 2009; Briceño-Méndez et al. 2014). The diet of white-lipped peccaries is 80 % fruit, while invertebrates are also important food source (Pérez-Cortez and Reyna-Hurtado 2008). These foods are found in greater quantities in the medium semi-perennial forest and in the low flooded forest of the Yucatán Peninsula (Reyna-Hurtado et al. 2009). This species is an assiduous visitor of *aguadas* during the dry season (Fig. 10.1b), and its foraging movements are strongly influenced by water availability in time and space, to such a degree that groups of this species have been classified as "central place foragers" with the central place being the *aguadas* (Reyna-Hurtado et al. 2012).

In a long-term study about white-lipped peccary movement patterns carried out in the Southern Yucatán Peninsula, Reyna-Hurtado et al. (2009) documented that groups move across areas larger than 100 km^2, and that movement patterns are strongly influenced by the availability of water and preferred forest types (Reyna-Hurtado et al. 2009). There is empirical evidence that the groups of this species can travel in a coordinated way and may visit places that they remember, inferring that spatial memory can play a large role in their movements (RRH pers. obs.). The forests of the Yucatán Peninsula hold one of the last and the largest population of white-lipped peccary in Mexico, after the reduction of its distribution in the last 40 years (Altrichter et al. 2012). The only documented estimate of density was 0.43 individuals per km^2 in the southern part of the Calakmul Biosphere Reserve, which would translate into approximately 1500 individuals in that specific area (assuming the forest maintains similar conditions) (Reyna-Hurtado et al. 2010). These findings contrast with research carried in communal forest where this species was very rare and its abundance very low in comparison with that in the protected area (Reyna-Hurtado 2009). These results also raise the possibility of this species becoming isolated in a few protected areas, and locally extinct in the near future in the communal forest where it still persists. This is not the best scenario for the conservation of this endangered wildlife species. In addition, the predicted reduction in rainfall in the center of the peninsula due to climate change (Magrin et al. 2007) will greatly affect the species' persistence in the dry areas of the Yucatán Peninsula. In the Calakmul region, it may force the species to migrate to adjacent communal forests where there is greater standing water availability, and a greater risk of being hunted. In South America, diseases are believed to be one potential cause of the large observed declines in this species, due to contact with domestic pigs (Fragoso 1997). In the Yucatán Peninsula we have no information on this topic, except for some evidence of skin diseases observed recently in wild individuals of white-lipped peccaries in the Maya forest (Reyna-Hurtado et al. 2014).

It will be essential to learn the following information on this species in the near future for conservation and management purposes:

1. Population estimates. Relative abundance and, when possible, density are very important pieces of information for the conservation of this endangered species. We need to document what is happening in terms of population dynamics. This is a species that has particular habitat requirements and is very sensitive to human perturbations, but we do not know the degree of impact on the species from hunting (see Chap. 13), or the effects of seasonal changes, and disease.
2. Movement ecology. Documenting movement patterns and the possible use of spatial memory could be an exciting research program that would contribute to understanding the long-distance movements this species performs, and the decision process in which these movements occur, as well as the ecological conditions that trigger them, providing some guidance for management plans.
3. Water dependence. We need to obtain basic information on the ecological strategies the species has developed to cope with water scarcity. This information may feed models for understanding or predicting the potential impact of climate change on the species in the future.
4. Diseases. An assessment of the health status of wild animals and the domestic animals in the areas where both types of animal overlap will provide sound information on this important issue for white-lipped peccaries.

*10.2.3 Collared Peccary (*Pecari tajacu*)*

In the Yucatán Peninsula collared peccary are locally know as *puerco de monte*, *puerco cinchado*, *coche de monte* and, in Mayan, *kitam*. The collared peccary belongs to the order Artiodactyla and, as mentioned previously, is one of the two peccary species of the Tayassuidae family that exist in Mexico. Collared peccaries are pig-like animals with grey-yellowish hair across their entire body except for the white "collar" that goes around the shoulder area and gives them their name. The collared peccary is significantly smaller than white-lipped peccary: its adult length ranges from approximately 90 to 100 cm, and they weigh between 15 and 28 kg, with a more rounded head and shorter legs than white-lipped peccary (Emmons and Feer 1997). Collared peccaries are also social animals, forming groups of 2–50, but less cohesive than groups of white-lipped peccaries (Sowls 1997). A comparative study of the two species of peccaries found that collared peccary had an average of six individuals per groups with a range of 1–44, for a total of 85 groups observed. Individuals usually remained close to each other, but can travel separated for periods of time, especially when they are in danger of predation and run in different directions, a behavioral strategy different than the stand-and-defend pattern of the larger white-lipped peccary (Sowls 1997).

The collared peccary has the largest distribution range in the Americas of any peccary species, occurring from northern Argentina to the southern United States. In the Yucatán Peninsula it is possible to find it in almost all regions except close to major cities or in the most disturbed areas (Reyna-Hurtado et al. 2014). The collared peccary is classified as No Risk by the Mexican list of species at risk

(NOM-059, Semarnat 2010), and as a species of Least Concern by the IUCN Red List (www.iucnredlist.org).

Collared peccaries use a diverse array of habitats, ranging from humid tall perennial forest to the desert. In the Yucatán Peninsula this species is found in tropical semi-perennial forest with some degree of tolerance to disturbed areas. For example, many reports of hunted collared peccaries come from highly disturbed areas of secondary habitat (*acahuales*) and *milpas* surrounding communities (Ramírez-Barajas and Naranjo 2007). The collared peccary is a common animal in the Yucatán Peninsula, even where it is hunted (Weber 2000; Gonzalez-Marin et al. 2008; Reyna-Hurtado and Tanner 2007; Urquiza-Haas et al. 2009). Reyna-Hurtado and Tanner (2005, 2007) found that collared peccary abundance is similar in a set of areas where they are hunted and within the Calakmul protected area. This finding seems surprising given that some reports point to the collared peccary as one of the most hunted animals in many communities (e.g., Escamilla et al. 2000; Ramírez-Barajas and Naranjo 2007; Weber 2000).

Reyna-Hurtado and Tanner (2005) also found that collared peccaries are habitat generalists, though they prefer tall humid habitats (medium sub-perennial forest) when available, while avoiding the dry forest to certain degree (Briceño-Méndez et al. unpubl. data). In the Yucatán Peninsula no major study has been conducted on this species, therefore it is worthwhile to pay closer attention to the following ecological aspects:

1. Population dynamics. It would be interesting to understand how the species adapt to hunting pressure and other human activities in the non-protected forest.
2. Social behavior. It would be interesting to conduct research on group size and fission-fusion social behavior in this species, and especially to compare behavior in protected versus non-protected areas.
3. Movement ecology. Movement patterns and home range are essential information for the development of management plans for this species, to ascertain the area needed in order to maintain a viable population on a given site.
4. Disease transmission. This is another relevant topic, as we know nothing about the potential transmission of diseases and parasites between collared peccaries and domestic pigs or other domestic species. This topic is becoming more relevant every day, as the borders between wild habitats and disturbed sites increase with the opening of previously forested areas.

*10.2.4 White-Tailed Deer (***Odocoileus virginianus***)*

White-tailed deer is known as *venado* or *venado real* in some areas of the Yucatán Peninsula or *Quej* in Mayan. The white-tailed deer, with a length of 1.1–2.2 m and weighing between 30 and 50 kg, is the largest of the three species of deer in the Yucatán Peninsula. They are brownish with grey in some areas, especially the head (Emmons and Feer 1997). The white-tailed deer is apparently distributed across almost the entire Yucatán Peninsula, with the exception of the most populated areas, near major cities. White-tailed deer are not classified as being at risk by the Mexican list of species at risk (NOM-059, Semarnat 2010), and are classified as Least Concern by the IUCN Red List (http://www.iucnredlist.org).

White-tailed deer is a very important species for the rural communities of the Yucatán Peninsula (Chap. 13) and despite being one of the most studied animals worldwide, it has been largely ignored by science in the Yucatán Peninsula, with only a handful of studies on diverse topics. For example, Mandujano and Rico-Gray (1991) studied the use of this species by the Mayan of the State of Yucatán and found that white-tailed deer were the most important game species for some Mayan communities there. They also found that hunters had five different methods to hunt white-tailed deer with *batida* (beat hunting) being the most common. In a 2-year study in the Calakmul region, Weber (2005) found that white-tailed deer are browsers, and that 70 % of their diet is comprised of leaves and stems year round, a high contrast to the more frugivorous brocket deer (*Mazama* spp). Also, white-tailed deer have the most diverse diet of the three deer species of the Yucatán Peninsula. Weber (2005) estimated that white-tailed deer were more common than brocket deer in agricultural and secondary forested areas. In addition, this author found that white-tailed deer was the most hunted of the three deer species in the Calakmul region, and provided the highest biomass of all species hunted in terms of kilograms of meat (Weber 2000).

In another study in El Eden Protected Area in the State of Quintana Roo, Gonzalez-Marin et al. (2008) found that white-tailed deer was the most abundant ungulate species with a density of 4.1–5.5 individuals per km^2. Reyna-Hurtado and Tanner (2007) found also that relative abundance of this species was higher in hunting and disturbed areas than in the Calakmul Biosphere Reserve; that white-tailed deer was a common species in secondary forest; and that they prefer low flooded forest in the hunting areas (Reyna-Hurtado and Tanner 2005). Weber (2005) found a much lower density, with 0.021 individuals per km^2, and did not observe variation in relative abundance between a community in which the deer were hunted and Calakmul Biosphere Reserve in the State of Campeche. These findings highlight the persistence of this species in highly disturbed areas, and its potential as a game species given a well-organized hunting and management plan that takes care to conserve populations.

Important information needed for this species for the development of management and conservation plans includes:

1. Movement ecology. It would be interesting to know how this species is able to move through and survive in highly disturbed areas, while minimizing encounters with hunters.
2. Population dynamics. As one of the preferred target species for hunters, this species is at risk of being overhunted despite its potential as a persistent species. We need to pay attention to its population status and any changes in that status, especially in highly disturbed areas in the State of Yucatán.

*10.2.5 Brocket Deer Species (*Mazama pandora *and* Mazama temama*)*

Brocket deer are known as *venados cabritos*, *cabritos* and *chac yuk* for the red brocket deer and *sac yuk* for the gray brocket deer in Mayan. Brocket deer include two species in the Yucatán Peninsula, the gray brocket deer (which was formerly classified as *M. gouzabira* but was renamed *M. pandora*, Medellín et al. 1998, Fig. 10.1c) and the red brocket deer, formerly classified as *M. americana* (Weber 2005) but recently renamed *M. temama*, based on genetic and morphometric criteria that distinguish it from the former species (Bello et al. 2010). Brocket deer are small deer, 90–120 cm in length and weighing no more than 20 kg. *M. pandora* is slightly heavier then *M. temama;* the weight of 21 individuals of *M. pandora* from Calakmul Biosphere Reserve averaged 17.5 kg for females and 20.5 kg for males (Weber and Medellin 2010), while *M. temama* averaged 16.3 kg for adult males hunted in the Calakmul region (Weber 2014). Brocket deer are also a very important species for subsistence hunters, especially in the southern areas of the Yucatán Peninsula (Escamilla et al. 2000; Weber 2000). Red brocket deer have been subjected to sport hunting in recent years in some communities where sport hunting is allowed under the UMA (Units for Wildlife Management and Conservation) scheme (Weber et al. 2006; RRH pers. obs.).

Red brocket deer inhabit humid tall and medium forests, so their distribution covers only the southern portion of the Yucatán Peninsula and extends to the north of the peninsula along the eastern forest reaching the coastal areas of Quintana Roo (Gonzalez-Marin et al. 2008). The gray brocket deer better tolerates dry areas and is a common inhabitant of most of the peninsula, although no signs of its presence (nor that of the red brocket deer) was found in a study carried out in the coastal ecosystems of the north of the Yucatán Peninsula (Hernández-Pérez unpubl. data). The gray brocket deer is a species endemic to most of the Yucatán Peninsula and the northern areas of Guatemala and Belize (Medellín et al. 1998; Weber et al. 2008). Red and grey brocket deer are not classified as at risk by the Mexican list of species at risk (NOM-059, Semarnat 2010), though the red brocket deer is classified as Data Deficient and the grey brocket deer as Vulnerable by the IUCN Red List (http://www.iucnredlist.org).

Brocket deer are small, shy animals that live in well-conserved forest rarely seen in disturbed areas (Reyna-Hurtado and Tanner 2005), although grey brocket deer can be found in forests with some degree of disturbance, especially in the north of the Yucatán Peninsula (RRH, S. Calmé pers. obs.). These facts make them difficult to study. The home range for the gray brocket deer was preliminarily estimated at less than 50 ha for a single animal during a three-month follow up (Weber and Reyna-Hurtado unpubl. data). Weber (2005) in an in-depth study on the feeding habits found that the gray brocket deer was a generalist, while the red brocket deer was a specialist, with more fruits consumed than the gray brocket deer and the white-tailed deer. This author found that gray brocket deer switch from frugivory to browsing during the year while red brocket deer specialize in fruits year-round (Weber 2014).

Estimates of the abundance of brocket deer using tracks are difficult to obtain because it is almost impossible to distinguish between the tracks of the two species. Some estimates of density using transects are between 0.90 and 1.5 deer/km^2 for both species of *Mazama* for the Calakmul region (Weber 2005). For El Eden, a protected area in the northern Quintana Roo, an estimate of 1.7 deer/km^2 was found for the two species combined (Gonzalez-Marin et al. 2008). In the Calakmul Biosphere Reserve, Reyna-Hurtado and Tanner (2007) found that brocket deer signs (again, the combination of tracks of the two species) were the most abundant of all ungulate species and that relative track abundance did not vary between hunting sites and the Calakmul protected area. The same authors also found that tracks of brocket deer were the most abundant tracks in dry forest of the protected area and that dry forest was used disproportionately for these species, while in the hunting sites the low flooded forest was used disproportionately (Reyna-Hurtado and Tanner 2005). The question remains as to which species is responsible for these habitat preferences, or whether low flooded forest could merely serve as refuge in hunted areas (see Chap. 13 for vegetation types used by hunters).

The research priorities for these two species include:

1. Population dynamics. The impact of hunting activities and deforestation on the population of the two species would be the most important information for conservation purposes.
2. Movement ecology. Home range size and movement patterns need to be investigated for both species in the protected areas and in sites with a high rate of hunting. If there are major differences in these variables between these two site types, it will have conservation implications, as we can assess the impact of human activities on the species.
3. Habitat use. Habitat preferences at the species level are needed to elucidate the differences between the species in the Yucatán Peninsula, the only place in the world where the two species live sympatrically. It would also be valuable to determine whether habitat preferences vary based on hunting regime.
4. Distribution. It is crucial to delimit the distribution range for the two species at the scale of the Yucatán Peninsula.

10.3 Carnivora

*10.3.1 Jaguar (***Panthera onca***)*

The jaguar is known in the Yucatán Peninsula as *tigre*, *onza* or *balam* in Mayan. This felid is the largest terrestrial predator in the Neotropics. It is unmistakable given its size and yellowish fur covered with black circular spots. The jaguar is the third largest cat in the world, with a length of 110–180 cm and a weight of 31–158 kg, with heavier individuals inhabiting open areas of Brazil (Emmons and Feer 1997). In the Yucatán Peninsula the average weight is 54 kg for males and 44 kg for females (Chávez 2010). The jaguar is listed as Endangered by the Mexican list of species at risk (NOM-059, Semarnat 2010), and as Near Threatened by the IUCN Red List (www.iucnredlist.org). The jaguar has been used as a charismatic species in the arguments for the conservation of the Mayan jungle, including issues such as impacts of road construction, the creation of biological corridors, and the impacts of infrastructure construction (Conde et al. 2010).

Research in the Calakmul region (southern Campeche and Quintana Roo) on feeding patterns and prey availability showed that jaguars prey on 76 % of the mammal species present in the area (Amín et al. 2004). Analysis of scats found prey belonging to 15 species of mammals, mostly weighing over 1 kg. The dominant prey in that area were coati (*Nasua narica*), white-tailed deer, collared peccary, and armadillo (*Dasypus novemcinctus*) (Chávez et al. 2011). The jaguar is opportunist in its choice of prey, except for collared peccary (*Tayassu tajacu*), paca (*Cuniculus paca*), and armadillo (*Dasypus novemcinctus*), on which it appears to feed selectively (Chávez et al. 2011; Chávez 2010).

Subsistence hunters in the Calakmul region compete heavily with jaguar for the same prey, because considerable overlap occurs between felids and local residents' hunting areas (Escamilla et al. 2000; Chávez et al. 2007, 2011; see Chap. 13 for list of hunted species). Some studies have documented a direct impact on jaguar population size and density, as well as that of other species, as a result of declining prey numbers (e.g. Peres 1990). In extreme cases, mature forests may not harbor large animals due to removal by hunters, a condition known as "empty forest syndrome" (Redford 1992).

In the Calakmul region, jaguars are mainly crepuscular and nocturnal; they prefer sites with good forest cover near water bodies or riparian habitat (Chávez 2006, 2010). In a 7-year study focused on wildlife associated with ponds in the Calakmul Biosphere Reserve using motion-triggered trap cameras, Reyna-Hurtado et al. (unpubl. data) have found that four different male jaguars periodically visited the same pond, and a single individual travelled to three different ponds in an area of approximately 20 km^2. The same researchers have found that jaguars visit ponds on a regular basis, and that the rate of these visits has remained constant since 2008.

Using GPS collars on jaguars, movement pattern estimation showed that they could move up to 20 km in a single night in search of prey, and that home ranges differ between males and females. In Calakmul, a male jaguar had an active range of 700 km^2; in contrast, females moved in areas of 180 km^2 (Chávez 2010). In El Eden Ecological Reserve a radio-collared male ranged across 800 km^2 (M. Lazcano and CC pers. obs.).

Jaguar's habitat use has been assessed using geographic information systems (GIS) and radio telemetry data in the Calakmul Biosphere Reserve. Jaguars favored medium semi-evergreen forest (60 % of the time), followed by tropical deciduous forest (25 %). These results were similar between males and females (Chávez 2006, 2010). In Ejido Caoba, Quintana Roo, jaguars used medium semi-evergreen forest

(49 %), followed by low semi-evergreen forest (25 %); usage rates by males and females were also similar (Chávez 2010).

Zarza et al. (2011) studied the habitat preferences and the use of human-transformed areas in radio-collared jaguars in the southern Yucatán Peninsula. He found that jaguars mostly ranged in areas far from human population settlements (at an average distance of 6.5 km) and roads (4.5 km). This analysis also showed that there were strong differences in habitat use between males and females, though both sexes preferred tall forests and avoided swamps and secondary vegetation. Males avoided low forests, while females used them according to their availability. In some instances, males used farmlands and grasslands less frequently than females (Chávez 2010; Conde et al. 2010; Colchero et al. 2011). A recent study, using satellite telemetry, is focusing on jaguars' movements and the impact of a major new road in the northeast Yucatán Peninsula. Researchers are assessing detailed information on the impact of the new road and its effect on habitat use and connectivity (Mircea-Hidalgo et al. unpubl. data).

Jaguars and humans engage in conflict when felids prey on livestock such as cattle or sheep. This problem is most acute when there is no cattle management and stock range freely. Several strategies to reduce predator-livestock conflicts are underway, including a predator compensation fund for dead animals to discourage ranchers from killing jaguars (*Fondo de Aseguramiento* run by SAGARPA). This strategy is difficult to apply, due to a series of hurdles such as a lack of personnel and bureaucratic delays before the livestock owner receives compensation (S. Calmé et al. unpubl. data; D. Sima pers. comm.). Other strategies include workshops to improve livestock management and keep domestic animals in enclosures instead of allowing them to range freely. The jaguar-human conflict is far from being solved in the Yucatán Peninsula, and will require several strategies for wildlife and livestock management, including political will and scientific attention.

Densities of jaguars in the peninsula have been estimated by various methods (genetic fingerprint, camera trapping, and radio telemetry) and have yielded densities from 1 to 7 individuals per 100 km^2 (Aranda 1998; Ceballos et al. 2002; Chávez 2010; Chávez et al. 2011). For example, in a long-term study on jaguars in the Calakmul Biosphere Reserve and southern Quintana Roo, Ceballos et al. (2002) obtained a density of 3–7 jaguars per 100 km^2 using radio telemetry.

Some potential distribution models in the Yucatán Peninsula (Chávez and Zarza 2009) have showed that nearly half (57,294 km^2; 42 %) of the peninsula is suitable habitat for the jaguar. At the state level, the States of Quintana Roo (24,409 km^2) and Campeche (22,763 km^2), encompass the largest proportion of this habitat, with about 90 % of the total area, while the rest lies in the State of Yucatán (7814 km^2). Taking into consideration the three major vegetation types: tall and medium forest (65 %), medium forest (19 %), and low forest (16 %) it has been estimated that about 1850 individuals inhabit in the peninsula, which accounts for half of the total population in Mexico (Ceballos et al. 2011). It was also estimated, that jaguars are distributed proportionally, with most of the individuals (84 %) in the Yucatán Peninsula dwelling in high and medium forests (1550 individuals) and 16 % (300 individuals) in the lowland forests (Chavez et al. unpubl. data).

The Calakmul region comprises a little more than 14,000 km^2 and includes four protected areas (Calakmul, Bala'an Ka'ax, Balam Ku and Balam Kin). It has an estimated population of 700 jaguars, the largest population in Mexico, which make this region very important for the species. The region is adjacent to the Maya Biosphere Reserve of Guatemala and Rio Bravo-Dos Milpas in Belize, which together encompass an extensive forest tract. They thus contain the largest estimated jaguar population in Mexico, and probably the largest population at the northern limit of the species' distribution.

In northern Yucatán Peninsula, the reserves of Ria Lagartos, Yum Balam, El Zapotal and El Eden Ecological Reserve harbor a population of around 200 individuals (Faller et al. 2007, 2011; Lazcano et al. unpubl. data). In the Yucatán Peninsula, population estimates have been based on criteria such as habitat size and quality, as well as spatial configuration. Six regional areas for jaguar conservation (ARCJ by its Spanish acronym) have been identified: (1) PY-ARCJ-Laguna-Centla Terms, (2) PY-ARCJ-Petenes-Palmar, (3) PY-ARCJ-Calakmul, (4) PY-ARCJ-Ticul-Bala'an Ka'ax, (5) PY-ARCJ-Sian Ka'an, and (6) PY-ARCJ-Dzilam-Yum Balam (CC unpubl. data). Those areas were selected based on their outstanding features and existing actions to conserve the largest populations within the limits of the species' distribution, together with Guatemala and Belize.

The Yucatán Peninsula is a crucial area for jaguar conservation in Mexico. Further scientific information is needed to develop conservation and management plans. Efforts must include:

1. Distribution and habitat change. Obtaining a detailed understanding of the jaguar's distribution range in the Yucatán Peninsula, especially in areas where it has not been intensively sampled, such as mangroves and lowland forests, is a priority. Additionally, the impact of changes in land use and its effect on the spatial distribution of the species must be assessed.
2. Population dynamics. The density and status of populations in the Mayan forest and in the coastal protected areas are necessary data for developing conservation plans, and for a monitoring program of jaguar and prey populations.
3. Corridors. Research in jaguars' use of existing corridors and determining connectivity areas between protected areas will be essential for the maintenance of the species.
4. Jaguar-Human conflict. Information on human-jaguar interactions and strategies to minimize conflicts is urgently needed to reduce losses to local residents' livestock, and prevent unnecessary killing of jaguars.
5. Movement ecology. Continuing investigation of home range and territoriality patterns will provide exciting insight on basic ecological aspects of the species that are only partially known at present.
6. Generate a regional strategy for jaguar conservation in the Yucatán Peninsula, in which various actors, institutions at different levels of government (ejido, municipal, state and regional), and other countries are involved. This strategy should be interagency and international, including especially Guatemala and

Belize, and could be made possible by reactivating the international program, "Jaguares sin Fronteras" (Jaguars Without Borders).

7. Strengthen the alliance, "Voice of the Jaguar", which connects jaguar conservation efforts by private and public reserves with natural protected areas.

*10.3.2 Puma (*Puma concolor*)*

The puma or cougar is locally known as *león* or *leoncillo* and is the second largest cat of America, after the jaguar. Pumas have a relatively small head and large tail; their fur is grey to reddish, and they reach 85–154 cm in length and weigh from 20 to 129 kg across their entire range (Emmons and Feer 1997). In the Calakmul region the average weight of eight males was 42 kg, and 38 kg for one female (Chávez and Ceballos 2014a, b). In the Yucatán Peninsula cougars are usually smaller than jaguars (Chávez 2010). Pumas have a wider distribution than jaguars: they inhabit temperate ecosystems across both American continents, from Argentina to Canada. Pumas show a flexible use of habitat, with records of the species from sea level to 4500 m a.s.l. (Emmons and Feer 1997). Due to their wide distribution, pumas are not included on the Mexican list of species at risk (SEMARNAT 2010) and are classified as of Least Concern by the IUCN Red List (www.iucnredlist.org).

Data on puma diets on the Yucatán Peninsula showed that they prey on 12 species of mammals, mainly *Cuniculus paca,* paca, followed by *Dasypus novemcinctus,* armadillo, and *Tayassu pecari*, white-lipped peccary (Chávez 2010). The same study found that they prey in equal proportion on *Dasyprocta punctata*, Central American Agouti, *Mazama* spp., *Pecari tajacu*, collared peccary, and *Odocoileus virginianus*, white-tailed deer (Chávez 2010). Pumas are common visitors of ponds within the Calakmul Biosphere Reserve (Reyna-Hurtado et al. unpubl data). A monitoring program of wildlife associated with ponds has recorded pumas for seven consecutive years with a relatively consistent visitation rate that is higher than that of jaguars, although visits decreased in the last years of the study (Reyna-Hurtado et al. unpubl. data).

In a radio-tracking study at the Costa Maya site in Calakmul, Campeche, the active range over a 3-year tracking study of a female cougar was 108 km^2, and 135 km^2 for a male (Ceballos et al. 2002; Chávez 2006). Male cougars tend to use dry deciduous forest, while females prefer semi-evergreen forest (Chávez 2010). In a long-term study (2008–2012) in El Eden Ecological Reserve using camera traps, it was found that the puma used medium forest and trails, and were mainly nocturnal (Ávila-Najera et al. 2014a, b).

We are not aware of any study that focuses exclusively in pumas in the Yucatán Peninsula, although one animal has been recently radio collared in the area near Playa del Carmen and is currently being followed by a team of researchers (M. Hidalgo-Mihart pers. comm.). In Belize, another team has studied movement of jaguars and pumas in a highly fragmented area in the central region of the country. Although the data are still unpublished, preliminary analyses suggest that home ranges of pumas are significant smaller than those of jaguars (O. Figueroa pers. comm.).

In the Private Reserve El Zapotal, Yucatán, the first photograph of a puma was taken in 2004 (Faller et al. 2007), and since then there has been a project to survey populations of jaguars and pumas in the area. A study on the effects of Hurricane Dean on the presence of carnivores in the ejidos X-Hazil and Felipe Carrillo Puerto found that jaguars were not present at the disturbed site but pumas were. In fact, pumas had a higher index of relative abundance in the disturbed site than in other four, non-disturbed sites (Hernández-Díaz et al. 2012). This highlights the puma's flexibility and ability to inhabit disturbed areas.

Cougar densities have been estimated using radio telemetry in an area of 3100 km^2 and it was extrapolated that perhaps 217 pumas inhabit the Calakmul Biosphere Reserve, and approximately 409 pumas may exist in the set of protected areas including Calakmul Biosphere Reserve, Balam Ku and Balam Kim State Reserves and Balam Ka'ax Protected Area (Chávez 2006). In El Eden Ecological Reserve, puma density was estimated at between 2.1 and 4.3 individuals per 100 km^2, using camera traps and spatially explicit models (Ávila-Najera et al. 2014a, b).

In the Calakmul Biosphere Reserve pumas have been observed in courtship during the dry season, and a den with two cubs was found in a seasonal pond in the middle of the reserve, also during the dry season (February–May) (J. Zuñiga pers. comm). There is not clear information on the role that pumas play in the human-large cat conflict due to livestock predation. However, pumas probably focus on different prey species than jaguars, and may prey on sheep or goats rather than calves or cows (CC pers. obs.). Attention needs to be paid to determine the role that pumas may be playing in human-large cats conflicts.

Important information needed on pumas include:

1. Distribution. We need to know precisely the current distribution of pumas, since there is not much detailed information on its distribution across the Yucatán Peninsula.
2. Population dynamics. The puma's population status on the peninsula is unknown, and little is known about variables that affect the population dynamics of the species. Ascertaining the basic parameters of this species in tropical forests is a priority. Population status information will be important for developing management and conservation plans for the species.
3. Movement ecology. Investigation of basic ecological issues such as home range, use of corridors, habitat use, and movement patterns is important for our knowledge of the species' spatial needs and use of connectivity areas.
4. Human-large cat conflicts. Is the puma responsible for livestock losses? Information regarding this point is important for preventing losses for local residents and reducing the unnecessary killing of pumas.

10.4 Primates

*10.4.1 Spider Monkey (*Ateles geoffroyi yucatanensis*)*

The spider monkey is one of the three primate species living in the Yucatán Peninsula; the subspecies *A. g. yucatanensis* is present in the forests of the region. Spider monkeys belong to the family Atelidae, and are the largest of the three species living in Mexico (Emmons and Feer 1997). In the area they are known as *mono araña*, *chango* or *Ma'ax* in Mayan [although see Stross (2008) for a set of names of monkeys at the Mayan archeological sites]. Spider monkeys have a distinctive prehensile tail and relatively long arms and legs in relation to their body and head. Their hair is long, with a dark brown and reddish color in the dorsal areas and white and yellowish color in the ventral areas. The size of the body ranges between 45 and 55 cm in length, but the tail alone ranges in length from 70 to 85 cm. Spider monkey can weigh up to 9.5 kg. This genus of primates has no thumb finger and females have a pink elongated clitoris (Napier and Napier 1967; Reid 1997).

Spider monkeys are diurnal primates inhabiting medium and tall tropical forests and feeding on fruits (Reid 1997). They also can live in dry or semi-dry tropical forest (Chapman 1990), and due to their specific feeding habits and habitat preferences, the presence of this species is used as an indicator of well-conserved ecosystems (Fedigan et al. 1988). Spider monkeys are social animals, forming territorial groups of from 8 to 78 individuals that during the day divide into small subgroups to forage (Fedigan and Baxter 1984; Nowak 1999; RRH pers. obs.).

Spider monkeys are inhabitant of the humid tall forest of the Yucatán Peninsula and their range is restricted to well-conserved forest of the southern areas of the peninsula, extending to the north through the humid forest of Quintana Roo (Ramos-Fernández et al. 2004) and the coastal ecosystem of the State of Yucatán (El Palmar State Reserve, E. Hernández-Perez pers. obs.), and Los Petenes Biosphere Reserve in northwest Campeche (Serio-Silva et al. 2006). Spider monkeys are listed as Endangered by the Mexican list of species at risk (SEMARNAT 2010) as well as by the IUCN Red List (www.iucnredlist.org).

Several studies have been conducted on spider monkeys in the Yucatán Peninsula. These studies detail their distribution (Serio-Silva et al. 2006; Vidal-García and Serio-Silva 2011; Vidal-García and Serio-Silva unpubl. data), and describe several localities where the presence of the species has been confirmed. These areas were localized in the southern and eastern areas of the peninsula and in some coastal ecosystems of the north of Yucatán and Campeche States (Vidal-García and Serio-Silva 2011; E. Hernandez-Perez and RRH pers. obs.). In another study about the movement patterns of this species in the north of the State of Quintana Roo, Ramos-Fernández et al. (2004) found that spider monkeys move through the landscape following a Levy walk pattern, a pattern consisting in several similar-size steps with occasional long steps that displace them to areas that have not been recently foraged (Viswanathan et al. 1996). This paper was the basis for developing another study that confirmed the same pattern in the Calakmul Biosphere Reserve (R. Reyna-Hurtado et al. unpubl. data.). In a protected area of Quintana Roo State (Punta Laguna Reserve) several studies have been carried out on behavior, including group fission-fusion strategies, the site fidelity this species exhibits (Ramos-Fernández and Morales 2014; Ramos-Fernández et al. 2013) and the male-female behavioral relationship (Slater et al. 2009).

One study aimed at investigating ecological aspects of spider monkeys in Calakmul Biosphere Reserve determined that they have a home range of 1.18–1.80 km^2 using the fixed kernel method at 95 % for a large group of this species inhabiting the forest of the ancient city of Calakmul (Sarabia-Hernandez et al. unpubl data). This group of spider monkeys is one of the largest recorded, with a sighting of 78 individuals heading to sleeping sites one evening (RRH pers. obs.).

Velázquez et al. (unpubl. data) investigate the impact of logged forest on patterns of seed dispersal. They found more types of seeds, and larger quantities, being dispersed by a community of spider monkeys living in unlogged forest than another community living in a logged forest of the Calakmul region, although the differences were not significant. However, it was shown that despite these findings, spider monkeys also disperse large quantities of seeds in logged forests. The same authors found that spider monkeys used significantly larger and taller than average trees as sleeping sites, and that some of the tree species being used as sleeping or feeding sites (*Diospyros* spp, *Bucidas buceras, Manilkara zapota, Swietenia macrophylla, Ficus* spp.) are subject to logging pressure by local residents. Velázquez et al. (unpubl. data) found 34 species that spider monkeys feed on, with *Diospyros, Cryosophila argentea, Brosimum alicastrum, Manilkara zapota* and *Ficus* spp. constituting important parts of the species' diet.

Spider monkeys live under pressure as their specific habitat is disappearing wherever humans colonize new areas. In disturbed forest, spider monkeys generally remain in larger forest blocks, because the species is very sensitive to primary forest fragmentation (Sorensen and Fedigan 2000). The main threats to the species are the loss of primary habitat, subsistence hunting, and the capture of newborns as pets or for circuses (Muñoz et al. 2008). Spider monkeys, in contrast to howlers (*Alouatta* spp.), do not tolerate fragmentation well and are generally absent from highly disturbed areas (Serio-Silva et al. 2006). In addition, in some sites, females with newborns are hunted to capture their offspring, so the juveniles can be sold on the black market as pets. This practice, although uncommon, is still practiced in some areas of the southern ejidos of Campeche and Quintana Roo (RRH pers. obs.).

The risk of disease and parasite transmission as well as human impacts on the stress level of spider monkey was studied by Deveaux (2014) in six groups of wild spider monkeys subjected to different levels of human contact. Deveaux (2014) found that increased blood cortisol levels corresponded to increased levels of human disturbance, and that multiple parasite infections as well as the intensity of the infections (number of eggs per gram of feces analyzed) had a direct positive relationship with degree of human disturbance (Deveaux 2014). This study showed that humans could affect spider monkeys via their presence for ecotourism, for example, the best conserved area in terms of tree coverage (the Calakmul archeological city) was also the area where spider monkeys presented the highest intensity and diversity of parasites (Deveaux 2014).

Research priorities for spider monkeys in the Yucatán Peninsula must include the following issues:

1. Distribution. An update on the species' distribution and the shifts or changes that have occurred in the last 10 years would be a very informative study.
2. Population dynamics. Monitoring the status of the spider monkey population and the groups living in areas subjected to hunting and logging will be essential to developing informed conservation plans.
3. Movement ecology. There are several aspects of movement that have not been explored in this species, such as the relationship of group size and movement patterns, or the movement of groups in highly disturbed areas where they still survive. There is no information on the use of corridors by this species. Therefore we know nothing about its dispersal between protected areas of the Yucatán Peninsula.

*10.4.2 Black Howler Monkey (*Alouatta pigra*)*

In the Yucatán Peninsula, black howlers are known as *mono aullador* or *saraguato*, or *ba'atz* in Mayan (J. Serio-Silva pers. obs.). The black howler is one of two species of howlers living in Mexico, and is the only one that inhabits most of the peninsula. In fact, the Yucatán Peninsula encompasses the majority of its world distribution, as this species lives only there, in the eastern parts of the States of Chiapas and Tabasco, and the northern forest of Belize and Guatemala (Emmons and Feer 1997). Black howlers are large, massive monkeys with entirely black bodies. The males have a large hyoid bone that they use to produce their characteristic howl. Black howlers' length varies from 52 to 63 cm and larger individuals, usually males, can weigh 9 kg, (Emmons and Feer 1997). Black howlers are listed as Endangered both by the Mexican list of species at risk (SEMARNAT 2010) and by the IUCN Red List (www.iucnredlist.org).

Black howlers prefer humid, tall, and well-conserved forest. However, they are tolerant of forest fragmentation and can survive in highly disturbed forest (i.e. Balancan Forest, Tabasco State; Pozo-Montuy et al. 2011; Serio-Silva et al. unpubl. data) better than spider monkeys. Black howlers inhabit different types of habitat, from semi-dry tropical forest to tall humid tropical forest and riparian vegetation (Cuarón et al. 2008; Marsh et al. 2008). They have been reported even in eucalyptus plantations, which demonstrates their great ecological flexibility (Bonilla-Sánchez et al. 2012). Black howlers are diurnal animals that live in the medium and upper strata of the forest and are generally considered folivorous, although they show high preference for fruits (Pavelka and Knopff 2004; Rivera and Calmé 2006).

In the Yucatán Peninsula black howlers are found in the forest of the southern protected areas, such as the Calakmul and Sian Ka'an Biosphere Reserves, and the communal forests that surround them, as well as in coastal ecosystems such as the Laguna de Términos Protected Area and Los Petenes Biosphere Reserve in Campeche State. Although black howlers tolerate forest fragmentation well, it is evident that this activity strongly impact populations of this species. There have been several documented cases in which howlers were hunted by domestic dogs, burned to death when ranchers set fires to convert forests into grasslands, or hit and killed by cars when crossing highways (G. Pozo-Montuy and JCSS pers. comm.).

Black howlers live in cohesive groups of 6 to 12, generally with a dominant male and two or three females, some subadult males, and newborns. In 158 groups of black howlers sighted in a survey of the entire Yucatán Peninsula, Serio-Silva and collaborators (2006) found an average of 5.5 and 6.4 individuals per group in Campeche and Quintana Roo State respectively, while the single troop found in Yucatán State had six individuals (Serio-Silva et al. 2006).

Black howlers are territorial and occasionally patrol the area, sometimes facing off against neighboring groups, although the fights consists of prolonged howling by both sides (K. Rizzo pers. comm.). Home ranges of groups of black howlers are significantly smaller than those of spider monkeys. For example, a study in Calakmul Biosphere Reserve found that over 8 months howlers' home range varied from 0.08 to 0.15 km^2 for two troops (estimated with the fixed kernel method; Hernández-Sarabia 2013). Rizzo studied 4 groups of black howler monkeys during the rainy season over 10 years and found similar home range size. Additionally, the same author found that black howlers show great site fidelity, and their home ranges shifted only slightly during these years (K. Rizzo pers. comm.).

In the most recent study about current distribution of *Alouatta pigra* (Vidal-García et al. 2015) for the State of Campeche, it was notable that black howler monkeys share their geographic range with *A. geoffroyi* (n = 37). This situation is likely because this is one of the sites with the largest extents of tropical forest (39 % of Campeche's territory is protected, Arguelles-Suarez et al. 2007), an area including the Calakmul Biosphere Reserve and the Laguna de Términos Protected Area. These two areas are without doubt of high importance for fostering the conservation of primate species in the Yucatán Peninsula. In the State of Quintana Roo, new data on *A. pigra* presence has been documented in sites with medium semi-evergreen seasonal tropical forest at the Sian Ka'an Biosphere reserve, and towards the western portion of the Calakmul Biosphere Reserve. Due to its location, this entire area is regularly impacted by natural phenomena such as hurricanes, which diminish habitat and food supply for primates (Pavelka et al. 2003, 2007). It is also relevant that the presence of the species was recorded in the northern portion of the State of Yucatán (Kantunilkin and Tizimin municipalities), in zones of high evergreen rainforest and low semi-evergreen seasonal rainforest (Durán-García and García-Contreras 2010). However these areas are especially vulnerable, as ranching and agriculture are the predominant land use activities (López-Castillo 2012).

Despite black howler monkeys status as the most studied species of primate in the Yucatán Peninsula, we still lack information on the following issues to assure the development of informed conservation and management plans for this species:

1. Population dynamics. The status of some populations are unknown, especially those in coastal ecosystems and highly fragmented areas. We do not know how many individuals are killed every year due to forest loss, road crossings, fires, or hunting.
2. Corridor use. Black howlers' potential use of corridors has been studied in Balancan, Tabasco State, but we need to obtain information from many areas about dispersal distances and the best design for a corridor, in order to design areas of connectivity among fragments with black howler populations.
3. Ecological interactions. The species' use of space and resources is a topic that has received scientific attention, but needs more research, as we still do not know if black howlers suffer from interspecific competition with spider monkeys as some observations suggest (Plante et al. 2014) or with mantled howlers (*Alouatta palliata*).
4. Distribution. There is a lack of knowledge of the species in Yucatán State, including whether or not there are viable populations in coastal ecosystems or in the forest that remains in the southern cone of the state.

10.5 Conclusions and Future Research Directions

The Yucatán Peninsula is a region that maintains large pieces of forest in good status of conservation. These forests are the hope for the conservation of large Neotropical fauna that need proportionally extensive habitat to maintain viable populations. In this chapter we have reviewed what is currently known about the ecological aspects of a set of ten species that encompass all large mammals of the Yucatán Peninsula, and are the largest of the Neotropical fauna found in Mexico. Although there are important and substantive advances in the knowledge of ecological relations and behavior of some of these species there is still much to do in terms of research with a focus on conservation goals. One issue, for nearly all species reviewed here, is the need to investigate the degree of human impact on populations. All kind of human activities are taking place in the forest daily: timber extraction, fires, forest fragmentation, hunting, pollution, and tourism, and we lack of information on how these activities are impacting wild populations of large fauna. For example, the synergistic effects of subsistence hunting associated with sport hunting is having a large impact on endangered species, for example the almost complete elimination of groups of white-lipped peccary in some communal forests surrounding the Calakmul Biosphere Reserve (Reyna-Hurtado et al. 2010). Also, in other instances, human impact can be subtle at first glance, as in the finding that tourism and human presence may have a large effect on cortisol level and parasite intensity in spider monkey populations, even in places where forest cover remains largely intact (the archeological zone of the ancient city of Calakmul; Deveaux 2014).

Another research focus that was highlighted as a priority was movement ecology. The fact that many of these species have large home ranges and high dispersal abilities presents unique opportunities to design research aimed at learning the extent of their movements, their home range, and defining or redefining what is known about their habitat preferences. While these topics are themselves exciting subjects of research, they also provide important information to assess the extent of areas that need protection if we want to conserve a viable population of a species. We also need to know how animals disperse, and what characteristics a potential corridor must have to function effectively for different species.

One pattern that was also consistently observed was that many species inhabit large protected areas (mainly in the southern areas of the States of Campeche and Quintana Roo), and there is an increasing risk of populations becoming isolated if connectivity among these protected areas is not maintained. Road construction, the expansion of cattle-ranching areas, and the establishment of monocrop agriculture (such as African oil palm) are threatening this connectivity in the Yucatán Peninsula. Isolation of populations of any species is not the best scenario for conservation at any level.

In summary, the Yucatán Peninsula is an amazing region that still hosts populations of large fauna. Jaguar, puma, deer, tapir and herds of peccaries still roam in the forest floor while primates find their way through the canopy. However, the entire region presents serious challenges to the maintenance of forest cover, forest connectivity, and to the continued intact functioning of ecological processes. We must assure that human communities living in the area meet all their basic needs without depleting forest resources and the species that live on them. This challenge is becoming urgent to attend as forests are being transformed every day. The task is one that needs to be tackled by all sectors of society: academia, NGOs, rural communities and governmental institutions all need to work closely to address these challenges. If we achieve these tasks successfully, we will assure that future generations can be amazed by the sighting of a jaguar or a puma, by observing a herd of white-lipped peccaries or a tapir visiting a pond, or by watching a spider monkey traveling on the trees, while howler monkeys are howling at the distance.

References

Altrichter M, Taber A, Beck H, Reyna-Hurtado R, Keuroghlian A. Range-wide declines of a key Neotropical ecosystem architect, the near threatened white-lipped peccary *Tayassu pecari*. Oryx. 2012;46:87–98.

Amín M, Ceballos, G, Chávez C. Patrones de alimentación y disponibilidad de presas del jaguar (*Panthera onca*) y del puma (*Puma concolor*) en la Reserva de la Biosfera Calakmul, Campeche, México. Doctoral dissertation, Facultad de Ciencias, Universidad Nacional Autónoma de México, México; 2004. 63 p.

Aranda M. Densidad y estructura de una poblacion del jaguar (*Panthera onca*) en la reserva de la biosfera Calakmul, Campeche, México. Acta Zoológica Mexicana. 1998;75:199–201.

Arguelles-Suarez LA, García Trujillo ZH, Foster R, Rodríguez Salazar J, Torres Pérez JA. Programa estratégico de desarrollo forestal sustentable del Estado de Campeche. Gobierno del Estado de Campeche, Secretaria de Desarrollo Rural, Campeche, Mexico; 2007.

Arroyo-Cabrales J, Álvarez T. A preliminary report of the late Quaternary mammal fauna from Lotún Cave, Yucatán, Mexico. In: Schubert BW, Mead JI, Russell WG, editors. Ice-age cave faunas of North America. Bloomington, IN: Indiana University Press; 2003.

Ávila-Najera DM, Chávez C, Pérez-Elizalde S, Alcántara JL, Lazcano-Barrero MA. Uso de hábitat y coexistencia del Jaguar (*Panthera onca*) y Puma (*Puma concolor*) en el norte de la Península de Yucatán, México. XII Congreso Nacional de Mastozoología, 27 al 31 de octubre de 2014, Puebla, Mexico; 2014.

Ávila-Najera DM, Chávez C, Pérez-Elizalde S, Alcántara JL, Figueroa V, Lazcano-Barrero MA. Densidad, uso de hábitat, patrones de actividad y especies asociadas del Puma (*Puma concolor*) en la Reserva Ecológica el Edén, México. XII Congreso Nacional de Mastozoología, 27 al 31 de octubre de 2014, Puebla, Mexico; 2014.

Bello J, Reyna-Hurtado R, Wilham J. Central American red brocket deer *Mazama temama* (Kerr 1792). In: Barbanti-Duarte JM, Gonzalez S, editors. Neotropical cervidology: biology and medicine of Latin American deer. Jaboticabal, Brasil/Gland, Switzerland: Funep/IUCN; 2010. p. 166–71.

Bonilla-Sánchez YM, Serio-Silva JC, Pozo-Montuy G, Chapman CA. Howlers are able to survive in eucalyptus plantations where remnant and regenerating vegetation is available. Int J Primatol. 2012;33:233–45.

Briceño-Méndez M, Reyna-Hurtado R, Calmé S, García-Gil G. Preferencias de hábitat y abundancia relativa de *Tayassu pecari* en un área con cacería en la región de Calakmul, Campeche, México. Rev Mex Biodivers. 2014;85:242–50.

Carrillo N, Reyna-Hurtado R, Schmook B. Abundancia relativa y selección de habitat de *Tapirus bairdii* en las reseras de Calakmul y Balam Ku, Campeche, Mexico. Rev Mex Biodivers. 2015;86:202–7.

Ceballos G, Chávez C, Rivera A, Manterola C. Tamaño poblacional y conservación del jaguar (*Panthera onca*) en la Reserva de la Biosfera Calakmul, Campeche, México. In: Medellín RA, Chetkiewicz C, Rabinowitz A, et al., editors. Jaguares en el nuevo milenio: Una evaluación de su estado, detección de prioridades y recomendaciones para la conservación de los jaguares en América. Mexico: Universidad Nacional Autónoma de México/Wildlife Conservation Society. México; 2002.

Ceballos G, Chávez C, List R, Zarza H and Medellín R, editors. Jaguar conservation and management in Mexico: case studies and perspectives. Alianza WWF-Telcel/Universidad Nacional Autonóma de México, México, D.F., Mexico; 2011.

Chávez C. Ecología poblacional y Conservación del Jaguar (*Panthera onca*) en la reserva de la Biósfera de Calakmul. Campeche. Master thesis. México, DF, Mexico: Universidad Nacional Autónoma de México; 2006.

Chávez C. Ecología y conservación del Jaguar (*Panthera onca*) y Puma (*Puma concolor*) en la región de Calakmul y sus implicaciones para la conservación de la Península de Yucatán. PhD thesis. Spain: Universidad de Granada; 2010.

Chávez C, Ceballos G, Amín M. Ecología poblacional del jaguar y sus implicaciones para la conservación en la Península de Yucatán. In: Ceballos G, Chávez C, List R, Zarza H, editors. Conservación y manejo del Jaguar en México estudios de caso y perspectivas. CONABIO-UNAM-Alianza WWF Telcel, México DF; 2007.

Chávez C, Zarza H. Distribución potencial del hábitat del jaguar y áreas de conflicto humano-jaguar en la Península de Yucatán. Revista Mexicana de Mastozoología. 2009;13:46–62.

Chávez C, Ceballos G and Amín M. Population ecology of jaguars and its implications for conservation in the Yucatán. In: Ceballos G, Chávez C, List R, Zarza H, editors. Conservación y manejo del Jaguar en México estudios de caso y perspectivas. CONABIO-UNAM-Alianza WWF Telcel, México, DF, Mexico; 2011.

Chávez C, Ceballos G. Puma (*Puma concolor*). In: Ceballos G, editor. Mammals of Mexico. Baltimore, MD: John Hopkins University Press; 2014a.

Chávez C, Ceballos G. Jaguar (*Panthera onca*). In: Ceballos G, editor. Mammals of Mexico. Baltimore, MD: John Hopkins University Press; 2014b.

Chapman CA. Association patterns of male and female spider monkeys: the influence of ecology and sex on social organization. Behav Ecol Sociobiol. 1990;26:409–14.

Colchero F, Conde DA, Manterola C, Chávez C, Rivera A, Ceballos G. Jaguars on the move: modeling movement to mitigate fragmentation from road expansion in the Mayan Forests. Anim Conserv. 2011;14:158–66.

Conde D, Colchero F, Christensen NL, Zarza H, Manterola C, Chávez C, Rivera A, Azuara D, Ceballos G. Sex matters: modeling male and female habitat differences for jaguar conservation. Biol Conserv. 2010;143:1980–8.

Cuarón AD, Shedden A, Rodríguez-Luna E, de Grammont PC, Link A. *Alouatta palliata* ssp. mexicana. The IUCN Red List of threatened species. Version 2014.3. 2008.

Deveaux N. Cortisol y endoparasitos de mono arana (*Ateles geoffroyi yucatanensis*) bajo diferente grado de presencia humana. Master thesis. Campeche, Mexico: El Colegio de la Frontera Sur; 2014.

Durán-García R, García-Contreras G. Distribución espacial de la vegetación. In: Durán-García R, Méndez-González M, editors. Biodiversidad y Desarrollo Humano en Yucatán. Mérida, Mexico: CICY, PPD-FMAM, CONABIO, SEDUMA; 2010.

Emmons LH, Feer F. Neotropical rainforest mammals: a field guide. Chicago, IL: The University of Chicago Press; 1997.

Escamilla A, Sanvicente M, Sosa M, Galindo-Leal C. Habitat mosaic, wildlife availability, and hunting in the tropical forest of Calakmul, México. Conserv Biol. 2000;14:1592–601.

Faller JC, Chávez C, Johnson S, Ceballos G. Jaguar population density and size in the northeastern of Yucatán Peninsula. In: Ceballos G, Chávez C, List R, et al., editors. Jaguar conservation and management in México: case studies and perspectives. México: Alianza WWF/Telcel UNAM; 2011.

Faller JC, Chávez C, Johnson S, Ceballos G. Densidad y tamaño de la población de jaguar en el noreste de la Península de Yucatán. In: Ceballos G, Chávez C, List R, Zarza H, et al., editors. Conservación y manejo del Jaguar en México estudios de caso y perspectivas. México: CONABIO-UNAM-Alianza WWF Telcel; 2007.

Fedigan LM, Baxter MJ. Sex differences and social organization in free-ranging spider monkeys (Ateles geoffroyi). Primates. 1984;25:279–94.

Fedigan LM, Fedigan L, Chapman C, Glander KE. Spider money home ranges: a comparison of telemetry and direct observation. Am J Primatol. 1988;16:19–29.

Fragoso JM. Desapariciones locales del baquiro labiado (*Tayassu pecari*) en la Amazonía: migración, sobre-cosecha o epidemia? In: Fang TG, Bodmer R, Aquino R, Valqui M, editors. Manejo de fauna silvestre en la Amazonía. La Paz, Bolivia: UNAP, University of Florida UNDP/GEF e Instituto de Ecologia; 1997.

Fragoso JM. A long-term study of white-lipped peccary (*Tayassu pecari*) population fluctuation in Northern Amazonia. In: Silvius K, Bodmer RE, Fragoso JMV, editors. People in nature, wildlife conservation in South and Central America. New York, NY: Columbia University Press; 2004. p. 286–96.

Gonzalez-Marin R, Gallina S, Mandujano S, Weber M. Densidad y distribucion de ungulados silvestres en la reserva ecológica El Eden, Quintana Roo, Mexico. Acta Zoológica Mexicana. 2008;24:73–93.

Hansen DM, Galetti M. The forgotten Megafauna. Science. 2009;324:42–3.

Hernández-Díaz M, Ramírez-Barajas PJ, Chávez C, Schmook B, Calmé S. Presencia y abundancia relativa de carnívoros en una selva dañada por el huracán Dean (2007). Rev Mex Biodivers. 2012;83:790–801.

Hernández-Sarabia RU. Estrategias de desplazamiento de *Ateles geoffroyi yucatanensis* y *Alouatta pigra* en la búsqueda de recursos tróficos durante la temporada seca y húmeda en la Reserva de la Biosfera de Calakmul, Campeche, México. Master thesis. Veracruz, Mexico: Universidad Veracruzana; 2013.

Hidalgo-Mihart M, Contreras-Moreno F. Registro de Pecarí labios blancos (*Tayassu pecari*) en la Región de la Laguna de Términos, Campeche, México. Rev Mex Biodivers. 2012;83:868–71.

Hildebrand AR, Penfield GT, Kring DA, Pilkington M, Camargo A, Jacobsen SB, Boynton WV. Chicxulub crater: a possible Cretaceous/Tertiary boundary impact crater on the Yucatán Peninsula, Mexico. Geology. 1991;19:867–71.

Jorgenson JP. Maya subsistence hunters in Quintana Roo, Mexico. Oryx. 1995;29:49–57.

López-Castillo D. Diagnóstico de la problemática social y económica asociada al vampiro común (*Desmodus rotundus*) en ranchos ganaderos de las zonas centro y oriente del estado de Yucatán, México. PhD Dissertation. Xalapa, Mexico: Instituto de Ecología AC; 2012.

Magrin G, Gay García C, Cruz-Choque D, Giménez JC, Moreno AR, Nagy GJ, Nobre C, Villamizar A. Latin América climate change: impacts, adaptability and vulnerability. In: Parry ML, Canziani OF, Palutikof JP, van der Linden PJ, Hanson CE, editors. Contribution of working group II to the fourth assessment report of the intergovernmental panel on climate change. Cambridge: Cambridge University Press; 2007.

Mandujano S, Rico-Gray V. Hunting, use, and knowledge of the biology of the white-tailed deer, *Odocoileus virginianus* (Hays), by the Maya of Central Yucatán, Mexico. J Ethnobiol. 1991;11:175–83.

Marsh LK, Cuarón AD, Cortés-Ortiz L, Shedden A, Rodríguez-Luna E, de Grammont PC. *Alouatta pigra*. The IUCN Red List of threatened species. Version 2014.3. 2008.

Mayer JJ, Wetzel RM. *Tayassu pecari*. Mamm Species. 1987;293:1–7.

McConkey K, O'Farrill G. Cryptic drivers of functional extinction in animal populations. Trends Ecol Evol. 2015;30:182–9.

Medellín RA, Gardner AL, Aranda JM. The taxonomic status of the Yucatán brown brocket, Mazama pandora (Mammalia: Cervidae). Proc Biol Soc Wash. 1998;111:1–14.

Muñoz D, Estrada A, García Y. Survey and conservation of a relict population of spider monkeys (*Ateles geoffroyi*) in the Sumidero Canyon, Mexico. Trop Conserv Sci. 2008;1:151–62.

Naranjo E. Ecology and conservation of Baird's tapir in Mexico. Trop Conserv Sci. 2009;2:140–58.

Naranjo E, Amador-Alcala S, Falconi-Briones F, Reyna-Hurtado R. Distribución, abundancia y amenazas a las poblaciones de tapir centroamericano (*Tapirus bairdii*) y pecarí de labios blancos (*Tayassu pecari*) en México. Therya. 2015;6:227–49.

Napier JR, Napier PH. A handbook of living primates. London: Academic Press; 1967.

Nowak RM. Walker's primates of the world. Baltimore, MD: John Hopkins University Press; 1999.

O'Farrill G, Gauthier-Schampaert K, Rayfield B, Bodin Ö, Calmé S, Gonzalez A, Sengupta R. The potential connectivity of waterhole networks and the effectiveness of a protected area under various drought scenarios. PlosOne. 2014;9, e95049.

O'Farrill G, Campos-Arceiz A, Galetti M. Frugivory and seed dispersal by tapirs: an insight on their ecological role. J Integr Zool. 2013;8:4–17.

O'Farrill G, Calmé S, Sengupta R, Gonzalez A. Effective dispersal of large seeds by Baird's tapir: a large-scale field experiment. J Trop Ecol. 2012;28:119–22.

O'Farrill G, Chapman CA, Gonzalez A. Origin and deposition sites influence seed germination and seedling survival of *Manilkara zapota*: implications for long-distance, animal-mediated seed dispersal. Seed Sci Res. 2011;21:305–13.

O'Farrill G, Calmé S, Gonzalez A. *Manilkara zapota*: a new record of a species dispersed by tapirs. Tapir Conserv. 2006;15:32–5.

Pavelka MSM, Brusselers OT, Nowak D, Behie AM. Population reduction and social disorganization in *Alouatta pigra* following a hurricane. Int J Primatol. 2003;24:1037–55.

Pavelka MSM, McGoogan K, Steffens T. Population size and characteristics of *Alouatta pigra* before and after a major hurricane. Int J Primatol. 2007;28:919–29.
Pavelka MSM, Knopff KH. Diet and activity in black howler monkeys (*Alouatta pigra*) in southern Belize: does degree of frugivory influence activity level? Primates. 2004;45:105–11.
Peres CA. Effects of hunting on Western Amazonian primate communities. Biol Conserv. 1990;54:47–59.
Pérez-Cortez S, Reyna-Hurtado R. La dieta de los pecaríes (*Pecari tajacu* y *Tayassu pecari*) en la región de Calakmul, Campeche, México. Revista Mexicana de Mastozoología. 2008;12:17–42.
Pérez-Cortez S, Enríquez PL, Sima-Panti D, Reyna-Hurtado R, Naranjo E. Influencia de la disponibilidad de agua en la presencia y abundancia de *Tapirus bairdii* en la selva de Calakmul, Campeche, Mexico. Rev Mex Biodivers. 2012;83:753–6.
Plante S, Colchero F, Calmé S. Foraging strategy of a Neotropical primate: how intrinsic and extrinsic factors influence destination and residence time. J Anim Ecol. 2014;83:116–25.
Pozo-Montuy G, Serio-Silva JC, Bonilla-Sánchez YM. Influence of the landscape matrix on the abundance of arboreal primates in fragmented landscapes. Primates. 2011;52:139–47.
Ramírez-Barajas PJ, Naranjo E. La cacería de subsistencia en una comunidad de la zona Maya, Quintana Roo. Mex Etnobiol. 2007;5:65–85.
Ramírez-Barajas PJ, Islebe GA, Calmé S. Impact of hurricane Dean (2007) on game species of the Selva Maya, Mexico. Biotropica. 2012;44:402–11.
Ramos-Fernández G, Mateos JL, Miramontes O, Cocho G, Larralde H, Ayala-Orozco B. Levy walk patterns in the foraging movements of spider monkeys (*Ateles geoffroyi*). Behav Ecol Sociobiol. 2004;55:223–30.
Ramos-Fernández G, Smith-Aguilar SE, Schaffner CM, Vick LG, Aureli F. Site fidelity in space use by spider monkeys (Ateles geoffroyi) in the Yucatán Peninsula, Mexico. PLoS One. 2013;8, e62813.
Ramos-Fernández G, Morales JM. Unraveling fission-fusion dynamics: how subgroup properties and dyadic interactions influence individual decisions. Behav Ecol Sociobiol. 2014;68:1225–35.
Redford K. The empty forest. BioScience. 1992;42:412–22.
Reid FA. A field guide to the mammals of Central America and Southeast México. New York, NY: Oxford University Press; 1997.
Reyna-Hurtado R. Conservation status of the white-lipped peccary (*Tayassu pecari*) outside the Calakmul Biosphere Reserve in Campeche, Mexico: a synthesis. Trop Conserv Sci. 2009;2:159–72.
Reyna-Hurtado R, Tanner G. Habitat preferences of an ungulate community in Calakmul Forest, Campeche, Mexico. Biotropica. 2005;37:676–85.
Reyna-Hurtado R, Tanner GW. Ungulate relative abundance in hunted and non-hunted sites in Calakmul Forest (Southern Mexico). Biodivers Conserv. 2007;16:743–57.
Reyna-Hurtado R, Rojas-Flores E, Tanner GW. Home range and habitat preferences of white-lipped peccary groups (*Tayassu pecari*) in a seasonal tropical forest of the Yucatán Peninsula, Mexico. J Mammal. 2009;90:1199–209.
Reyna-Hurtado R, Naranjo E, Chapman CA, Tanner GW. Hunting and the conservation of a social ungulate: the white-lipped peccary *Tayassu pecari* in the Calakmul, Mexico. Oryx. 2010;44:88–96.
Reyna-Hurtado R, Chapman CA, Calme S, Pedersen E. Searching in heterogeneous environments: foraging strategies in the white-lipped peccary (*Tayassu pecari*). J Mammal. 2012;93:124–33.
Reyna-Hurtado R, March I, Naranjo EJ, Mandujano S. Pecaríes en México. In: Valdez R, Ortega JA, editors. Ecología y manejo de fauna silvestre en México. México: Colegio de Postgraduados; 2014.
Rivera A, Calmé S. Forest fragmentation and the changes in the feeding ecology of the black howler monkey (*Alouatta pigra*) in Calakmul region. In: Estrada A, Garber P, Pavelka M, Luecke L, editors. New perspectives on the distribution, ecology and conservation of Meso-american primates. New York, NY: Springer; 2006. p. 189–213.

Serio-Silva JC, Rico-Gray V, Ramos-Fernández G. Mapping primate populations in the Yucatán Peninsula, Mexico: a first assessment. In: Estrada A, Garber PA, Pavelka MSM, Luecke L, editors. New perspectives in the study of Mesoamerican primates: distribution, ecology, behavior, and conservation. New York, NY: Springer; 2006.
SEMARNAT. Norma Oficial Mexicana NOM-059-SEMARNAT-2010. Protección ambiental-especies nativas de México de flora y fauna silvestres- categorías de riesgo y especificaciones para su inclusión, exclusión o cambio- lista de especies en riesgo. Diario Oficial de la Federación, 30 de diciembre de 2010. 2010.
Slater KY, Schaffner CM, Aureli F. Sex differences in the social behavior of wild spider monkeys (*Ateles geoffroyi yucatanensis*). Am J Primatol. 2009;71:21–9.
Sorensen TC, Fedigan LM. Distribution of three monkey species along a gradient of regenerating tropical dry forest. Biol Conserv. 2000;92:227–40.
Sowls LK. Javelinas and the other peccaries: their biology, management and use. 2nd ed. Collage Station, TX: Texas A&M University Press; 1997.
Stross B. K'U: the divine monkey. J Mesoam Lang Linguist. 2008;1:1–34.
Urquiza-Haas T, Peres CA, Dolman PM. Regional scale effects of human density and forest disturbance on large-bodied vertebrates throughout the Yucatán Peninsula, Mexico. Biol Conserv. 2009;142:134–48.
Vazquez-Dominguez E, Arita H. The Yucatán Peninsula, biogeographical history 65 millions years in the making. Ecography. 2010;33:212–9.
Vidal-García F, Serio-Silva JC. Potential distribution of Mexican primates: modeling the ecological niche with the maximum entropy algorithm. Primates. 2011;52:261–70.
Vidal-García F, Serio-Silva JC, Pineda E. Evaluating the accuracy of species distribution models through field validation: an example of two threatened primate species. Am J Primatol. 2015.
Viswanathan GM, Afanasyev V, Buldyrev SV, Murphy EJ, Prince PA, Stanley HE. Levy flight search patterns of wandering albatrosses. Nature. 1996;381:413–5.
Weber M. Effects of hunting on tropical deer populations in Southeastern México. Master thesis. London: University of London; 2000.
Weber M. Ecology and conservation of sympatric tropical deer populations in the Greater Calakmul Region, South-Eastern, Mexico. PhD thesis. Durham: University of Durham; 2005.
Weber M, García-Marmolejo G, Reyna-Hurtado R. The tragedy of the commons Mexican style: a critique to the Mexican UMAs concept as applied to wildlife management and use in South-Eastern Mexico. Wildl Soc Bull. 2006;34:1480–8.
Weber M, de Grammont PC, Cuarón AD. *Mazama pandora*. The IUCN Red List of threatened species. Version 2014.3. 2008.
Weber M, Medellin R. Yucatán brown brocket deer *Mazama pandora* (Merriam 1901). In: Barbanti-Duarte JM, Gonzalez S, editors. Neotropical cervidology biology and medicine of Latin American deer. Jaboticabal: Funep/IUCN; 2010.
Weber M. Temazates y venados cola blanca tropicales. In: Valdez R, Ortega JA, editors. Ecología y Manejo de Fauna Silvestre en Mexico. México: Colegio de Posgraduados; 2014.
Zarza H, Chávez C, Ceballos G. Jaguar habitat use in human dominated landscape in the Yucatán Peninsula, Mexico. In: Ceballos G, Chávez C, List R, Zarza H, Medellín R, editors. Jaguar conservation and management in México: case studies and perspectives. México: Alianza WWF/Telcel UNAM; 2011.
Zuñiga-Fuentes HE. Distribución y conocimiento ecológico local del tapir centroamericano (*Tapirus bairdii*) en el estado de Campeche. BSc thesis. Campeche, Mexico: Universidad Autónoma de Campeche; 2009.

Chapter 11
Amphibians and Reptiles

Pierre Charruau, José Rogelio Cedeño-Vázquez, and Gunther Köhler

Abstract The three Mexican states of the Yucatán Peninsula have been relatively well explored for herpetofauna, when compared with other states of the country. However, most studies on the herpetofauna of the Yucatán Peninsula have focused on their diversity, taxonomy, and species distribution, and less on their ecology, behavior or conservation status. The major conservation efforts have focused on sea turtles. Although some conservation programs exist locally for crocodiles in the north of the peninsula, to date conservation strategies have mostly been restricted to the designation of protected areas. With 24 species of amphibians and 118 species of reptiles, the Yucatán Peninsula harbors 11.5 % of national herpetofauna diversity, and 19 % of species are endemic to the peninsula. Reptiles and amphibians are two major globally threatened groups of vertebrates, with amphibians being the most threatened vertebrate class. Both groups face the same threats, namely habitat loss and modification, pollution, overharvest for food and pet trade, introduction of exotic species, infectious diseases, and climatic change. Unfortunately, almost none of these issues have been investigated for key populations in the Yucatán Peninsula. For amphibians, studies exploring the presence of the chytrid fungus (*Batrachochytrium dendrobatidis*) and the effects of climatic change are badly needed to understand the specific factors that negatively affect populations in this area. In general, conservation efforts for reptiles and amphibians in the Yucatán Peninsula need to include environmental education, scientific investigation, and law enforcement and application.

P. Charruau (✉)
Centro del Cambio Global y la Sustentabilidad en el Sureste A.C., Calle Centenario del Instituto Juárez s/n, CP 86080 Villahermosa, Tabasco, Mexico
e-mail: charruau_pierre@yahoo.fr

J.R. Cedeño-Vázquez
Departmento Ecología y Sistemática Acuática, El Colegio de la Frontera Sur Unidad Chetumal, Apartado Postal 424, CP 77014 Chetumal, Quintana Roo, Mexico
e-mail: rcedenov@ecosur.mx

G. Köhler
Senckenberg Forschungsinstitut und Naturmuseum Frankfurt, Senckenberganlage 25, 60325 Frankfurt, Germany
e-mail: gkoehler@senckenberg.de

G.A. Islebe et al. (eds.), *Biodiversity and Conservation of the Yucatán Peninsula*,
DOI 10.1007/978-3-319-06529-8_11

Keywords Herpetofauna • Amphibians • Reptiles • Diversity • Ecology • Natural History • Conservation • Yucatán Peninsula

11.1 Introduction

The herpetofauna of Mexico, with 376 species of amphibians and 864 species of reptiles, corresponds to the fifth and second worldwide positions of diversity of these groups, respectively (Flores-Villela and García-Vázquez 2014; Parra-Olea et al. 2014). The earliest European explorations of the herpetofauna of Mexico go back to the sixteenth century, and although they continued during the following centuries, they were mainly made in southern (e.g., Veracruz, Oaxaca), central (e.g., Mexico City, Puebla), western (e.g., Guanajuato) and northern (e.g., Coahuila) states of Mexico. Published scholarship continues to be conducted mainly by Europeans (Flores-Villela et al. 2004). At the end of the nineteenth century, the states of the Yucatán Peninsula were herpetologically almost unknown, perhaps due to a less attractive diversity of species in comparison with other southern states such as Veracruz, Chiapas, or Oaxaca (Flores-Villela and García-Vázquez 2014; Parra-Olea et al. 2014), or to more difficult access. Nevertheless, at the beginning of the twentieth century the majority of the species of amphibians and reptiles occurring in the Yucatán Peninsula had already been described based on specimens collected in other regions from Mexico, Guatemala, and Belize (Lee 1996). In fact, the first explorations and descriptions of species from specimens of the Yucatán Peninsula started in the late nineteenth and early twentieth centuries and increased between 1920 and 1930, mainly due to an increase in archeological explorations in the state of Yucatán (Lee 1996). In the first half of the twentieth century, Yucatán was thus herpetologically the best-known state of the peninsula, whereas records and collections from Campeche and Quintana Roo remained scarce until the mid-twentieth century, when the first studies of herpetofauna began there (Lee 1996; Flores-Villela et al. 2004). Although Campeche, Yucatán and Quintana Roo are now better known in terms of diversity of amphibians and reptiles species, research on herpetofauna in the Yucatán Peninsula is still limited compared to other states (Flores-Villela et al. 2004). In this chapter, the term Yucatán Peninsula refers to the three Mexican states of Quintana Roo, Campeche and Yucatán (i.e., the geopolitical definition) when speaking of species diversity. When speaking of endemism we refer to the species native to this peninsula, but not necessarily restricted to the Mexican portion of the peninsula, since several of these species extend their range into Belize and/or Guatemala (i.e., the geological definition of the peninsula).

The herpetofauna of the Yucatán Peninsula is composed of 142 species (118 reptiles and 24 amphibians) representing 94 genera and 33 families (Table 11.1), corresponding to 11.5 % of the national species count (1240 species; Parra-Olea et al. 2014; Flores-Villela and García-Vázquez 2014). The state of Quintana Roo shows the highest diversity (130 species: 107 reptiles and 23 amphibians) followed by Campeche (121 species: 100 reptiles and 21 amphibians), and Yucatán

Table 11.1 Taxonomic composition of the herpetofauna of the Yucatán Peninsula and number (%) of endemic and introduced species

Groups	Orders	Families	Genera	Species	Endemic species	Introduced
Reptiles	Crocodilia	1	1	2	0 (0)	0 (0)
	Testudines	6	14	16	2 (12.6)	0 (0)
	Squamata					
	Sauria	10	21	44	9 (20.5)	3 (6.8)
	Serpentes	6	40	56	13 (23.2)	1 (1.8)
	Total	23	76	118	24 (20.3)	4 (3.4)
Amphibians	Caudata	1	1	3	1 (33.3)	0 (0)
	Anura	9	17	21	2 (9.5)	1 (4.8)
	Total	10	18	24	3 (12.5)	1 (4.2)
Total		33	94	142	27 (19.0)	5 (3.5)

(101 species: 84 reptiles and 17 amphibians). Among the 142 species, 24 species of reptiles (20.3 %) and three species of amphibians (12.5 %) are endemic to the peninsula (Table 11.1). These endemics include two turtles, nine lizards, 13 snakes, one salamander and two anurans (Table 11.2). Four species of reptiles and one amphibian species have been introduced: *Hemidactylus frenatus*, *H. turcicus*, *Anolis sagrei*, and *Rhamphotyplops braminus* for reptiles (Lee 1996), and *Eleutherodactylus planirostris* for amphibians (Cedeño-Vázquez et al. 2014).

The peninsula effect (a decrease in species richness from the base to the tip of a peninsula), was originally interpreted as a result of potential colonization/extinction dynamics, attributable to the isolating effects of peninsulas (Ricklefs 1973). Despite the fact that this phenomenon has been documented in Florida for amphibians and reptiles (Keister 1971), and in Florida, Yucatán, and Baja California for birds (Mac-Arthur and Wilson 1967) and mammals (Simpson 1964), according to Lee (1980), in the Yucatán Peninsula this effect is observed among herpetofauna only for anurans (species diversity diminishes dramatically from south to northwest). Rather than an isolation process, this pattern could be explained by climatic effects, as the base is much more moist than the tip of the peninsula. For snakes and lizards, species richness is highest at the base, lowest at the center, and intermediate at the northern end of the region. The number of endemic species is greatest at the northern end and diminishes rapidly to the south (Lee 1980). This can be attributed to the fact that the original vegetation of the northwestern area is dry tropical forest, which is isolated from other such forests, leading to increased endemism. In the following sections, we present a synthesis of information on the diversity and ecology of each group of herpetofauna present in the Yucatán Peninsula. We also describe the various threats that currently face this diversity in the peninsula, and offer prescriptions for the direction of the future research on the herpetofauna in this region.

Table 11.2 Endemic species of amphibians and reptiles to the Yucatán Peninsula

Groups	Order	Species
Amphibians	Caudata	*Bolitoglossa yucatana*
	Anura	*Craugastor yucatanensis*
		Triprion petasatus
Reptiles	Testudines	*Cryptochelys creaseri*
		Terrapene yucatana
	Sauria	*Ctenosaura alfredschmidti*
		Ctenosaura defensor
		Sceloporus chrysostictus
		Sceloporus cozumelae
		Sceloporus lundelli
		Mesoscincus schwartzei
		Aspidoscelis angusticeps
		Aspidoscelis cozumela
		Aspidoscelis rodecki
	Serpentes	*Amerotyphlops microstomus*
		Coniophanes meridanus
		Coniophanes schmidti
		Dipsas brevifacies
		Imantodes tenuissimus
		Sibon sanniolus
		Symphimus mayae
		Tantilla cuniculator
		Tantilla moesta
		Tantillita canula
		Agkistrodon russeolus
		Crotalus tzabcan
		Porthidium yucatanicum

11.2 Amphibians

The amphibian community of the Yucatán Peninsula consists of 24 species, distributed in 18 genera, 10 families, and two orders. The order Caudata is represented by three species of lungless salamanders (Plethodontidae) of the genus *Bolitoglossa*, the northern banana salamander (*B. rufescens*), the Mexican mushroomtongue salamander (*B. mexicana*), and the endemic Yucatán mushroomtongue salamander (*B. yucatana*) (Fig. 11.1a). Like other plethodontid salamanders, they presumably exhibit direct development (there is no free-living aquatic stage involved). The eggs are deposited in moist terrestrial substrates, and the larvae complete their development within the egg and hatch into miniature replicas of the adults. There is no specific information on the reproduction of *B. rufescens* and *B. yucatana*, but clutch size in *B. mexicana* may be up to 63 eggs (Wake and Lynch 1976). Gas exchange is effected through skin and tissues lining the mouth and pharynx (Lee 1996). *Bolitoglossa rufescens* is small (reaching

Fig. 11.1 (**a**) The endemic Yucatán Mushroomtongue Salamander (*Bolitoglossa yucatana*) by G. Köhler. (**b**) The endemic Yucatán Casque-Headed Tree Frog (*Tripion petasatus*) by J.R. Cedeño Vázquez. (**c**) The Black-Backed Frog (*Leptodactylus melanonotus*) by G. Köhler.

a maximum snout-vent length [SVL] of 28–33 mm) compared to the mid-sized *B. yucatana* (49–55 mm SVL) and *B. mexicana* (55–80 mm SVL). They are uncommon, both arboreal and terrestrial, and found beneath surface debris, limestone rocks, logs, and on roads at night. During the dry season they likely inhabit leaf axils of bananas, bromeliads, and other water-retaining epiphytes (Wake and Lynch 1976; Calderón-Mandujano et al. 2003). The terrestrial activity of these salamanders is probably restricted to the rainy season. They feed on a variety of small invertebrates, including ants and termites (JRCV pers. obs.); the examination of droppings of a specimen of *B. yucatana* collected close to Chetumal city revealed that ants were the main source of food (Cedeño-Vázquez et al. 2006a).

In toads and frogs (Anura), fertilization is usually external, although internal fertilization occurs in a few species. Most of the species of the Yucatán Peninsula deposit their eggs in water, which hatch into aquatic larvae that subsequently metamorphose into froglets. Some frogs place their eggs in foam nests (genus *Leptodactylus* and *Engystomops*), while other species deposit eggs on vegetation overhanging water (*Agalychnis callidryas* and *Dendropsophus ebraccatus*). Still others (the endemic Yucatán rainfrog *Craugastor yucatanensis* as well as the introduced greenhouse frog *Eleutherodactylus planirostris*) lay their eggs in moist terrestrial sites, where they undergo direct development. During the breeding season, adult males call from temporary or permanent water bodies to attract females for mating. In general, adult females are larger than males. The anurans in the Yucatán Peninsula possess external vocal sacs, except *Rhinophrynus dorsalis,* which has internal sacs. Most species have unpaired vocal sacs in subgular positions (e.g., *Incilius valliceps*). Paired vocal sacs may be in subgular (e.g., *Smilisca baudinii*) or lateral positions, that is, they lie posterior to the angle of the jaws (e.g., *Trachycephalus typhonius* and *Lithobates brownorum*).

The burrowing toad (*Rhinophrynus dorsalis*), the only living species of the family Rhinophrynidae, is relatively large in size (63–80 mm SVL) and its body is globose, flaccid, and covered with loose skin. This fossorial toad is common and is generally found in savannas and seasonal forests (Duellman 1971). Surface activity of adults is restricted to the beginning of the summer rainy season, when males call from temporary bodies of water, females produce clutches of several thousand eggs (Lee 1996). Adults feed on insects, especially ants and termites (McCoy 1966), and during the dry season pass long periods in subterranean chambers of their own construction (Foster and McDiarmid 1983).

Fig. 11.1 (continued) (**d**) The Tungara Frog (*Engystomops pustulosus*) by J.R. Cedeño Vázquez. (**e**) The American Crocodile (*Crocodylus acutus*), note the well-developed preorbital elevation, by P. Charruau. (**f**) The critically endangered Central American River Turtle (*Dermatemys mawii*) by G.A. González-Desales. (**g**) The Yucatán Banded Gecko (*Coleonyx elegans*) by G. Köhler. (**h**) The Allison's Anole (*Anolis allisoni*) by P. Charruau. (**i**) The endemic Yucatecan Cantil (*Agkistrodon russeolus*) by G.A. González-Desales. (**j**) The Snail-Eating Thirst Snake (*Dipsas brevifacies*) by G. Köhler

Two bufonid toads are present in the peninsula: the Gulf Coast toad (*Incilius valliceps*, formerly known as *Bufo valliceps*) and the marine toad (*Rhinella marina*, formerly known as *Bufo marinus*). Both are stout, have large parotid glands on the neck, and a thick, glandular and wart-covered dry skin. The parotoid glands produce a toxic secretion with pharmacological properties. Dogs have been known to die from biting *R. marina*. *Incilius valliceps* is moderate-sized (73–100 mm SVL) compared to *R. marina*, which is the largest (150–200 mm SVL) anuran in the Yucatán Peninsula. These nocturnal and terrestrial toads are abundant in most habitats within the Yucatán Peninsula, including open habitats (common in disturbed areas and in association with human habitation). In contrast, *R. marina* is uncommon or absent from closed-canopy forests (Lee 1996). Breeding may occur throughout the year in both species, but in *I. valliceps* individuals are cued by the first heavy rains at the onset of the summer rainy season. Females of both species release eggs in long strings into water, where they hatch and where tadpoles complete their development. While *I. valliceps* feeds predominantly on insects, *R. marina* consumes a wide variety of invertebrate and vertebrate prey, including smaller members of their own species (Easteal 1986). Despite their presumably noxious skin secretions, *I. valliceps* members are common items in the diet of snakes like *Drymobius margaritiferus*, *Xenodon rabdocephalus*, and *Drymarchon melanurus* (Lee 1996; Cedeño-Vázquez et al. 2006a; JRCV pers. obs.).

The arboreal hylid frogs are the most diverse family in the region, with nine species from seven genera. They are nocturnal and feed on a variety of invertebrates, especially insects and spiders. Larger-sized species (e.g., *Trachycephalus typhonius* [formerly known as *Phrynohyas venulosa*] and *Triprion petasatus*) may also eat small vertebrates such as other frogs. Based on their size we can distinguish two groups of hylid frogs in the region. The first group is composed of four small (18–30 mm SVL) and slender species (*Dendropsophus microcephalus* [formerly known as *Hyla microcephala*], *D. ebraccatus* [formerly *Hyla ebraccata*], *Scinax staufferi*, and *Tlalocohyla picta* [formerly *Hyla picta*]). The second group is integrated by five mid-sized to large (38–102 mm SVL) species (*Agalychnis callidryas*, *Smilisca baudinii*, *Tlalocohyla loquax* [formerly known as *Hyla loquax*], *Trachycepahlus typhonius*, and *Triprion petasatus*). Most species are relatively common in forests and open areas (i.e., savannas, pastures), but some (e.g., *S. baudinii* and *T. typhonius*) are frequently more widespread, including urban zones, especially during the rainy season (Lee 1996). During daylight hours and during the dry season, some species (e.g., *S. staufferi*, *S. baudinii*, and *T. typhonius*) hide beneath the bark on standing trees, within tree holes and crevices, or in arboreal vegetation such as bromeliads and other water-retaining plants. The endemic Yucatán casque-headed treefrog (*T. petasatus*) (Fig. 11.1b) is particularly abundant in the arid northwestern portion of the peninsula. During the day these frogs seek refuge in the recesses of tree trunks and in rock crevices, where they plug the openings with their bony head, making them nearly impossible to extract (Lee 1996). This behavior probably helps to prevent desiccation and predation.

Hylid frogs are common prey of a variety of colubrid snakes like *Leptophis mexicanus*, *L. ahaetulla*, and *Coniophanes imperialis* (Cedeño-Vázquez et al. 2006a; JRCV pers. obs.). Breeding activity is associated with summer rains, when anywhere from few to hundreds of males congregate at water bodies and call from trees, shrubs, and emergent grasses. Small clutch sizes are typical for the two species that deposit egg masses on vegetation overhanging the water: *D. ebraccatus* generally deposits clutches of 20–80 eggs, while *A. callidryas* produces multiple clutches of 20–50 eggs per breeding episode (Lee 1996). The other hylid species produce masses of hundreds of eggs in the water, where the tadpoles complete their development (Lee 1996).

Two frogs have direct development, the endemic Yucatán rainfrog (*Craugastor yucatanensis* [formerly known as *Eleutherodactylus yucatanensis*], family Craugastoridae) and the invasive greenhouse frog (*Eleutherodactylus planirostris,* family Eleutherodactylidae). The biology of *C. yucatanensis* is poorly understood. The type specimen was collected in a cave, suggesting that these small (30–34 mm SVL) frogs are terrestrial and inhabit sinkholes (*cenotes*) and caverns (Lee 1996), but Calderón-Mandujano et al. (2008) state that these frogs are arboreal. These authors observed adult males calling (the call is like the cheep of a young chicken) in the forests of Sian Ka'an Biosphere Reserve, where the species is abundant and more active at night, especially after rainfall episodes. The tiny (16–32 mm SVL) greenhouse frog is native to the Caribbean islands of Cuba, Bahamas, and Cayman Islands (Díaz and Cádiz 2008; Olson et al. 2012), but has been introduced in several countries of the Neotropics. It was recently recorded in the Mexican Caribbean (Cedeño-Vázquez et al. 2014). Direct development facilitates human-mediated colonization of this frog (Christy et al. 2007) through transportation in potted plants (Kraus et al. 1999; Kraus and Campbell 2002). It feeds mostly on leaf litter invertebrates; in Hawaii, Olson and Beard (2012) estimated a mean consumption of 129,000 invertebrates/ha/night by these little frogs. Research to assess the invasion and its possible ecological impacts in the Mexican Caribbean is required to determine if control is necessary (Cedeño-Vázquez et al. 2014).

Three species of the genus *Leptodactylus* (family Leptodactylidae) occur in the peninsula. These are nocturnal, semiaquatic or terrestrial, small to mid-sized frogs (30–50 mm SVL) that construct foam nests within which they deposit their eggs; these nests provide protection against desiccation and predation. They breed during the rainy season in permanent ponds (*aguadas*) and temporary bodies of water, including roadside ditches, flooded pastures, small puddles, and even hoof prints of horses and cattle, often in close association with human settlements (Lee 1996). The white-lipped frog (*Leptodactylus fragilis*) and the black-backed frog (*L. melanonotus*) (Fig. 11.1c) are widespread and common species that occur in a variety of habitats across the peninsula. Whereas foam nests of *L. fragilis* are usually placed in sites subject to flooding to ensure the release of the larvae from the nest, *L. melanonotus* constructs foam nests at the water's edge. Tungara frogs (*Engystomops pustulosus* [formerly known as *Physalaemus pustulosus*], family Leiuperidae) (Fig. 11.1d) live in open areas (savannas) and deciduous forests of southern Campeche and Quintana Roo. They are rarely found, except at breeding congregations. Multiple clutches (200–300 eggs per clutch) are deposited in foam nests in shallow water (Rand 1983). Tadpoles hatch after 6 days and complete metamorphosis within 4–6 weeks (Galindo-Leal 2003). Their calls attract predators, such as the bat *Trachops cirrhosus,* and it is reported that tadpoles of *A. callidryas* feed on tadpoles of *E. pustulosus* when their nests have been broken apart by heavy rain (Ryan 1985).

Two small-sized (24–38 mm SVL) species of narrow-mouthed frogs (family Microhylidae) occur in the peninsula, the elegant narrow-mouthed toad (*Gastrophryne elegans*) and the sheep toad (*Hypopachus variolosus*). They are characterized by their chubby shape, short limbs, pointed heads, and a fold of skin across the back of the head. During the rainy season males call from temporary bodies of water, usually from the surface, and eggs are deposited directly into the water. The uncommon and inconspicuous *G. elegans* lives in leaf litter and crevices in moist soil (Cedeño-Vázquez et al. 2006a) in humid lowland forests, although individuals are occasionally found on the forest floor at night. As with most species of *Gastrophryne,* it feeds on small insects, particularly ants, termites, and small beetles (Nelson 1972; Köhler 2011). *Hypopachus variolosus*, in contrast, is widespread and common in both forested and more open areas (Lee 1996). Terrestrial and fossorial, it feeds on small arthropods, particularly insects such as ants and termites (Cedeño-Vázquez et al. 2006a; Köhler 2011). *Hypopachus variolosus* is often found at night on roads after heavy rains (Lee 1996). Typical of an explosive breeder, these frogs emerge after heavy downpours, mostly during the early rainy season, and use temporary waters bodies such as flooded pastures, roadside ditches, and marshes for reproduction (Lee 1996; Savage 2002). In addition, they occasionally breed in tree hollows above ground (McDiarmid and Foster 1975).

Finally, there are two mid-sized (60–114 mm SVL) species of true frogs (family Ranidae) of the genus *Lithobates.* These are classic frogs with smooth skin, long legs, and semi-aquatic habits, and are excellent jumpers (Galindo-Leal 2003). They feed primarily on invertebrates, but also on small vertebrates like fish, frogs, and mid-sized lizards (Lee 1996; Terán-Juárez 2011). They breed during the summer rainy season; the tadpoles develop in ponds where the eggs are laid. *Lithobates brownorum* (formerly known as *Rana berlandieri*) is terrestrial, diurnal and nocturnal. It is widely distributed throughout the Yucatán Peninsula (Zaldívar-Riverón et al. 2004), and is commonly found in and around freshwater bodies, reaching high densities in open, disturbed settings. During the reproduction period, males call from the surface of permanent or temporary bodies of water. The Vaillant's frog *L. vaillanti* (formerly known as *Rana vaillanti*) is terrestrial and lives primarily in humid lowland forests throughout the base of the peninsula, generally in lakes, *aguadas*, woodland pools, and slow-moving stretches of streams and rivers, but individuals have also been found at night on the forest floor (Lee 1996). Males call from the water's edge or surface during the breeding season.

11.3 Crocodiles

Two species of crocodiles occur in the Yucatán Peninsula, the American crocodile *Crocodylus acutus* and the Morelet's crocodile *C. moreletii*. Both species were intensively hunted for their skin during the last century, and their populations have decreased drastically. Today, populations of *C. moreletii* seem to have recovered, but continental populations of *C. acutus* remain in low numbers and exhibit poor recruitment (Cedeño-Vázquez et al. 2006b). However, island populations of *C. acutus* appear relatively well preserved but exhibit a male-biased sex ratio (Charruau et al. 2005; González-Cortés 2007).

Crocodylus moreletii inhabits inland freshwater systems on the mainland, while *C. acutus* occur in coastal saltwater habitats including offshore atolls and islands (Cedeño-Vázquez et al. 2006b; Charruau et al. 2005; González-Cortés 2007). Both species occur sympatrically in several areas of brackish mangrove swamp on the mainland (Cedeño-Vázquez et al. 2006b), where hybridization occurs (Cedeño-Vázquez et al. 2008; Rodriguez et al. 2008; Machkour M'rabet et al. 2009). Continental populations of *C. acutus* show high levels of hybridization and introgression with *C. moreletii*, while island populations remain apparently genetically pure (Machkour M'rabet et al. 2009). Hybridization may also complicate the identification of individuals in sympatric areas, as hybrids are cryptic or present characteristics of both species (Cedeño-Vázquez et al. 2008).

Crocodylus acutus and *C. moreletii* differ only by subtle morphological characteristics, such that the correct identification of the species can pose problems even to experienced herpetologists (Platt and Rainwater 2006). Differences in scale arrangements and skull morphology are the principal useful characteristics for identification of the species. *Crocodylus moreletii* is a mid-sized crocodile with males presumably reaching a maximum total length of 4.5 m (Platt et al. 2009), although individuals of more than 3.5 m are rare in the peninsula. This species presents a broad snout, weakly keeled osteoderms on the back, and irregular scales on the ventral and lateral surfaces of the tail (Platt and Rainwater 2006). In contrast, *C. acutus* is a large species, with males reaching maximum total lengths of more than 6 m and females generally reaching maximum total lengths of 3.5 m with exceptional individuals of 4.4 m reported (Thorbjarnarson 2010). However, individuals of more than 4 m are now rare in the peninsula. This species has a longer and more slender snout than *C. moreletii* and shows a well-developed preorbital elevation (Fig. 11.1e) (Platt and Rainwater 2006). *Crocodylus acutus* has also the most reduced and irregular pattern of dorsal osteoderms of any crocodilian (Platt and Rainwater 2006). Moreover, *C. acutus* generally presents many fewer caudal irregular scales than *C. moreletii* and usually never shows irregular scales on ventral surface of the tail (Platt and Rainwater 2006).

Crocodiles are opportunistic predators that eat a large array of prey. Studies in the Yucatán Peninsula show that the diet of *Crocodylus acutus* and *C. moreletii* includes insects, arachnids, mollusks, crustaceans, fish, amphibians, reptiles, birds and mammals (Gómez-Hernández 2004; Platt et al. 2002, 2006, 2007, 2013a). Moreover, frugivory, necrophagy, and kleptoparasitism have been observed, as well as the presence of stones in the stomach of some individuals (i.e., gastroliths; Platt et al. 2002, 2006, 2007, 2013b). In both species, the diet varies as a function of ontogenic development (Platt et al. 2006, 2013a). Hatchlings and small juveniles are the most specialized classes, feeding mainly on insects and arachnids; larger juveniles and subadults have the most diverse diet, increasing their consumption of fish and non-fish vertebrates. The diet of adults differs between species: *C. acutus* feeds principally on crustaceans, whereas *C. moreletii* has a broader diet including gastropods, crustaceans, fish, mammals, birds and reptiles (Platt et al. 2006, 2013a). Dietary overlap is greatest between adjacent size classes and least between the smallest and largest size classes (Platt et al. 2006, 2013a).

All crocodilians are oviparous and lay a single clutch of hard-shelled eggs per year. However, *Crocodylus acutus* and *C. moreletii* show clear differences in their reproductive ecology. *Crocodylus moreletii* is a mound nester: the female builds a mound nest of fresh and decomposing vegetation, sticks and mud, or soil, into which 20–50 eggs are deposited at the end of the dry season (Platt et al. 2008). The incubation period lasts 75–85 days; hatching occurs in August and September during the peak of the wet season (Platt et al. 2008). Females are very protective and remain in the vicinity of the nest during incubation, attacking potential egg predators. At hatching, the female excavates the nest, helps the hatchlings to emerge, and transports them to the water's edge, where she protects them from predators (Platt et al. 2010). *Crocodylus acutus*, on the other hand, is a hole nesting species. The female excavates a hole in open sandy areas into which 9–60 eggs are laid, and then re-covers them with substrate, sometime creating slightly elevated mounds (Charruau et al. 2010a; Thorbjarnarson 2010). Nesting occurs generally during the dry season and hatching during the beginning of the rainy season (Charruau et al. 2010a; Thorbjarnarson 2010). Maternal behavior of this species seems variable but on Banco Chinchorro atoll, females have been observed to visit their nests, rebuild them when disturbed, help hatchlings to emerge from the nest, and transport them to the water (Charruau and Hénaut 2012).

Crocodylus acutus and *C. moreletii*, like all crocodilians, have temperature-dependent sex determination (TSD), in which sex of embryos is determined during incubation by the action of temperature on the sexual differentiation system during a thermo-sensitive period (Charruau 2012; Lang and Andrews 1994). Both species show a FMF (Female-Male-Female) TSD pattern where low and high incubation temperatures produce a majority of females (up to 100 %) and intermediate temperatures produce a majority of males (Charruau 2012; Lang and Andrews 1994).

11.4 Turtles

The turtle fauna of the Yucatán Peninsula consists of 16 species in 14 genera and six families. This count includes five marine and 11 continental forms. The later includes freshwater (five species), semi-aquatic (four species), semi-terrestrial (one species), and terrestrial (one species) turtles.

As with crocodilians and other reptiles, temperature-dependent sex determination (TSD) is common in turtles, but no data exist on this for the species of the peninsula. Sea turtles (families Dermochelydae and Cheloniidae) are strictly marine, but normally come ashore to lay eggs in sandy beaches. As adaptations to aquatic life, they have forelimbs that have evolved into paddle-shaped fins, and salt glands, which expel superfluous salt that has been ingested through feeding (Köhler 2008). Adult sea turtles are among the largest living reptiles and the only reptiles that exhibit long-distance migrations, comparable only with those of terrestrial and avian vertebrates (Plotkin 2003).

Five species of sea turtles, all endangered, inhabit the waters surrounding the Yucatán Peninsula. These are the leatherback (*Dermochelys coriacea*), which is the only species of the family Dermochelydae, and four species of hard-shelled sea turtles (family Cheloniidae): the loggerhead (*Caretta caretta*), green turtle (*Chelonia mydas*), hawksbill (*Eretmochelys imbricata*) and Kemp's ridley (*Lepidochelys kempii*).

The leatherback is the largest of all living turtles. The largest specimen ever measured had a carapace (shell) length of 256.5 cm and a mass of 916 kg (Márquez 1990). They usually stay in the open ocean and are observed only occasionally near the coast (Köhler 2008). They are mostly carnivorous and feed on a variety of invertebrates, mainly tunicates and jellyfish (and the fish and crustaceans associated with jellyfish), though they do occasionally consume seaweeds (Mortimer 1982). Females undertake reproductive migrations to nesting beaches every 2–3 years, where they emerge from the sea at night to oviposit five or six clutches at 9-day intervals (Boulon et al. 1996; Steyermark et al. 1996). Females do not display strong beach fidelity and may travel among adjacent (Steyermark et al. 1996) or distant beaches (Keinath and Musick 1993) over the course of a nesting season. The nesting season in the Yucatán Peninsula lasts from April to October (Köhler 2008). Leatherbacks are thought to nest on Arrecife Alacrán, Yucatán (Carranza 1959), and are known to nest at several coastal locales in Campeche and Quintana Roo (Lee 1996).

The loggerhead (*Caretta caretta*) is the largest species of the family Cheloniidae (Lee 1996); the adults attain a carapace length of up to 230 cm and a mass of 540 kg (Pritchard 1967). Loggerheads are abundant and spend most of their time in nearshore and inshore shallow waters, sometimes associated with reefs and other natural and artificial hard substrates (Dodd 1988). They feed on a variety of invertebrates including, but not limited to: crabs, barnacles, gastropods, bivalve mollusks, shrimps, squids, and jellyfish, as well as fish (Lee 1996). The eggs are especially vulnerable to predation by a variety of terrestrial animals, including insects (e.g., larvae of the beetle *Lanelater sallei*) and mammals such as the raccoon *Procyon lotor* (Talbert et al. 1980; Donlan et al. 2004), while hatchlings suffer high mortality from numerous invertebrate and vertebrate predators Dodd (1988). They nest in large numbers on the coast of Quintana Roo, from Punta Allen to Isla Contoy, and along the Yucatán and Campeche beaches to Isla del Carmen, Campeche (Lee 1996). The females crawl up onto the nesting beaches at night, laying eggs from April to July (Lee 1996; Köhler 2008). After a short mating phase that occurs offshore, females produce 2–5 clutches, each of 40–190 eggs at intervals of about 2 weeks (Köhler 2008). Young hatch after 55–68 days of incubation (Magnuson et al. 1990; Márquez 1990).

The green turtle (*Chelonia mydas*) is a mid-sized sea turtle: the carapace can measure up to 150 cm in length (Lee 1996; Köhler 2008). This turtle inhabits relatively shallow coastal waters with abundant marine grasses (Lee 1996). Seaweeds and sea grass are the major components of its diet, but it also feeds upon sponges, mollusks, jellyfish, and fish (Mortimer 1982; Lee 1996; Köhler 2008). The waters off the northwest and southeast coasts of the Yucatán Peninsula are important feeding areas for this species (Carr 1952). During the nesting season (April to October), solitary females come to land at night and produce two to five clutches at intervals of 12–14 days, each with 38–150 eggs, from which the young hatch after 50–75 days (Magnuson et al. 1990; Márquez 1990; Lee 1996). Females nest every 2–5 years (Köhler 2008). Along the peninsula, females nest on Islas Contoy, Isla Mujeres and Cozumel Island, Quintana Roo (Smith and Smith 1979), and on Cayo Arcos, Isla Arena, Arrecife Alacrán, and Arrecife Triángulos on the Campeche Bank (Hildebrand 1982).

The hawksbill (*Eretmochelys imbricata*) is a rather small sea turtle, with a maximum carapace length of 100 cm, although most specimens today are considerably smaller. Hawksbills exhibit a pronounced sexual dimorphism, with females growing larger than males (Lee 1996). It inhabits shallow coastal waters, especially rocky substrates and coral reefs, but also occurs in regions with muddy bottoms, and sometimes enters the lower reaches of streams (Lee 1996). This turtle is mostly carnivorous, feeding especially on encrusting organisms such as sponges, sea anemones, bryozoans, tunicates, mollusks, and algae, which it scrapes off reef faces (Mortimer 1982). It also consumes crabs, jellyfish, squid, and fish (Lee 1996; Köhler 2008). There is an argument for the appearance of geographical differences in regard its range of foods, since in some regions individuals feed primarily or exclusively upon sponges (Carr and Stancyk 1975; Meylan 1988). In the Yucatán Peninsula it can be found at any coastal locality; its nesting distribution includes the beaches of southwestern Campeche and northeastern Yucatán (Carranza 1959; Fuentes 1967), but it is particularly abundant along the Caribbean coast (Hildebrand 1982; Meylan 1989). Nesting specimens have been reported elsewhere only sporadically (Köhler 2008). Females breed every 2–3 years, and nest at night, usually at high tide (Lee 1996) once or twice during the breeding season (April to August) to bury clutches consisting of 50–223 eggs each (Lee 1996; Köhler 2008). The incubation period lasts 47–75 days (Márquez 1990). Hatchlings and juveniles are apparently pelagic, but at a carapace length of about 23–25 cm they become benthic feeders in coastal habitats (Meylan 1988).

Kemp's ridley (*Lepidochelys kempii*) is the smallest sea turtle. The carapace length of adults averages 60–65 cm, and there is little or no sexual dimorphism (Lee 1996). This sea turtle is found throughout the Gulf of Mexico with confirmed records from Campeche and Yucatán, but apparently not on the Caribbean coast of the Yucatán Peninsula (Lee 1996). It prefers relatively shallow waters and the crustacean-rich banks of the Campeche—Tabasco area, which represent an important feeding ground (Hildebrand 1982; Byles 1989). This turtle is primarily carnivorous, feeding mostly on crabs, especially those of the genera *Ovalipes* and *Callinectes* (Mortimer 1982), but occasionally they may eat other invertebrates (e.g., sponges, sea anemones, jellyfish, and squids) and plant material (Lee 1996; Köhler 2008). In contrast with the other sea turtles described, the females of *L. kempii* usually lay eggs on beaches during the day (Lee 1996; Köhler 2008). Nesting is generally synchronized, and large numbers of females emerge from the ocean at the same time, producing huge aggregations (Lee 1996; Köhler 2008). These groups once numbered many thousands of individuals (Magnuson et al. 1990), but now are reduced to a few hundred (Lee 1996). Also remarkable is the fact that females nest almost every year, producing 2–3 clutches from April to August, each with 42–167 eggs; the incubation period is about 50–66 days (Casas-Andreu 1978).

The 11 species of continental turtles belong to three families, the most diverse being Kinosternidae, with six species in four genera, followed by Emydidae with three species in three genera, and Dermatemydidae and Chelydridae with one species each. The family Kinostenidae contains two musk turtles (*Claudius angustatus* and *Staurotypus triporcatus*) and four mud turtles (genus *Kinosternon* and *Cryptochelys*), including one that is endemic (*Cryptochelys creaseri*). Kinosternid turtles inhabit slow-moving bodies of water such as lakes, swamps, seasonally flooded marshes, temporary ponds, roadside ditches, *aguadas*, *cenotes*, lagoons in large rivers, and small streams, especially those with muddy bottoms in open or forested areas (Lee 1996; Legler and Vogt 2013). Frequently, kinosternids (*Claudius* and *Kinosternon*) can be encountered on land as they wander from one body of water to another (Köhler 2008; Legler and Vogt 2013). During the dry season, they bury themselves in the loose soil or mud at the site of temporary bodies of water and remain in aestivation until the beginning of the rainy season (Lee 1996; Legler and Vogt 2013).

The monotypic narrow-bridged musk turtle (*Claudius angustatus*) is characterized by a small plastron, which is connected to the carapace by ligaments (Lee 1996; Köhler 2008). It attains a maximum carapace length of about 17 cm, and males generally are 10 mm bigger than females (Lee 1996). This turtle is carnivorous and preys on invertebrates (e.g., snails, earthworms, crustaceans, insects and arachnids) and small vertebrates (Lee 1996; Köhler 2008; Legler and Vogt 2013). It reproduces after the annual floods, and oviposition occurs in November (Pritchard 1979). Clutch size varies from 1 to 8 eggs, which hatch after an incubation period of

94–150 days depending on the environmental conditions (Ernst and Barbour 1989; Vogt and Flores-Villela 1992; Köhler 1997).

The Mexican musk turtle (*Staurotypus triporcatus*) is the largest kinosternid in the Yucatán Peninsula; the adults may attain a maximum carapace length of 40 cm and weight 10 kg (Lee 1996; Legler and Vogt 2013). When disturbed they can be extremely aggressive, so they need to be handled with great caution, because their strong sharp jaws can cause serious wounds (Lee 1996; Köhler 2008). The strongly developed dorsal keels help to stabilize the carapace, thereby making it more difficult for predators (e.g., *Crocodylus moreletii*) to break (Köhler 2008). This turtle is strictly aquatic, and it is most abundant in rivers with slow current at depths of 1–2 m, generally along shorelines where they forage (Legler and Vogt 2013). They are carnivorous, feeding on crabs, worms, snails, bivalves, insects, amphibians, fish, aquatic birds and reptiles such as snakes and mud turtles (Pritchard 1979; Campbell 1998; Köhler 2008; Legler and Vogt 2013). During the reproductive season, the female buries several clutches, each with three to ten eggs (Köhler 2008). Incubation lasts 120–207 days (Ernst and Barbour 1989; Köhler 1997).

The mud turtles were recently divided in two genera: *Kinosternon* and *Cryptochelys* (Iverson et al. 2013). Their members are characterized by having one or two moveable hinges in the plastron that enable complete closure of the shell (Lee 1996; Köhler 2008). The four species that occur in the Yucatán Peninsula are small to mid-sized: *Cryptochelys acuta* are the smallest (9–12 cm of carapace length) and *Kinosternon scorpiodes* the largest (15.7–17.5 cm of carapace length). In *C. acuta*, females are larger than males, while in the remaining species (*Cryptochelys creaseri*, *Cryptochelys leucostoma* and *K. scorpiodes*) males are larger than females (Lee 1996). The most common species locally is *K. scorpiodes*, which can be found in abundance (Lee 1996), followed by the endemic *C. creaseri*, which is most abundant in eastern Yucatán and northern Quintana Roo (Iverson 1988). The diet of *C. acuta* in the Yucatán Peninsula is poorly known, but like other species of the genus it probably feeds primarily on aquatic invertebrates, such as snails, worms, and insects (Köhler 2008). *C. leucostoma* and *K. scorpiodes* consume large quantities of plant material (e.g., leaves, stems, seeds, fruits, and flowers), but are facultative omnivores, also preying prey on small vertebrates (e.g. fish, frogs, lizards, and snakes), including carrion (Iverson 1976, 1988; Vogt and Guzman-Guzman 1988; Köhler 2008), when these are available. The reproduction of these kinosternids is poorly known in the Yucatán Peninsula, but in other places nesting occurs from March to April, and females produce multiple clutches, which, depending on the species, contain from one to ten eggs (Köhler 1997).

The family Emydidae is represented in the Yucatán Peninsula by two mid-sized (15–20 cm of carapace length) species of predominantly terrestrial turtles, *Rhinoclemmys areolata* and the endemic *Terrapene yucatana*, and a large-sized (24–60 cm carapace length), predominantly aquatic slider turtle (*Trachemys venusta*). Although in the Yucatán Peninsula the furrowed wood turtle (*R. areolata*) typically inhabits savannas and other open areas, it also occurs in forests and marshy grounds (Lee 1996). It is frequently found during the day, especially during the summer rainy season (Lee 1996). Little is known about its diet in the Yucatán Peninsula itself, but based on analysis of feces of two specimens from northern Belize examined by Platt (1993), it feeds on plant material (e.g., herbs, legumes, and fruits), beetles and other insects, and turtle eggs. In a recent study, Vogt and co-author (2009) showed that the plants consumed by *R. areolata* in Belize vary seasonally. Data collected in Cozumel, Quintana Roo indicate that sexual maturity in this turtle occurs at about 14.5–15 cm of carapace length and at ages of 9–10 years (Legler and Vogt 2013). Nesting occurs from May to July, and females may lay up to 4–5 clutches per year, each with one large egg, which hatches in 80–90 days (Vogt et al. 2009; Legler and Vogt 2013). The Yucatán box turtle (*T. yucatana*) has been considered a subspecies (*T. carolina yucatana*), but following Legler and Vogt (2013), we treat this taxon as a full species. It inhabits wet savannas and pastureland, as well as thorn forest and evergreen rain forest, with preference for open habitats (Lee 1996). Although it exhibits a terrestrial lifestyle, individuals may occasionally enter shallow water (Lee 1996; Köhler 2008). No reports of the diet of this turtle are available, but according to available data from other species of *Terrapene*, omnivory can be assumed (Legler and Vogt 2013). Information on the reproduction of the Yucatán box turtle is also absent. However the dissection of three females by Legler (see Legler and Vogt 2013) in Yucatán suggests that clutches are small (2–3 eggs), with the first clutch is laid in July; that follicular potential for a second clutch exists, and that mature females may skip seasons of reproduction. The common slider (*T. venusta*) has often been placed in the genera *Pseudemys* and *Chrysemys*, and much of the literature referring to this turtle is found under those names (Lee 1996). Formerly, *T. venusta* was treated as a subspecies of *T. scripta* (*T. scripta venusta*), but we follow Seidel (2002) who elevated this subspecies to full specific status. This large turtle is diurnal and commonly occupies freshwater habitats, with a preference for quiet waters with soft bottoms, such as ponds, rivers, streams, lakes, *aguadas* and *cenotes* (Lee 1996; Köhler 2008). Occasionally individuals do wander some distance from water, particularly during the rainy season (Lee 1996). *Trachemys venusata* usually basks on partially submerged rocks, roots, and logs (Köhler 2008). They feed predominantly on aquatic vegetation (leaves, stems), aquatic invertebrates (mollusks, insects), and fish (Lee 1996). This species is commonly preyed upon by humans and crocodiles (Lee 1996; Köhler 2008). Himmelstein (1980) described a specimen from Quintana Roo that had punctures on the carapace and plastron, suggesting attempted predation by a crocodile. During the egg laying period (December–May) females will construct a flask-shaped cavity about 15 cm deep into which they place 9–25 eggs; a female may produce up to six clutches per season, and juveniles (carapace length 30–33 mm) hatch after an incubation period of 65–85 days (Lee 1996; Köhler 2008). Females are sexually mature at an age of 5–7 years and a carapace length of 20–24 cm (Lee 1996).

The family Dermatemydidae comprises a single living genus and species, the Central American river turtle (*Dermatemys mawii*) (Fig. 11.1f). This very large aquatic turtle may reach a carapace length of 65 cm and a mass up to 22 kg (Lee 1996). This turtle commonly inhabits large deep rivers, and associated oxbow freshwater lakes and lagoons throughout the year (Legler and Vogt 2013), though occasionally it enters brackish water (Lee 1996; Viveros-León 1996; Calderón-Mandujano 2008; Köhler 2008). These nocturnal turtles are able to perform prolonged dives (Köhler 2008). *Dermatemys* is completely herbivorous from hatching to adulthood (Vogt and Flores-Villela 1992), feeding almost exclusively on plant material such as algae, aquatic grasses, leaves, flowers and fruits that fall into the water (Holman 1963; Álvarez del Toro et al. 1979; Lee 1996; Köhler 2008; Legler and Vogt 2013). During the rainy season they enter flooded forest, and also travel up smaller tributaries to nest (Legler and Vogt 2013). The egg-laying period lasts from April through September (Smith and Smith 1979), and females bury their clutches of 2–20 eggs on sandy banks near the shoreline (Polisar 1992; Köhler 2008). Females produce up to three clutches per year (Lee 1996). In Belize the incubation period lasts 217–300 days from oviposition to hatching, and a maximum of four clutches and 47 eggs per year was observed (Polisar 1992). There are many animals (e.g. coati, otter, herons and other large wetland birds, and crocodiles) that prey on this turtle during all its life stages (see references in Legler and Vogt 2013), but humans are clearly the major predator of *Dermatemys*, and hunter-gatherers have probably exploited this food source for centuries (Legler and Vogt 2013).

The Central American snapping turtle (*Cheydra rossignoni*) was previously recognized by Gibbons et al. (1988) as a subspecies of *C. serpentina* (*C. s. rossignoni*). Here we follow Legler and Vogt (2013) who treat this taxon as a full species. This turtle species has the smallest distribution range within in the Yucatán Peninsula of any turtle species described here, occupying only the southwestern part of Campeche (though its complete range includes the coastal plain from central Veracruz and adjacent Oaxaca southeastern Tabasco, southwestern Campeche and Chiapas to northern Guatemala, southern Belize, and the extreme northwestern part of Honduras; see Legler and Vogt 2013). According to Lee (1996), *C. rossignoni* rivals *Dermatemys mawii* as the largest freshwater turtle in Mexico, attaining a carapace length of at least 47 cm and a mass of 20 kg. Based on observations, trapping, and information from fishermen, this species is not found in abundance anywhere (Legler and Vogt 2013). No information exists on the biology of this turtle for Campeche, thus we present data on its habitat, diet, reproduction and predation from other areas. In Veracruz and Chiapas this snapping turtle inhabits the marshy borders of shallow lakes and associated slow, meandering, muddy streams covered by water hyacinth; it does not appear to utilize the deep portions of lakes or inhabit large rivers, and basking is uncommon (Legler and Vogt 2013). Examinations of stomach contents of individuals from Veracruz suggest an opportunistic diet including small invertebrates (mostly crustaceans), fruits, seeds, and aquatic vegetation (Legler and Vogt 2013). In Chiapas Álvarez del Toro (1974) reported *C. rossignoni* feeding on hatchling crocodiles, and found that younger individuals feed on duckweed, aquatic invertebrates, and terrestrial invertebrates that fall into the water, as well as small fish and carrion. This turtle is known for its aggressive, pugnacious temperament; even hatchlings will attempt to bite when confronted (Legler and Vogt 2013). Limited observations and dissections suggest that females of *C. rossignoni* produce a single clutch of a massive number of eggs

(20–55) from April to July each year (Álvarez del Toro 1982; Legler and Vogt 2013). Álvarez del Toro (1974, 1982) reported that small individuals are consumed by *Staurotypus triporcatus*, *Crocodylus moreletii*, otters (*Lontra longicaudis*), and herons. Hatchlings are preyed upon by snakes of the genus *Drymarchon* (Ruthven 1912 in Legler and Vogt 2013). Adults are probably consumed only by crocodiles and humans (Lee 1996; Legler and Vogt 2013).

11.5 Lizards

With 45 species in 23 genera and 10 families, lizards are well represented in the Yucatán Peninsula. The Least geckos (genus *Sphaerodactylus*) are the smallest lizards known to occur in the area, reaching a maximum SVL of 25–30 mm. Least geckos are typical leaf litter inhabitants; they dart through the leaves, skillfully hunting small arthropods, such as springtails, cricket larvae and small spiders. Most gecko species are characterized by broadened adhesive lamellae beneath fingers and toes (allowing these animals to climb on smooth, vertical surfaces), as well as by the absence of moveable eyelids (their eyes are covered by a transparent membrane or "spectacle" instead). Some gecko species are crepuscular or nocturnal (e.g., *Hemidactylus* and *Phyllodactylus*), whereas others such as *Sphaerodactylus* are active during the day. Two species of Old World house geckos (genus *Hemidactylus*) have been introduced to the Yucatán Peninsula. These geckos are usually associated with human habitation. At night, they can be observed in houses, where they mainly prey on arthropods but also on small vertebrates such as smaller conspecifics. They are capable of vocalization; their call sounds like "gec-gec-gec." All gecko species in the region are oviparous, usually producing one or two eggs per clutch. In contrast to most other geckos, *Coleonyx elegans* (Fig. 11.1g) has moveable eyelids. This terrestrial nocturnal lizard hides beneath rocks and fallen tree trunks, and in hollow tree trunks during daytime. Its diet consists mostly of insects and spiders. The movements of this attractively patterned gecko are very graceful, almost cat-like. When alarmed, it expands its throat and assumes a stiff posture with the body held high, well off the ground.

At any given locality in the region, anoles are among the most common lizards and often several species of anoles occur together. Niche partitioning is evident in anoles, with different species found in bushes and lower parts of trees (e.g., *Anolis lemurinus*, *A. rodriguezii*, *A. sericeus*), on the trunks of large trees (e.g., *A. biporcatus*), or on the leaf litter (e.g., *A. tropidonotus*). One species, *A. beckeri*, even spends most of its time on twigs in the canopy, although it can also be observed also on the trunks of large trees. Most *Anolis* species have a brown to gray ground color, except for species such as *A. allisoni*, which is bright green when relaxed (Fig. 11.1h) but capable of rapid color change (metachrosis) to brown when stressed. Male anoles usually have large colorful dewlaps that are used during display behavior such as courtship and territorial defense. Displays consist of species-specific sequences of head bobbing with the dewlap extended. All anole species reproduce by laying eggs, and, during the breeding season, females will produce several clutches, each consisting of a single egg, at intervals of 1–2 weeks. The ovaries and oviducts on each side of the body are alternately active. For example, as soon as one egg is laid from the right oviduct, a new follicle is produced by the left ovary, which passes through the left oviduct, where it is fertilized and shelled within a few days, before being laid. In the meantime, the right ovary has produced a new follicle, which is ready to ovulate. As long as there is enough food available and weather conditions are suitable, this cycle will continue and females will lay eggs at regular intervals (Fitch 1982).

Three species of spiny-tailed iguanas (genus *Ctenosaura*) occur on the peninsula, with *C. similis* being the most common and largest—it can reach a total length of over 100 cm. While *C. similis* is mostly a ground dweller that also climbs on walls, roofs, and trees, the two smaller species endemic to the Yucatán Peninsula (i.e., *C. alfredschmidti* and *C. defensor*) have a largely arboreal lifestyle. However, because of the scarcity of trees with hollow branches in the rocky terrain of the northern Yucatán Peninsula, the beautifully colored *C. defensor* mostly uses limestone crevices as hiding places. All spiny-tailed iguanas lay eggs, with clutch sizes ranging from 2–12 (for smaller species) and 20–88 (*C. similis*). Suitable nesting sites for *C. similis* are open, sun exposed areas where females will dig a burrow to lay their eggs (Fitch and Hackforth-Jones 1983). The smaller species usually lay their eggs in the back portion of their hiding hole, typically a hollow branch or rock crevice. The eggs require an incubation of approximately 3 months (Köhler 2002). Hatchlings of *C. similis* are predominantly green in color with green markings. Spiny-tailed iguanas reach sexual maturity at 2–3 years of age.

The green iguana (*Iguana iguana*) is certainly the best-known lizard in Central and South America. The preferred habitat of this magnificent giant lizard is lowland forest, near streams, rivers, or lakes. While adult iguanas spend a lot of their time in the treetops, juveniles are often found closer to the ground in bushes. Green iguanas are impressive swimmers and divers, leaping into the water from as high as 10 m to escape predation. Green iguanas consume an almost exclusively vegetarian diet, consisting mainly of leaves. At the beginning of the dry season, the female will excavate a nesting site, which, depending on ground hardness, will be from 30 to 200 cm long (Rand 1968; Rand and Dugan 1983). Clutch size varies from 20 to 60 (rarely as many as 80) eggs, primarily depending upon the size of the female. The young will hatch following an incubation period of approximately 3 months, which usually corresponds to the onset of the rainy season. This ensures hatchlings a ready supply of tender leaves as nourishment.

Basilisks (genus *Basiliscus*), with head and dorsal crests that give them a truly dragon-like appearance, are among the most impressive and bizarre lizards of Central America. Their ability to run on their hind legs even across rapidly flowing water has earned them the local name "*Iguano Jesucristo*" (Jesus Christ Iguana). Two important adaptations that allow these animals to run on the water's surface are broad fringes of skin on the toes, which distribute their body weight over a larger area, and the relatively heavy tail, which acts as a counter-balance to the upraised body during bipedal locomotion. These lizards are found in a number of habitats, wherever water in the form of streams, ponds, or swamps is present, and are quite abundant in many areas. Basilisks feed on insects, spiders, and small vertebrates, as well as occasional vegetable matter. Females basilisks produce three to seven clutches of eggs per season, each with 4–18 eggs, which hatch in 2–3 months (Köhler 1999).

Helmeted iguanas (genus *Corytophanes*) are relatives of the basilisks and are characterized by having a laterally flattened body and large occipital crests, or "helmets". Helmeted iguanas are tree-dwellers that spend most of their time in partial shade, waiting for their predominantly insect prey (Köhler 1999). Casquehead iguanas (genus *Laemanctus*) are remarkably well adapted for life in the trees, which they leave only to lay eggs. These iguanas are difficult to spot amongst the leaves. Several times a year, *Laemanctus* females will produce a clutch of four to nine eggs, which hatches after only 2 months (Köhler 1999).

In the dry regions of Mexico and Central America, next to spiny-tailed iguanas, spiny lizards (genus *Sceloporus*) are the dominant lizard group. In the Yucatán Peninsula, six species of *Sceloporus* are known to occur. Spiny lizards are diurnal small to mid-sized lizards, most of which are covered with extremely spiny scales. They feed on arthropods, as well as small vertebrates. Among *Sceloporus* species, some are ground-dwelling and saxicolous species, as well as primarily arboreal species (e.g., *S. lundelli*). The ground-dwelling *S. chrysostictus*, *S. cozumelae*, *S. teapensis*, and *S. variabilis* can frequently be seen on tree stumps or rocks. *Sceloporus serrifer* can be seen in great numbers at several archeological sites, although the animals are quite shy and will flee at the slightest disturbance. These attractive lizards bask on trees, as well as on rocks and brick walls, and will disappear into the nearest crack or opening when threatened. *Sceloporus* males are highly territorial and attempt to intimidate intruders with push up-like movements of their upper body. Most *Sceloporus* species in the region reproduce by laying eggs, with the exception of *S. serrifer*, which is viviparous (Köhler and Heimes 2002).

Several species of the Teiidae family occur in the Yucatán Peninsula, the Common Ameiva, *Holcosus undulatus* (formerly known as *Ameiva undulata*), being the largest. Five species of whiptails (genus *Aspidoscelis*) also belong to this group of fast diurnal lizards. To the human observer, these lizards seem tremendously restless, as they are in virtually constant motion. As heat-loving animals, ameivas and whiptails are out in the open primarily on hot, sunny days, whereas they retreat to their underground burrows when the sky is cloudy. Cool and rainy days are spent in burrows as well. Ameivas and whiptails are tenacious hunters that will wander around their territory looking for prey, scratching in the leaves, digging in the ground, and searching in the undergrowth and under bark, primarily consuming insects and arachnids. These lizards are constantly flicking their tongues, checking the scents in their surroundings. Although they are ground dwellers, these lizards will also climb trees and low vegetation to bask or to reach prey. Notably, several *Aspidoscelis* species, including *A. cozumela* and *A. rodecki* reproduce parthenogenetically. In these two species, the populations consist exclusively of females, which produce fertile eggs without any previous copulation—males are unknown in these two species (Köhler 2008).

The four species of the family Scincidae in the region are small to moderate-sized diurnal lizards with shiny, smooth scales and short limbs. They feed primarily on arthropods, but will also eat small vertebrates. These animals are generally found on the ground, but they will climb tree trunks in order to bask. *Mabuya unimarginata* is viviparous, with litter size varying from two to seven young. The other scincid lizards in the region are oviparous, producing clutches of one to three (seldom four) eggs (*Sphenomorphus cherriei*), or even up to 11 eggs per clutch (*Eumeces sumichrasti*). While adult *E. sumichrasti* have a pale brown color with black longitudinal stripes, young individuals are somewhat more brightly colored. The latter have a bright blue tail, and several orange spots on the head, as well as three orange to yellow longitudinal stripes on the body (Köhler 2008).

A single species of the family Anguidae occurs in the Yucatán Peninsula, *Celestus rozellae*. It is a diurnal tree-dweller that spends time at a considerable height above ground, but can also be observed basking on tree trunks only 50–250 cm above the ground (Lee 1996; McCranie and Wilson 1996). *Celestus rozellae* is viviparous, producing litters of three to five juveniles from May to July (Álvarez del Toro 1982).

Lepidophyma flavimaculatum (family Xantusiidae) is a bizarre-looking mid-size lizard (SVL to 95 mm) that has no eyelids. As with many gecko species, they possess so-called "spectacles". The sides of the body are decorated with countless, irregularly arranged tubercular scales, while the tail is equipped with whorls of slightly enlarged scales. During the day, these ground-dwelling lizards hide under rocks and fallen logs, which provide a moist microclimate. These nocturnal animals are sit-and-wait predators whose prey consists largely of insects and spiders. The remarkable thing about this genus is its viviparous method of reproduction with a fully developed placenta. There are also populations composed entirely of females that reproduce parthenogenetically. From April to July, the females of *L. flavimaculatum* bear two to five (seldom as many as eight) young, which have a SVL of 35–38 mm at birth (Álvarez del Toro 1960; Telford and Campbell 1970).

11.6 Serpentes

The known snake fauna of the Yucatán Peninsula consists of 55 species, encompassing 40 genera and six families. This count includes six venomous species, one of which is the elapid *Micrurus diastema*, while the other species are in the family Viperidae. *M. diastema*, the only venomous coral snake in the Yucatán Peninsula, is highly variable in its color pattern, from regularly tricolored individuals to almost uniformly red animals. In view of the highly potent neurotoxic venom of this species, extreme caution is advised in dealing with them. Coral snakes feed predominantly on other snakes, but will also eat eels and caecilians.

Five viperid species are known to occur in the region with three endemics, *Crotalus tzabcan*, *Porthidium yucatanicum* and *Agkistrodon russeolus* (Fig. 11.1i). The latter species is a small terrestrial inhabitant of thorn forest, tropical deciduous forest, and tropical semi-deciduous forest. It feeds on small vertebrates such as lizards, frogs, snakes, and mammals. The largest and most dangerous venomous snake in the region is *Bothrops asper*, which can reach a total length of more than 2 m. It lives not only in forested areas, but also in secondary undergrowth near human habitation, preferably close to water. There, this nocturnal pit viper will hunt frogs, lizards, and small mammals. All viperids in the region are live-bearing, producing litters of four to ten (*P. yucatanicum*) or even up to 86 (*B. asper*; usually 20–50), neonates (Solórzano and Cerdas 1989; McCoy and Censky 1992).

The small snakes of the families Typhlopidae and Leptotyphlopidae are represented by three species in the region, one of which (*Ramphotyphlops braminus*) is introduced. It is assumed that this small snake was transported from its native Southeast Asia, Australia, and the South Pacific, in potted plants. As it is parthenogenetic, only a single female is required to establish a new population (Nussbaum 1980). *Amerotyphlops microstomus* is endemic to the Yucatán Peninsula. This secretive burrower feeds mainly on termites, ants and other small insects.

The largest snake in the region is *Boa constrictor*. Although it is predominantly crepuscular and nocturnal, it is occasionally encountered outside of its hiding places during the day. Typical daytime hiding places include hollow trees, as well as piles of leaf litter among the buttress roots of large trees. Boas are not too particular about their vertebrate prey and will consume any lizard, bird, or mammal they can swallow. *B. constrictor* is viviparous and produces litters of 10–64 young (Köhler 2008).

With 45 species, the family Colubridae is well represented in the Yucatán Peninsula. For convenience, we have grouped these species according to their lifestyles, temperament, and size. These categorizations, of course, do not reflect phylogenetic relationships.

A large proportion of the snakes in the Yucatán Peninsula can be referred to as terrestrial diurnal racers. These are medium-sized to large, diurnal, mostly terrestrial, fast-moving, and easily irritated snakes such as *Conophis lineatus*, *Masticophis mentovarius*, *Mastigodryas melanolomus*, and *Drymobius margaritiferus*. These snakes have large eyes, reach more than 1 m in total length when adult (except for *C. lineatus* with adults in the 70–90 cm range), and feed mostly on small vertebrates such as frogs, lizards, and rodents. When caught or handled, these racers promptly deliver a bite. *Conophis lineatus* is a rear-fanged snake with a pair of greatly enlarged, erectile, grooved teeth in the posterior upper jaw and a mildly venomous secretion from its Duvernoy's glands, which are primitive venom glands situated posterior to the eyes. The clinical signs of a bite by *C. lineatus* can include severe pain at the bite site and local swelling. These racers are often seen crossing a trail or road before quickly disappearing into the vegetation. Although it is never a good idea to grab the tail of a snake, in the case of *D. margaritiferus*, the tail easily comes off because these snakes are capable of tail autotomy. In fact, a good proportion of individuals of this species have incomplete tails (Köhler 2008).

Another group of large snakes are the diurnal arboreal racers, which share similar characteristics to the species mentioned above, except that these are arboreal and therefore slenderly-built climbing species. Some of them are predominantly green in color such as *Leptophis ahaetulla*, *L. mexicanus*, and *Oxybelis fulgidus*. *Leptophis ahaetulla* is notorious for its irritable temperament. When handled, it will respond with a threatening gape, expanding the lower jaw, hence their Latin name. They will often remain in this pose for an extended period of time, but rarely make use of their numerous pointed teeth. It feeds primarily on frogs, as well as the occasional salamander, lizard, snake or bird egg (Oliver 1948; Sexton and Heatwole 1965). *Pseustes poecilonotus* lives primarily in old growth rainforest, where it can be found both on the forest floor, as well as in shrubs and trees. If it feels threatened, it will inflate its upper body, open its mouth wide and bite without hesitation. This large snake feeds primarily on birds, small tree-dwelling mammals, and bird eggs (Álvarez del Toro 1982; Campbell 1998). The large, green *Oxybelis fulgidus* and the smaller, gray to brown *O. aeneus* are both rear-fanged snakes, with mild venom that does not pose a serious threat to humans, but can lead to localized swelling and itching (Crimmins 1937). When a vine snake is handled, it will react with a threatening gape, exposing the blue-black interior of its mouth; it will also bite without any warning. As a defensive reaction, this colubrid will continuously hold its outstretched tongue rigid with the two tips of the tongue pressed together. A large part of their diet is made up of anoles (*Anolis*), but other lizards are also eaten along with frogs, rodents, birds and insects. *Spilotes pullatus* is extremely agile at moving through vegetation, thus its Spanish name, "*voladora*" (the flying one). It feeds primarily on small mammals, birds, and bird eggs (Henderson and Hoevers 1977; Lee 1996). All these arboreal racers have in common that they reproduce oviparously, producing clutches of 4–11 (*D. corais*), 3–5 (*O. aeneus*), 8–10 (*O. fulgidus*), 7–14 (*P. poecilonotus*), and 7–25 (*S. pullatus*) eggs, respectively (Goode 1989; Campbell 1998; Köhler 2005). *Drymarchon melanurus* is a fast, diurnal colubrid that feeds on a variety of vertebrates (e.g., fish, frogs, turtles, lizards, rodents, and small birds).

Several other diurnal snakes of small to moderate size occur in the Yucatán Peninsula, with *Coniophanes* being the most diverse genus in the area. The four species of these secretive, rear-fanged (opisthoglyphous), but usually inoffensive snakes are known to feed on frogs, salamanders, lizards, and snakes, as well as reptile and bird eggs; they will also consume invertebrates, such as earthworms and insect larvae (Minton and Smith 1960; Landy et al. 1966; Seib 1985; Platt and Rainwater 1998). The species of this genus are oviparous and lay 1–10 eggs per clutch, usually under rotting logs or roots (Campbell 1998). *Symphimus mayae* is a diurnal and semi-arboreal Yucatán endemic, inhabiting mostly dry areas. It is known to feed on insects (mostly crickets and grasshoppers) (Rossman and Schaefer 1974). The species is oviparous and produces clutches of two to four eggs during the wet season (Stafford 2005). *Stenorrhina freminvillei* is a harmless rear-fanged snake, which will not attempt to bite. As a nocturnal ground-dweller, it feeds almost exclusively on spiders and scorpions (Duellman 1963; Sexton and Heatwole 1965; GK pers. obs.) It is oviparous and produces clutches of four to nine eggs during the dry season (Censky and McCoy 1988).

Six species of snail-eating snakes occur in the Yucatán Peninsula, distributed among the genera *Dipsas* (*D. brevifacies*) (Fig. 11.1j), *Sibon* (*S. nebulatus*, *S. sanniolus*), and *Tropidodipsas* (*T. fasciatus*, *T. fischeri*, *T. sartorii*). The morphological adaptations that enable these snakes to pull their slimy prey out of their shells include particularly long teeth in the lower jaw and a conspicuously short snout. These are small to mid-sized, harmless colubrids that can be found at dusk and at night crawling around in vegetation, particularly during rain. They feed on snails and slugs, and have developed specific strategies to pull out the soft bodies of snails. Some hold the shell of the snail firmly in the coils of their body; others seek help of a Y-shaped twig as an anchor point (A. Hertz pers. comm.) The snake will grab the foot of the snail with the teeth of the lower jaw, press the shell against the upper jaw and then use the lower jaw to pull the soft body parts out of the shell. Up to a third of the snake's lower jaw is shoved into the shell during this procedure. After the snail is pulled out, the snake will drop the shell and swallow the soft body parts. Slugs are grabbed in the middle, pulled out of the substrate and usually swallowed tail first. All these species are oviparous and produce clutches of two to six eggs towards the end of the dry season or at the beginning of the rainy season (Kofron 1987; Campbell 1998).

Two species of Garter snakes (genus *Thamnophis*) occur in the region (i.e., *T. marcianus* and *T. proximus*). Garter snakes are semiaquatic, harmless snakes that will expel a vile-smelling secretion from the cloaca as their first line of defense before they will bite. They live in ponds, lakes, swamps, streams, and rivers, and feed on tadpoles, frogs, fish, and earthworms (Rossman et al. 1996). They are predominantly diurnal, but are occasionally also seen at night (Campbell 1998). All species are livebearers, producing litters of 6–22 young (Rossman et al. 1996).

Finally, there is a group of small, secretive, terrestrial colubrids with representatives in the genera *Ficimia*, *Ninia*, *Tantilla*, and *Tantillita*. These are without exception harmless snakes that will rarely attempt to bite when they are handled. They are most commonly found by turning over rotten logs or rocks in the forest, as well as in meadows. They feed on invertebrates, such as earthworms, beetle larvae, and centipedes (Van Devender and Cole 1977; Lee 1996; Campbell 1998).

11.7 Conservation

The principal threats to amphibians and reptiles in the Yucatán Peninsula are habitat loss/modification, pollution, overharvest for food and the pet trade, exotic species, infectious diseases, and climate change (Duellman 1999; Whitfield-Gibbons et al. 2000; Köhler 2011). Among the 142 species of reptiles and amphibians present in the states of Campeche, Yucatán, and Quintana Roo, 18 (13 %) appear as endangered (i.e., as Near Threatened and higher categories) in the Red List

Table 11.3 Number (%) of species of amphibians and reptiles listed in an endangered category (i.e., near-threatened and above) or appendix of the IUCN red list, CITES and Mexican NOM-59-SEMARNAT-2010

Groups	Orders	IUCN	CITES	NOM
Reptiles	Crocodilia	1 (50)	2 (100)	2 (100)
	Testudines	12 (75)	7 (43.8)	14 (87.5)
	Squamata			
	Sauria	3 (6.8)	1 (2.3)	18 (40.9)
	Serpentes	1 (1.8)	3 (5.4)	18 (32.1)
	Total	17 (14.4)	13 (11.0)	52 (44.1)
Amphibians	Caudata	0 (0)	0 (0)	3 (100)
	Anura	1 (4.8)	1 (4.8)	5 (23.8)
	Total	1 (4.2)	1 (4.2)	8 (33.3)
Total		18 (12.7)	14 (9.9)	60 (42.3)

of threatened species of the International Union for the Conservation of Nature (IUCN). Fourteen (9.9 %) appear in the appendices of the Convention on International Trade in Endangered Species of Wild Fauna and Flora (CITES), and 60 (42 %) are listed as threatened species under Mexican law (NOM-059-SEMARNAT-2010) (Table 11.3). Of the 18 species appearing as endangered in the IUCN Red List, eight are listed as Near Threatened, five as Vulnerable, two as Endangered, and three as Critically Endangered. The three critically endangered species are the freshwater Central American river turtle (*Dermatemys mawii*), the Hawksbill Turtle (*Eretmochelys imbricata*) and Kemp's Ridley turtle (*Lepidochelys kempii*). Among the 14 species appearing in the CITES appendices, six are listed in Appendix I (*Crocodylus acutus* and the five sea turtles), another six appear in Appendix II (*Crocodylus moreletii, Dermatemys mawii, Terrapene carolina, Iguana iguana, Boa constrictor, Agalychnis callidryas*), and two appear in Appendix III (*Micrurus diastema* y *Crotalus simus*). Of the 60 species listed by the NOM-059-SEMARNAT-2010, 41 appear with the status of Under Special Protection, ten are Threatened and nine Endangered. The nine endangered species are the five sea turtles, the freshwater turtles *Dermatemys mawii* and *Claudius angustatus* and the lizards *Ctenosaura defensor* and *Aspidoscelis rodecki*.

Habitat loss and degradation are the main factors driving the decline of both amphibian and reptile populations (Whitfield-Gibbons et al. 2000; Köhler 2011). Populations of coastal species are particularly vulnerable to these threats because of a higher rate of modification and destruction of coastal habitat. This is due to accelerated urban development along the Caribean coast of the Yucatán Peninsula, accompanying the mass tourism occurring in Quintana Roo. Coastal habitats of the Yucatán Peninsula are important for a great number of species of amphibians, lizards, freshwater and marine turtles, snakes and crocodiles, including endemic species such as *Sceloporus cozumelea* and *Aspidoscelis rodecki* (Escobedo-Galván et al. 2011). Populations of the latter species are restricted to a small area in northern Quintana Roo where urban development is increasing, and the destruction of its remaining habitat could drive this endemic species to extinction. Development-associated loss of vegetation cover involves increased soil temperature, reduced connectivity among subpopulations, and decreased availability of food/prey, refuges and reproduction sites, among other factors. Reptiles are ectotherms that depend on the ambient temperature for many aspects of their biology, ethology and physiology; thus, changes in temperatures could lead to important impact on coastal reptile populations. Concerning connectivity loss, the fragmentation of Quintana Roo's coastal habitat is the major threat to the recovery of the American crocodile, which presents high rates of inbreeding (Machkour M'rabet et al. 2009). Furthermore, these populations of *Crocodylus acutus* show high levels of hybridization and introgression with *C. moreletii*, especially in the southern coastal areas, which could lead to local extinctions of small, isolated, genetically pure populations (Rhymer and Simberloff 1996; Cedeño-Vázquez et al. 2008; Machkour M'rabet et al. 2009). However, island populations remain genetically pure and conservation efforts are needed in these islands to preserve the last reservoirs of *C. acutus* (Machkour M'rabet et al. 2009). Moreover, the modification of beaches, coastal dunes and nearby brackish lagoons due to tourist development is reducing the potential area for nesting sites and nursery habitat of marine turtles and crocodiles (Platt and Thorbjarnarson 2000; Escobedo-Galván et al. 2011). The increased human colonization of crocodile habitat could also lead to increased crocodile/human encounters, which generally end with the death of crocodiles (Escobedo-Galván et al. 2011). The number of crocodile attacks on humans in Quintana Roo increased between 2005 and 2011, after almost a decade without incidents (Charruau et al. 2012).

The increase in urban and agriculture areas in the Yucatán Peninsula also raises the level of environmental pollution by chemicals or plastic debris, which have adverse effects in amphibians and reptiles. While amphibians have been the subjects of many ecotoxicological studies, reptiles have received little attention; most studies on contaminant effects have focused on turtles and crocodiles (Whitfield-Gibbons et al. 2000; Gardner and Oberdörster 2005). Reptiles are particularly sensitive to organochlorine pesticides, which are endocrine system disruptors and can cause severe abnormalities such as sex reversal, feminization of males and lowered reproductive success (Guillette and Milnes 2000). Heavy metals and polychlorinated biphenyls (PCBs) can also cause abnormalities in reptiles and amphibians (Gardner and Oberdörster 2005; Gonzalez-Jauregui et al. 2012; Adlassnig et al. 2013). In the Yucatán Peninsula, the few studies on ecotoxicology in herpetofauna have shown the presence of organochlorine pesticides, PCBs, and heavy metals, which have been detected in crocodile eggs and tissues, and also in sea turtle eggs (Cuevas et al. 2003; Gonzalez-Jauregui et al. 2012; Charruau et al. 2013; Buenfil-Rojas et al. 2015). PCBs affect the testosterone levels in female *C. moreletii* (Gonzalez-Jauregui et al. 2012). Although a case of amelia that could be due to contamination has recently been observed in *C. moreletii* (Charruau and Niño-Torres 2014), no other effects of contamination on the herpetofauna have been detected in the Yucatán Peninsula. Plastic debris is also a serious threat to amphibians and reptiles, either by ingestion or entanglement (Gregory 2009).

Entanglement in a polyvinyl chlorine ring by *Ctenosaura similis* has been observed (Fig. 11.1k) on the atoll of Banco Chinchorro, as well as the regurgitation of 85 plastic bags weighing a total of 5 kg by an American crocodile in the Nichupté lagoon in Cancun (J. Carballar-Osorio pers. comm.)

Introduced species affect native species through competition, predation, disease transmission, habitat modification, and alteration of the trophic structure (Pitt et al. 2005; Bucciarelli et al. 2014). Some invasive reptiles and amphibians are known to affect species of their own group (Pitt et al. 2005; Bucciarelli et al. 2014). As mentioned previously, four species of reptiles and one species of amphibians have been introduced in the Yucatán Peninsula. These species are *Hemidactylus frenatus* (Fig. 11.1l), *H. turcicus*, *Anolis sagrei*, and *Rhamphotyplops braminus* for reptiles and *Eleutherodactylus planirostris* for amphibians. *E. planirostris* is a highly successful invasive species that was recently detected in Playa del Carmen, Quintana Roo (Cedeño-Vázquez et al. 2014). *Hemidactylus* spp. and *A. sagrei* are known successful invasive species and are well established in the Yucatán Peninsula (Case et al. 1994; Lee 1996; Campbell 1999; Echternacht 1999; Cole et al. 2005; Dame and Petren 2006). However, other introduced invasive vertebrates (e.g., janitor fish, tilapias, rats, feral pigs), invertebrates (e.g., fire ants) and plants (e.g., sheoak *Casuarina equisetifolia*) present in the peninsula could threaten native species of amphibians and reptiles (Whitfield-Gibbons et al. 2000).

Certain diseases can seriously affect species of reptiles and amphibians, causing population decline and potentially species extinction (Whitfield-Gibbons et al. 2000; Köhler 2011). The chytrid fungus *Batrachochytrium dendrobatidis* is causing a wave of extinction of amphibians in different parts of the world and is present in the state of Chiapas, near the Yucatán Peninsula (Lips et al. 2004). Another disease, fibropapilloma (Herbst 1994), is a viral tumor-causing disease that can potentially kill loggerhead sea turtles, as the tumors grow and inhibit movement, feeding, and vision. It has yet to be been detected on the peninsula, but due to the wandering behavior of sea turtle species it is likely only a matter of time before it arrives. Parthenogenetic species, such as the lizards *Aspidoscelis cozumela* and *A. rodecki* are more vulnerable to disease, as all individuals are females and share the exact same genome (Taylor et al. 2014).

Global climate change is a major threat to herpetofauna in the peninsula, through potentially increasing intensity and frequency of tropical cyclones, alterations of rainfall and temperatures patterns and increasing sea levels (Michener et al. 1997; Whitfield-Gibbons et al. 2000; Webster et al. 2005). Hurricanes and tropical storms are frequent in the Yucatán Peninsula and an increase of the number of category 3, 4 and 5 hurricanes has been observed over the last decades, while tropical storms and category 1 and 2 hurricanes have decreased (Charruau 2010). These cyclones destroy extensive areas of reptiles and amphibians habitat, resulting in the same changes described in the paragraph on habitat loss. A key impact of tropical storms is the destruction of turtle and crocodile nests (Michener et al. 1997; Charruau et al. 2010a). One study on the effects of tropical storms on American crocodiles nesting in Quintana Roo showed that hurricanes dramatically reduced nesting success by cooling and flooding clutches (Charruau et al. 2010a). However, it is likely that heavy rainfall from tropical cyclones can benefit hatchling survival by decreasing salinity in hypersaline nurseries. Tropical cyclones have negative short-term impacts on crocodile nesting but likely have an overall beneficial effect by creating and maintaining open sandy nesting areas used by crocodiles (Charruau et al. 2010a). However, rapid sea level rise would cause a reduction of coastal habitat including nesting sites of turtles and crocodiles. This would especially affect species that are only distributed on the coast and are already declining such as the endemic *Aspidoscelis rodecki*.

Scenarios of climate change in the Yucatán Peninsula predict an increase in temperature and a decrease in rainfall for the year 2020 (Orellana et al. 2009). Changes in rainfall and temperature patterns would affect amphibians and reptiles because of their tight dependence on these parameters. A change in temperature regime is also likely to affect reptiles with temperature-dependent sex determination (TSD), where the sex of embryos is determined by the incubation temperature and not by sexual chromosomes (Valenzuela and Lance 2004). An increase in nest temperature of TSD species could cause highly biased sex ratios, and affect population survival (LeGalliard et al. 2005). The crocodiles and several species of lizards and turtles in the Yucatán Peninsula that show this type of sex determination could be affected (Valenzuela and Lance 2004; Charruau 2010, 2012). Moreover, incubation temperature determines many physiological and morphological characteristics in amphibians and reptiles that are important for the survival of individuals and thus of their populations (Valenzuela and Lance 2004). Thus, the alteration of temperature and rainfall patterns in the peninsula in the near future is likely to have an important impact on the herpetofauna.

The illegal or unsustainable use of species, which includes illegal hunting, capture, collection, and transport of specimens, may also threaten some species of amphibians and reptiles (Whitfield-Gibbons et al. 2000). Although no study exists on the level of this threat in the Yucatán Peninsula, data on wildlife use and hunting (including data on reptiles) have been recorded in rural communities of the peninsula (Lee 1996; Santos-Fita et al. 2012). One example of the consequences of unsustainable use of a reptile in the Yucatán Peninsula is the case of the American crocodile. This species was highly over-exploited for its skin from the beginning to the middle of the twentieth century. Populations remain depleted in most areas, and show a difficult recovery in current conditions of global change (Cedeño-Vázquez et al. 2006b). Opportunistic killing of crocodiles is still occurring in the peninsula, and although it is difficult to assess the number of killings, the loss of even a few individuals, especially adult females, can have a dramatic impact on the depleted and fragmented populations of *C. acutus* in the peninsula (Cedeño-Vázquez et al. 2006b).

One option for the herpetofauna to adapt to changes in habitat modification is shifting their distributional range. However, amphibians and reptiles have a limited ability to disperse, and the destruction, fragmentation and modification of their habitat further reduces their dispersal ability.

Regarding conservation measures, major efforts have focused on sea turtles (Calderón-Mandujano and Cedeño-Vázquez 2011; Herrera-Pavón 2011). Although some conservation programs exist locally for crocodiles in the northern portion of the peninsula, the major conservation strategy for the herpetofauna is the declaration of protected natural areas (Cedeño-Vázquez and Calderon-Mandujano 2011; Calderón-Mandujano and Cedeño-Vázquez 2011). The peninsula hosts 45 protected natural areas (25 federal, 16 state reserves and four municipal reserves), some of them ranking among the largest protected areas in the country. The principal protected areas are Laguna de Terminos, Calakmul, Los Petenes, Sian Ka'an, Ría Lagartos and Ría Celestun. Calakmul Biosphere Reserve is preserving 723,185 ha of tropical forest in the south-center of the peninsula, while the other reserves are coastal and located all around the peninsula. The 45 reserves together cover an area of 4,838,553 ha which represent 34.2 % of the total area (14,136,157 ha) of the Yucatán Peninsula.

11.8 Perspectives

The diversity of amphibians and reptiles is relatively well known in the Yucatán Peninsula (Lee 1996; Köhler 2008, 2011), but knowledge on the ecology of the species remains scarce. For instance, the majority of the studies on turtles has been centered on the marine species, and conducted in the state of Quintana Roo (e.g., Isla Contoy, Isla Mujeres, Cozumel, X'Cacel, and other coastal zones from La Riviera Maya, Sian Ka'an Biosphere Reserve, and Costa Maya). Most of our knowledge comes from studies conducted on post-nesting females because these turtles are relatively easy to capture and tag. Topics that have been addressed about marine turtles to date in the peninsula include conservation and management (Viveros-León 1991; Zurita et al. 1993; Herrera-Pavón and Zurita 2005; Ortiz de Montellano 2010), reproduction (Prezas-Hernández 1991; Espinosa 1992; Pérez-Castañeda et al. 2007), feeding ecology (Plotkin et al. 1993), interactions (Sentíes-Granados et al. 1999; Cárdenas-Palomo and Maldonado-Gasca 2005), phylogeography, the population's genetic structure (Bowen et al. 1992; Encalada et al. 1996, 1998) and toxicology (Talavera-Sáenz 2010). Crocodiles are the most studied reptiles in the Yucatán Peninsula. However, studies have been mainly carried out in the state of Quintana Roo, and little work has been done on crocodiles in other regions of the peninsula. Studies to date have dealt with species morphology and distribution, population status and ecology, reproductive ecology, maternal behavior, population genetics (including hybridization), toxicology, health, archaeology and ethnozoology (Cedeño-Vázquez et al. 2006b; Machkour M'rabet et al. 2009; Charruau 2011; Thurston 2011; Zamudio-Acedo et al. 2011; Charruau et al. 2010a, b, 2012; Charruau and Hénaut 2012).

Future studies on the herpetofauna in the peninsula should focus on the determination of their ecological needs for reproduction, feeding and thermoregulation, among other issues. Studies on factors threatening the region's herpetofauna are urgent given the current conditions of global change, and coastal species are likely the most threatened due to accelerated development in coastal areas.

References

Adlassnig W, Sassmann S, Grawunder A, Puschenreiter M, Horvath A, Koller-Peroutka M. Amphibians in metal-contaminated habitats. Salamandra. 2013;49:149–58.

Álvarez del Toro M. Los Reptiles de Chiapas. 1st ed. Tuxtla Gutiérrez: Inst Zool del Estado; 1960.

Álvarez del Toro M. Los crocodylia de Mexico (estudio comparativo). Mexico, DF: Inst Rec Nat Renov; 1974.

Álvarez del Toro M. Los Reptiles de Chiapas. 3rd ed. Tuxtla Gutiérrez: Inst Zool del Estado; 1982.

Álvarez del Toro M, Mittermeier RA, Iverson JB. River turtle in danger. Oryx. 1979;15:170–3.

Boulon Jr RH, Dutton PH, McDonald DL. Leatherback turtles (*Dermochelys coriacea*) on St. Croix, U.S. Virgin Islands: fifteen years of conservation. Chelonian Conserv Biol. 1996; 2:141–7.

Bowen BW, Meylan AB, Ross JP, Limpus CJ, Balazs GH, Avise JC. Global population structure and natural history of the green turtle (*Chelonia mydas*) in terms of matriarchal phylogeny. Evolution. 1992;46:865–81.

Bucciarelli GM, Blaustein AR, Garcia TS, Kats LB. Invasion complexities: the diverse impacts of nonnative species on amphibians. Copeia. 2014;2014:611–32.

Buenfil-Rojas AM, Álvarez-Legorreta T, Cedeño-Vázquez JR. Metals and metallothioneins in Morelet's crocodile (*Crocodylus moreletii*) from a transboundary river between Mexico and Belize. Arch Environ Contam Toxicol. 2015;68:265–73.

Byles RA. Satellite telemetry of Kemp's ridley sea turtle, *Lepidochelys kempii*, in the Gulf of Mexico. In: Eckert SA, Eckert KL, Richardson TH, comps. Proceedings of the Ninth Annual Workshop on Sea Turtle Conservation and Biology. NOAA Tech Memo NMFS-SEFC-232. Miami; 1989. p. 25–6.

Calderón-Mandujano RR. Conocimiento y uso de la tortuga blanca (*Dermatemys mawii* Gray, 1847) en diez ejidos en el sur de Quintana Roo, México. Etnobiología. 2008;6:42–55.

Calderón-Mandujano RR, Cedeño-Vázquez JR. Reptiles. In: Pozo C, editor. Riqueza biologica de Quintana Roo, Un analisis para su conservación, Tomo 2. El Colegio de la Frontera Sur, Comisión Nacional para el Conocimiento y el Aprovechamiento de la Biodiversidad, Gobierno del Estado de Quintana Roo, Programa de Pequeñas Donaciones, México, DF; 2011. p. 247–51.

Calderón-Mandujano R, Cedeño-Vázquez JR, Pozo C. New distributional records for amphibians and reptiles from Campeche, México. Herpetol Rev. 2003;34:269–72.

Calderón-Mandujano R, Bahena Basave H, Calmé S. Amphibians and reptiles of the Sian Ka'an Biosphere Reserve and surrounding zones. Global Environment Facility, El Colegio de la Frontera Sur, Comisión Nacional para el Conocimiento y el Aprovechamiento de la Biodiversidad, Community Management of Protected Areas Conservation Programme, Sociedad Herpetologica Mexicana, Programa de las Naciones Unidas para el Desarrollo, Quintana Roo, México; 2008.

Campbell JA. Amphibians and reptiles of northern Guatemala, the Yucatán, and Belize. Norman: University of Oklahoma Press; 1998.

Campbell TS. Consequences of the Cuban brown anole invasion: it's not easy being green. Anolis Newsl. 1999;V:12–21.

Cárdenas-Palomo N, Maldonado-Gasca A. Epibiontes de tortugas de carey juveniles *Eretmochelys imbricata* en el santuario de tortugas marinas de Río Lagartos, Yucatán, México. CICIMAR Oceánides. 2005;20:29–35.

Carr AF. Handbook of turtles. Ithaca: Cornell University Press; 1952. Reprint 1995, Comstock Classic Handbooks, Cornell University Press.

Carr A, Stancyk S. Observations on the ecology and survival outlook of the hawksbill turtle. Biol Conserv. 1975;8:161–72.

Carranza J. Pesca y recursos pesqueros. Tortugas. In: Beltrán E, editor. Los recursos naturales del sureste y su aprovechamiento. México, DF: Inst Mex Rec Nat Renov; 1959. p. 227–38.

Casas-Andreu G. Análisis de la anidación de las tortugas marinas del genero *Lepidochelys* en México. An Centro Cienc Mar Limnol Univ Nac Autón Mex. 1978;5:141–58.

Case TJ, Bolger DT, Petren K. Invasions and competitive displacement among house geckos in the tropical Pacific. Ecology. 1994;75:464–77.

Cedeño-Vázquez JR, Calderon-Mandujano RR. Anfibios. In: Pozo C, editor. Riqueza biologica de Quintana Roo, Un analisis para su conservación, Tomo 2. El Colegio de la Frontera Sur, Comisión Nacional para el Conocimiento y el Aprovechamiento de la Biodiversidad, Gobierno del Estado de Quintana Roo, Programa de Pequeñas Donaciones, México, DF; 2011. p. 242–6.

Cedeño-Vázquez JR, González-Vázquez J, Martínez-Arce A, Canseco-Márquez L. First record of the invasive greenhouse frog (*Eleutherodactylus planirostris*) in the Mexican Caribbean. Rev Mex Biodiv. 2014;85:650–3.

Cedeño-Vázquez JR, Calderón-Mandujano R, Pozo C. Anfibios de la región de Calakmul, Campeche, México. Global Environment Facility, El Colegio de la Frontera Sur, Comisión Nacional para el Conocimiento y el Aprovechamiento de la Biodiversidad, Comision Nacional de Areas Naturales Protegidas. Reserva de la Biosfera Calakmul, Sociedad Herpetologica Mexicana, Programa de las Naciones Unidas para el Desarrollo, Quintana Roo, México; 2006a.

Cedeño-Vázquez JR, Ross JP, Calmé S. Population status and distribution of *Crocodylus acutus* and *C. moreletii* in southeastern Quintana Roo. Herpetol Nat Hist. 2006b;10:17–30.

Cedeño-Vázquez JR, Rodriguez D, Calmé S, Ross JP, Densmore III LD, Thorbjarnarson JB. Hybridization between *Crocodylus acutus* and *Crocodylus moreletii* in the Yucatán Peninsula: I. Evidence from mitocondrial DNA and morphology. J Exp Zool. 2008;309A: 661–73.

Censky EJ, McCoy CJ. Female reproductive cycles in five species of snakes (Reptilia: Colubridae) from the Yucatán Peninsula. Biotropica. 1988;20:326–33.

Charruau P. Ecología y etología de anidación del cocodrilo americano (*Crocodylus acutus*): Un estudio para su conservación. Tesis de Doctorado, El Colegio de la Frontera Sur, Chetumal, Quintana Roo, México; 2010.

Charruau P. Estimación de la edad de los cocodrilos (*Crocodylus acutus*) de Banco Chinchorro, Quintana Roo, México. Quehacer Científico en Chiapas. 2011;1:36–43.

Charruau P. Microclimate of American crocodile nests in Banco Chinchorro biosphere reserve, Mexico: effect on incubation length, embryos survival and hatchlings sex. J Therm Biol. 2012;37:6–14.

Charruau P, Hénaut Y. Nest attendance and hatchling care in wild American crocodiles (*Crocodylus acutus*) in Quintana Roo, Mexico. Anim Biol. 2012;62:29–51.

Charruau P, Niño-Torres CA. A third case of amelia in Morelet's crocodile from the Yucatán Peninsula. Dis Aquat Org. 2014;109:263–7.

Charruau P, Hénaut Y, Álvarez-Legorreta T. Organochlorine pesticides in nest substratum and infertile eggs of American crocodiles (Reptilia, Crocodylidae) in a Mexican Caribbean atoll. Caribb J Sci. 2013;47:1–12.

Charruau P, Cedeño-Vázquez JR, Calmé S. Status and conservation of the American crocodile (*Crocodylus acutus*) in Banco Chinchorro Biosphere Reserve, Quintana Roo, Mexico. Herpetol Rev. 2005;36:390–5.

Charruau P, Thorbjarnarson JB, Hénaut Y. Tropical cyclones and reproductive ecology of *Crocodylus acutus* Cuvier, 1807 (Reptilia: Crocodilia: Crocodylidae) on a Caribbean atoll in Mexico. J Nat Hist. 2010a;44:741–61.

Charruau P, Cedeño-Vázquez JR, Villegas A, González-Cortés H. Tasas de crecimiento del cocodrilo Americano (*Crocodylus acutus*) en estado silvestre en la Península de Yucatán, México. Rev Lat Cons. 2010b;1:63–72.

Charruau P, Pérez-Flores J, Pérez-Juárez JG, Cedeño-Vázquez JR, Rosas-Carmona R. Oral and cloacal microflora of wild crocodiles *Crocodylus acutus* and *C. moreletii* in the Mexican Caribbean. Dis Aquat Org. 2012;98:27–39.

Christy MT, Clark CS, Gee II DE, Vice D, Vice DS, Warner MP, Tyrrell CL, Rodda GH, Savidge JA. Recent records of alien anurans on the Pacific Island of Guam. Pac Sc. 2007;61: 469–83.

Cole NC, Jones CG, Harris S. The need for enemy-free space: the impact of an invasive gecko on island endemics. Biol Conserv. 2005;125:467–74.

Crimmins ML. A case of *Oxybelis* poisoning in man. Copeia. 1937;1937:233.

Cuevas E, Maldonado A, Cobos V. Determinación de DDT y DDE en huevos de tortuga blanca (*Chelonia mydas*) y de tortuga carey (*Eretmochelys imbricata*) en la costa de Yucatán, México. Oceánides. 2003;18:87–92.

Dame EA, Petren K. Behavioral mechanisms of invasion and displacement in Pacific island geckos (*Hemidactylus*). Anim Behav. 2006;71:1165–73.

Díaz LM, Cádiz A. Guía taxonómica de los anfibios de Cuba, Abc Taxa, vol. 4. Brussels: Royal Belgium Institute of Natural Sciences; 2008.

Dodd CK Jr. Synopsis of the biological data on the loggerhead sea turtle *Caretta caretta* (Linnaeus 1758). US Fish Wildl Serv Biol Report 88–14. Washington, DC; 1988.

Donlan EM, Townsend JH, Golden EA. Predation of *Caretta caretta* (Testudines: Cheloniidae) eggs by larvae of *Lanelater sallei* (Coleoptera: Elateridae) on Key Biscayne, Florida. Caribb J Sci. 2004;40:415–20.

Duellman WE. Amphibians and reptiles of the rainforests of southern El Petén, Guatemala. Univ Kan Publ Mus Nat Hist. 1963;15:205–49.

Duellman WE. The burrowing toad, *Rhinophrynus dorsalis*, on the Caribbean lowlands of Central America. Herpetologica. 1971;27:55–6.

Duellman WE. Global distribution of Amphibians: Patterns, conservation, and future challenges. In: Duellman WE, editor. Patterns of distribution of amphibians, a global perspective. Baltimore: The Johns Hopkins University Press; 1999. p. 1–30.

Easteal S. *Bufo marinus*. Cat Am Amphib Reptiles. 1986;395:1–4.

Echternacht AC. Possible causes for the rapid decline in population density of green anoles, *Anolis carolinensis* (Sauria: Polychrotidae) following invasion by the brown anole, *Anolis sagrei*, in the southeastern United States. Anolis Newsl. 1999;V:22–7.

Encalada SE, Lahanas PN, Bjorndal KA, Bolten AB, Miyamoto MM, Bowen BW. Phylogeography and population structure of the Atlantic and Mediterranean green turtle Chelonia mydas: a mitochondrial DNA control region sequence assessment. Mol Ecol. 1996; 5:473–83.

Encalada ES, Bjorndanl KA, Bolten AB, Zurita JC, Schoroeder B, Possar E, Sear CJ, Bowen BW. Population structure of loggerhead turtle (Caretta caretta) nestings colonies in the Atlantic and Mediterranean as inferred from mitochondrial DNA control region sequences. Mar Biol. 1998;130:567–75.

Ernst CH, Barbour RW. Turtles of the world. Washington, DC: Smithsonian Inst Press; 1989.

Escobedo-Galván AH, López-luna MA, Diaz de la Vega Pérez AH. Anfibios y reptiles en la zona costera de México. In: de la Lanza Espino G, Hernández Pulido S, compositors. Ambiente, biología, sociedad, manejo y legislación de sistemas costeros mexicanos. Universidad Michoacana de San Nicolás de Hidalgo, WWF-Fundación Gonzalo Río Arronte, Plaza y Valdés S.A. de C.V. Plaza y Valdés editores, Madrid, Spain; 2011.

Espinosa DM. Anidación de las tortugas Marinas (*Caretta caretta*) Linn. (1758) y Blanca (*Chelonia mydas*) Linn (1758) en la isla de Cozumel, Q. Roo, México. Tesis de Licenciatura en Biología. Universidad Veracruzana. Xalapa; 1992.

Fitch HS. Reproductive cycles in tropical reptiles. Occ Pap Mus Nat Hist Univ Kansas. 1982;96: 1–53.

Fitch HS, Hackforth-Jones J. *Ctenosaura similis* (Garrobo, Iguana Negra, Ctenosaur). In: Janzen DH, editor. Costa Rican Natural History. Chicago and London: University of Chicago Press; 1983. p. 394–6.

Flores-Villela O, García-Vázquez UO. Biodiversity of reptiles in Mexico. Rev Mex Biodiv. 2014; Suppl 85:S467–75.

Flores-Villela OA, Smith HM, Chiszar D. The history of herpetological exploration in Mexico. Bonn Zool Beitr. 2004;52:311–35.

Foster MS, McDiarmid RW. *Rhinophrynus dorsalis*. In: Janzen GH, editor. A Costa Rica natural history. Chicago: University of Chicago Press; 1983. p. 419–20.

Fuentes CD. Perspectivas de cultivo de tortugas marinas en el Caribe mexicano. Inst Nac Invest Biol Pesq Bol Prog Nac Marc Tort Mar. 1967;1:1–9.

Galindo-Leal C. De dos mundos/of two worlds: Las ranas, sapos y salamandras de la Península de Yucatán, México/The frogs, toads and salamanders of the Yucatán Peninsula. Saint Paul, MN: Pangaea; 2003.

Gardner SC, Oberdörster E. Toxicology of reptiles. Boca Raton, FL: CRC Press/Taylor & Francis; 2005.

Gibbons JW, Novak SS, Ernst CH. *Chelydra serpentina*. Cat Am Amphib Reptiles. 1988;420:1–4.

Gómez-Hernández Y. Hábitos alimentarios del cocodrilo de ría (*Crocodylus acutus*) y del cocodrilo de pantano (*Crocodylus moreletii*) en la reserva de la biosfera Sian Kaán Q. Roo, México. Tesis de Licenciatura, Universidad Autónoma Metropolitana, México, DF; 2004.

González-Cortés H. Estudio poblacional de *Crocodylus acutus* (Cuvier, 1807) en el Refugio Estatal de Flora y Fauna Laguna de Colombia, Cozumel, Quintana Roo, Mexico. Tesis de Licenciatura, Universidad Veracruzana, Córdoba, Veracruz, Mexico; 2007.

Gonzalez-Jauregui M, Valdespino C, Salame-Méndez A, Aguirre-León G, Rendón-vonOsten J. Persistent organic contaminants and steroid hormones levels in Morelet's crocodiles from the southern Gulf of Mexico. Arch Environ Contam Toxicol. 2012;62:445–54.

Goode M. *Pseustes sulphureus*. Reproduction. Herpetol Rev. 1989;20:73.

Gregory MR. Environmental implications of plastic debris in marine settings-entanglement, ingestion, smothering, hangers-on, hitch-hiking and alien invasions. Philos Trans R Soc B. 2009;364:2013–25.

Guillette Jr LJ, Milnes MR. Recent observations on the reproductive physiology and toxicology of crocodilians. In: Grigg GC, Seebacher F, Franklin CE, editors. Crocodilians biology and evolution. Chipping Norton, NSW: Surrey Beatty & Sons; 2000. p. 199–213.

Henderson RW, Hoevers LG. The seasonal incidence of snakes at a locality in northern Belize. Copeia. 1977;1977:349–55.

Herbst LH. Fibropapillomatosis of marine turtles. Annu Rev Fish Dis. 1994;4:389–425.

Herrera-Pavón RL. La Tortuga marina, omnipresente en la cultura Maya. In: Pozo C, Armijo-Canto N, Calmé S, editors. Riqueza biológica de Quintana Roo, Un análisis para su conservación, Tomo 1. El Colegio de la Frontera Sur, Comisión Nacional para el Conocimiento y el Aprovechamiento de la Biodiversidad, Gobierno del Estado de Quintana Roo, Programa de Pequeñas Donaciones, México, DF; 2011. p. 241–7.

Herrera-Pavón RL, Zurita JC. La protección de las tortugas marinas y el desarrollo turístico en la Rivera Maya, Quintana Roo, México. Revista de Medio Ambiente, Turismo y Sustentabilidad. 2005;1:35–8.

Hildebrand H. A historical review of the status of sea turtle populations in the western Gulf of Mexico. In: Bjorndal K, editor. Biology and conservation of sea turtles. Washington, DC: Smithsonian Institution Press; 1982.

Himmelstein J. Observations and distributions of amphibians and reptiles in the state of Quintana Roo, Mexico. Herpetol Bull New York Herpetol Soc. 1980;16:18–34.

Holman JA. Observations on dermatemyid and staurotypine turtles from Veracruz, Mexico. Herpetologica. 1963;19:277–9.

Iverson JB. The genus *Kinosternon* in Belize (Testudines: Kinosternidae). Herpetologica. 1976;32:258–62.

Iverson JB. Distribution and status of Creaser's mud turtle, *Kinosternon creaseri*. Herpetol J. 1988;1:285–91.

Iverson JB, Le M, Ingram C. Molecular phylogenetics of the mud and musk turtle family Kinosternidae. Mol Phylogenet Evol. 2013;69:929–39.

Keinath JA, Musick JA. Movements and diving behavior of a leatherback turtle, *Dermochelys coriacea*. Copeia. 1993;1993:1010–7.

Keister AR. Species density of North American amphibians and reptiles. Syst Zool. 1971;20: 127–37.
Köhler G. Inkubation von Reptilieneiern—Grundlagen, Anleitungen, Erfahrumgen. Offenbach: Herpeton; 1997.
Köhler G. Basilisken, Helmleguane, Kronenbasilisken. 2nd ed. Offenbach: Herpeton; 1999.
Köhler G. Schwarzleguane—Lebensweise, Pflege, Zucht. Offenbach: Herpeton; 2002.
Köhler G. Incubation of reptile eggs. Malabar, FL: Krieger; 2005.
Köhler G. Reptiles of Central America. 2nd ed. Offenbach: Herpeton; 2008.
Köhler G. Amphibians of Central America. Offebach: Herpeton; 2011.
Köhler G, Heimes P. Stachelleguane—Lebensweise, Pflege, Zucht. Offenbach: Herpeton; 2002.
Kofron CP. Systematics of Neotropical gastropod-eating snakes: the *fasciata* group of the genus *Sibon*. J Herpetol. 1987;21:210–25.
Kraus F, Campbell EW. Human-mediated escalation of a formerly eradicable problem: the invasion of Caribbean frogs in the Hawaiian Islands. Biol Invasions. 2002;4:327–32.
Kraus F, Campbell EW, Allison A, Pratt T. *Eleutherodactylus* frog introductions to Hawaii. Herpetol Rev. 1999;30:21–5.
Landy MJ, Langebartel DA, Moll EO, Smith HM. A collection of snakes from Volcan Tacana, Chiapas, Mexico. J Ohio Herpetol Soc. 1966;5:93–101.
Lang JW, Andrews HV. Temperature-dependent sex determination in Crocodilians. J Exp Zool. 1994;270:28–44.
Lee JC. An ecogeographic analysis of the herpetofauna of the Yucatán Peninsula. Univ Kansas Misc Publ Mus Nat Hist No. 1980;17.
Lee JC. The amphibians and reptiles of the Yucatán Peninsula. Ithaca and London: Cornell University Press; 1996.
LeGalliard JF, Fritze PS, Ferriere R, Clobert J. Sex ratio bias, aggression, and population collapse in lizards. Proc Natl Acad Sci USA. 2005;102:18231–6.
Legler JM, Vogt RC. The turtles of Mexico. Land and freshwater forms. Berkeley, Los Angeles, CA: University of California Press; 2013.
Lips KR, Mendelson III JR, Muñoz-Alonso A, Canseco-Márquez L, Mulcahy DG. Amphibians population declines in montane southern Mexico: resurveys of historical localities. Biol Conserv. 2004;119:555–64.
Mac-Arthur RH, Wilson EO. The theory of island biogeography. Monog Pop Biol. 1967;1:1–203.
Machkour M'rabet S, Hénaut Y, Charruau P, Gevrey M, Winterton P, Legal L. Between introgression events and fragmentation, islands are the last refuge for the American crocodile in Caribbean Mexico. Mar Biol. 2009;156:1321–33.
Magnuson JJ, Bjorndal KA, DuPaul WD, Graham GL, Owens DW, Peterson CH, Pritchard PCH, Richardson JI, Saul G, West CW. Decline of the sea turtles: causes and prevention. Washington, DC: National Academic Press; 1990.
Márquez MR. Sea turtles of the world. FAO Species Catalogue. 1990;11:1–81.
McCoy Jr CJ. Additions to the herpetofauna of southern El Petén, Guatemala. Herpetologica. 1966;22:306–8.
McCoy Jr CJ, Censky EJ. Biology of the Yucatán hognosed pitviper, *Porthidium yucatanicum*. In: Campbell JA, Brodie ED, editors. Biology of the Pitvipers. Tyler: Selva; 1992. p. 217–22.
McCranie JR, Wilson LD. A new arboreal lizard of the genus *Celestus* (Squamata: Anguidae) from northern Honduras. Rev Biol Trop. 1996;44:259–64.
McDiarmid RW, Foster MS. Unusual sites for two Neotropical tadpoles. J Herpetol. 1975;9: 264–5.
Meylan A. Spongivory in hawksbill turtles. A diet of glass. Science. 1988;239:393–5.
Meylan A. Status report of the hawksbill turtle. In: Ogren L, Berry F, Bjorndal K, Kumpf H, Mast R, Medina G, Reichart H, Withman R, editors. Proceedings of the Second Western Atlantic Turtle Symposium. NOAA Technical Memorandum NMFS-SEDC-226. National Oceanic and Atmospheric Administration. Washington, DC; 1989. p. 101–15.
Michener WK, Blood ER, Bildstein KL, Brinson MM, Gardner LR. Climate change, hurricanes and tropical storms, and rising sea level in coastal wetlands. Ecol Appl. 1997;7:770–801.

Minton Jr SA, Smith HM. A new subspecies of *Coniophanes fissidens* and notes on Central American amphibians and reptiles. Herpetologica. 1960;16:103–11.

Mortimer JA. Feeding ecology of sea turtles. In: Bjorndal K, editor. Biology and conservation of sea turtles. Washington, DC: Smithsonian Institution Press; 1982. p. 103–9.

Nelson CE. *Gastrophryne elegans*. Cat Am Amphib Reptiles. 1972;121:1–2.

Nussbaum RA. The Brahminy blind snake (*Ramphotyphlops braminus*) in the Seychelles Archipelago: Distribution, variation, and further evidence for parthenogenesis. Herpetologica. 1980;36:215–21.

Oliver JA. The relationships and zoogeography of the genus *Thalerophis* Oliver. Bull Am Mus Nat Hist. 1948;92:157–280.

Olson CA, Beard KH. Diet of the introduced greenhouse frog in Hawaii. Copeia. 2012;1:121–9.

Olson CA, Beard KH, Pitt WC. Biology and impacts of Pacific Island Invasive Species. 8. *Eleutherodactylus planirostris*, the Greenhouse Frog (Anura: Eleutherodactylidae). Pac Sci. 2012;66:255–70.

Orellana R, Espadas C, Conde C, Gray C. Atlas escenarios de cambio climático en la península de Yucatán. Centro de Investigación Científica de Yucatán y Centro de Ciencias de la Atmosfera de la Universidad Nacional Autónoma de México, Mérida, Yucatán, México; 2009.

Ortiz de Montellano AM. La tortuga marina: conservación en Campeche. Fomix Campeche. Revista Año. 2010;2(3):24–8.

Parra-Olea G, Flores-Villela O, Mendoza-Almeralla C. Biodiversity of amphibians in Mexico. Rev Mex Biodiv. 2014;Suppl 85:S460–6.

Pérez-Castañeda R, Salum-Fares A, Defeo O. Reproductive patterns of the hawksbill turtle *Eretmochelys imbricata* in sandy beaches of the Yucatán Peninsula. J Mar Biol Assoc UK. 2007;87:815–24.

Pitt W, Vice D, Pitzler M. Challenges of invasive reptiles and amphibians. In: Proceedings of the 11th Wildlife Damage Management Conferences, May 16–19, Traverse City, Michigan; 2005. p. 112–119.

Platt SG. Life history notes. *Rhinoclemmys areolata* (Furrowed wood turtle). Diet Herpetol Rev. 1993;24:32.

Platt SG, Rainwater TR. Distribution records and life history notes for amphibians and reptiles in Belize. Herpetol Rev. 1998;29:250–1.

Platt SG, Thorbjarnarson JB. Nesting ecology of the American crocodile in the coastal zone of Belize. Copeia. 2000;2000:869–73.

Platt SG, Rainwater TR. A review of morphological characters useful for distinguishing Mortelet's crocodile (*Crocodylus moreletii*) and American crocodile (*Crocodylus acutus*) with an emphasis on populations in the coastal zone of Belize. Bull Chicago Herp Soc. 2006;40:25–9.

Platt SG, Rainwater TR, McMurry ST. Diet, gastrolith acquisition and initiation of feeding among hatchling Morelet's crocodiles in Belize. Herpetol J. 2002;12:81–4.

Platt SG, Rainwater TR, Finger AG, Thorbjarnarson JB, Anderson TA, McMurry ST. Food habits, ontogenetic dietary partitioning and observations of foraging behaviour of Morelet's crocodile (*Crocodylus moreletii*) in northern Belize. Herpetol J. 2006;16:281–90.

Platt SG, Rainwater TR, Snider S, Garel A, Anderson TA, McMurry ST. Consumption of large mammals by *Crocodylus moreletii*: field observations of necrophagy and interspecific kleptoparasitism. Southwest Nat. 2007;52:310–7.

Platt SG, Thorbjarnarson JB, Rainwater TR, Martin DR. Diet of the American crocodile (*Crocodylus acutus*) in marine environments of coastal Belize. J Herpetol. 2013a;47:1–10.

Platt SG, Elsey RM, Liu H, Rainwater TR, Nifong JC, Rosenblatt AE, Heithaus MR, Mazzotti FJ. Frugivory and seed dispersal by crocodilians: an overlooked form of saurochory? J Zool. 2013b;291:87–99.

Platt SG, Rainwater TR, Thorbjarnarson JB, McMurry ST. Reproductive dynamics of a tropical freshwater crocodilian: Morelet's crocodile in Northern Belize. J Zool. 2008;275:177–89.

Platt SG, Rainwater TR, Thorbjarnarson JB, Finger AG, Anderson TA, McMurry ST. Size estimation, morphometrics, sex ratio, sexual size dimorphism, and biomass of Morelet's crocodile in northern Belize. Caribb J Sci. 2009;45:80–93.

Platt SG, Sigler L, Rainwater TR. Morelet's Crocodile *Crocodylus moreletii*. In: Manolis SC, Stevenson C, editors. Crocodiles. Status Survey and Conservation Action Plan. 3rd ed. Darwin: Crocodile Specialist Group; 2010. p. 79–83.
Plotkin P. Adult migrations and habitat use. In: Luts PM, Musick JA, Wyneken J, editors. The biology of sea turtles, vol. II. New York, NY: CRC Press; 2003. p. 225–41.
Plotkin PT, Wicksten MK, Amos AF. Feeding ecology of the loggerhead sea turtle *Caretta caretta* in the Northwestern Gulf of México. Mar Biol. 1993;115:1–5.
Polisar J. Reproductive biology and exploitation of the Central American river turtle *Dermatemys mawii* in Belize. MSc thesis, University of Florida; 1992.
Prezas Hernández B. Aspectos de la anidación de tortuga caguama *Caretta caretta* [linnaeus 1758], en la Costa Central del Estado de Quintana Roo, México. Tesis de Licenciatura en Biología. Universidad Veracruzana. Xalapa; 1991.
Pritchard PCH. Living turtles of the world. Neptune, NJ: TFH Publishing; 1967.
Pritchard PCH. Encyclopedia of turtles. Neptune, NJ: TFH Publishing; 1979.
Rand AS. *Physalaemus pustulosus*. In: Janzen DH, editor. A Costa Rica natural history. Chicago: University of Chicago Press; 1983. p. 412–5.
Rand AS. A nesting aggregation of iguanas. Copeia. 1968;1968:552–61.
Rand AS, Dugan BA. Structure of complex iguana nests. Copeia. 1983;1983(3):705–11.
Rhymer JM, Simberloff D. Extinction by hybridization and introgression. Annu Rev Ecol Syst. 1996;27:83–109.
Ricklefs RE. Ecology. Portland: Chiron Press; 1973.
Rodriguez D, Cedeño-Vázquez JR, Forstner MRJ, Densmore LD. Hybridization between *Crocodylus acutus* and *Crocodylus moreletii* in the Yucatán Peninsula: II. Evidence from microsatellites. J Exp Zool. 2008;309A:674–86.
Rossman DA, Ford NB, Seigel RA. The garter snakes—Evolution and ecology. Norman and London: University of Oklahoma Press; 1996.
Rossman DA, Schaefer GC. Generic status of *Opheodrys mayae*, a colubrid snake endemic to the Yucatán Peninsula. Occ Pap Mus Zool Louisiana State Univ. 1974;45:1–12.
Ruthven AG. The amphibians and reptiles collected by the University of Michigan-Walker expedition in southern Veracruz, Mexico. Zool Jahrb. 1912;32:295–332.
Ryan MJ. The Tungara frog, a study in sexual selection and communication. Chicago: University of Chicago Press; 1985.
Santos-Fita D, Naranjo EJ, Rangel-Salazar JL. Wildlife uses and hunting patterns in rural communities of the Yucatán Peninsula, Mexico. J Ethnobiol Ethnomed. 2012;8:38.
Savage JM. The amphibians and reptiles of Costa Rica: a herpetofauna between two continents, between two seas. Chicago: University of Chicago Press; 2002.
Seib RJ. Euryphagy in a tropical snake, *Coniophanes fissidens*. Biotropica. 1985;17:57–64.
Seidel ME. Taxonomic observations on extant species and subspecies of slider turtles, genus *Trachemys*. J Herpetol. 2002;36:285–92.
Sentíes-Granados A, Espinosa-Avalos J, Zurita JC. Epizoic algae of nesting sea turtles *Caretta caretta* and *Chelonia mydas* from the Mexican Caribbean. Bull Mar Sci. 1999;64:185–8.
Sexton OJ, Heatwole H. Life history notes on some Panamanian snakes. Caribb J Sci. 1965;5: 39–43.
Simpson GG. Species density of North American recent mammals. Syst Zool. 1964;13:57–73.
Smith HM, Smith RB. Synopsis of the herpetofauna of Mexico, Guide to Mexican turtles. Bibliographic addendum III, vol. 6. North Bennington, VT: John Johnson; 1979.
Solórzano A, Cerdas L. Reproductive biology and distribution of the terciopelo, *Bothrops asper* Garman (Serpentes: Viperidae) in Costa Rica. Herpetologica. 1989;45:444–50.
Stafford PJ. Diet and reproductive ecology of the Yucatán cricket-eating snake *Symphimus mayae* (Colubridae). J Zool. 2005;265:301–10.
Steyermark AC, Williams K, Spotila JR, Paladino FV, Rostal DC, Morreale SJ, Koberg MT, Arauz-Vargas R. Nesting leatherback turtles at Las Baulas National Park, Costa Rica. Anidamiento de las tortugas baula en el Parque Nacional Marino Las Baulas, Costa Rica. Chelonian Conserv Biol. 1996;2:173–83.

Talavera-Sáenz AL. Evaluación de metales pesados en hembras y crías de tortuga blanca *Chelonia mydas* (Reptilia, Cheloniidae) de las costas de Quintana Roo. México Tesis de maestría El Colegio de la Frontera Sur. 2010;45:444–50.

Talbert Jr OR, Stancyk SE, Dean JM, Will JM. Nesting activity of the loggerhead turtle (*Caretta caretta*) in South Carolina I: a rookery in transition. Copeia. 1980;1980:709–19.

Taylor HL, Cole CJ, Cooley CR. Origins and evolution in the *Aspidoscelis cozumela* complex of parthenogenetic Teiid lizards: morphological and karyotypic evidence and paradoxes. J Herpetol. 2014;48:343–54.

Telford SR, Campbell HW. Ecological observations on an all female population of the lizard *Lepidophyma flavimaculatum* (Xantusiidae) in Panama. Copeia. 1970;1970(2):379–81.

Terán-Juárez SA. Helmintos parásitos en cinco especies de anuros en el sur de Quintana Roo, México. Tesis de licenciatura. Instituto Tecnológico de Chetumal; 2011.

Thorbjarnarson JB. American Crocodile *Crocodylus acutus*. In: Manolis SC, Stevenson C, editors. Crocodiles. Status Survey and Conservation Action Plan. 3rd ed. Darwin: Crocodile Specialist Group; 2010. p. 46–53.

Thurston EB. Crocodiles and the ancient Maya: an examination of the iconographic and zooarchaeological evidence. MA thesis, Trent University, Ontario, Canada; 2011.

Valenzuela N, Lance V. Temperature dependent sex determination in vertebrates. Washington, DC: Smithsonian Institution Press; 2004.

Van Devender RW, Cole CJ. Notes on a colubrid snake, *Tantilla vermiformis,* from Central America. Am Mus Novit. 1977;2625:1–12.

Viveros-León P (1991) Manejo y protección de las tortugas marinas en Quintana Roo, México. Tesis de licenciatura (Médico Veterinario Zootecnista). Universidad Nacional Autónoma de México. México.

Viveros-León P. Uso tradicional y estatus de la tortuga blanca de río (*Dermatemys mawii*) en el complejo lagunar Guerrero, Quintana Roo, México. Tesis de maestría. El Colegio de la Frontera Sur. México; 1996.

Vogt RC, Guzman-Guzman S. Food partitioning in a Neotropical freshwater turtle community. Copeia. 1988;1988:37–47.

Vogt RC, Flores-Villela O. Effects of incubation temperature on sex determination in a community of Neotropical freshwater turtles in southern Mexico. Herpetologica. 1992;48:265–70.

Vogt RC, Platt SG, Rainwater TR. Rhinoclemmys areolata (Duméril and Bibron 1851)-Furrowed wood turtle, black-bellied turtle, mojina. In: Rhodin AGJ, Pritchard PCH, van Dijk PP, Saumure RA, Buhlmann KA, Iverson JB, Mittermeier RA, editors. Chelonian Research Mongraphs No. 5; 2009. p. 22.1–22.7.

Wake DB, Lynch JF. The distribution, ecology, and evolutionary hustory of plethodontid salamanders in tropical America. Nat Hist Mus Los Angeles Cty Sci Bull. 1976;25:1–65.

Webster PJ, Holland GJ, Curry JA, Chang HR. Changes in tropical cyclone number, duration, and intensity in a warming environment. Science. 2005;309:1844–6.

Whitfield-Gibbons J, Scott DE, Ryan TJ, Buhlmann KA, Tuberville TD, Metts BS, Greene JL, Mills T, Leiden Y, Poppy S, Winne CT. The global decline of reptiles, déjà vu amphibians. BioScience. 2000;50:653–66.

Zaldívar-Riverón A, León-Regagnon V, Nieto-Montes de Oca A. Phylogeny of the Mexican coastal leopard frogs of the *Rana berlandieri* group based on mtDNA sequences. Mol Phylogenet Evol. 2004;30:38–49.

Zamudio-Acedo F, Bello-Baltazar E, Estrada-Lugo EIJ. Integrando conocimientos mayas y científicos sobre el lagarto (*Crocodylus moreletii*) en el ejido de Xhazil Sur y anexos, Quintana Roo, México. In: Bello-Baltazar E, Estada-Lugo EIJ, compilers. Cultivar el territorio maya. Conocimiento y organización social en el uso de la selva. El Colegio de la Frontera Sur. Mexico; 2011. p. 161–188.

Zurita JC, Herrera R, Prezas B. Tortugas Marinas del Caribe. In: Salazar-Vallejo SI, González NE, editors. Biodiversidad Marina y Costera de México. Comisión Nacional para el Conocimiento y Uso de la Biodiversidad (CONABIO)/Centro de Investigaciones de Quintana Roo (CIQROO). Mexico; 1993. p. 735–50.

Chapter 12
Birds

Sophie Calmé, Barbara MacKinnon-H, Eurídice Leyequién, and Griselda Escalona-Segura

Abstract In this chapter, we present the current situation faced by 30 species of the Yucatán Peninsula which conservation status is currently at stake. We focus primarily on endemic species as well as resident and migrant species of conservation concern. Understanding how these species respond to their changing environment is vital to foreseeing their future in the Yucatán Peninsula. The information available for each species is highly variable, and we report many unpublished data and rely on personal observations to provide species accounts as complete as possible to assess the situation. In general coastal birds are at great risk due to rampant development threatening many coastal ecosystems, especially to the North and East of the peninsula, and because the Caribbean coast is regularly beaten by strong hurricane winds which frequency tends to increase with climate change. Small forest birds so far do not appear vulnerable, but large birds have declined due to hunting pressure or to the pet trade. Natural protected areas cover large tracts of the peninsula and represent most ecosystems, thus ensuring the protection of some species, but many species or populations are found outside these areas and require other strategies.

S. Calmé (✉)
El Colegio de la Frontera Sur, Avenida Centenario km. 5.5, CP 77014 Chetumal, Quintana Roo, Mexico

Université de Sherbrooke, 2500 boul. de l'Université, Sherbrooke, J1K 2R1 QC, Canada
e-mail: sophie.calme@usherbrooke.ca

B. MacKinnon-H
Amigos de Sian Ka'an A.C., Cancún, Quintana Roo, Mexico
e-mail: barbaramackinnonde@gmail.com

E. Leyequién
Centro de Investigación Científica de Yucatán, Unidad de Recursos Naturales, Calle 43 No. 130, Colonia Chuburná de Hidalgo, CP 97200 Mérida, Yucatán, Mexico
e-mail: leyequien@cicy.mx

G. Escalona-Segura
El Colegio de la Frontera Sur, Av. Rancho Poligono 2-A, Ciudad Industrial, CP 24500 Campeche, Campeche, Mexico
e-mail: gescalon@ecosur.mx

G.A. Islebe et al. (eds.), *Biodiversity and Conservation of the Yucatán Peninsula*,
DOI 10.1007/978-3-319-06529-8_12

Keywords Avifauna • Conservation • Endemic species • Distinct populations • Large raptors • Game birds • Pet trade • Natural history

12.1 Introduction

Birds are among the most conspicuous and attractive wildlife species, and as such one of the best known taxa. This explains the long history of ornithological research in the Yucatán Peninsula relative to other taxonomic groups. The first scientific expeditions included scientists dedicated exclusively to the discovery of bird diversity, with reports in scientific outlets starting in the mid-1800 (e.g. Cabot 1845; Lawrence 1869; Ridgway 1873). The number of publications on birds of the peninsula peaked in the 1980s (Fig. 12.1), after the opening of the last forest frontiers (permitted by road construction) and its concomitant surge of deforestation started in the late 1960s (Turner et al. 2001). Later, and despite the presence of a large number of protected areas and several new research centers, the number of ornithological publications decreased. In this regard, the region lags well behind other Neotropical areas despite its 555 bird species (MacKinnon 2013).

It is only by the second half of the last century that ornithology moved from the discovery of species to formal surveys on their ecology in the Yucatán Peninsula. The turn started in the 1950s, when bird research was still oriented towards discovery and natural history as illustrated by Paynter's notes from 1950 to 1957, and his famous "Ornithogeography of the Yucatán Peninsula" (Paynter 1955), but some ecological or behavioral research on migrant birds started to emerge (e.g., Eisenmann 1955). Understanding the ecology of species is particularly relevant in the Yucatán Peninsula as it is an important wintering area and migration corridor

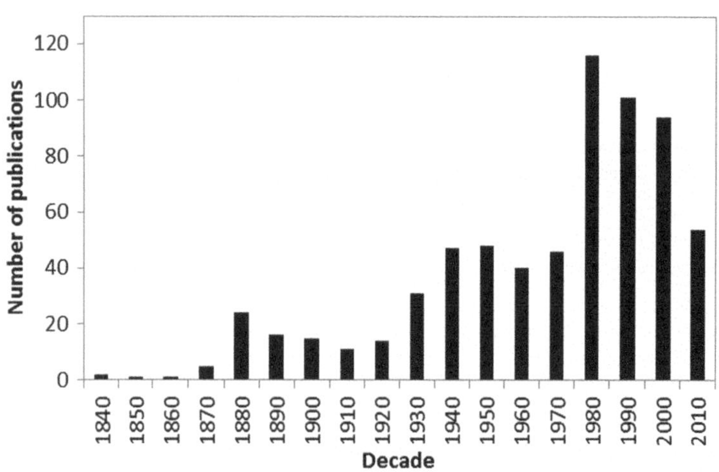

Fig. 12.1 Number of publications related to birds of the Yucatán Peninsula published per decade between 1840 and 2014

for many Nearctic-Neotropic bird species. More than 200 migratory bird species use the peninsula, 106 of which overwinter there, with another 26 species that have both resident and migrant populations, 12 species that are mainly transient of which some individuals overwinter in the peninsula, one species that is both transient and resident, 35 transient species, and 47 vagrant species (<5 records). The Yucatán Peninsula also hosts a large number of resident species, 17 of which are endemics or near-endemics, even when land cover has been changing considerably and climate change is presenting further challenges for species.

The status of the Yucatán Peninsula as bird habitat has changed dramatically over recent time. For the one part, population growth—along with urbanization—has increased pressure on land, while for the other part, new activities (e.g. cattle ranching) have promoted land use change. The peninsula remained scarcely populated as late as the 1960s, but both the government-promoted colonization of Quintana Roo and Southern Campeche, and natural demographic growth due to tourism development that began with the island of Cancún in 1969 have contributed to increased population (Fondo Nacional de Fomento al Turismo 1980). Between 1930 and 2010, the population of the Yucatán Peninsula has increased by more than 752 %, with the present population slightly above four million (Fig. 12.2). Currently three quarters of the peninsula land cover has been transformed or disturbed (see Chaps. 8 and 15), leading to changes in the composition and abundance of bird communities. Even if no long-term data exist, reviewing the localities from which earlier ornithologists reported their findings provides a clear idea of how great that change has been. For instance, of the 15 localities with known coordinates from which Ocellated Turkeys (*Meleagris ocellata*) were collected prior to 1960, only less than half hosted the species in 2000 (Calmé and Sanvicente 2000).

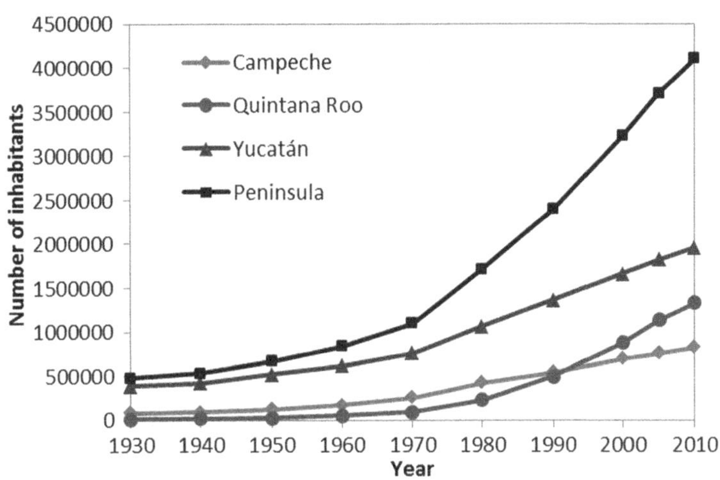

Fig. 12.2 Human Population change in the Yucatán Peninsula between 1930 and 2010. Data from Instituto Nacional de Estadistica y Geografia

In this chapter, we aim to present the current situation on the risks faced by a select number of species which conservation status is currently at stake. For detailed accounts on species we refer the reader to reference books such as The Handbook of the Birds of the World (del Hoyo et al. 1992–2013). We focus primarily on endemic species as well as resident and migrant species of conservation concern according to BirdLife International and the International Union for the Conservation of Nature (Table 12.1). Understanding how these species respond to their changing environment is vital to foreseeing their future in the Yucatán Peninsula. For this reason, we organize the chapter around the main physical features of the peninsula: the coastal areas, Cozumel Island, and the forested interior. We conclude tentatively by looking at the environmental history of the Yucatán Peninsula and reflecting on how it has shaped its avifauna.

12.2 Coastal Birds

Because of its peninsular nature, the Yucatán has a large proportion of coastal areas, some of which go far inland due to its landform (Chap. 2). The coast also borders two seas, the Gulf of Mexico to the west and north, and the Caribbean to the east. Coastal habitat as we define it in this chapter includes shallow salt waters, mudflats, beaches, dunes, mangroves, and scrublands found within the first kilometers inland. Following that definition, we include eight species in this section: Reddish Egret, Snowy Plover, Piping Plover, White-crowned Pigeon, Mexican Sheartail, Yucatán Vireo, Yucatán Wren, and Black Catbird. Seven are considered near-threatened by the International Union for the Conservation of Nature (IUCN); two overwinter in the peninsula (both plovers), one is endemic to Mexico (Mexican Sheartail), two others are endemic to the peninsula (Yucatán Wren and Black Catbird), and one is near-endemic (Yucatán Vireo).

Reddish Egret In the peninsula this medium-sized heron (Fig. 12.3a) is rather rare, and information about it is sparse. However, recent surveys have allowed recording the largest known colony. This colony of 670 pairs is located close to Sisal, Yucatán (Palacios et al. 2013). A total of 897 pairs living in ten colonies are estimated to reside in the whole peninsula (Palacios et al. 2013); however, three quarters of that number live in the largest colony mentioned above, while half of the other colonies have less than 25 pairs. While the Reddish Egret is confirmed to breed in numbers on the northern and eastern coasts, as well as in small colonies in Ría Celestún estuary and on a mangrove island at Isla Aguada, future research may confirm the presence of more colonies on the west coast of the peninsula (MacKinnon pers. obs.). Colonies are sometimes found among those of other birds such as herons and spoonbills (Kushlan and Hancock 2005). The species nests on all kinds of substrate, usually mangrove trees (Wilson et al. 2014), while it feeds on mudflats, preying mainly upon small minnows (Ramo and Busto 1993).

Table 12.1 Bird species included in the chapter, with their distribution and conservation status

English name	Scientific name	Distribution status	Conservation status
Coastal birds			
Reddish Egret	*Egretta rufescens*	R/M	NT
Snowy Plover	*Charadrius nivosus*	R/M	NT
Piping Plover	*Charadrius melodus*	M	NT
White-crowned Pigeon	*Patagioenas leucocephala*	R	NT
Mexican Sheartail	*Doricha eliza*	NE	NT
Yucatan Vireo	*Vireo magister*	NE	LC
Yucatan Wren	*Campylorhynchus yucatanicus*	E	NT
Black Catbird	*Melanoptila glabrirostris*	E	NT
Cozumel birds			
Cozumel Emerald	*Chlorostilbon forficatus*	E	LC
Cozumel Vireo	*Vireo bairdi*	E	LC
Cozumel Thrasher	*Toxostoma guttatum*	E	CR
Forested interior birds			
Great Tinamou	*Tinamus major*	R	NT
Great Curassow	*Crax rubra*	R	VU
Black-throated Bobwhite	*Colinus negrogularis persiccus*	NE	LC
Ocellated Turkey	*Meleagris ocellata*	E	NT
Agami Heron	*Agamia agami*	R	VU
Ornate Hawk-Eagle	*Spizaetus ornatus*	R	NT
Yucatan Poorwill	*Nyctiphrynus yucatanicus*	E	LC
Yucatan Nightjar	*Antrostomus badius*	E	LC
Yucatan Woodpecker	*Melanerpes pygmaeus*	E	LC
Orange-breasted Falcon	*Falco deiroleucus*	R	NT
Yellow-headed Parrot	*Amazona oratrix*	R	EN
Yellow-lored Parrot	*Amazona xantholora*	E	LC
Northern Mealy Parrot	*Amazona guatemalae*	R	NT
Yucatan Flycatcher	*Myiarchis yucatanensis*	E	LC
Yucatan Jay	*Cyanocorax yucatanicus*	E	LC
Wood Thrush	*Hylocichla mustelina*	M	NT
Rose-throated Tanager	*Piranga roseogularis*	E	LC
Painted Bunting	*Passerina ciris*	M	NT
Orange Oriole	*Icterus auratus*	E	LC

Distribution status is defined according to residence status and endemism with *R* resident; *M* migrant; *E* endemic; and *NE* near-endemic. Endemism to the Yucatán Peninsula here corresponds to a distribution encompassing the three Mexican states of the Yucatán Peninsula (Yucatan, Campeche and Quintana Roo), Northern Belize and Northern Guatemala, i.e., the peninsula in its geographical rather than geopolitical definition. Conservation status follows up-to-date Birdlife International Red List. *LC* Least Concern; *NT* Near Threatened; *VU* Vulnerable; *EN* Endangered; *CR* Critically Endangered

Fig. 12.3 (**a–g**) Some of the remarkable species of birds present in the Yucatán Peninsula. (**a**) Reddish Egret, which largest known colony (>600 pairs) is located in Sisal, Yucatán. Photo by Alexander Dzib Chay; (**b**) Cozumel Emerald, an endemic of Cozumel Island, on a rare sighting of

Despite no data existing for the Yucatán Peninsula, a recent genetic analysis of populations from Baja California, Northwestern Gulf coast, and Bahamas showed a strong differentiation among these regions, suggesting that dispersal is limited (Hill et al. 2012). These authors also suggest that given the scattered distribution and rather small size of its habitat, colonies tend to be small, increasing the risks of extinction caused by stochastic events, demographic or environmental.

The main environmental stochastic events on the coast of the peninsula are hurricanes, especially on its Caribbean coast (see also Chap. 7). In fact, the Yucatán Peninsula is the region of the Eastern American continent with the highest records of hurricane strikes. Some argue that hurricane probability or strength will further rise with climate change. If so, this could be of great concern for Reddish Egrets of the peninsula. Although there was no detected effect on breeding, colonies of the species were more clustered after Hurricane Ike (2008) swept the Great Inagua in the Bahamas on its path, leaving mangrove islands severely damaged (Green et al. 2011). Another study looking at the effect of hurricanes Katrina (2005) and Rita (2005) found that wind speed was associated with waterbird colony abandonment or with the reduction of the number of breeding pairs following these events (Leberg et al. 2007). The authors suggested these effects were caused by the heavy damages suffered by the vegetation serving as breeding support, although they could not demonstrate it.

Because feeding habitat (mudflats) is located in areas improper for tourist development, we believe that the sprawling urban development on the Yucatán and Quintana Roo coasts should not affect the Reddish Egret in the short term. In fact, the mudflats around known colonies are Ramsar sites protected both in Yucatán (El Palmar State Reserve) and in Quintana Roo (Sian Ka'an Biosphere Reserve). However, recreational activities (e.g. kayaking) in the lagoons where the species is nesting have the potential to disturb the birds if not undertaken with care. The access to feeding grounds might also be disturbed. The majority of the large colony of Sisal, for instance, flies over areas that will be subject to development to reach feeding grounds (B. MacKinnon pers. obs.). Development will also surely include the filling (at least partial) of the lagoon behind Sisal, leading to the loss of feeding habitat and to potential contamination. We should note, moreover, that drought may reduce the availability of mudflats or increase their salinity, leading to a higher mortality of organisms living in them. However, no information currently exists on the effect of drought on the Reddish Egret or its feeding habitat.

Fig. 12.3 (continued) an adult feeding nestlings. Photo by Rafael Chacón; (**c**) Yucatan Woodpecker, a little known endemic of the peninsula. Photo by José Antonio Linage. (**d**) Yucatan Wren, an endemic of the dry scrub of northern Yucatán. Photo by José Antonio Linage; (**e**) Orange Oriole, an endemic of the peninsula. Photo by Alejandro Pacheco Moreno; (**f**) Black Catbird, an endemic mainly nesting on the Caribbean coast. Photo by Jorge Machado Castro; (**g**) Ocellated Turkey, an endemic, emblematic and heavily hunted bird. Photo by Alexander Dzib Chay

Snowy Plover This shorebird was recently recognized as a New World species when it was officially split from the Kentish Plover (*C. alexandrinus*) by the American Ornithologists' Union (Chesser et al. 2011), after it was found to be genetically distinct (Küpper et al. 2009). It is a relatively uncommon bird that can be found in small flocks on sandy beaches, salt ponds and salt flats of the Yucatán Peninsula (Howell and Webb 1995; Thomas et al. 2012). However, year-long residents and breeders are found on the northern coast of the peninsula; the birds encountered along the Western Gulf coast and Eastern Caribbean coast are rather considered winter migrants. The breeding population is probably quite small given the number of birds (180 in 11 locations) counted during the North American-wide sampling realized by Thomas et al. (2012).

The Snowy Plover is very vulnerable to coastal tourist development, which in turn causes both habitat destruction and creates a lot of disturbance to feeding and nesting birds. On the one hand, dune vegetation is usually removed by beach-front property owners (pers. obs.), while the species nests at the limit of the beach and the front dune (Lafferty et al. 2006). Increasing tourist development of the peninsula northern coast is thus likely to destroy a significant part of Snowy Plover's nesting habitat in that region, even if development remains mostly concentrated along a 120-km stretch between Sisal and Dzilam de Bravo, in Yucatán State, and around Holbox, in Yum Balam Biosphere Reserve, in Quintana Roo. This latter area is of particular concern since the development of Holbox Island has literally exploded over the past few years. Tourists are also a main source of disturbance to birds feeding or nesting on beaches (Lafferty et al. 2006). Snowy Plover are greatly disturbed by people at distances of 10 m (>50 % of the events cause disturbance; Lafferty 2001). This latter study also found that dogs contributed more to disturbance than people, especially when unleashed. In Mexico, lots of stray dogs wander around towns and villages in search for food and may therefore cause further disturbance and mortality (to eggs and chicks) than reported in the USA by Lafferty et al. (2006), where dogs were accompanying their owners.

Salt flats and ponds, important feeding grounds where Snowy Plovers prey on small invertebrates, are common features in several parts of the peninsula northern coast. They remain generally little disturbed. However, there exists locally salt-works and saltpans built for the production of brine shrimp (*Artemisia* sp.) cysts. Those artificial basins are located in areas not suitable for tourism. In consequence, overall habitat modification tends to expand over most of the peninsula northern coast, at the exception of the protected areas of Ría Lagartos Special Biosphere Reserve, and Dzilam de Bravo State Reserve. In these protected areas, development is limited and better controlled than elsewhere, in the latter one because the coast can only be reached by boat. One other important reserve, Ría Celestún Special Biosphere Reserve, is failing to provide long-term protection on its Yucatán side, as it is being slowly developed.

Piping Plover This relatively rare migrant species uses beaches and sandflats in the same area on the peninsula Northern coast as resident Snowy Plovers in addition to Isla Blanca north of Cancún. Like this latter species, it also feeds on small invertebrates (Wiersma et al. 2014). It is therefore subject to the same kind of disturbance on the beaches it frequents, i.e., people and dogs. Burger (1994) found that the time Piping Plovers spent foraging in dunes was halved in the presence of people within 100 m, while vigilance increased. Because overwintering migrants need to acquire enough energy to be able to fly back to their breeding grounds, the cost of vigilance is potentially high. Furthermore, the kind of short flights initiated to flee a source of disturbance are also very costly to birds (Nudds and Bryant 2000), and may decrease their probability of survival during spring migration.

White-Crowned Pigeon This is a relatively common Caribbean bird that is therefore present only on the Caribbean coast of the Yucatán Peninsula (Howell and Webb 1995), except when individuals get blown off course and reach the Río Lagartos area. With the exception of small winter populations on Cozumel Island and Cayo Centro and Cayo Norte on Banco Chinchorro, the species arrives in late spring and leaves in the fall (MacKinnon and Acosta Aburto 2003). Along the coast, the White-crowned Pigeon uses two distinct vegetation types for feeding and breeding. The species breeds on mangrove islets, cays and large islands off the coast, but feeds in seasonal tropical forest on the coast (Howell and Webb 1995). Although no study on the diet of the White-crowned Pigeon has been conducted in the Yucatán Peninsula, the bird appears to be almost strictly frugivorous elsewhere, most probably feeding on trees present in Quintana Roo forests such as *Metopium* and *Ficus* (Bancroft and Bowman 1994), and *Brosimum alicastrum* (Baptista et al. 2014). In Florida, nesting phenology, in terms of timing and number of nests initiated, seems strongly associated with the fruiting phenology (timing and abundance of fruits) of *Metopium* (Bancroft et al. 2000).

Even when habitat destruction is the main threat to White-crowned Pigeon in other parts of its distribution (Baptista et al. 2014), we do not believe it to strongly affect the species in the Yucatán Peninsula. The Northern Caribbean coast has been subject to intensive tourist development leading to deforestation over the past 40 years, and to the South a similar development is looming, despite being aimed to be less intensive. However, White-crowned Pigeon has the ability to fly over distances in the tens of kilometers. For instance, MacKinnon and Acosta Aburto (2003) observed 18 individuals leave Banco Chinchorro prior to sunset in direction of the mainland, located 31 km from the cay; and five individuals later coming from the mainland. This flight ability should allow the species to reach feeding habitat in the less developed forested interior, as development remains usually limited to a relatively narrow strip along the coast. Also, the Yum Balam, Sian Ka'an, and Banco Chinchorro Biosphere Reserves, as well as the Isla Contoy National Park protect more than 150 km of coast and offshore islands with suitable White-crowned Pigeon feeding and breeding habitat. The species is sensitive, however, to human disturbance, and is known to abandon nesting sites when disturbed (e.g. in Sian Ka'an; B. MacKinnon pers. obs.)

As noted before, hurricanes are common events on the Caribbean coast of the Yucatán Peninsula, and cause extensive damage to forests (Whigham et al. 2003) and mangroves. Trees in the region are adapted to this type of disturbance by sprouting on stems after having been snapped or uprooted, allowing post-hurricane forest composition to remain stable (Vandermeer et al. 1997; Whigham et al. 2003). The trees on which the White-crowned Pigeon feeds are mainly fast-growing species that recover after hurricane damage but also after burning (e.g. *Metopium*, *Bursera*). This is an important characteristic since hurricanes are often followed by wildfires due to the large accumulation of dead organic material (see Chap. 7). Even if trees have the ability to recover from hurricane disturbance, fruiting is reduced or even suppressed immediately after, and many trees change their fruiting cycle for a few years (Wunderle 1999). Since nesting in White-crowned Pigeon is synchronized with fruit-peaking, and varies quantitatively with fruit availability (Bancroft et al. 2000), hurricane damages have obvious consequences on its population dynamics within the area.

We believe the combined effect of habitat destruction (via land use change) and habitat disturbance (via hurricanes and humans) to be the most serious threat to the White-crowned Pigeon on the Yucatán Peninsula. Birds might have a hard time finding a new feeding ground following a hurricane if too much forest has been cleared for tourism. This could be especially worrying if the area affected by the climatic event is a Biosphere Reserve. Given the current rate of development in coastal Quintana Roo and the possible increase in the frequency or strength of hurricanes in the face of climate change, this threat might materialize sooner than later.

Black Catbird The distribution of this endemic (Fig. 12.3f) encompasses the Yucatán Peninsula; however it is locally common only on the Eastern Yucatán Peninsula and Cozumel Island, including Northern Belize (Morgenthaler 2003). The density of nests found in some areas on the Caribbean coast (up to 3.9 per ha; Roldán-Clarà et al. 2013), as well as territory clustering (Morgenthaler 2003; Roldán-Clarà 2009), suggest that the Black Catbird could be colonial. Inland Yucatán, B. MacKinnon (pers. obs.) found the species completely colonial when nesting. To that point the reasons for such aggregation remain unknown (see Danchin and Wagner 1997 for a review of hypotheses). The species uses a variety of vegetation types for breeding that are typical of Caribbean coastal vegetation: scrubby vegetation, sometimes dominated by mangrove, littoral forest and mangrove (Morgenthaler 2003; Roldán-Clarà et al. 2013). Inland the species has been observed nesting in chukum trees, *Havardia albicaus*, but is also frequently found in thick scrub, and almost always near a water source (B. MacKinnon pers. obs.). According to Morgenthaler (2003) Black Catbirds forage on arthropods, especially during the breeding season, and a variety of fruits from trees (e.g. *Manilkara zapota*, *Metopium brownei*), palm (*Thrinax radiata*), coastal scrub (*Coccoloba uvifera*) and bush. It also feeds on the fruit of the Black Mangrove (*Avicennia germinans*) and Black Sage (*Lantana involucrata*) (MacKinnon 2005).

Like plovers and White-Crowned Pigeons, Black Catbirds are at risk of sprawling tourist development along the Caribbean coast. Even in Sian Ka'an Biosphere Reserve, a small amount (<3 %) of the reserve is private property but it is mostly located along the coast (Roldán-Clarà et al. 2013). Small properties usually result in the almost complete loss of native vegetation on the dune and behind the dune (S. Calmé pers. obs.). We are also concerned by the fact that the species appears to require a matrix of different vegetation types as shown by Morgenthaler (2003), when tourist development favors homogenization of the habitat (Gunn 1997).

Nest predation is another potential threat to the species. Roldán-Clarà et al. (2013) observed several events of predation or intents of it by the Great-tailed Grackle (*Quiscalus mexicanus*), and they believed that this species was responsible for a large fraction of depredation in Sian Ka'an Biosphere Reserve. Contrary to the Black Catbird, this species thrives in areas subjected to development (Fraga 2011), with the risk of increasing predation pressure on Black Catbird in an unsustainable way.

Mexican Sheartail It is an endemic species in Mexico with two allopatric populations, one in central Veracruz and the other one on the northern extreme of the Yucatán Peninsula. This species is listed as near-threatened by the IUCN (Birdlife International 2014) and as in danger of extinction by the Mexican government (SEMARNAT 2010). Both populations have a small geographical range and are locally threatened by habitat loss and degradation (Ortiz-Pulido et al. 2002). In the Yucatán Peninsula, the Mexican Sheartail hummingbird occurs mainly in the coastal dune and the edge of the mangrove, with at least a geographic range of 4 km from the coastline (Santamaría-Rivero et al. 2013). However, a few observations have been reported in the dry tropical deciduous forest within a distance of 5 km from the coastline (Leyequién unpubl. data) and in gardens in Mérida City during the summer months, 36 km inland (MacKinnon unpubl. data). López-Ornat and Lynch (1990) reported an isolated observation on Isla Cancun, which represents its most northeastern range. This species forages and constructs nests on native vegetation found in its habitat. It feeds on at least 23 plant species, i.e., *Malvaviscus arboreus*, *Agave angustifolia*, *Ernodea littoralis*, *Opuntia stricta*, *Tillandsia dasyliriifolia*, *Caesalpinia vesicaria*, *Cannavalia rosea*, *Cordia sebestena*, *Dicliptera sexangularis*, *Sophora tomentosa*, *Gossypium hirsutum*, *Ipomoea pes-caprae*, *Lycium carolinianum*, *Passiflora foetida*, *Suriana marítima*, *Tournefortia gnaphalodes*, *Scaevola plumieri*, *Myrmecophila cristinae*, *Euphorbia personata*, *E. heterophylla*, *Turnera diffusa*, *Bougainvillea spectabilis*, and *Cocos nucifera*. For nesting, Mexican Sheartails use Black Mangrove (*Avicenia germinans*), Button Mangrove (*Conocarpus erectus*), Bay Cedar (*Suriana maritima*), and White Mangrove (*Laguncularia racemosa*), among others (Leyequién unpubl. data). This species presents two breeding peaks during the year: the first from January to April, and the second from June to November (Leyequién unpubl. data).

Current threats to the Mexican Sheartail include potentially low recruitment rates and loss of feeding habitat. The probability of nest survival, i.e. probability that a nest fledges at least one young with a nesting period of 48 days, is as low as 16.5 % (based on Mayfield, 1975 from Leyequién unpubl. data). Causes of recorded failure are predation by ants, praying mantis, strong winds, parasites, and infertility (Leyequién unpubl. data). We believe that habitat loss and degradation is another important risk factor for the population, as native vegetation is replaced by homogenous and/or non-native plants in the gardens of hotels and summer houses (an important breeding ground has been found in Telchac Puerto surrounded by summer houses, hotels, and roads; E. Leyequién pers. obs.). Also the potential construction of wind farms in ecologically vulnerable areas such as in Dzilam de Bravo, Yucatán State, could severely threaten the nest success and population dynamics of this species through collision risk, displacement due to disturbance, barrier effect acting upon local flight paths, and habitat change and loss (Drewitt and Langston 2006). A number of studies recommend the precautionary avoidance of construction of wind turbines close to any important bird breeding colony and within the most frequent foraging flight paths (Everaert and Stienen 2007).

Yucatán Vireo This near endemic species is distributed for the most part along the east coast of the Yucatán Peninsula, Belize and adjacent islands, including Honduras islands of Guanaja—abundant, Utila and Roatán—not abundant (Howell 1991), and Grand Cayman. There are at least three records of individuals photographed or netted in the northeastern corner of Yucatán (San Felipe, Ría Lagartos and El Cuyo) between 2004 and 2013 that can be considered vagrants (Gómez de Silva 2004, 2005, 2014), and one vagrant was documented in Texas in 1984 (Morgan et al. 1985). Although found principally in coastal zones, it has been recorded in the past 45 km inland at the archeological zone of Cobá (MacKinnon pers. obs.). Its habitat consists of humid to semi-humid scrubby woodland and edge and mangroves (Howell and Webb 1995).

Vireo magister magister probably forms a super-species with *V. olivaceus*, *V. gracilirostris*, *V. flavoviridis* and *V. altiloquus*. Four subspecies are recognized, including: *V. m. magister* (S. F. Baird, 1871)—SE Mexico (Quintana Roo, including islands of Mujeres and Cozumel) and mainland Belize; *V. m. decoloratus* (A. R. Phillips, 1991)—islands off Belize (Ambergris Cay S to Turneffe Is); *V. m. stilesi* (A. R. Phillips, 1991)—Glover's Reef (off C Belize) and Bay Is (off N Honduras); and *V. m. caymanensis* Cory, 1887—Grand Cayman I (Brewer 2010). Although it may have a small range, it is not believed to approach the thresholds for "Vulnerable" under the range size criterion of the IUCN and consequently is listed as a species of Least Concern. Partners in Flight estimated the population to figures fewer than 50,000 individuals (BirdLife International 2015).

The species nests between April and July on the peninsula in a nest similar to others of its family, a cup suspended from a forked, horizontal branch at medium height. It feeds on insects and fruits, including those of the chaká (*Bursera simaruba*), as do all vireos, and abounds in the very dense dune vegetation that provides both food and refuge (MacKinnon 2013).

Unfortunately, no in-depth study has been made on the ecology of this species, which should be a priority due to the rapid development of tourism infrastructure and expanding urbanization along the majority of the east coast of the peninsula, affecting the species major distributional habitat. As seen with the Black Catbird, even within the Sian Ka'an Biosphere Reserve, private coastal properties are cleared of native dune vegetation upon which the species depends.

Yucatán Wren Campylorhynchus yucatanicus is an endemic species of the Yucatán Peninsula, whose distribution is highly restricted to the waterfront in the northwestern portion of the Peninsula, in the states of Campeche and Yucatán. It lives exclusively in coastal scrub (Fig. 12.3d) and adjacent black mangrove forest, with the first one regarded as rare and restricted to the shoreline of the above mentioned area (Flores and Espejel 1994). Sixty five percent of the Yucatán Wren nests found by Vargas-Soriano et al. (2010) were in coastal scrub, while the remaining 35 % were in flooded lowland forest (between the coastal scrub and mangrove forest). Only three species, Button Mangrove (*Conocarpus erectus*), Peccary Tree (*Sideroxylon americanum*), and Blackbead (*Pithecellobium pallens*), supported 75 % of all nests, suggesting that the Yucatán Wren is highly dependent on these tree species (Vargas-Soriano et al. 2010). According to their observations, nest construction began in early April and finalized in mid-September; nest building ranged from 4 to 12 days. Despite early nest construction in April, the first egg was not laid until June; the reason why egg laying was delayed by 67 days (the time span between the first nest built and the first egg laid) is unknown. The last nest was recorded at the end of July, covering a nesting period of 48 days from the first brood. Second attempts after failure or success of the nests were recorded. Clutch size varied from 1 to 5 eggs that were incubated for 15–17 days. Brood hatching was synchronous. Parental care was provided by two adult individuals, presumably the parents, which fed the chicks in turn. In 4 % of the population, three adults participated, hinting at a cooperative breeding system in this species (Vargas-Soriano et al. 2010).

Predation is the main natural cause of failure of Yucatán Wren nests. However, the Yucatán Wren and its habitat, the coastal scrub and the black mangrove are being threatened primarily by the development of hotel infrastructure, and to a lesser extent by charcoal production and logging. Given the extremely restricted range of the species, an assessment of its population should be performed, and strict protection measures should be enforced.

12.3 Cozumel Birds

Cozumel avifauna is unique in that the island is not all that distant from the mainland (17.5 km), separated by a channel approximately 400 m deep, but has three endemic species and 16 resident endemic subspecies (Clements 2014). However, rich as it is in endemics it is limited in extension (478 km^2) and habitats, being a coral based ocean island with a maximum height of 10 m and with no ground level water source (Cuarón 2009). In this section we present the cases of the three

Cozumel endemics: Cozumel Emerald, Cozumel Vireo, and Cozumel Thrasher. The first two are not considered of conservation concern, while the Cozumel Thrasher is listed as Critically Endangered. Threats to the endemic species and subspecies include the annual reduction and fragmentation of habitat due to increased tourism infrastructure and urbanization, as well as introduction of invasive species (both competitors and predators) that include *Boa constrictor*, feral dogs (*Canis lupus familiaris*) and cats (*Felis catus*), rats (*Rattus rattus*) and mice (*Mus musculus*), and more recently feral bovine cattle (*Bos* sp.) (Cuarón et al. 2004; Cuarón 2009; McFadden et al. 2010). In addition, hurricane impacts on habitat have taken their toll over the past few decades. Of concern for all bird species on Cozumel, but even more so for endemics, is that population trends and natural history have not been studied in depth or are completely unknown. Of recent creation, the Cozumel Birding Club, in collaboration with the Commission for Natural Protected Areas (CONANP) are carrying out monitoring of bird species on the island, which over the long term will help assess the population trends for all endemics on Cozumel.

Cozumel Emerald Formally considered a subspecies of *Chlorostilbon canivetii* it was recognized as a separate species in the early 1990s (Howell 1993). Its distribution is limited to the island of Cozumel, with an individual found on nearby Isla Mujeres being considered a vagrant. It is fairly common on the island where it is found in "scrub and low deciduous insular forest" (Paynter 1955) and in vegetation described as "brushy woodland and scrub, second growth, etc." (Howell and Webb 1995). In more recent times it has been commonly found feeding and nesting in the yards of private homes and even restaurants in the town of San Miguel (MacKinnon pers. obs.). The species has been documented by local biologist Rafael Chacón reproducing from January through May, building nests mostly at low to mid-level in tree branches near food sources as well as in man-made objects such as broken lamp fixtures and even cables. Although hummingbirds are known for laying just two white eggs, there are some examples of three offspring in a nest, albeit not necessarily from the same female. A picture was taken of two eggs in a nest built atop a frond of a potted xiat palm (*Chamaedorea* sp.), but at a later date, a picture was taken of three featherless chicks in this nest (Fig. 12.3b); however, only two survived. Casual observations of five nests between January and April of 2015 produced estimates of between 4 and 5 weeks for incubation and 17 days between hatching and when fledglings left the nest. Nest destruction observed was a result of inclement weather in which strong winds caused nests to plummet from their location. In one case, the female stayed out all night, returning to the nest the next morning, without any harm caused to the chicks (R. Chacón pers. com).

The IUCN Red List has assessed the Cozumel Emerald of Least Concern (BirdLife International 2012d) as even though the species has a very small range, it is not believed to approach the thresholds for vulnerable under the range size criterion; nor is the species considered threatened under Mexican law. However, its population trend is not known. There was concern for the species after the passing of Hurricane Wilma (4 on Saffir-Simpson scale when it made landfall) in October of 2005 as the species was not visible to observers the following January, but appeared normal again in March. The species therefore seems quite resilient to hurricanes, even as their frequency or intensity is expected to increase with current climate change.

Cozumel Vireo Considered common just 20 years ago (Macouzet 1997), this endemic bird to the island of Cozumel currently is reported fairly common as evidenced by a recent monitoring project initiated by CONANP and the Cozumel Birding Club. However, since its population trend has not been scientifically studied, it is currently considered a species of Least Concern (BirdLife International 2012c) and is not considered threatened under Mexican law.

The Cozumel Vireo belongs to the 'eye-ring' group of vireos that includes eight species endemic to the West Indies or the adjacent Caribbean Coast of Mexico and Central America, of which few studies exist. The most range-restricted species among these is the Cozumel Vireo (LaPergola et al. 2012). The species inhabits "scrubby woodland and edge, and second growth" (Howell and Webb 1995), which would seem to protect it, but not if considering the increasing threats due to the growth of the island's human population, both resident and visitor, as well as the damage created to habitat as a result of category five hurricanes passing over the island (Gilbert in 1988; Wilma in 2005). Only recently a description of the nest and eggs of this species was provided (LaPergola et al. 2012), leaving much of its breeding biology completely unknown. In 2012 a wind farm project was proposed for the island which would have impacted negatively on the habitat of this species as well as other endemics, but it eventually was turned down by authorities.

Cozumel Thrasher Once known as a fairly common to common bird species endemic to Cozumel Island (AOU 1998), the Cozumel Thrasher seemed to almost disappear from view after Hurricane Gilbert pounded the island in September 1988, followed by Roxanne in 1995. Howell (pers. com.) only saw one individual during 2 days in July 1988 vs. 3–4 a day in early to mid-1980s, and saw none in February 1989 or thereafter. It would seem then that the species was already beginning to wane before Hurricane Gilbert hit the island. Prior to Roxanne, there were only four records of the species obtained during monthly visits to the island spanning August 1994–August 1995. The last definitive scientific record was in June 1995, when a single bird was captured in a mist net (Macouzet and Escalante Pliego 2001). Since then, intensive searches have been very disappointing with only four observations of what may well have been the same individual during 2004, with another report from a different area the same year. There have been other possible sightings in 2004 and 2006, but none have been positively confirmed (BirdLife 2015). During 2013 and 2014, a program of intensive search was performed without a single sighting being reported (Martinez-Morales 2013, 2014). Although it is listed as Critically Endangered by the IUCN Red List and in danger of extinction under Mexican law (it is considered the bird most at risk of extinction in Mexico, and perhaps globally), the species may in fact already be extinct. Sadly, no systematic studies were previously made to estimate population parameters or learn its natural history.

At one time the Cozumel Thrasher was considered a subspecies of the Long-billed Thrasher by some authorities (Paynter 1955; Mayr and Short 1970). However, the Cozumel Thrasher is currently listed as a distinct species in the AOU Checklist, 7th edition (1998); this assessment is based on Zink et al. (1999), who found *T. guttatum* to be about 5 % genetically different from both *T. longirostre* and *T. rufum*. According to R. L. Curry (www.homepage.villanova.edu/robert.curry/Cozumel), (the genetic uniqueness of the Cozumel Thrasher raises the question of its origin based on different theories. One is that it is a relic species that once inhabited the Yucatán Peninsula. Another refutes this saying that relatively rapid divergence in morphology and plumage is common in mimids which have colonized even more remote islands, such as the Galápagos Islands and Socorro Island in the Revillagigedo Archipelago. Others speculate that the species may have originated from stray migrants from much further away than the mainland peninsula.

The Cozumel Thrasher has been described as a shy and secretive bird found in dense vegetation, preferring to run rather than fly when disturbed. It is a brown and white bird with black streaking on chest and sides; a long, slightly decurved black bill with amber eyes; juveniles have never been described (Howell and Webb 1995). Due to it being an endemic species it has always been of interest to visiting birdwatchers, but was not noticed by resident islanders, whose interest in birdwatching is of recent development National and international researchers, with few exceptions, only began to pay serious attention to the conservation of the avifauna of the island in the late 1980's. In any case, the story of the Cozumel Thrasher is truly a tragic one...a species on the verge of extinction and its life history described very little.

12.4 Birds of the Forested Interior

The interior of the Yucatán Peninsula is characterized by two landforms, flat lowlands to the North and East, and low rolling hills to the South and West (see Chap. 2). Even if there is a strong rainfall gradient and edaphic factors that allow heterogeneity, the dominant ecosystem is tropical forest. Forests vary from dry (and low) to humid (and tall), and also include the very particular low seasonally flooded forests (Chap. 3). In this forest matrix, a number of wetlands are found, and the ecotone between forest and these wetlands are also important areas for many birds. Anthropogenic land use change increased the quantity of edge habitat between forest and open habitat, even if their nature is obviously different. The 19 bird species we document in this section occupy these different types of forest and wetlands found inland. More than half of them are endemic or near-endemic to the Yucatán Peninsula (nine endemics, one near-endemic): Black-throated Bobwhite, Ocellated Turkey, Yucatán Poorwill, Yucatán Nightjar, Yucatán Woodpecker, Yellow-lored Parrot, Yucatán Flycatcher, Yucatán Jay, Rose-throated Tanager, and Orange Oriole. The same number of species is listed in some category of threat in the IUCN Red List (seven as Near-Threatened, NT, two as Vulnerable, VU, and one as Endangered, EN): Great Tinamou (NT), Great Curassow (VU), Ocellated Turkey (NT), Agami Heron (VU), Ornate Hawk-Eagle (NT), Orange-breasted Falcon (NT), Yellow-headed Parrot (EN), Northern Mealy Parrot (NT), Wood Thrush (NT), and Painted Bunting (NT). It is noteworthy that none of the Yucatán Peninsula endemics or near-endemics but one (Ocellated Turkey) are of conservation concern, providing some insight on the degree of conservation of their original habitat.

Great Tinamou This species reaches its northernmost distribution in the Yucatán Peninsula, where it is found only in the South. It is a forest-dweller that prefers undisturbed forests (Cabot et al. 2014). Contrary to what Howell and Webb (1995) report, it does not use only evergreen tropical forest, but also mature seasonal tropical forest (Calmé et al. unpubl. data). The Great Tinamou is rather uncommon in the peninsula. Over one complete year of sampling (372 km of line-transect walked) in three large (>20,000 ha) community forests, the species was detected only in two of these forests, once in one, and 10 times in another (Calmé et al. unpubl. data). Because some of the records belonged to the same transects, they likely corresponded to the same individuals. In a same 3-km transect, individuals detected on a same day were separated by at least 450 m. Most individuals were detected while vocalizing between mid-May and mid-July. All individuals were found at more than 12 km from human settlements, reflecting the fact than farther from villages forests are less disturbed, and hunting pressure is lower (Manzón Che 2010). In fact, only one Great Tinamou was hunted of a total of 647 harvested animals in the same villages and over the same period (Calmé et al. unpubl. data).

The low abundance of the Great Tinamou in the Yucatán Peninsula is probably linked to several factors, namely its position within its distribution range and habitat quality, and subsistence hunting. As mentioned in Chap. 3, evergreen forest, the common habitat of the Great Tinamou, is rare in the peninsula. Evergreen forest has also been subject to land use change disproportionately to its abundance (Vester et al. 2007). As a consequence, the best habitat for the species is extremely reduced. Even without accounting for global climate change effects, increased deforestation in the area already induces reduced rainfall during the dry season (roughly by 15 %; Ray et al. 2006). This is enough to alter forest species composition toward drier forests as suggested by Ray et al. (2006), as well as increase the frequency of fires (Cochrane 2003), likely reducing Great Tinamou habitat. A further threat comes from hunting. The species is considered a treat (del Hoyo et al. 1992) and hunted whenever possible (S. Calmé pers. obs.). However, few locals have hunted Great Tinamou over the past 10 years in the Southern Yucatán, both because it is uncommon and because it occupies areas difficult to access. Hunting in the region is done mainly for self-consumption of the meat, and is carried out along with other activities, especially agriculture (see Chap. 13). Therefore, hunters do not travel to remote parts of the forest other than for large prey, typically during timber harvest, which is very limited spatially and temporally, and chicle gum tapping (González-Abraham et al. 2007), an activity that has sharply declined.

Great Curassow This cracid is found in almost all the forest types of the Yucatán Peninsula, including *peten* (tree island hammocks) of Los Petenes and Ría Celestun Biosphere Reserves in Campeche and Yucatán (Calmé unpubl. data; MacKinnon pers. obs.). It is only absent from the dry low forests found both in Yucatán and Campeche states (Calmé unpubl. data). The Great Curassow distribution in the peninsula is therefore different from what is reported elsewhere (e.g., del Hoyo and Kirwan 2013). An endangered, small population of an endemic subspecies, *C. r. griscomi*, is also found in the forests of Cozumel Island at low densities of approximately 1.0 ind./km^2 (Martínez-Morales et al. 2009). On the mainland the species is rare to common, with documented densities varying from 1.2 ind./km^2 in the Eastern Yucatán Peninsula (Ramírez-Barajas et al. 2006; Ramírez-Barajas and Naranjo 2007) to 14.5 ind./km^2 in the Southern Yucatán Peninsula (Calmé unpubl. data). It uses well preserved forests dominated by late successional species (Martínez-Morales 1999, all co-authors pers. obs.) such as *Manilkara zapota* and *Brosimum alicastrum*, which fruits it consumes. Almost 80 % of the Great Curassow diet (expressed in dry weight) consists of fruits or seeds (del Hoyo and Kirwan 2013). The species is also positively associated with water bodies within the forest, with an influence of water bodies on its abundance up to 2 km (Martínez-Morales 1999). That importance is also well illustrated by the observation of a party of 40 Great Curassows, all males, at one *aguada* in Calakmul Biosphere Reserve during the dry season of 2002 (S. Calmé pers. obs.).

The Great Curassow presents large variations in density in the Yucatán Peninsula, linked to habitat suitability, but also to subsistence hunting pressure. As a result, the species is much rarer in areas with fragmented, secondary forests, which dominate much of Yucatán and Northern Campeche (Calmé unpubl. data). Elsewhere, it can still be relatively abundant even when it is one of the two most hunted large bird species along with the Ocellated Turkey. The Great Curassow represented up to 24 % of all the annual kills in one Mayan community of the Southern Yucatán Peninsula (Calmé et al. 2008). More importantly, the Great Curassow was actively selected by hunters in three of five communities studied in the Southern and Eastern Yucatán Peninsula (Calmé et al. 2006, 2008), portending a bleak future.

On Cozumel Island, the situation of the Great Curassow is likely to become critical. Martínez-Morales et al. (2009) estimated the population at 282 birds in 1994–1995, and 372 birds in 2005, which may appear to indicate a stable population. However, these estimates were made previous to the passing of, respectively, hurricane Roxanne (3 on Saffir-Simpson Hurricane Wind Scale) in 1995, and hurricanes Emily and Wilma (both 4 on Saffir-Simpson Scale) in 2005. Martínez-Morales et al. (2009) thus consider that the population had been recovering over the 10 years between the two estimates. The last two hurricanes also strongly damaged the forest vegetation (preventing these authors to continue fieldwork), which has an impact on fruit production as seen earlier in this chapter (Wunderle 1999). On the mainland, Ramírez-Barajas et al. (2012) showed that the Great Curassow abundance had sharply declined (−46 %) 8–14 months after Hurricane Dean (4 on Saffir-Simpson scale) struck in 2007. Two years and a half after the impact, the species still showed no sign of recovery (Ramírez-Barajas et al. 2012), even if in this instance the area impacted by the hurricane was well connected to other parts of the landscape so that recolonization was possible. On Cozumel, recolonization is very unlikely if not assisted, and population models show that the subspecies could become extinct by the mid-century or even faster if hurricane frequency increases (Martínez-Morales et al. 2009).

Ocellated Turkey This gregarious, endemic species of the Yucatán Peninsula, including northern Guatemala and Belize down to the Maya Mountains (Howell and Webb 1995; Fig. 12.3g), has declined substantially or been extirpated from many areas of its range over the past 35 years (Calmé and Sanvicente 2000). The species is a habitat generalist, using a wide range of vegetation types ranging from seasonally flooded savannas (during the dry season) to tall humid forests, including secondary vegetation (Calmé and Sanvicente 2000); however it depends on forest for the reproductive season, and specifically for nesting (González et al. 1998). The Ocellated Turkey is also quite generalist in its diet, and feeds opportunistically (McRoberts 2014). It is mainly granivore (almost 80 %, expressed in dry weight) without regards for the habitat (Baur 2008: forest, McRoberts 2014: agricultural landscape). Its adaptability is such that in an agricultural landscape McRoberts (2014) found that more than 92 % of the diet was composed of cultivated plants (mainly sorghum *Sorghum bicolor* and maize *Zea mays*). The only study presenting annual demographic data indicates a low annual survival rate of 0.5 (McRoberts 2014). This author also estimated a high juvenile:adult ratio of 2.89, and a male: female ratio of 0.44; these data correspond to an area where the population has been stable and even increasing for years (Calmé et al. 2010). However, Kampichler et al. (2010) found that the species was likely engaged in an extinction vortex, suggesting that the population is highly fragmented with low dispersal rates among sub-populations. Dispersal did not appear strongly limited in McRoberts' study (2014), with distances travelled by Ocellated Turkeys of approximately 20 km between non-breeding and breeding habitat; but since turkeys' movements were not analyzed, we are still limited in our understanding of how the species moves across the landscape.

Kampichler et al. (2010) showed that the decline of Ocellated Turkey between 1980 and 2000 was associated with decreasing forest cover and increasing human populations. In fact the species did not persist when forest cover was below 50 %, and abundance remained stable only when forest cover was over 80 % (Kampichler et al. 2010). Increasing human population is linked to declines in forest cover through land use change, but also to hunting pressure, as the species is prized by subsistence hunters. Ocellated Turkey represented up to 28 % of all the animals bagged annually by hunters in a community of the Southern Yucatán Peninsula (Calmé et al. 2006). It is the bird most commonly hunted along with the Great Curassow in the forests of the Yucatán Peninsula (see Chap. 13). The Ocellated Turkey tends to be an easy prey, particularly males during the breeding season. Their vocalizations are heard at several hundreds of meters and when fighting they do not pay attention to potential predators (Calmé pers. obs.). When individuals of all sexes and ages gather during the fall, they also often move to agricultural areas to feed on crops, increasing the probability of encounters with hunters. Predation other than by humans is also high. González et al. (1998) estimated that only 15 % of the chicks made it to the juvenile stage in Tikal National Park, and in Yucatán where the species has become rare interviewees mentioned that they do not see females with chicks anymore (Calmé and Sanvicente 2000). It comes to no surprise as smaller predators like foxes, known to predate on nests, flourish in anthropogenic landscapes after large predators' numbers are reduced.

Despite the grim figures provided by Kampichler et al. (2010), the recent study by McRoberts (2014) in an area already considered a model for conservation in agricultural settings (Calmé et al. 2010) brings the prospect of a better future than anticipated. In that area some huge flocks of >100 individuals gather at field edges (photo provided by J. Guerrero) even when it was very rare a decade earlier. The case of Calakmul Biosphere Reserve also provides another example where the ending of subsistence hunting was enough to increase flock size (Calmé et al. 2010). The Ocellated Turkey adaptability to a large array of vegetation types and food plants weights in its favor. The species even appears to be resilient to huge disturbances such as hurricanes. Ramírez-Barajas et al. (2012) found that despite abundance was reduced by 55 % in the 8–14 months after the passing of hurricane Dean in 2007, the species quickly recovered two years and a half after: its abundance increased by 66 % and habitat occupancy by 28 %. Should hunting pressure on the Ocellated Turkey decrease significantly, there are good reasons to be optimistic regarding the fate of the species; otherwise, a steady decline linked to local extinctions will continue.

Black-Throated Bobwhite The species, *Colinus nigrogularis*, which possibly is conspecific with *C. virginianus*, is composed of four recognized subspecies distributed in the following manner: *C. n. persiccus* Van Tyne and Trautman, 1941—Progreso area of Yucatán; *C. n. caboti* Van Tyne and Trautman, 1941—N Campeche, Yucatán (except Progreso area) and N Quintana Roo; *C. n. nigrogularis* (Gould, 1843)—Belize and N Guatemala; and *C. n. segoviensis* Ridgway, 1888—E Honduras and NE Nicaragua. *C. n. persiccus* is distinctive in that it is paler than the other three subspecies and its limited distribution may merit further investigation (Carroll and Boesman 1994). This subspecies "has its center of distribution at Progreso and extends to Celestún on the west, Mina de Oro (Dzilam de Bravo) on the east, and inland about 20 km from Progreso, tapering to less than 10 km at the eastern extremity and confined to the narrow projection of land at Celestún on the west" (Paynter 1955).

The species as a whole is not globally threatened by any means, in spite of being a ground nester; to the contrary, its distribution has expanded considerably since the 1950s (Paynter 1955; Leopold 1959) with the most increase taking place during the past 20 years due to the expansion of cattle ranching in the southern portion of the peninsula (B. MacKinnon pers. obs.). Paynter (1955) found the species only as far south in Quintana Roo as Laguna Chichancanab and in Campeche to Pixoyal (between Champotón and Laguna de Términos), along a longitude comparable to slightly north of Noh Bec, Quintana Roo. Leopold (1959), on the other hand, described the species as being from the arid northwestern part of the Yucatán Peninsula, with disjunct populations in Guatemala and British Honduras (Belize), being separated by uninhabitable humid forest. Today, the species is found in appropriate habitat almost continuously to Caobas in Southern Quintana Roo. If one is to consider the impact of the ancient Maya civilization on the ecosystems of what is today Central America prior to its collapse in the ninth century (Schele and Freidel 1990), it is not difficult to imagine that at one time the species may have been distributed continuously throughout its range.

The endemic subspecies of *C. n. persiccus* may well be at risk in the future due to the elimination of native vegetation surrounding coastal properties and increased urbanization of coastal villages which may impact dramatically on the Northern Coast of the peninsula. However, for the time being the species is presently benefiting from the replanting of sisal fields which have found new markets for its products.

Ornate Hawk-Eagle This species inhabits humid forest from central Mexico to Northern Argentina with three main distributional regions. The first includes mostly Southeastern Mexico, the Caribbean coast of Central America and the northern part of Colombia and Ecuador; the second region encloses Southern Venezuela, Colombia and Ecuador, Eastern Peru, Guyana, Suriname, French Guiana, Bolivia, and Northern Brazil; while the third region includes Southeastern Brazil, Paraguay and a reduced area in Argentina. However, in Mexico its broad distribution has been fragmented and restricted to large clumps of forest such as the Chimalapas and the south of the Yucatán Peninsula, with at least one record up to Laguna Ocom to the south of Felipe Carrillo Puerto, Quintana Roo (MacKinnon pers. obs.). The Ornate Hawk-Eagle usually inhabits the most extensive primary forests compared to Black Hawk-Eagle that sometimes is found in open areas (Escalona et al. 2006). Territory size is modest for such a large bird, at almost 7 km^2 (in French Guiana; Thiollay 1989). It is a powerful predator that primarily hunts large prey such as birds including macaws, parrots, toucans, doves, great curassow, guans, tinamous, herons, and even backyard chickens. It also feeds on several mammals like tayras, agoutis, rats, bats, and squirrels. Occasionally it will eat Black Vultures (*Coragyps atriceps*) and remains of monkeys, iguanas, and snakes (Brown and Amadon 1968; Lyon and Kuhnigk 1985).

In the peninsula, the incubation period lasts from January to May, which coincides with the dry season. The species has a long period of courtship prior to egg laying (Klein et al. 1988). Nests were previously reported to range between 100 and 125 cm wide and 45 cm high, but Balán Medina (2009) recorded larger dimensions, with nests varying between 109 and 160 cm in diameter and on average 93 cm high (n = 6). Nests are often built on long branches far from the main trunk in Central and South America; however, in the peninsula the nests are built on the principal axis of the shaft, which confers easy access to these raptors, but makes it difficult for potential terrestrial predators to reach the egg or chick. Ornate Hawk-Eagles nest high, close to the top of large trees, but not necessarily in tall forest

(Balán Medina 2009; MacKinnon pers. obs.); many of these nests were also on the edge of small, not much frequented roads. Although the incubation is mostly done by the female, the male does share incubation and bring prey to the female (Lyon and Kuhnigk 1985). Incubation time in Brazil was estimated to last 40 days (Brown 1977), and the nestling period 80–90 days. Once fledged, parental dependency is long (Klein et al. 1988). The species has a low productivity, with only one young per year if successful (Klein et al. 1988).

As Ornate Hawk-Eagles rely on big emergent trees for nesting, large extensions of well-preserved humid forests are required to sustain a viable population. Some of the tree species most used as support for the nest in the peninsula, for instance *Bucida buceras* (Balán Medina 2009; Calmé pers. obs.), are subject to logging outside protected areas. For the time being, it does not appear to be a big threat, as protected areas in the Southern peninsula cover a huge territorial extension. Outside reserves, however, the species requirements make it vulnerable to habitat loss, which is one of the most prominent threats.

Yucatán Nightjar This species is a year-round resident of the Yucatán Peninsula, including Cozumel Island and the northern half of Campeche (BirdLife International 2015). Mostly sedentary, it is a winter visitor in Belize, northern Honduras (near Tela), and Half Moon Cay in the Bahamas during December to February (Cleere and Nurney 2010). This species is insectivorous, solitary and inhabits mostly open, lowland forests, and forest edge (Cleere 1999). A recent study carried out in the northeast of Yucatán state, close to the city of Mérida, found much greater abundance of this species in earlier successional forests (i.e., between 4 and 15 years old), than in older secondary forest (i.e., between 30 and 40 years old) (Avilés-Peraza 2014). Aside from this latter information, which indicates the tolerance of the Yucatán Nightjar to anthropogenic landscapes, knowledge about this species remains scarce.

Yucatán Poorwill This species is endemic to the Yucatán Peninsula ranging from the Mexican states to northern Guatemala and Belize (Cleere and Nurney 2010). Although the northernmost limit of its distribution has been reported in the city of Mérida, Yucatán (BirdLife International 2015), Avilés-Peraza (2014) recently reported systematic observations of the Yucatán Poorwill in the municipality of Hunucmá. These observations extend the geographical range of this species to the north for nearly 10 km. It is a poorly known species that inhabits low deciduous forest and open woodland in the Yucatán peninsula; also found in secondary forests and farmland. In the Hunucmá municipality, the Yucatán Poorwill has been found in early secondary forests (i.e., between 4 and 15 years) but not in older secondary forest (i.e., between 30 and 40 years). However, the landscape surrounding the sample sites in Hunucmá corresponds to the older successional age, which contrasts with the dominance of early succession and semi-rural areas surrounding the samples of Temozón Norte where there are no records. The latter raises the question of whether the Yucatán Poorwill requires older forest for refuge and breeding, whereas it prefers early successional forest for foraging. The breeding season is from April to June in Guatemala (Cleere and Nurney 2010) but there is no systematic study regarding the breeding period in the Yucatán Peninsula.

Yucatán Woodpecker This endemic species to the Yucatán Peninsula (except the Southwest) and central Belize has three subspecies, *Melanerpes p. pygmaeus* which resides on the island of Cozumel, *M. p. tysoni* found on the island of Guanaja off the northern coast of Honduras, and *M. p. rubricomus* resident on the mainland (Clements et al. 2014). The Yucatán Woodpecker inhabits deciduous forest, forest edges, clearings, secondary forests and coastal scrub (Winkler et al. 2013; Fig. 12.3c). The species forages solitary "usually lower than other Centurus" (= Melanerpes) (Howell and Webb 1995), feeding, for instance, on dead coconut palm trunks (B. MacKinnon pers. obs.). However, Philips (2007) reported Yucatán Woodpecker foraging high on tall trees with big diameters, in closed-canopy forests, with a preference for branches rather than trunks. The breeding season includes April to June (Clements et al. 2014; B. MacKinnon pers. obs.), and there are reliable observations of adults bringing food to the nest (years 1975 and 1977), and an adult in a nest (year 1978) on Isla Cancún (B. MacKinnon pers. obs.). In the years 1977 and 1979, this species had been observed on numerous occasions nesting in live or dead narrow trunks of the Chit palm (*Thrinax radiata*) in the dune vegetation on Isla Cancún.

The ecology of this species is poorly known, and its population trend and breeding ecology have not been systematically studied. In the Yucatán Peninsula this species appears not to be as near as common as it was 35 years ago, whereas on Cozumel Island, the opposite is true in that the species is more common there than the Golden-fronted Woodpecker (*Melanerpes auratus*) (B. MacKinnon pers. obs.). Land use change from forest to pasture and urban development might pose a threat, as well as logging since they frequently use big tall trees for foraging. However, its use of a wide array of vegetation types and substrate for foraging and nesting suggest adaptability to varying habitat conditions.

Orange-Breasted Falcon This species reaches its northernmost geographical limit in the south of Mexico, including the base of the Yucatán Peninsula (del Hoyo et al. 1994). The remaining sparse distribution is throughout Central America, and outstretches to South America, but the occurrence of Orange-breasted Falcon is rare everywhere. At the present time, this species appears to be virtually extirpated from most of its distribution area in southern Mexico and Central America, and there are indications that its range in South America is contracting (Berry et al. 2010). The first reported observation of the Orange-breasted Falcon in the Yucatán Peninsula was in 1973, approximately 15 km north of Xpujil, Campeche (Hardy et al. 1975), in what is now called the Calakmul region. Since then there has been a number of isolated observations of the species in the Calakmul region (e.g., in the 1990s by J. Salgado pers. com.), the last reported by Balán Medina (2009).

The Orange-breasted Falcon is ecologically highly specialized, with low reproductive rate, which, added to its rarity and sparse distribution, make it of high risk for population decline and local extirpation (Kruger and Radford 2008; Berry et al. 2010). The habitat of this species is mature humid tropical forest (Baker et al. 2000). It is not known to nest apart from mature forest, where it tends to use cliffs to build its nest, but it can traverse a diverse landscape matrix with different types of habitat such as farmed fields, orchards, pasture lands (Berry et al. 2010). This species preys on small and medium-sized birds, and specializes on primarily hunting above the canopy (Baker et al. 2000). In the Yucatán Peninsula there are no systematic studies about the biology or ecology of the Orange-breasted Falcon, but in the near region of Belize, Berry et al. (2010) found a steep decline in occupancy and the mean number of fledged young per territorial pair, with a high proportion of nest failure (i.e., 55 %). The imminent threat for this species is the expansion of rural and urban development which causes habitat loss; there are clear examples of the loss of territories in Belize with the construction of hydroelectric dams along the Macal River, or in Ecuador with logging activities (Berry et al. 2010). These authors also report the expansion of the Black Vulture as a threat as that species tends to compete for the same nesting sites as the Orange-breasted Falcon.

Yellow-Headed Parrot Amazona oratrix was formerly distributed in both coastal slopes of Mexico from the Tres Marías Islands in Nayarit to Oaxaca and from Nuevo León to northern Chiapas and southern Tabasco and southwestern Campeche. Also there are disjunct populations in Belize, Guatemala, and Honduras. According to Monterrubio-Rico et al. (2010) its distribution in the Mexican Pacific coast has shrunk by 79 %. This is one reason for which the Yellow-headed Parrot is the second most endangered species in the region after the Cozumel Thrasher, and the situation is critical because of habitat loss and the high value of the species in the illegal pet trade, given its ability to mimic sounds and words (Cantú et al. 2007). In the Yucatán Peninsula, the occurrence of *Amazona oratrix* is related to areas where small forest fragments are interspersed in a matrix of open habitat (Plasencia-Vázquez et al. 2014). Plasencia-Vázquez and Escalona-Segura (2014) also confirmed that the Yellow-headed Parrot still maintains a high abundance in Palizada, Campeche, despite the loss of much of its vegetation cover and high fragmentation of forests. It is interesting that the species, which is characterized by a low reproductive rate and a preference for high and medium tropical forests, is present in large numbers within an area of significant human activity, making it highly vulnerable to illegal trafficking and deforestation (Enkerlin-Hoeflich 2000). The historical component has not been taken into account but may be a determining factor in explaining the observed patterns. Very little is known about this species in the Yucatán Peninsula and no data are available on its abundance prior to the deforestation of most of the forests that dominated the distribution area of this species in the southwest of the peninsula.

The area encompassing the Palizada population of Yellow-headed Parrot has been identified as a priority for conservation in Mexico (Macías-Caballero et al. 2000; Macías-Caballero and Iñigo-Elías 2003). In fact, parrots have high longevity (Munshi-South and Wilkinson 2006), thus many of the individuals observed in this region of the Yucatán Peninsula are adults that have managed to survive in a suboptimal environment. Furthermore, adult parrots normally remain in the forest canopy, making them difficult to capture, while parrot chicks suffer the highest capture rates by poachers as they are more easily found and removed from the nest. Therefore, despite low reproduction rates and high predation, large numbers of Yellow-headed Parrots continue to be observed in that portion of their habitat in the Yucatán Peninsula (Plasencia-Vázquez and Escalona-Segura 2014). If recruitment is too low, however, in the near future the population will be composed of mainly old individuals with very low productivity and on the verge of a potentially catastrophic population crash. Unfortunately, illegal pet trade remains important in the area as reported by local people, a large number of which kept Yellow-headed Parrots as pets. Because of this situation, we recommend that surveillance be strengthened and sites with high parrot abundance be better controlled during nesting, particularly outside the boundaries of the Laguna de Términos protected area.

Yellow-Lored Parrot Amazona xantholora is endemic to the Yucatán Peninsula biotic province including Northeastern Belize, very much similar to the range of the Yucatán Woodpecker. It inhabits deciduous to evergreen forest (Collar and Boesman 2014), and mangroves in *petenes*, as well as forest edges near agricultural crops (Plasencia-Vázquez and Escalona-Segura 2014). The Yellow-lored Parrot is present in landscapes characterized by large fragments of tropical semi-evergreen medium-height forest, and its distribution probability decreases with increasing fragment irregularity and the isolation of forest fragments (Plasencia-Vázquez et al. 2014). Hence, in forested areas their relative abundance can be as high as 37 individuals in Sian Ka'an, compared to deforested areas such as Palizada with 0 out of 100 parrots in each area (Plasencia-Vázquez and Escalona-Segura 2014). In Southwestern Campeche, Gómez-Garduño (2006) found that this species nests more frequently in areas with less human activity. According to this author, it is selective in its choice of nesting cavities, and nests are usually distant 450 m to 1 km from each other.

Although considered stable by Birdlife International (2015), the Yellow-lored Parrot population is perceived to have declined by local ornithologists (B. MacKinnon pers. obs.). It may well have to do with its need for large extensions of forest. White-fronted Parrots (*Amazona albifrons*) on the other hand have increased with forest fragmentation. Historically (1970s) the White fronted Parrot moved into the coastal coconut groves on Cancún and along the east coast to nest, showing their resilience to coastal habitats, something the Yellow-lored did not seem to do. Actually, these species present a contrary controversy.

Northern Mealy Parrot Amazona guatemalae has two subspecies, *A. g. guatemalae* and *A .g. virenticeps*. This Mesoamerican parrot reaches its northernmost range in the south of the Yucatán Peninsula, where it has been observed only near the border with Guatemala and Belize, mainly in Calakmul and Escárcega municipalities. Bjork (2004) suggested that this parrot makes use of these areas during the non-reproductive stage, when performing migrations from the Petén in Guatemala to certain areas of Belize, Campeche and Chiapas. However, in 2006 O. Gómez-Garduño (pers. comm.) recorded a nest in southeast Calakmul. No other study has shown movements in other areas of its distribution. However, movements can be linked to food availability, because the Mealy Parrot only feeds in primary evergreen forest, mainly in the canopy, and has disappeared from certain parts of its range, probably because the conditions in disturbed areas do not meet its ecological requirements (Ridgely 1981). Indeed, the habitat of this species is mostly closed-canopy humid lowland forest, but it can be found near edges or clearings (del Hoyo et al. 2014). There are no studies in Mexico about the reproduction of the Northern Mealy Parrot that we are aware of, but in nearby Guatemala the breeding season encloses April to May (Harrison and Holyoak 1970). The two biggest threats to this species is its capture for the pet trade, national and international, and the loss and fragmentation of its habitat.

Yucatán Flycatcher There are three subspecies of this endemic flycatcher, *Myiarchus yucatanensis yucatanensis* found in eastern Mexico from extreme eastern Tabasco to northern and central Yucatán Peninsula; *M. y. lanyoni* restricted to Cozumel Island, where it is a year-round resident; and *M. y. navai* resident in southern Quintana Roo to northern Belize and Guatemala. Due to its large range, and despite the fact that the population trend appears to be decreasing, the IUCN does not believe that the decline is sufficiently rapid to approach the thresholds for Vulnerable under the population trend criterion.

Unlike the other two *Myarchus* species in the region, the Yucatán Flycatcher is not migratory and is found year-round on Cozumel and throughout its range (Parks 1982), but in very low numbers. It tends to stay in the "shadows" of medium height tropical forest and edge, along with the Dusky-capped Flycatcher (*M. tuberculifer*), making its presence more difficult to ascertain than that of the Brown-crested Flycatcher (*M. tyrannulus*), which prefers more open perches. However, it is very much in the open when feeding on the fruit of the chaká (*Bursera simaruba*), or when declaring its territory during its reproductive period in March–April. It nests in natural tree holes as do the other two resident species of its genus, and although its nest hole has not been described it may be assumed that its height is similar to those of the Dusky-capped and Brown-crested Flycatchers, varying from 1.5 to 3 m from the ground and may even substitute a tree for a former wooden fence post (B. MacKinnon pers. obs.). There are no published accounts of this endemic bird's reproductive biology, which in light of the fact that its population appears to be declining, makes it all the more urgent that the species be studied in order to ascertain its habitat needs, and hence the risk of decline caused by habitat loss.

Yucatán Jay The distribution of this species is throughout the Yucatán Peninsula, the adjacent Mexican state of Tabasco, the north of Belize, and in the Petén district of Guatemala (BirdLife International 2015). In the Yucatán Peninsula it is a common bird in the edges and clearings of tropical deciduous forest and dense scrubby woodland (dos Anjos 2009). The Yucatán Jay has a sedentary character that most likely favours the kinship of breeding flock members, and is characterised by delayed maturity, low reproductive rate, low dispersal, and nonmigratory habits (Raitt and Hardy 1976). The diet of this species is omnivorous, consisting of seeds, e.g., from *Bursera simaruba*, *Ehretia tinifolia*, and maize (Scott and Martin 1984), and arthropods such as beetles (Coleoptera), Lepidoptera larvae, Orthoptera, spiders, ants, dipterans, and also mollusks (Raitt and Hardy 1976). The mating system of this species is cooperative, and the nest success in the Yucatán Peninsula has been estimated at 42 %, where predation by snakes and small mammals is an important cause of nest losses (Raitt and Hardy 1976). The species thrives in relatively disturbed habitats, however due to its low reproductive rate and low dispersal, the population dynamics could be disturbed by rapid land use change, especially the fragmentation of its habitat.

Wood Thrush Although the Wood Thrush is still common throughout the forests of eastern North America, populations suffered a cumulative decline of 55 % between 1966 and 2010, according to the North American Breeding Bird Survey. The species is a prime example of what is happening to forest songbirds in North America. The species is presently listed as Near Threatened on the IUCN Red List, but not listed as threatened in Mexico. The major reason for concern over the species internationally is the impact of habitat loss and fragmentation in both breeding and wintering grounds. On migration to Mexico and Central America each winter, the Wood Thrush will settle into undisturbed to moderately disturbed wet primary tropical forests. Its preferred habitat includes interior understory of tropical primary, closed-canopy, semi-evergreen, broad-leaved, and mixed palm forests at 50–1000 m elevation from humid lowland to arid or humid mountain forest, as well as scrub and thickets (Rappole et al. 1989; Winker et al. 1990; Blake and Loiselle 1992; Petit et al. 1992).

Researchers tracking these birds to their wintering grounds have discovered that they stay in the same area, even though it has been destroyed, and usually die within a short period from starvation or predation. The Wood Thrush is one of the songbird species selected by researchers from the Universities of Illinois and Alabama, as well as from Canada to attach automated radio telemetry transmitters on south-bound migrants in order to learn about their migration routes. The result is that some will go down the east coast to Florida, across to Cuba and over water to Central America whereas others will fly over the Gulf of Mexico, entering land along the north coast of the Yucatán Peninsula before heading further south where they will spend the winter. Some will remain in the humid forests of Quintana Roo and southern Campeche, but most will move further south into Central America. Because of the large cover of mature tropical forest still remaining in this region, we consider that the situation in the peninsula does not pose any threat to the species.

Rose-Throated Tanager This endemic tanager has three subspecies: *Piranga roseogularis roseogularis* found in the more arid regions of the northern portion of the peninsula; *P. r. tincta* that inhabits the central-southern portion of the peninsula; and *P.r. cozumelae* endemic to Cozumel. This latter subspecies is much darker and less colorful than its mainland relatives. The species was removed from the family of Thraupidae and placed in that of Cardinalidae in recent years (Chesser et al. 2009). It is considered a species of least concern by the IUCN (BirdLife International 2012a, b) and is not of special concern under Mexican law as there is no indication that its population is diminishing.

With the advent of more trained observers and ability to enter more areas, this species has been recorded further west in Northern Yucatán than shown in Howell and Webb (1995), including Chichén Itza where Paynter collected a specimen in 1949 (MacKinnon 2013). However, distribution in the dryer semi-deciduous forests of Yucatán near Espita and Tizimín appears to be very local and of very low density (MacKinnon pers. obs.). Contrary to its distribution in Yucatán State, the species is quite common in humid forest in the states of Quintana Roo and Campeche. It is found in forests at medium to high levels but rarely on an exposed limb. Although it is known to nest from March to July, there is no published account that describes its nest and eggs, and nothing is known of its ecology, thus putting it at risk in the future due to a lack of information.

Painted Bunting The breeding distribution of the Painted Bunting is comprised of two allopatric populations separated by a distance of 550 km in southeastern US. It is unknown if the Atlantic Coast and interior birds are also isolated during migration and on wintering grounds. Two subspecies are recognized, *Passerina ciris pallidor*, distributed in the southwest US and northern Mexico, migrating to western Panama, and *P. ciris ciris* occupying the coastal area of southeastern US and migrating to Bahamas, Cuba, Jamaica and Yucatán (Herr et al. 2011), where it is present between September and beginning of May (MacKinnon 2013). Although there is no conclusive proof that the eastern population of the Painted Bunting winters in the Yucatán Peninsula, it is supposed based on the fact that the species is very common in Cuba during spring and fall migrations but not during winter. To confirm scientifically the distribution of the Atlantic Coast population in the Yucatán would require that birds be examined genetically or by stable isotope analysis (Sykes et al. 2007). The species is listed as Near Threatened by the IUCN (BirdLife International 2012a, b) and is ranked a "species at risk" and a species of Continental Importance by Partners in Flight (Rich et al. 2004), but at present is not listed as threatened under Mexican law. On the peninsula, the species is more apt to be found feeding on grasses of the genus *Cenchrus* and *Eragrostis* situated in open areas within forests than in open fields where most seedeaters are prominent.

The concern over the species has to do with its long-term decline of 1.6 % per year, with the eastern population declining 3.5 % annually between 1966 and 1996. The major factors contributing to the overall decline of the species include loss of habitat and the cage bird trade (Herr et al. 2011). In the peninsula, loss of habitat is not nearly as much of a threat as is the illegal capture and commercialization of the colorful male of the Painted Bunting. In the past, the species could be captured legally in the region as it was assumed incorrectly that all the birds found on the peninsula were from the population that breeds in the north of Mexico. It is estimated that at least 100,000 Painted Buntings were trapped in Mexico between 1984 and 2000. The present law does not allow for their legal capture but illegally, it is only second in demand after the Cardinal in the illegal trade market.

A great deal of emphasis is presently being put on environmental education in general, and bird conservation specifically, in the region, in which the Painted

Bunting is more often than not the popular figurehead for attracting attention to the need to conserve habitat for migratory birds.

Orange Oriole This is an endemic species (Fig. 12.3e) whose distribution is restricted to the three states of the Yucatán Peninsula with only a small population that winters in northeastern Belize and Ambergris Cay from October to April (Jones 2003). The species is resident in Campeche south to the community of Las Carolinas, 5 km southwest of Zoh Laguna. IUCN lists it as a species of Least Concern as it has a large range and appears to have a stable population; nor is it considered threatened under Mexican law. Partners in Flight estimated the population to number fewer than 50,000 individuals (BirdLife international 2015).

The little that is known of the breeding biology of the Orange Oriole points to it rarely breeding individually, but rather in colonies that will vary from just two or three nests in close proximity in the same tree, camouflaged by leaves in the upper canopy which are usually located near a water source, or low down in seasonally inundated scrub vegetation in colonies with as many as 35 nests (Howell et al. 1991). The species of trees apparently is not as important as the location near water, which may have to do with the availability of a very strong, black fiber of a basidiomycete fungus (Ortega Jiménez 2002) that grows on the bark of trees (MacKinnon pers. obs.). Under different circumstances the nest may be constructed of Silver Saw Palm (*Acoellorraphe wrightii*) or something similar (Ortega Jiménez 2002). Colonial nesting sites are located in seasonally inundated depressions as well as in open savannas, containing islands of scrub vegetation, surrounding a spring fed pond, all within a larger forest habitat. It has been reported laying 4–6 eggs in its pouch-like nest (Ortega Jiménez 2002), although for other *Icterus* species the norm is 2–3 or 3–4 eggs. The nest is a bit deeper than that of the Hooded Oriole (*I. cucullatus*) and it reproduces for the most part between April and August; nesting is not synchronous.

The Orange Oriole falls into the same category as other endemic and near-endemic species of the region, including the Yucatán Woodpecker, Rose-throated Tanager, and Yucatán Vireo, in that very little has been studied of its selection of habitat both for food and reproduction.

12.5 Conservation Perspective

The Yucatán Peninsula harbors a very rich avifauna that includes several endemic and near-endemic species, a very special fact given that the peninsula does not present any geographical barrier. This richness and uniqueness are threatened to different degrees and for different reasons that the species accounts in this chapter showcase almost completely. Broadly, threats can be divided into those related to anthropogenic activities and those linked to climate change.

Anthropogenic activities are by far the most important threats to birds in the Yucatán Peninsula. These activities foster land use change, be it for agriculture or for urban or tourist development. The ecosystems most at risk currently in the peninsula are those with very limited distribution found on or near the coasts, namely the dunes, the dry forest of Yucatán State and, close to urban areas, mangroves (Chap. 3). The vegetation of these ecosystems is being removed, and it the case of wetlands, landfills allow for urban spreading. These ecosystems are the habitat of a number of bird species for which the peninsula is key to ensure their conservation, either because they are endemic (e.g., Yucatán Wren) or because the peninsula sustains important and probably distinct populations (e.g., Reddish Egret). Given the rate of urban growth along the coast or around main cities, especially in Yucatán and Quintana Roo, we are let but with a dismal prospective for some of the species depending of these rare ecosystems.

Deforestation for agriculture has greatly increased over the past decades, but most forest types remain well represented in the Yucatán Peninsula, both within and outside reserves. Cavity nesters or large raptors, however, are most at risk as small scale logging for domestic purposes usually targets the largest trees, those that are or would become most adequate for nesting. This problem is acute outside reserves, and even in large natural protected areas where ranging patrols are rare (e.g., Balamku, Campeche).

Several bird species are also directly affected by human activities leading to their extraction from their habitat, e.g., hunting or trapping. Hunting is a pervasive activity reaching almost the farthest reaches of the forests of the Yucatán Peninsula, even if activities such as gum-tapping that required working within the forest have severely declined. This threat concerns mainly large, terrestrial game birds (e.g., Great Curassow and Ocellated Turkey), but their decline will certainly affect smaller birds currently considered common (e.g., Plain Chachalaca *Ortalis vetula*). Trapping for the pet trade has already greatly impacted the populations of the largest psittacids of the peninsula (i.e., Yellow-headed and Northern Mealy Parrots), and of the colored Painted Bunting. Even if these species are now legally protected, and some awareness exists among the population, trading remains important due to demand and to weak law enforcement.

A last concern regards the current ongoing climate change, which has the potential to grossly change the face of the peninsula's ecosystems. The vegetation distributional changes are too speculative to be discussed in terms of their impact on the avifauna, but the expected change in the intensity or frequency of the hurricanes striking the peninsula is a far more concrete fact. The birds of Cozumel Island have experienced several strong hurricanes over the past 25 years that have led to the presumed extinction of the endemic Cozumel Thrasher and to a strong decline of the endemic subspecies of Great Curassow. Other species have recovered, both on the island (e.g. Yucatán Emerald) and on the mainland (e.g. Ocellated Turkey). In fact, one would expect many bird species of the Yucatán Peninsula, especially those more common on the Caribbean coast, to be resilient to hurricane impacts. However, these impacts may become not only stronger or more frequent, but synergetic effects with habitat fragmentation may well emerge (Brook et al. 2008) with dire consequences for those species with limited dispersal ability or too specialized in their habitat or feeding habits.

Even if natural protected areas remain a tool of choice for conservation (see Chaps. 14 and 15), our ability to protect effectively any species depends largely on how well their natural history and ecology are known. In face of all the recent changes in the landscapes of the peninsula, more studies are badly needed for some species without regards for their conservation status. The state of Yucatán has encouraged birdwatching tourism, which may have the double benefit of providing information and raising awareness among the population benefited from the activity. Some NGOs have also been very active in that regard, by encouraging the conservation of the resident American Flamingo (*Phoenicopterus ruber*) in Ría Lagartos area. Due to the large number and size of the protected areas in the Yucatán Peninsula, we believe that the National Commission of Natural Protected Areas (CONANP) has a major role to play that it has yet to assume.

References

AOU. Checklist of North American birds. New York, NY: American Ornithologists Union; 1998.

Avilés-Peraza GC. Abundancia y composición de especies de caprimúlgidos (Caprimulgiformes) en parches de diferente edad sucesional en una selva baja caducifolia de Yucatán. Honor's thesis. Instituto Tecnológico de Conkal, Mérida, México; 2014.

Baker AJ, Whitacre DF, Aguirre-Barrera OA, White C. The Orange-breasted Falcon *Falco deiroleucus* in Mesoamerica: a vulnerable, disjunct population? Bird Conserv Int. 2000;10:29–40.

Balán Medina GI. Anidación de aves rapaces (Órdenes: Falconiformes y Strigiformes, Clase: Aves) en el centro y sur del estado de Campeche, México. BSc thesis. Universidad Autónoma de Campeche, Campeche, Mexico; 2009.

Bancroft GT, Bowman R. Temporal patterns in diet of nestling white-crowned pigeons: implications for conservation of frugivorous columbids. Auk. 1994;111:844–52.

Bancroft T, Bowman R, Sawicki RJ. Rainfall, fruiting phenology, and the nesting season of white-crowned Pigeons in the Upper Florida Keys. Auk. 2000;117:416–26.

Baptista LF, Trail PW, Horblit HM, Boesman P. White-crowned Pigeon (*Patagioenas leucocephala*). In: del Hoyo J, Elliott A, Sargatal J, Christie DA, de Juana E, editors. Handbook of the Birds of the World Alive. Barcelona: Lynx Edicions; 2014. Retrieved from http://www.hbw.com/node/54126 on 6 Apr 2015.

Baur EH. Structure of a lowland neotropical galliform bird guild. Master's thesis. University of Florida, Gainesville, USA; 2008.

Berry RB, Benkman CW, Muela A, Seminario Y, Curti M. Isolation and decline of a population of the Orange-breasted Falcon. Condor. 2010;112:479–89.

BirdLife International. *Passerina ciris*. In: The IUCN Red List of Threatened Species. Version 2014.3; 2012a. Retrieved from http://www.iucnredlist.org on 30 Mar 2015.

BirdLife International. *Piranga roseogularis*. In: The IUCN Red List of Threatened Species. Version 2014.3; 2012b. Retrieved from http://www.iucnredlist.org on 30 Mar 2015.

BirdLife International. *Vireo bairdi*. In: The IUCN red list of threatened species. Version 2015.2; 2012c. Retrieved from http://www.iucnredlist.org on 9 July 2015.

BirdLife International. *Chlorostilbon forficatus*. In: The IUCN red list of threatened species. Version 2015.2; 2012d. Retrieved from http://www.iucnredlist.org on 9 July 2015.

BirdLife International. Doricha eliza. In: The IUCN red list of threatened species. Version 2015.2; 2014. Retrieved from http://www.iucnredlist.org on 9 July 2015.

BirdLife International. IUCN Red List for birds; 2015. Retrieved from http://www.birdlife.org on 17 Apr 2015.
Blake JG, Loiselle BA. Habitat use by neotropical migrants at La Selva Biological Station and Braulio Carrillo National Park, Costa Rica. In: Hagan J, Johnston DW, editors. Ecology and conservation of neotropical landbird migrants. Washington, DC: Smithsonian Institution Press; 1992. p. 257–72.
Brewer D. Yucatán Vireo (*Vireo magister*). In: del Hoyo J, Elliott A, Sargatal J, Christie DA, de Juana E, editors. Handbook of the Birds of the World Alive. Barcelona: Lynx Edicions; 2010. Retrieved from http://www.hbw.com/node/61270 on 20 Apr 2015.
Brook BW, Sodhi NS, Bradshaw CJ. Synergies among extinction drivers under global change. Trends Ecol Evol. 2008;23:453–60.
Brown L. Eagles of the world. New York, NY: Universe Books; 1977.
Brown L, Amadon D. Eagles, hawks and falcons of the world. New York, NY: McGraw-Hill; 1968.
Bjork RD. Delineating pattern and process in tropical lowlands: mealy parrot migration dynamics as a guide for regional conservation planning. PhD dissertation. Oregon State University, Corvallis, Oregon; 2004.
Burger J. The effect of human disturbance on foraging behavior and habitat use in Piping Plover (*Charadrius melodus*). Estuaries. 1994;17:695–701.
Cabot Jr S. Descriptions and habitats of some of the birds of Yucatán. Boston Soc Nat Hist J. 1845;4:460–7.
Cabot J, Jutglar F, Garcia EFJ, Boesman P, Sharpe CJ. Great Tinamou (Tinamus major). In: del Hoyo J, Elliott A, Sargatal J, Christie DA, de Juana E, editors. Handbook of the birds of the world alive. Barcelona: Lynx Edicions; 2014. Retrieved from http://www.hbw.com/node/52412 on 9 July 2015.
Calmé S, Sanvicente M. Distribución actual, estado poblacional y evaluación del estado de protección del pavo ocelado (*Agriocharis ocellata*). Final report of Project R114 to Comisión Nacional para el Conocimiento y Uso de la Biodiversidad. México, DF, Mexico; 2000.
Calmé S, Guerra Roa M, Xijun Kantun M, Manzón Che M. Cacería. In: Uso y monitoreo de los recursos naturales en el corredor biológico mesoamericano (áreas focales X'Pujil-Zoh Laguna y Carrillo Puerto). Final report of Project BJ002 to Comisión Nacional para el Uso y Conocimiento de la Biodiversidad (CONABIO), Mexico, DF, Mexico; 2006.
Calmé S, Guerra Roa M, Armijo Canto N. Manejo comunitario de la fauna silvestre en comunidades rurales de Calakmul. Final report of Project CAMP-2005-C01-053 to Consejo Nacional de Ciencias y Tecnologías (CONACYT) and Government of Campeche, Campeche, Mexico; 2008.
Calmé S, Sanvicente M, Weissenberger H. Estudio de caso: manejo del pavo ocelado. In: Villalobos-Zapata GJ, Mendoza Vega J, editors. La biodiversidad en Campeche: estudio de estado. México: Comisión Nacional para el Conocimiento y Uso de la Biodiversidad; 2010. p. 507–9.
Cantú JC, Sánchez ME, Grosselet M, Silva J. Tráfico ilegal de pericos en México. Una evaluación detallada. Defenders of Wildlife, Mexico; 2007.
Chesser RT, Banks RC, Barker FK, Cicero C, Dunn JL, Kratter AW, Lovette IJ, Rasmussen PC, Remsen JV, Rising JD, Stotz DF, Winker K. Fiftiest supplement to the American Ornithologists' Union check-list of North American birds. Auk. 2009;126:705–14.
Chesser RT, Banks RC, Barker FK, Cicero C, Dunn JL, Kratter AW, Lovette IJ, Rasmussen PC, Remsen JV, Rising JD, Stotz DF, Winker K. Fifty-second supplement to the American Ornithologists' Union check-list of North American birds. Auk. 2011;128:600–13.
Cleere N. Family Caprimulgidae (Nightjars). In: del Hoyo J, Elliott A, Sargatal J, editors. Handbook of the Birds of the World, Barn Owls to Hummingbirds, vol. 5. Barcelona: Lynx Edicions; 1999. p. 302–86.
Cleere N, Nurney D. Nightjars. A guide to nightjars and related nightbirds. London: A&C Black Publishers Ltd. 2010.

Clements JF, Schulenberg TS, Iliff MJ, Roberson D, Fredericks TA, Sullivan BL, Wood CL. The eBird/Clements checklist of birds of the world: Version 6.9; 2014. http://www.birds.cornell.edu/clementschecklist/download/

Cochrane MA. Fire science for rainforests. Nature. 2003;421:913–9.

Collar N, Boesman P. Yellow-lored Amazon (*Amazona xantholora*). In: del Hoyo J, Elliott A, Sargatal J, Christie DA, de Juana E, editors. Handbook of the Birds of the World Alive. Barcelona: Lynx Edicions; 2014. Retrieved from http://www.hbw.com/node/54735 on 24 Apr 2015.

Cuarón AD, Martínez-Morales MA, McFadden KW, Valenzuela D, Gompper ME. The status of dwarf carnivores on Cozumel Island, Mexico. Biodivers Conserv. 2004;13:317–31.

Cuarón AD. Cozumel. In: Gillespie R, Clague DA, editors. Encyclopedia of islands. Berkeley, CA: University of California Press; 2009. p. 203–6.

Danchin E, Wagner R. The evolution of coloniality: the emergence of new perspectives. Trends Ecol Evol. 1997;12:342–7.

dos Anjos L. Yucatán Jay (*Cyanocorax yucatanicus*). In: del Hoyo J, Elliott A, Sargatal J, Christie DA, de Juana E, editors. Handbook of the Birds of the World Alive. Barcelona: Lynx Edicions; 2009. Retrieved from http://www.hbw.com/node/60710 on 17 Apr 2015.

del Hoyo J, Elliot A, Sargatal J. Handbook of the Birds of the World, Ostrich to Ducks, vol. 1. Barcelona: Lynx Edicions; 1992.

del Hoyo J, Elliot A, Sargatal J. Handbook of the Birds of the World, New World Vultures to Guineafowl, vol. 2. Barcelona: Lynx Edicions; 1994.

del Hoyo J, Elliot A, Sargatal J. Handbook of the Birds of the World, Hoatzin to Auks, vol. 3. Barcelona: Lynx Edicions; 1996.

del Hoyo J, Elliot A, Sargatal J. Handbook of the Birds of the World, Sandgrouse to Cuckoos, vol. 4. Barcelona: Lynx Edicions; 1997.

del Hoyo J, Elliot A, Sargatal J. Handbook of the Birds of the World, Barn-howls to Hummingbirds, vol. 5. Barcelona: Lynx Edicions; 1999.

del Hoyo J, Elliot A, Sargatal J. Handbook of the Birds of the World, Mousebirds to Hornbills, vol. 6. Barcelona: Lynx Edicions; 2001.

del Hoyo J, Elliot A, Sargatal J. Handbook of the Birds of the World, Jacamars to Woodpeckers, vol. 7. Barcelona: Lynx Edicions; 2002.

del Hoyo J, Elliot A, Christie DA. Handbook of the Birds of the World, Broadbills to Tapaculos, vol. 8. Barcelona: Lynx Edicions; 2003.

del Hoyo J, Elliot A, Christie DA. Handbook of the Birds of the World, Cotingas to Pipits and Wagtails, vol. 9. Barcelona: Lynx Edicions; 2004.

del Hoyo J, Elliot A, Christie DA. Handbook of the Birds of the World, Cuckoo-shrikes to Thrushes, vol. 10. Barcelona: Lynx Edicions; 2005.

del Hoyo J, Elliot A, Christie DA. Handbook of the Birds of the World, Old World Flycatchers to Old World Warblers, vol. 11. Barcelona: Lynx Edicions; 2006.

del Hoyo J, Elliot A, Christie DA. Handbook of the Birds of the World, Picathartes to Tits and Chickadees, vol. 12. Barcelona: Lynx Edicions; 2007.

del Hoyo J, Elliot A, Christie DA. Handbook of the Birds of the World, Penduline-tits to Shrikes, vol. 13. Barcelona: Lynx Edicions; 2008.

del Hoyo J, Elliot A, Christie DA. Handbook of the Birds of the World, Bush-shrikes to Old World Sparrows, vol. 14. Barcelona: Lynx Edicions; 2009.

del Hoyo J, Elliot A, Christie DA. Handbook of the Birds of the World, Weavers to New World Warblers, vol. 15. Barcelona: Lynx Edicions; 2010.

del Hoyo J, Elliot A, Christie DA. Handbook of the Birds of the World, Tanagers to New World Blackbirds, vol. 16. Barcelona: Lynx Edicions; 2011.

del Hoyo J, Elliot A, Christie DA. Handbook of the Birds of the World, New species and global index, vol. special volume. Barcelona: Lynx Edicions; 2013.

del Hoyo J, Kirwan GM. Great Curassow (*Crax rubra*). In: del Hoyo J, Elliott A, Sargatal J, Christie DA, de Juana E, editors. Handbook of the Birds of the World Alive. Barcelona: Lynx Edicions; 2013. Retrieved from http://www.hbw.com/node/53311 on 14 Apr 2015.

del Hoyo J, Collar N, Kirwan GM. Northern Mealy Amazon (*Amazona guatemalae*). In: del Hoyo J, Elliott A, Sargatal J, Christie DA, de Juana E, editors. Handbook of the Birds of the World Alive. Barcelona: Lynx Edicions; 2014. Retrieved from http://www.hbw.com/node/467504 on 24 Apr 2015.

Drewitt AL, Langston RHW. Assessing the impacts of wind farms on birds. Ibis. 2006;148:29–42.

Eisenmann E. The species of Middle American birds. Transactions of the American Museum of Natural History. Linnaean Society of New York, NY; 1955.

Enkerlin-Hoeflich EC. Loro cabeza amarilla. In: Ceballos G, Marquez Valdelamar L, editors. Las Aves de México en Peligro de Extinción. México, DF: CONABIO/UNAM; 2000.

Escalona G, Puga I, Moo L. El imperio del águila ornada. Ecofronteras. 2006;29:9–10.

Everaert J, Stienen EWM. Impact of wind turbines on birds in Zeebrugge (Belgium): significant effect on breeding tern colony due to collisions. Biodivers Conserv. 2007;16:3345–59.

Flores S, Espejel I. Tipos de vegetación de la Península de Yucatán. Fascículo de la Etnoflora Yucatánense. Mérida: Universidad Autónoma de Yucatán; 1994.

Fondo Nacional de Fomento al Turismo. La Costa Turquesa, El Caribe Mexicano, Cancún, Quintana Roo, Mexico; 1980.

Fraga R. Great-tailed Grackle (*Quiscalus mexicanus*). In: del Hoyo J, Elliott A, Sargatal J, Christie DA, de Juana E, editors. Handbook of the Birds of the World Alive. Barcelona: Lynx Edicions; 2011. Retrieved from http://www.hbw.com/node/62287 on 7 Apr 2015.

Gómez de Silva H. N Am Birds. 2004;58(1):154.

Gómez de Silva H. N Am Birds. 2005;59(1):161.

Gómez de Silva H. N Am Birds. 2014;69(1). *in litt.*

Gómez-Garduño JO. Disponibilidad y ocupación de cavidades para tres especies de loros del género *Amazona* en Calakmul, Campeche. Master's thesis. El Colegio de la Frontera Sur, Campeche, Campeche, Mexico; 2006.

González MJ, Quigley HB, Taylor CI. Habitat use and reproductive ecology of the ocellated turkey in Tikal National Park. Wilson Bull. 1998;110:505–10.

González-Abraham A, Schmook B, Calmé S. Distribución espacio-temporal de las actividades extractivas y su relación con la conservación de los recursos naturales. El caso del ejido Caoba al Sur de Quintana Roo. Investigaciones Geográficas. 2007;62:69–86.

Green MC, Hill A, Troy JR, Holderby Z, Geary B. Status of breeding Reddish Egrets on Great Inagua, Bahamas with comments on breeding territoriality and the effects of hurricanes. Waterbirds. 2011;34:213–7.

Gunn CA. Vacationscape: developing tourist areas. 3rd ed. Washington, DC: Taylor and Francis; 1997.

Hardy JW, Raitt RJ, Orejuela J, Webber T, Edinger B. First observation of the Orange-breasted Falcon in the Yucatán Peninsula of Mexico. Condor. 1975;77:512.

Harrison CJO, Holyoak DT. Apparently undescribed parrot eggs in the collection of the British Museum (Natural History). Bull Br Ornithol Club. 1970;90:42–6.

Herr CA, Sykes Jr PW, Klicka J. Phylogeography of a vanishing North American songbird: the Painted Bunting – (*Passerina ciris*). Conserv Genet. 2011;12:1395–410.

Howell SNG. Recent Observations of Birds in the Honduras Bay Islands. Internal report. Point Reyes Bird Observatory, California; 1991.

Howell SNG. Taxonomy and distribution of the hummingbird genus *Chlorostilbon* in Mexico and northern Central America. Euphonia. 1993;2:25–37.

Howell SNG, Webb S, MacKinnon de Montes B. Colonial nesting of the Orange Oriole. Wilson Bull. 1991;104:189–90.

Howell SNG, Webb S. A guide to the birds of Mexico and Northern Central America. New York, NY: Oxford University Press; 1995.

Hill A, Green C, Palacios E. Genetic diversity and population structure of North America's rarest heron, the reddish egret (*Egretta rufescens*). Conserv Genet. 2012;13:535–43.

Jones HL. Birds of Belize. Austin, TX: University of Texas Press; 2003.

Kampichler C, Calmé S, Weissenberger H, Arriaga-Weiss SL. Indication of a species in an extinction vortex: the ocellated turkey on the Yucatán Peninsula, Mexico. Acta Oecol. 2010;36:561–8.

Klein BC, Harper LH, Bierregaard RO, Powell GVN. The nesting and feeding behavior of the ornate hawk-eagle near Manaus. Brazil Condor. 1988;90:239–41.

Kruger O, Radford AN. Doomed to die? Predicting extinction risk in the true hawks Accipitridae. Anim Conserv. 2008;11:83–91.

Küpper C, Augustin J, Kosztolányi A, Burke T, Figuerola J, Székely T. Kentish versus Snowy Plover: phenotypic and genetic analyses of *Charadrius alexandrinus* reveal divergence of Eurasian and American subspecies. Auk. 2009;126:839–52.

Kushlan JA, Hancock JA. The Herons. Oxford: Oxford University Press; 2005.

Lafferty KD. Disturbance to wintering western snowy plovers. Biol Conserv. 2001;101:315–25.

Lafferty KD, Goodman D, Sandoval CP. Restoration of breeding by snowy plovers following protection from disturbance. Biodivers Conserv. 2006;15:2217–30.

LaPergola JB, Marina Hipolito JG, Martínez-Gómez JE, Curry RL. First description of the nest and eggs of the island-endemic cozumel vireo, *Vireo bairdi*. Wilson J Ornithol. 2012;124:743–9.

Lawrence GN. List of a collection of birds from Northern Yucatán. Lyceum Nat Hist NY. 1869;9:198–210.

Leberg PL, Green MC, Adams BA, Purcell KM, Luent MC. Response of waterbird colonies in southern Louisiana to recent drought and hurricanes. Anim Conserv. 2007;10:502–8.

Leopold AS. Wildlife of Mexico: the game birds and mammals. Berkeley: University of California Press; 1959.

López-Ornat A, Lynch JF. Landbird communities of the coastal dune scrub in the Yucatán Peninsula: species composition, ecology and zoogeographic affinities. Vida Silvestre Neotropical. 1990;2:21–31.

Lyon B, Kuhnigk A. Observations on nesting ornate Hawk-Eagles in Guatemala. Wilson Bull. 1985;92:141–264.

Macías-Caballero C, Íñigo Elías E, Enkerlin Hoeflich E. Proyecto de recuperación de especies prioritarias: proyecto nacional para la conservación, manejo y aprovechamiento sustentable de los psitácidos de México. Instituto Nacional de Ecología/SEMARNAP, México, DF; 2000

Macías-Caballero C, Íñigo Elías E. Evaluación del estado de conservación actual de las poblaciones de loro cabeza amarilla (*Amazona oratrix*) en México. Informe final del proyecto AS002. Apoyado por la Comisión Nacional para el Conocimiento y Uso de la Biodiversidad (CONABIO); 2003.

Mayr E, Short LL. Species Taxa of North American Birds: a contribution to comparative systematics. Cambridge, MA: Nuttall Ornithological Club; 1970.

MacKinnon HB, Acosta Aburto J. Critical Habitat for Migratory Land Birds, Banco Chinchorro, Quintana Roo, Mexico. Bull Mar Sci. 2003;73:171–86.

MacKinnon HB. Plantas Costeras que Conservan las Playas y Alimentan las Aves. Amigos de Sian Ka'an A.C., Cancún, Q. Roo; 2005.

MacKinnon HB. Sal a pajarear Yucatán. México, DF, Mexico: La Vaca Independiente; 2013.

Macouzet T. Diversidad avifaunistica de la Isla Cozumel, Quintana Roo, Mexico. B.Sc. thesis. UNAM, México, DF; 1997

Macouzet T, Escalante Pliego P. Registros del Cuitlacoche de Cozumel *Toxostoma guttatum* posteriores al Huracán Gilberto. Cotinga. 2001;15:32–3.

Manzón Che MJ. Uso de fauna silvestre en relación con las actividades productivas en dos comunidades forestales de Calakmul, Campeche, México. B.Sc. thesis. Instituto Tecnológico de Chetumal, Chetumal, Mexico; 2010.

Martínez-Morales MA. Conservation status and habitat preferences of the Cozumel Curassow. Condor. 1999;101:14–20.

Martínez-Morales MA, Cruz PC, Cuarón AD. Predicted population trends for Cozumel Curassows (*Crax rubra griscomi*): empirical evidence and predictive models in the face of climate change. J Field Ornithol. 2009;80:317–27.

Martinez-Morales MA. Estudio diagnóstico del estado poblacional del Mapache Enano y el Cuitlacoche de Cozumel. Final report to PROCER. El Colegio de la Frontera Sur; 2013.
Martinez-Morales MA. Acciones de conservación y monitoreo de cuatro vertebrados endémicos críticamente amenazados de Isla Cozumel. Final report to PROCER. El Colegio de la Frontera Sur; 2014.
McFadden KW, García-Vasco D, Cuarón AD, Valenzuela D, Medellín RA, Gompper ME. Vulnerable island carnivores: the endangered endemic dwarf procyonids from Cozumel Island. Biodivers Conserv. 2010;19:491–502.
McRoberts JT. Investigations into the ecology and management of ocellated turkeys in Campeche, Mexico. Ph.D. dissertation. Texas Tech University, Lubbock, USA; 2014.
Monterrubio-Rico T, Renton C, Ortega-Rodríguez JM, Pérez-Arteaga A, Cancino-Murillo A. The endangered Yellow-headed parrot *Amazona oratrix* along the Pacific coast of Mexico. Oryx. 2010;44:602–9.
Morgan JG, Eubanks Jr TL, Eubanks V, White LN. Yucatán Vireo appears in Texas. Am Birds. 1985;39:245–6.
Morgenthaler A. Ecology of the Black Catbird, *Melanoptila glabrirostris*, at Shipstern Nature Reserve (Belize), and distribution in Yucatán. Master's thesis. University of Neuchâtel, Neuchâtel, Switzerland; 2003.
Munshi-South J, Wilkinson GS. Diet influences life span in parrots (Psittaciformes). Auk. 2006;123:108–18.
Nudds RL, Bryant DM. The energetic cost of short flights in birds. J Exp Biol. 2000;203:1561–72.
Ortega Jiménez VM. Estrategias de anidación de la Yuca Naranja (Icteridae: *Icterus auratus*). Master's thesis. El Colegio de la Frontera Sur, Campeche, Campeche, Mexico; 2002.
Ortiz-Pulido R, Townsend A, Robbins M, Díaz-Valenzuela R, Navarro A, Escalona G. The Mexican Sheartail (*Doricha eliza*): morphology, behavior, distribution and endangered status. Wilson Bull. 2002;114:153–60.
Palacios E, Amador E, Green MC. Assessment of the reddish egret population in Yucatán and postbreeding dispersal studies in west Mexico. Centro de Investigación Científica y de Educación Superior de Ensenada (CICESE), Centro de Investigaciones Biológicas del Noroeste, Texas State University – San Marcos. La Paz, Baja California Sur; 2013.
Parks KC. Parallel geographic variation in three Myarchus flycatchers in the Yucatán Peninsula and adjacent areas (Aves: Tyrannidae). Ann Carnegie Museum. 1982;51:1–16.
Paynter Jr RA. The ornithogeography of the Yucatán Peninsula. Bulletin 9. Peabody Museum of Natural History, Yale University, New Haven, CT; 1955.
Petit DR, Petit LJ, Smith KG. Habitat associations of migratory birds overwintering in Belize, Central America. In: Hagan J, Johnston DW, editors. Ecology and conservation of neotropical landbird migrants. Washington, DC: Smithsonian Institution Press; 1992. p. 247–56.
Philips ME. The foraging ecology and habitat use of two Melanerpinae woodpeckers in the Yucatán Peninsula. Master's thesis. Stephen F. Austin State University, Austin, TX; 2007.
Plasencia-Vázquez AH, Escalona-Segura G. Relative abundance of parrots throughout the Yucatán Peninsula: implications for their conservation. Wilson J Ornithol. 2014;126:759–66.
Plasencia-Vázquez AH, Escalona-Segura G, Esparza-Olguín LG. Landscape variables interaction on parrots' potential geographical distribution in the Yucatán Peninsula, Mexico. Anim Biodivers Conserv. 2014;37:191–203.
Raitt RJ, Hardy JW. Behavioral ecology of the Yucatán Jay. Wilson Bull. 1976;88:529–721.
Ramírez-Barajas PJ, Naranjo E. La cacería de subsistencia en una comunidad de la Zona Maya, Quintana Roo, México. Etnobiología. 2007;5:65–85.
Ramírez-Barajas PJ, Torrescano-Valle N, Chan-Rivas C. Diagnóstico del aprovechamiento de flora y fauna por los mayas del Ejido Petcacab y evaluación de la cacería y pesca, Felipe Carrillo Puerto, Quintana Roo. Final report of Project BJ008 to Comisión Nacional para el Uso y Conservación de la Biodiversidad. México, DF, Mexico; 2006.
Ramírez-Barajas PJ, Islebe GA, Calmé S. Impact of Hurricane Dean (2007) on game species of the Selva Maya, Mexico. Biotropica. 2012;44:402–11.

Ramo C, Busto B. Resource use by herons in a Yucatán wetland during the breeding season. Wilson Bull. 1993;105:573–86.

Rappole JH, Ramos MA, Winker K. Wintering wood thrush movements and mortality in southern Veracruz. Auk. 1989;106:402–10.

Ray DK, Welch RM, Lawton RO, Nai US. Dry season clouds and rainfall in northern Central America: implications for the Mesoamerican Biological Corridor. Global Planet Change. 2006;54:150–62.

Rich TD, Beardmore CJ, Berlanga H, Blancher PJ, Bradstreet MSW, Butcher GS, Demarest DW, Dunn EH, Hunter WC, Iñigo-Elias EE, Kennedy JA, Martell AM, Panjabi AO, Pashley DN, Rosenberg KV, Rustay CM, Wendt JS, Will TC. Partners in Flight North American Landbird Conservation Plan. Ithaca, NY: Cornell Lab of Ornithology; 2004.

Ridgely R. The current distribution and status of mainland neotropical parrots. In: Pasquier RF, editor. Conservation of new world parrots. ICBP Technical Publication No. 1. Washington, DC: Smithsonian Institution Press; 1981.

Ridgway R. On some new forms of American birds. Am Nat. 1873;7:602–19.

Roldán-Clarà B. Morfometría, historia natural, abundancia y éxito reproductivo de *Melanoptila glabrirostris* y el efecto antropogénico en la Reserva de la Biosfera de Sian Ka'an, México. Master's thesis, El Colegio de la Frontera Sur, Chetumal, Quintana Roo, Mexico; 2009.

Roldán-Clarà B, LaPergola JB, Chapa-Vargas L, Calmé S. Nest survival in the Neotropical Black Catbird (Melanoptila glabrirostris). J Ornithol. 2013;154:491–9.

Santamaría-Rivero W, MacKinnon HB, Leyequién E. Registros de anidación del colibrí tijereta mexicano (*Doricha eliza*) en el estado de Yucatán, México. Huitzil. 2013;14:139–45.

Schele L, Freidel D. A forest of kings. New York, NY: Wm. Morrow & Co.; 1990.

Scott PE, Martin RF. Avian consumers of *Bursera*, *Ficus*, and *Ehretia* fruit in Yucatán. Biotropica. 1984;16:319–23.

SEMARNAT. Norma Oficial Mexicana NOM-059-SEMARNAT-2010, Protección Ambiental – Especies nativas de México de flora y fauna silvestres – Categorías de riesgo y especificaciones para su inclusión, exclusión o cambio – lista de especies en riesgo. Diario Oficial de la Federación. 30 de diciembre de 2010, Segunda Sección. México, DF; 2010.

Sykes Jr PW, Holzman S, Iñigo-Elias E. Current range of the eastern population of Painted Bunting (*Passerina ciris*) – Part II: Winter range. N Am Birds. 2007;61:378–406.

Thiollay JM. Area requirements for the conservation of rain forest raptors and game birds in French Guiana. Conserv Biol. 1989;3:128–37.

Thomas SM, Lyons JE, Andres BA, Elliot-Smith E, Palacios E, Cavitt JF, Royle JA, Fellows SD, Maty K, Howe WH, Mellink E, Melvin S, Zimmerman T. Population size of Snowy Plovers breeding in North America. Waterbirds. 2012;35:1–14.

Turner II BL, Cortina Villar S, Foster D, et al. Deforestation in the southern Yucatán peninsular region: an integrative approach. For Ecol Manage. 2001;154:353–70.

Vargas Soriano J, Salgado Ortíz J, Escalona Segura G. Breeding phenology and nesting success of the Yucatán Wren in the Yucatán Peninsula, Mexico. Wilson J Ornithol. 2010;122:439–46.

Vandermeer J, Granzow de la Cerda I, Boucher D. Contrasting growth rate patterns in eighteen tree species from a post-hurricane forest in Nicaragua. Biotropica. 1997;29:151–61.

Vester HFM, Lawrence D, Eastman JR, Turner II BL, Calmé S, Pozo C, Dickson R, Sangermano F. Land change in the southern Yucatán and Calakmul Biosphere Reserve: implications for habitat and biodiversity. Ecol Appl. 2007;17:989–1003.

Whigham DF, Olmsted I, Cabrera Cano E, Curtis AB. Impacts of hurricanes on the forests of Quintana Roo, Yucatán Peninsula, Mexico. In: Gomez-Pompa A, Allen M, Fedick S, Jiménez-Osornio J, editors. The Lowland Maya area: three Millennia at the Human-Wildland interface. New York, NY: CRC Press; 2003.

Wiersma P, Kirwan GM, Christie DA, de Juana E, Boesman P. Piping Plover (*Charadrius melodus*). In: del Hoyo J, Elliott A, Sargatal J, Christie DA, de Juana E, editors. Handbook of the Birds of the World Alive. Barcelona: Lynx Edicions; 2014. Retrieved from http://www.hbw.com/node/53828 on 5 Apr 2015.

Wilson TE, Wheeler J, Green MC, Palacios E, editors. Reddish Egret Conservation Action Plan. Reddish Egret Conservation Planning Workshop, October 2014. Corpus Christi, TX; 2014.

Winker K, Rappole JH, Ramos MA. Within-forest preferences of wood thrushes wintering in the rainforest of southern Veracruz. Wilson Bull. 1990;102:715–20.

Winkler H, Christie DA, Kirwan GM. Yucatán Woodpecker (*Melanerpes pygmaeus*). In: del Hoyo J, Elliott A, Sargatal J, Christie DA, de Juana E, editors. Handbook of the Birds of the World Alive. Barcelona: Lynx Edicions; 2013. Retrieved from http://www.hbw.com/node/56169 on 17 Apr 2015.

Wunderle Jr JM. Pre- and post-hurricane fruit availability: Implications for Puerto Rican parrots in the Luquillo Mountains. Caribb J Sci. 1999;35:249–64.

Zink RM, Dittman DL, Klicka J, Blackwell-Rago RC. Evolutionary patterns of morphometrics, allozymes, and mitochondrial DNA in thrashers (genus *Toxostoma*). Auk. 1999;116:1021–38.

Chapter 13
Subsistence Hunting and Conservation

Pablo Jesús Ramírez-Barajas and Sophie Calmé

Abstract The Maya civilization was one of the most important throughout the Americas. Today, Mayans still exist as a cultural group in Mexico, Belize, Honduras and Guatemala. The contemporary Mayan people of the Yucatán Peninsula live in a modern world; however, they still retain their language, customs and ancestral knowledge. Several studies have highlighted how this cultural group uses its natural environment and transforms it according to a complex knowledge of nature and natural systems. This knowledge is expressed in the use they make not only of plants and animals, but also of entire ecosystems. These relationships are the result of thousands of years of Mayan co-habitation with their environment. The aim of this chapter is to characterize subsistence hunting by Mayan people in the Yucatán Peninsula as a form of biodiversity utilization. Subsistence hunting is not an activity isolated from other forms of resource use, which creates a dynamic that has rarely been studied. The complex interplay of natural, social and economic conditions makes it difficult to assess the intensity of extraction and its impact on wildlife populations. Given that challenge, this chapter emphasizes the importance of sociocultural context and the multiple use character of animal and plants resources to effectively assess sustainable extractive use of fauna in future studies of the region.

Keywords Bushmeat • Game species • Harvest rates • Hunting techniques • Mayans • Sustainability • Wildlife use

---

P.J. Ramírez-Barajas (✉)
Departamento Conservación de la Biodiversidad, El Colegio de la Frontera Sur Unidad Chetumal, Apartado Postal 424, CP 77014 Chetumal, Quintana Roo, Mexico
e-mail: pjrbarajas@gmail.com

S. Calmé
Université de Sherbrooke, 2500, boul. de l'Université, Sherbrooke, QC, Canada
e-mail: Sophie.Calme@usherbrooke.ca

G.A. Islebe et al. (eds.), *Biodiversity and Conservation of the Yucatán Peninsula*,
DOI 10.1007/978-3-319-06529-8_13

13.1 Introduction

The Maya stand out as one of the cultural groups that use their natural surroundings and transform it based on their highly complex knowledge of the natural systems with which they live. This knowledge is not only expressed in their use of plants and animals, but also of entire ecosystems, which results from thousands of years of coexistence with nature (Gómez-Pompa 2003; Gill 2008; Estrada-Belli 2011). The main cultural group in the Yucatán Peninsula is the "peninsular Maya" (Bartolomé 2001). Since the 1970s, however, the federal government has promoted a new settlement policy in the tropical areas of Mexico, allowing other Mayan people (e.g. *choles*, *tzeltales*, *tzotziles*) to settle in the region, along with other ethnic groups and mestizos. Currently, Mayas from the Yucatán Peninsula live in a modern world; nevertheless, many of them maintain their language, customs and ancestral ecological knowledge (Estrada-Belli 2011).

Detailed studies reported contemporary Maya use of between 300 and 500 species of plants and animals from agricultural and natural ecosystems (Ramírez-Barajas et al. 2001; Barrera-Bassols and Toledo 2005; García-Frapolli and Toledo 2008; Toledo et al. 2008; Toledo 2009). These species have been extracted, domesticated, tolerated, managed and cultivated for generations in anthropic systems like *milpas*, family orchards, backyards gardens and plantations, as well as in natural ecosystems such as tropical forests, savannahs (*sabanas*), lagoons, sinkholes (*cenotes*), mangroves and other coastal ecosystems (Ramírez-Barajas et al. 2001; Gómez-Pompa 2003; García-Frapolli et al. 2008; Estrada-Belli 2011). For instance, Ramírez-Barajas et al. (2001, 2006) reported up to 15 productive activities for a Maya community in Quintana Roo, of which 11 were related with their natural surroundings, and the rest, with transformed systems. Eight of these 15 activities did not provide an economic benefit, but rather were subsistence oriented. Among them, subsistence hunting stands out as an activity with strong cultural, religious, and economic dimensions (Mandujano and Rico-Gray 1991; Terán and Rassmusen 1994; Quijano-Hernández and Calmé 2002; Ramírez-Barajas and Naranjo 2007; Santos-Fita et al. 2012).

Subsistence hunting is defined here as the extraction of wild fauna by communities or individuals for self-consumption and local trade without the influence of an external market to make it a lucrative activity (Fa 2000; Pérez-Gil et al. 1995; Ramírez-Barajas and Naranjo 2007; Santos-Fita et al. 2013). In such contexts, the sale of meat is proportional to the satisfaction of basic needs of hunters and their economic dependents (Montiel et al. 1999; Ojasti 2000). Subsistence hunting is a complementary activity situated within a multiple-use resource strategy (Ramírez-Barajas et al. 2001; Ramírez-Barajas and Naranjo 2007; Santos-Fita et al. 2013). Its importance transcends its subsistence role of providing animal protein, as it is fully integrated into the Mayan system of understanding the world, their *cosmovision* (Villa-Rojas 1987; Terán and Rassmusen 1994). Hunting remains culturally significant, even if it has lost its past status and scope, when it generated regional products for exchange and trade among the ancient Maya, as well as instruments, elite costumes (e.g., plumes), clothing, ceremonial offerings and objects of civic-religious character with their denotation of social status as prestige goods (Götz 2014).

In this chapter we describe subsistence hunting in the Yucatán Peninsula, placing hunting in the context of Maya culture and comparing it to *mestizo* hunting group practices. We address subsistence hunting by first characterizing hunters, and how they practice the activity, showing that few differences now exist between Maya and mestizos. The second part of the chapter addresses with the products of hunting, i.e., the species that are subject to extraction, and their different uses. We show that throughout the peninsula meat is the prime reason for hunting; however, in some cases selling it locally is much more widespread among Maya than mestizos. Finally, we discuss the implications of the current forms of hunting for the conservation of wildlife.

13.2 Hunters and Hunting Practices

Hunters in the Yucatán Peninsula are men, with very few exceptions, e.g., Calmé and Sanvicente (2000), found only one woman among the 246 hunters they interviewed across the whole peninsula. Typically, the learning process to become a hunter starts as young boys between the ages 8–12 learn how to track with their father or uncle. Between the ages of 13 and 18 they learn to use firearms (Calmé et al. 2006; Manzón-Che 2010). Knowledge is thus still transmitted from one generation to the next and also to other hunters-to-be, who are not yet experienced yet (Lechuga 2001; Ramírez-Barajas and Naranjo 2007). In the Calakmul area, however, a surprisingly large number of hunters were self-taught (Manzón-Che 2010). Specialized hunters have developed a deep knowledge of the wild fauna, habitats and food resources needed by these animals. For example, outstanding Maya knowledge-holders, among them hunters and elders, recognize around 65 % of tree species in tropical forests as a food source for animals (Ramírez-Barajas et al. 2012a). This knowledge is used to hunt game using different techniques described below (spying, *batida*, hunting in cultivated areas, etc.). Hunters are usually deeply involved in activities based on the use of natural resources. Overwhelmingly (97 %), they are traditional *milpa* farmers, but half of them are also beekeepers, loggers, tree markers, gum-tappers, and seed collectors, among other occupations (Calmé et al. 2006).

Subsistence hunters can be classified as specialized or opportunistic; in the latter case they hunt when on their way to (or around in the vicinity of) their *milpa* or apiary. Hunting trips may be solitary or group activities. Whether or not a hunter prefers to hunt alone, it is very common for one or more hunters to join the hunting group for security and company. In the Calakmul area, one or two hunters performed 86 % of the hunts, and hunting parties never exceeded four men. In the Zona Maya (e.g., Felipe Carrillo Puerto and José María Morelos municipalities), parties included up to 12 men, but a single hunter or pair of hunters conducted most hunts (72 %) (Calmé et al. 2006).

Specialized hunters are often a small and select group (Mandujano and Rico-Gray 1991; Quijano-Hernández and Calmé 2002; Ramírez-Barajas and Naranjo 2007). They express a particular taste for hunting and for being in the "wild", which often implies spending several days in the forest. Besides the knowledge acquired from the persons with whom they learn to hunt, their knowledge results from their constant and close contact with nature (Ramírez-Barajas and Naranjo 2007). The hunting effort of these men usually accounts for a large proportion of the game harvest in their communities. For instance, five hunters were responsible for 50 % of the harvest in each of three communities of Calakmul and the Zona Maya studied by Calmé et al. (2006), and four hunters harvested 79 % of the game in Tres Reyes (Quijano-Hernández and Calmé 2002). In Petcacab, 10 out of 79 different hunters were responsible for providing 67 % of the total biomass extracted over 3 years (Ramírez-Barajas et al. 2006; Ramírez-Barajas and Naranjo 2007). From the above, hunting can be described both as a generalized activity, because it is practiced by many people, and a specialized activity, because it is performed more intensively by a few hunters with a sophisticated knowledge of species and the environment.

Traditional devices used by hunters, such as traps, have practically disappeared in the Yucatán Peninsula. Most hunters now use firearms to obtain their game; for example, in 97 % of the cases reported in the community of Petcacab hunters shot their game using guns (Ramírez-Barajas and Naranjo 2007). Other devices may include machete, stick, dog, and even bare hands when dealing with very young animals (Manzón-Che 2010). Among the hunting techniques used and reported by several authors (Mandujano and Rico-Gray 1991; Ramírez-Barajas and Naranjo 2007; León and Montiel 2008) those that stand out are described in detail, below.

Active Searching or Walking This technique consists of day and night walks, the latter conducted using flashlights. The route follows paths and trails or by getting deep into the forest and *sabana* (seasonally flooded vegetation composed primarily of Cyperaceae and Poaceae) using *picaderos* (small access-openings) made with machetes. The walking distance covered varies from 1 to 6 km and is determined by the time spent to find game (Morales 2000; Quijano-Hernández and Calmé 2002). According to Ramírez-Barajas and Naranjo (2007), in the Maya community of Petcacab, Quintana Roo, the time invested can be 0.5 to 6 h, with an average of 1.7 h. This technique was used in 33 % of all recorded hunting events during the study year and accounted for extraction of 36 % of all hunted animals. In terms of biomass, it represented 49 % of the total extracted (545 kg), mainly because 11 white-tailed deer (totaling 327 kg) were hunted in this manner (Ramírez-Barajas and Naranjo 2007). In other communities, active search was the most common technique: it corresponded to 48 % of game in the culturally mixed Nuevo Becal, Campeche (Manzón-Che 2010), and to 83 % of the hunts in the Maya community of Tres Reyes, Quintana Roo (Quijano-Hernández and Calmé 2002). In Nuevo Becal, Campeche, active searching was the technique of choice for large game birds (great curassow and ocellated turkey), but also for ungulates (collared and white-lipped peccaries, and white-tailed deer) (Manzón-Che 2010).

Watching or Stalking This technique is performed day or night (using a lamp), normally when the hunter has previously observed some indications of the presence of game. In this places the hunter put *espiaderos*, where hung a hammock between two trees at a height of 2 m or more, or build a *tapesco* (small base of sticks), which waits for the game. The average waiting time is 3.5 h, but varies between 1 and 12 h (Ramírez-Barajas and Naranjo 2007). This technique was performed in 21 % of the hunting events reported by these latter authors and corresponded to 22 % of the extracted biomass (248 kg). In the same study the number of individual animals hunted with this technique represented 20 % of the total of recorded game numbers. However, it is a technique that *mestizo* hunters tend to deprecate when game abundance is high; in Nuevo Becal it was used to capture only 10 % of recorded game (Manzón-Che 2010). Hunters commonly perform this technique in tropical forest and *milpas*, although sometimes, in sabanas and *huamiles* (young secondary vegetation), and report that it generates a 70 % success rate (Ramírez-Barajas and Naranjo 2007). *Espiaderos* are effective particularly for paca, ocellated turkey, collared peccary and white-tailed deer, but hunters can also harvest armadillos, brocket deer and Central American agouti during the wait.

Opportunistic Encounters It is important to mention that in this technique, hunting is not the main purpose. Rather, hunters carry firearms while engaged in other activities such as *milpa* cultivation, gum-tapping, farm maintenance, beekeeping, and fishing, among others. Thus, hunters find game in route to their destination or while at work. Respectively 35 % and 38 % of all hunted individuals were harvested in opportunistic encounters in Petcacab and Nuevo Becal (Ramírez-Barajas and Naranjo 2007; Manzón-Che 2010). In Petcacab it corresponded to 40 % of hunting events and to 28 % of the biomass obtained (Ramírez-Barajas and Naranjo 2007). The species most frequently hunted using this technique are white-nosed coati, ocellated turkey, great curassow, agouti, collared peccary, white-tailed deer, armadillo, both brocket deer, and white-lipped peccary.

Use of Dogs As with opportunistic hunting, this technique is usually performed while dogs are accompanying their owners in other activities. In the study of Ramírez-Barajas and Naranjo (2007), only 6 % of the hunts were obtained by this technique; it represented 9 % of game and barely 1 % of the biomass. The main animal species obtained in this way are nine-banded armadillo, paca and white-nosed coati. According to interviews with hunters of Petcacab, when dogs detect game, they cornered in a tree or find its burrow, and there the hunter takes the game (Ramírez-Barajas and Naranjo 2007). For this reason, it was forbidden by general agreement during an assembly of Petcacab, because it was considered a form of disturbance to the forest that could drive animals away from their territories. This technique may contribute more to capture rates and hunted biomass in communities where the practice is not banned by the assembly or conservation agreements.

Batida The *batida* is a group hunting technique performed regularly in Maya communities from the northern Yucatán Peninsula, where there is low tropical dry forest or is deforested (Mandujano and Rico-Gray 1991; León and Montiel

2008; Rodríguez et al. 2012). It is much rarer in the forest-dominated areas of Calakmul, Campeche and Zona Maya, Quintana Roo (Manzón-Che 2010; Ramírez-Barajas and Naranjo 2007). Participants form two hunting groups that drive an animal, most frequently a white-tailed deer, towards the shooters' group (for more details see Mandujano and Rico-Gray 1991). The technique has more of a cultural and organizational purpose than an economic one, because it is regularly practiced for religious events or town festivities and the meat is shared among all the participants (Mandujano and Rico-Gray 1991; Rodríguez et al. 2012). We believe, however, that there are also ecological drivers to this practice: when game availability (and ease of capture) is high, *batida* is usually not practiced (Ramírez-Barajas and Naranjo 2007; Quijano-Hernández and Calmé 2002). León and Montiel (2008) report that each hunter gets about 2.5 kg of meat in a *batida*, compared to 20 kg when hunting using active search. In those communities where *batida* is not practiced, group hunts may be occasionally organized by three or four hunters to drive an animal when it is found by chance in the forest or in a *milpa* (Ramírez-Barajas and Naranjo 2007).

Traps and Other Methods Only a few hunters use traps, nonetheless, some hunters report hunting of lesser species (gophers, squirrels, doves) with traps, mousetraps, and even homemade boxes. Also, birds and squirrels are hunted with slingshots, and hunters use calls to imitate some animals (mainly agouti and brocket deer), by placing the leaves of certain plants or plastic bags between their lips and blowing.

In the Yucatán Peninsula, hunting is mostly carried out in mature forest. In a study in four communities of the Calakmul-Sian Ka'an section of the Mesoamerican Biological Corridor, between 46 and 70 % of game was harvested in mature forest; however secondary forest were also very important, accounting for 27–37 % of harvested game (Calmé et al. 2006). Third came the *milpa*, in which 8–26 % of game was hunted. Hunting in *milpas* is a dual-purpose activity: producers protect vegetables, corn, beans, and squash crops, among other products from the *milpa* from depredation; and bring wild meat to the table via hunting of animals attracted by these crops (Jorgenson 1995). Fauna associated with crops are either considered a pest species when they are not hunted (e.g. raccoon, tapir, parrots, and chachalacas), or a reward if they are prized species (e.g. white-tailed deer, white-nosed coati, paca, Central American agouti, ocellated turkey, and great curassow). According to Santos-Fita et al. (2013), there are variants in this kind of hunting, and they report that some hunters intentionally cultivate small areas (0.4–1 ha) to lure and hunt particular vertebrate species in a system the authors call feeder-trap *milpa*.

Many communities of the Yucatán Peninsula have rules for hunting agreed on during their meetings. These agreements include restricting the use of dogs, respecting others' plots, limiting the removal of females and young, as well as respect for closed seasons and community conserved areas (Quijano-Hernández and Calmé 2002; Ramírez-Barajas and Naranjo 2007; Calmé et al. 2006). Some rules are also established ad hoc; for example, after Hurricane Dean, the community of Petcacab agreed to a temporally hunting ban to protect wildlife and allow population recovery (Ramírez-Barajas et al. 2012a). Enforcement, however, is often difficult, as the communities are small and most inhabitants have close ties to one another (Calmé unpublished data).

Despite differences in hunting practices in the Yucatán Peninsula, the studies performed in Mayan communities concur that cultural conceptions, syncretism and symbology still guide hunting practices, which tends to define the importance of the activity, its persistence, and even extraction intensity (Quijano-Hernández and Calmé 2002; Ramírez-Barajas and Naranjo 2007; León and Montiel 2008). The hunter incorporates magical and religious beliefs that limit the number of game that can be taken, and maintains implicit respect for the hunted species and certain places in the forest (Mandujano and Rico-Gray 1991; Terán and Rasmussen 1994; Quijano-Hernández and Calmé 2002; Ramírez-Barajas and Naranjo 2007; Rodríguez et al. 2012). However, as beliefs lose importance among Maya hunters, so does the control they place on extraction (Quijano-Hernández and Calmé 2002). Moreover, external market forces can exert a strong pressure on hunters by increasing the demand for wild meat (Calmé et al. 2006; Ramírez-Barajas and Naranjo 2007). Thus, depending on their own status in the community (rights-owner or not) and on the opportunities for selling the meat, *mestizo* and Maya hunters alike may find themselves involved in the commercialization of wild meat (Lechuga 2001; Quijano-Hernández and Calmé 2002; Calmé et al. 2006; Ramírez-Barajas and Naranjo 2007).

Extraction rates (estimations of the number of animals extracted per square kilometer) from mixed Mayan and *mestizo* communities vary widely among communities, and do not reflect any clear cultural pattern (Fig. 13.1). In fact, those communities where extraction rates are high are communities where wildlife products are mostly sold rather that consumed for subsistence. For instance, in Laguna Kana and Tres Reyes 57 % of hunted game was sold outright, and hunters and their families consumed most of the rest, with some portion shared and sold. In contrast, hunters in Nuevo Becal and Conhuas fully consumed 73 % of hunted game, while they sold 25 % of these products whole (Calmé et al. 2006; Quijano-Hernández and Calmé 2002). Hunting products (mostly wild meat, but sometimes live animals) are sold either on the highway or in the municipal market. Although a more detailed study on commercialization of wildlife products and its drivers is required to better assess the situation, several factors such as access to economic opportunities, location relative to a market, as well as location relative to a police or army checkpoint seem to determine the fate of hunted goods (Calmé et al. 2006). It is noteworthy that both extraction rates and estimates of abundance (density) of species are necessary elements to assess the sustainability of hunting; therefore it is of great importance to address both aspects simultaneously for conservation, either by species or species group.

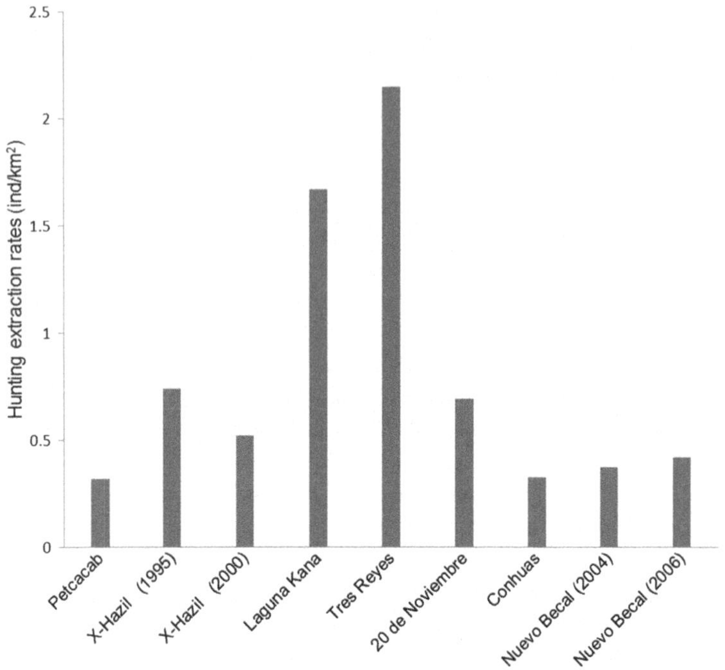

Fig. 13.1 Rates of extraction of wildlife in Mayan and *mestizo*-mixed communities of the Yucatán Peninsula. Mayan communities: Petcacab (Ramírez-Barajas and Naranjo 2007), X-Hazil (Jorgenson 1995; Morales 2000), Laguna Kana (Calmé et al. 2006) and Tres Reyes (Quijano-Hernández and Calmé 2002), in Quintana Roo; 20 de Noviembre (Calmé et al. 2006), in Campeche. *Mestizo*-mixed communities: Conhuas (Calmé et al. 2006) and Nuevo Becal (Calmé et al. 2006), in Campeche

13.3 Wildlife Uses and Diversity of Vertebrates Used

More than 50 species are reportedly hunted in the Yucatán Peninsula (Table 13.1). The main species hunted are mammals, followed by birds and, in lower proportion, reptiles (turtles and crocodiles). The number of species in the Yucatán Peninsula identified as being used with some purpose differs among regions and study objectives. For example, Chablé-Santos and Delfín-González (2010) reported through interviews the use (for alimentary, medicinal, and as pets, among other purposes) of 28 mammal species, 38 bird species and 15 reptile species in the state of Yucatán. Ramírez-Barajas et al. (2001) and Ramírez-Barajas et al. (2006), also through interviews, reported up to 26 wildlife species (9 mammals, 10 birds, 2 turtles, and 5 fishes) extracted solely for food consumption in subsistence hunting and fishing in a Maya community in Quintana Roo.

Preferences are based on the taste of different species' meat, and to affective and aesthetic conceptions for some species, mainly mammals. For example, hunting of

Table 13.1 Terrestrial vertebrates used in the Yucatán Peninsula. Species are organized by inferior to superior groups (reptiles, birds and mammals) and then alphabetically by family

Common name	Species	Family	Use[a]	Part used[b]	Site[c]
Reptiles					
Mojina, Furrowed Wood Turtle	*Rhinoclemmys areolata*	Bataguridae	P		MSF, ML, HG, R, LSF
Imperial boa	*Boa constrictor*	Boidae	D, M, O	M, P	MSF, LSF
Morelet's crocodile	*Crocodylus moreletii*	Crocodylia	F, H	M, Sk	L, S, Ce
American crocodile	*Crocodylus acutus*	Crocodylia	F, H	M, Sk	L, S, Ce
Jicotea, Slider	*Trachemys venusta*	Emydidae	F	M	L, S, HG, Ce
Black Iguana	*Ctenosaura similis*	Iguanidae	F	M	
Morocoy, Tres Lomos	*Staurotypus triporcatus*	Kinosternoidae	F	M	L, S, HG, Ce
Pochitoque tres quillas, Scorpion Mud Turtle	*Kinosternon scorpioides*	Kinosternoidae	P, M		L, S, Ce
Casquito, Creaser's Mud Turtle	*Cryptochelys creaseri*	Kinosternoidae	P, M		L, S, HG, Ce
Pochitoque, White-lipped Mud Turtle	*Cryptochelys leucostomum*	Kinosternoidae	P, M		L, S, HG, Ce
Terciopelo, Fer-de-lance	*Bothrops asper*	Viperidae	D		MSF, ML, R, HG
Cascabel, Rattlesnake	*Crotalus durissus*	Viperidae	D, M	M	MSF, ML, R, HG
Birds					
Muscovy duck	*Cairina moschata*	Anatidae	F, B	M	L, Ce, S
Gray-necked Wood-rail	*Aramides cajanea*	Aramidae	F, P	M	LSF, S, Ce, L
Dove	*Claravis pretiosa*	Columbidae	F, P	M	MSF
Dove	*Columba flavirostris*	Columbidae	F, P	M	MSF
Dove	*Zenaida asiatica*	Columbidae	F, P	M	MSF
Plain chachalaca	*Ortalis vetula*	Cracidae	F, P	M	MSF, ML, S
Crested guan	*Penelope purpurascens*	Cracidae	F	M	MSF
Great curassow	*Crax rubra*	Cracidae	F, P	M	MSF, ML, S
Northern cardinal	*Cardenalis cardenalis*	Fringillidae	P		MSF
Blue bunting	*Cyanocompsa parellina*	Fringillidae	P		MSF
Ocellated turkey	*Meleagris ocellata*	Phasianidae	F, O	M, Fe	MSF, ML, S
Aztec parakeet	*Eupsittula astec*	Psittacidae	F, P	M	ML, MSF

(continued)

Table 13.1 (continued)

Common name	Species	Family	Use[a]	Part used[b]	Site[c]
White-fronted/ Yellow-lored parrot	*Amazona albifrons/ xantholora*	Psittacidae	F, P	M	ML, MSF
Collared aracari	*Pteroglossus torquatus*	Ramphastidae	F, P, T	M, Bi	MSF
Keel-billed toucan	*Ramphastos sulfuratus*	Ramphastidae	F, P, T	M, Bi	MSF
Eastern thicket tinamou	*Crypturellus cinnamomeus*	Tinamidae	F	M	MSF
Mammals					
Paca	*Cuniculus paca*	Agoutidae	F, P	M	MSF, ML, S
Red/Brown brocket deer	*Mazama americana/ pandora*	Cervidae	F, P, O	M, Sk	MSF, ML, S
White-tailed deer	*Odocoileus virginianus*	Cervidae	F, P, H, O, D	M, An	MSF, ML, S, LSF
Mexican black howler monkey	*Alouatta pigra*	Cebidae	F	M	MSF
Central American spider monkey	*Ateles geoffroyi*	Cebidae	P, F	M	MSF
Nine-banded armadilo	*Dasypus novemcinctus*	Dasypodidae	F	M	MSF, ML
Central American agouti	*Dasyprocta punctata*	Dasyproctidae	F	M	MSF, ML
Common opossum	*Didelphis marsupialis*	Didelphidae	D, M		S
Mexican hairy porcupine	*Coendu mexicanus*	Erethizontidae	F, M	M	MSF
Puma	*Puma concolor*	Felidae	F, D, O	Sk, Fa	MSF, R
Jaguar	*Panthera onca*	Felidae	D, O	Sk, Fa	MSF, R
Jaguarundi	*Puma yagouaroundi*	Felidae	O	Fa	MSF
Margay	*Leopardus wiedii*	Felidae	O	Fa, Sk	MSF
Ocelot	*Leopardus pardalis*	Felidae	O	Fa, Sk	MSF
Hispid pocket gopher	*Orthogeomys hispidus*	Geomydae	F, D	M	ML
Long-tailed weasel	*Mustela frenata*	Mustelidae	D		MSF, HG
Skunk	*Conepatus semistriatus*	Mustelidae	M, D	M	MSF
Tayra	*Eyra barbara*	Mustelidae	D		Ap
White-nosed coati	*Nasua narica*	Procyonidae	F, P, D	M	MSF, ML
Kinkajou	*Potos flavus*	Procyonidae	F	M	MSF
Northern raccoon	*Procyon lotor*	Procyonidae	D		ML
Deppe's/Yucatán squirrel	*Sciurus deppei/ yucatanensis*	Sciuridae	F	M	MSF

(continued)

Table 13.1 (continued)

Common name	Species	Family	Use[a]	Part used[b]	Site[c]
Baird's tapir	*Tapirus bairdii*	Tapiridae	D, F	M	LSF, ML, S
Collared peccary	*Pecari tajacu*	Tayassuidae	F, P, D	M	MSF, ML
White-lipped peccary	*Tayassu pecari*	Tayassuidae	F, P, D	M	MSF, ML, LSF, S
Eastern cottontail	*Sylvilagus floridanus*	Leporidae	F	M	ML

Sources: Escamilla et al. (2000), Ramírez-Barajas et al. (2001), Quijano-Hernández and Calmé (2002), Ramírez-Barajas et al. (2006), Chablé-Santos and Delfín-González (2010), León and Montiel (2008), Segovia-Castillo et al. (2010)

[a]Use: *F* Food, *P* pet, *D* damage control, *O* ornamental, *H* handcraft, *B* bait, *T* tool, *M* medicinal

[b]Part used: *M* meat, *Sk* skin, *An* antlers, *Fa* fangs, *Fe* feathers, *Bi* bill

[c]Site: *MSF* medium statured forest, *ML* milpa, *R* ranch, *Ap* apiary, *HG* homegardens, *LSF* low statured forest, *S* sabana, *L* lagoon, *Ce* cenote

primates (spider and howler monkeys), carnivores (Felidae and Canidae), kinkajou (*Potos flavus*), tapir (*Tapirus bairdii*) and marsupials (Didelphidae) is rare (Jorgenson 1995; Quijano-Hernández and Calmé 2002; Ramírez-Barajas and Naranjo 2007; Manzón-Che 2010). In general the diversity of game hunted by Mayas is higher than that of *mestizos*, as they consume some species that mestizos do not, such as the hispid pocket gopher and white-nosed coati (see, e.g., Escamilla et al. 2000; Jorgenson 1995).

Wild fauna in the Yucatán Peninsula is used for diverse purposes, but primarily for the meat it provides (e.g. 78 % in two communities in the Calakmul area; Manzón-Che 2010). Other uses are medicinal, decorative, baiting, ritual, keeping as pets (sometimes with the intent to rear the animal to later consume it), and killings to eliminate nuisance animals (Jorgenson 1995; Ramírez-Barajas and Naranjo 2007; Santos-Fita et al. 2013). Similar patterns of use have been reported in archaezoological research on the Maya (Götz 2008; Masson and Peraza-Lope 2008). For example, animal remains found in excavations from the ancient city of Mayapán, in Yucatán, suggest that white-tailed deer were raised in captivity or carefully managed on the outskirts of cities (Masson and Peraza-Lope 2008). Similar records show that the wild turkey *Meleagris gallopavo* was raised or domesticated, along with the ocellated turkey *Meleagris ocellata*, a wild species endemic to the Yucatán Peninsula (Thornton et al. 2012).

The former presupposes not only an extractive use, but also a management strategy that implies a well-developed knowledge of both species as well as of the ecosystem. This is known in ethnoecology as the kosmos-corpus-praxis (k-c-p) complex, generated through the intellectual and material appropriation of the landscape and natural resources [for further details see Toledo (1992) and Barrera-Bassols and Toledo (2005)]. It is represented as a system of beliefs, knowledge and practices developed via millennia of use of natural resources. This systemic understanding is expressed in the diversity of wild vertebrates used by the Maya, mostly including mammals and birds, but also reptiles with specific uses.

Despite the existence of more available potential game species, the number of species used for subsistence hunting in the peninsula oscillates between 12 and 25 (Ramírez-Barajas and Naranjo 2007; Santos-Fita et al. 2012). However, no more than ten species are regularly used with high intensity, and these are medium- and large-sized mammal and bird species (>2 kg weight), which meat is consumed (Calmé et al. 2006; Jorgenson 1995; Manzón-Che 2010; Montiel et al. 1999; Escamilla et al. 2000; Ramírez-Barajas and Naranjo 2007; León and Montiel 2008; Segovia-Castillo and Hernández-Betancourt et al. 2003; Segovia-Castillo et al. 2010; Weber and Retana-Guiascón 2010). In general terms, this large mammals and birds make up respectively 48–74 % and 18.38 % of the prey (Manzón-Che 2010), but large mammals comprise between 66 and 97 % of total biomass of species extracted by hunters (Jorgenson 1995; Manzón-Che 2010; Ramírez-Barajas and Naranjo 2007). Götz (2008) suggests this hunting pattern was already present among the ancient Maya (600–1050 AD). He found that between 82 and 95 % of the bone remains of fauna used for food consumption in ancient cities corresponded with large birds and mammals.

Ungulates like the white-tailed deer, the brown and red brocket deer (*Mazama pandora*, *M. temama*), and the collared and white-lipped peccaries (*Tayassu tajacu, T. pecari*) are the species providing most of the biomass in the harvest of fauna for three main reasons: they are large in size (>10 kg), provide higher return when sold than medium sized species, and are preferred by hunters because of the taste of their meat (Jorgenson 1995; Quijano-Hernández and Calmé 2002; Ramírez-Barajas and Naranjo 2007; Escamilla et al. 2000; Hernández-Betancourt and Segovia-Castillo 2010; Ramírez-Barajas et al. 2012a; Santos-Fita et al. 2012). Two large rodent species, paca and Central American agouti, are also generally heavily hunted, as are white-nosed coati and nine-banded armadillo (*Dasypus novemcinctus*). As for birds, the crested guan (*Penelope purpurascens*), the ocellated turkey (*Meleagris ocellata*) and the great curassow (*Crax rubra*) are the large species frequently extracted (Table 13.2).

There are some local exceptions related, for instance, to available ecosystems (e.g., iguana is the second most hunted species close to Los Petenes Biosphere Reserve; León and Montiel 2008), and to availability or taste (e.g., Central American agouti and white-nosed coati are hunted as pests and then given to dogs in one community of the Calakmul region; Manzón Che 2010). Smaller species are hunted infrequently and under specific circumstances, for example, while *campesinos* are working on-farm, in *milpas*, or developing some extractive activity (Jorgenson 1995; Ramírez-Barajas and Naranjo 2007). As an example of extraction of fauna, Table 13.2 presents the hunting performed in a Maya community of Petcacab in the state of Quintana Roo (Ramírez-Barajas et al. 2006; Ramírez-Barajas and Naranjo 2007). The monetary income reported in this study corresponded solely to wild meat intended for sale (between 50 and 60 % of meat biomass extracted). The authors also observed that several lesser species did not contribute economically.

The behavior of subsistence hunters in the Yucatán Peninsula is similar to other tropical regions of South America and Africa (e.g., Amazonas and Congo; Nasi

Table 13.2 Game captured in the Maya community of Petcacab over 3 years (2003–2005). Species are listed in decreasing order of individuals captured

Species	Number of individuals	Estimated biomass (kg)	Income (USD)
White-nosed coati	243	897	981
Paca	122	643	1375
White-tailed deer	108	3490	6010
Central American agouti	105	349	66
Collared peccary	99	1670	2521
Great curassow	76	343	406
Plain chachalaca	35	18	0
Nine-banded armadillo	34	1578	72
Red brocket deer	18	276	304
Ocellated turkey	8	40	50
Deppe's squirrel	5	3	0
White-fronted parrot	5	2	0
Crested guan	3	9	0
Thicket tinamou	3	2	0
Keel-billed toucan	3	1	0
Yellow-lored parrot	2	1	0
Puma	1	40	14
Pigeon	1	<1	0
Total	871	7941	11,800

et al. 2011), because of the preference for large mammals, mainly ungulates, as they provide an important source of meat and income (Milner-Gulland et al. 2003). However, it differs largely in that primates are never hunted for meat, both among Mayans and *mestizos*, probably for cultural reasons. Moreover, as elsewhere, there is no single type of hunting. The types of hunting practices corresponds to multiple circumstances, making it difficult to address in research and policy formulation. Those hunting practices is related to the biological characteristics of the species, to the ecological features of each community (e.g., the presence of large forest remnants where white-lipped peccary may still be found), as well as to differences in culture, religion and economic status of hunters.

13.4 Implications of Subsistence Hunting for Conservation

The intense extraction of one a particular group of species has serious implications for its conservation, since species are also subject to other, non-hunting driven pressures affecting their populations (e.g., habitat loss and degradation), leading sometimes to negative growth rates. Some species are more vulnerable because of intrinsic aspects of their life histories, like large home ranges, low reproductive rates, and high body biomass (Dirzo et al. 2014). For example, in the specific case of the white-lipped peccary, a species that needs large extents of unbroken forest and is sensitive to habitat fragmentation and hunting, data indicate that populations remain in the two largest natural protected areas, the Biosphere Reserve of Calakmul in Campeche (Reyna-Hurtado et al. 2014) and the Biosphere Reserve of Sian-Ka'an in Quintana Roo (Ramírez-Barajas and Naranjo 2007; Ramírez-Barajas unpublished data). However, in the forests of Maya communities located in the central portion between these reserves, the white-lipped peccary has not been reported by local inhabitants for the last 40 years, and has not been recorded in biodiversity monitoring for at least 10 years (Ramírez-Barajas and Naranjo 2007; Ramírez-Barajas et al. 2012a), which indicates a possible local extinction of the species as a consequence of hunting and loss of connectivity in their historical dispersion routes due to highways, habitat transformation, and human influence (Ramírez-Barajas and Naranjo 2007).

Unsustainable over-exploitation of vertebrates has led to what is known as "empty forests syndrome" (Redford 1992; Wilkie et al. 2011; Camargo-Sanabria et al. 2015). This "defaunation" process results in the absence of large vertebrates (mainly top predators and large herbivores) in apparently intact forests (Wilkie et al. 2011). This phenomenon is of such magnitude that it has been called the defaunation of the Anthropocene period, similar in speed and magnitude to previous mass extinctions (Dirzo et al. 2014). The degradation of habitat, loss of species and populations of wild fauna, and even decreasing population abundance have consequences in ecosystem processes. This degradation may unleash extinction cascades or the decline of plants and animals, as well as having serious implications for the structure, function and resilience of forests (Wilkie et al. 2011).

Hunting is consistently mentioned in the literature as one promoter of defaunation in the tropics; strongly suggesting it is an important factor in current extinction processes. Even so, unlike deforestation, which can be identified through rigorous, quantitative information, defaunation remains a relatively cryptic phenomena that requires intensive field sampling (Galetti et al. 2013; Dirzo et al. 2014). Without such intensive efforts, it is complicated to prove its presence and effects, because it is almost undetectable with remote sensing techniques (Peres et al. 2006; Wilkie et al. 2011). For those reasons, research on hunting is a priority for conservation in the Yucatan Peninsula. A better understanding of the underlying social factors is much needed, but the trends of game species' populations, both locally and regionally, are critical.

The negative impact of hunting also depends on the species and the habitat conditions. Game species show differences in vulnerability according to biological attributes and life traits. Some species are able to sustain populations in secondary habitats with intense hunting, while others require intact forests with minimal extraction to maintain healthy populations (Nasi et al. 2011). Therefore, species' overall vulnerability, and susceptibility to the effects of hunting, should not only be studied in the cultural context in which hunting is developed, but in conjunction with the natural conditions of the habitat type, and co-occurring threats like habitat loss, fragmentation and natural disturbances. For example, frugivores and herbivores show more vulnerability to high-impact natural disturbances like hurricanes than omnivores and carnivores (Hernández-Díaz et al. 2012; Ramírez-Barajas et al. 2012a, b). Herbivores and frugivores (e.g. white-tailed deer, brocket deer, paca, Central American agouti) regularly show higher rates of extraction (Ramírez-Barajas and Naranjo 2007), raising the possibility of deleterious and synergic effects in the presence of increasingly frequent disturbances in the region. These potential synergies have yet to be evaluated, and should be integrated into studies of populations dynamics.

According to Peres (2001), subsistence hunting has strong effects on species diversity, biomass, size and structure of the assembly of terrestrial vertebrates; those effects are increasing heavily with habitat degradation and forest fragmentation. It is known that the loss of medium and large herbivores and frugivores (predators and seed dispersers) affects dispersal patterns, especially those of large-seeded plants (Dirzo and Miranda 1990; Redford 1992). Evidence of these and others effects of species extirpation and their consequences over the ecosystems dynamics have increased, to the extent that some species or groups can be considered as ecosystem engineers and keystone species (Dirzo and Miranda 1990; Wright et al. 2000; Nasi et al. 2011; Wilkie et al. 2011).

The consequence of uncontrolled overexploitation of wild fauna is obviously the extinction of animal species, but also a loss of sustainability for millions of people depending on these resources (Fa et al. 2007). This applies not just to the Yucatán Peninsula, but also to the entire tropical region of worldwide. However, we must emphasize that there is no extant monitoring of wildlife (of population or harvest rates) to implement strategies for conservation that take into account the cultural conditions of the Yucatán Peninsula. As in the past, subsistence hunting is an activity deeply rooted in culture, especially in contemporary Maya population; hence, its cultural, and economic value is a topic of great scientific interest because of its implications for conservation and the understanding of the ecological repercussions of the extraction of wild fauna. Subsistence hunting is developed within a spatio-temporal logic that also contains the rest of the suite of productive activities. These accord it certain characteristics, not yet evaluated, based on and contributing to the resilience of human communities. Not being an activity isolated from other forms of resource exploitation, subsistence hunting is characterized by a rarely-studied dynamics. However, the complexity of conditions surrounding hunting complicates the evaluation of the intensity of fauna extraction. Therefore, it is necessary to consider the socio-cultural context of hunting, the multiple-use character of the resource, as well as the importance of this information in evaluating the following studies focused on population estimates and sustainability of the extraction of wild fauna in the region.

Assessing the sustainability of hunting in the Yucatán Peninsula is not entirely possible yet, because of the paucity of available information on the population status of species, extraction rates, and other aspects of the life history of hunted species. Although several studies have separately addressed some of these issues (Calmé et al. 2006; Ramírez-Barajas and Naranjo 2007; León and Montiel 2008; Rodríguez et al. 2012), the information in their descriptions of hunting patterns indirectly suggests the risk of unsustainable extraction for various species. For example, in some instances the hunters choose less preferred species, and select smaller species in the absence of large-sized species (Stearman 2000; Ramírez-Barajas and Naranjo 2007), or indices of selectivity show strong selection of some species (Calmé et al. 2006). Increasing distances that hunters need to travel (Bennett and Robinson 2000), which is in turn is related to an increase in hunting effort (Stearman 1999), also indicates reduced availability of game species. Such changes have been documented through interviews (e.g., León and Montiel 2008), but hard data remain scarce.

Hunting strategies indirectly show their relation to game availability and pressure on some species. The technique of driving game or "*batida*", even if it is a traditional Mayan technique, may indicate a low availability of wildlife, requiring increasingly more group effort to obtain larger game like white-tailed deer and peccary (Mandujano and Rico-Gray 1991; Quijano-Hernández and Calmé 2002; Rodríguez et al. 2012). In turn, increasing average time and travel distance spent on individual hunting serves as an indirect indicator of lower abundance of game species (Quijano-Hernández and Calmé 2002; Ramírez-Barajas and Naranjo 2007). The sale of bushmeat at the community level may not affect game species populations. When pursued on a regional scale and in response to increasing demand from external markets, however, it also increases the intensity of extraction levels, often exceeding reproduction rates in the local environment (Bennett and Robinson 2000; Calmé et al. 2006; Ramírez-Barajas and Naranjo 2007). Elsewhere, this has given rise to the so-called "bushmeat crisis"; which involves influencing factors such as poverty, food insecurity, low levels of development, economic market failure, and lack of political and institutional understanding (Fa et al. 2007; Nasi et al. 2011).

Future studies on subsistence hunting in the Yucatán Peninsula, in addition to considering sustainability of extraction and population estimates of game species, must face the challenge of considering the synergies of factors that interact to impact wildlife. Researchers must collaborate across disciplinary boundaries to elucidate the cultural and economic factors that define hunting and its intensity, as well as loss and fragmentation of habitat, human population growth, changes in the availability of water and food due to climate change, and biological and ecological characteristics of game species.

References

Barrera-Bassols N, Toledo VM. Ethnoecology of the Yucatec Maya: symbolism, knowledge and management of natural resources. J Lat Am Geogr. 2005;4:9–41.

Bartolomé MA. El derecho a la autonomía de los mayas macehualob. Alteridades. 2001;11:97–110.

Bennett EL, Robinson JG. Hunting for sustainability: the start of a synthesis. In: Robinson JG, Bennett EL, editors. Hunting for sustainability in tropical forests. New York, NY: Columbia University Press; 2000.

Calmé S, Sanvicente M. Distribución actual, estado poblacional y evaluación del estado de protección del pavo ocelado (*Agriocharis ocellata*). Final report of Project R114 to Comisión Nacional para el Conocimiento y Uso de la Biodiversidad. México, DF, Mexico; 2000.

Calmé S, Guerra RM, Xijun KM, Manzón-Che M. Cacería. In Uso y monitoreo de los recursos naturales en el corredor biológico mesoamericano (áreas focales X'Pujil-Zoh Laguna y Carrillo Puerto). Final report for Comisión Nacional para el Uso y Conocimiento de la Biodiversidad (CONABIO), Mexico; 2006.

Camargo-Sanabria AA, Mendoza E, Guevara R, Martínez-Ramos M, Dirzo R. Experimental defaunation of terrestrial mammalian herbivores alters tropical rainforest understorey diversity. Proc R Soc B Biol Sci. 2015;282:1–6.

Chablé-Santos J, Delfín-González H. Uso tradicional de fauna silvestre. In: Durán R, Méndez M, editors. Biodiversidad y Desarrollo Humano en Yucatán. CICY, PPD-FMAM, CONABIO, SEDUMA. 2010.

Dirzo R, Miranda A. Contemporary neotropical defaunation and forest structure, function and diversity a sequel to John Terborgh. Conserv Biol. 1990;4:444–7.

Dirzo R, Isaac NJB, Young HS, Collen B, Galetti M, Ceballos G. Defaunation in the Anthropocene. Science. 2014;345:401–6.

Escamilla A, Sanvicente M, Sosa M, Galindo-Leal C. Habitat mosaic, wildlife availability, and hunting in the tropical forest of Calakmul, México. Conserv Biol. 2000;14:1592–601.

Estrada-Belli F. The first Maya civilization: ritual and power before the classic period. New York, NY: Routledge/Taylor And Francis Group; 2011.

Fa JE. Hunted animals in Bioko Island, West Africa: sustainability and future. In: Robinson JG, Bennett EL, editors. Hunting for sustainability in tropical forests. New York, NY: Columbia University Press; 2000.

Fa JE, Albrechtsen L, Brown D. Bushmeat: the challenge of balancing human and wildlife needs in African moist tropical forests. In: Macdonald D, Service K, editors. Key topics in conservation biology. Oxford: Blackwell; 2007.

Galetti M, Guevara R, Côrtes MC, Fadini R, Von Matter S, Leite AB, Labecca F, Ribeiro T, Carvalho CS, Collevatti RG, Pires MM, Guimarães Jr PR, Brancalion PH, Ribeiro MC, Jordano P. Functional extinction of birds drives rapid evolutionary changes in seed size. Science. 2013;340:1086–90.

García-Frapolli E, Toledo VM. Evaluación de sistemas socioecológicos en áreas protegidas: un instrumento desde la economía ecológica. Argumentos. 2008;21:103–16.

García-Frapolli E, Toledo VM, Martínez-Alier J. Apropiación de la naturaleza por una comunidad Maya yucateca: un análisis económico-ecológico. Rev Iberoam Econ Ecol. 2008;7:27–42.

Gill RB. Las grandes sequías mayas. Agua, vida y muerte. Fondo de Cultura Económica. Mexico, DF; 2008.

Gómez-Pompa A. Chapter 1. Research challenges for the lowland Maya area: an introduction. In: Gómez-Pompa A, Allen MF, Fedick SL, Jiménez-Osornio JJ, editors. The lowland maya area: Three millennia at the human-wildland interface. Binghamton, NY: Haworth Press; 2003.

Götz CM. Coastal and inland patterns of faunal exploitation in the prehispanic northern Maya lowlands. Quat Int. 2008;191:154–69.

Götz CM. La alimentación de los mayas prehispánicos vista desde la zooarqueología. An Antropol. 2014;48:167–99.

Hernández-Betancourt S, Segovia-Castillo A. La cacería de subsistencia en el sur del estado de Yucatán. In: Guerra M, Calmé S, Gallina S, Naranjo E, Coords. Uso y manejo de fauna silvestre en el norte de Mesoamérica. México: Gobierno del Estado de Veracruz, Instituto de Ecología A.C, El Colegio de la Frontera Sur; 2010.

Hernández-Díaz M, Ramírez-Barajas PJ, Chávez C, Schmook B, Calmé S. Presencia y abundancia relativa de carnívoros en una selva dañada por el huracán Dean (2007). Rev Mex Biodivers. 2012;83:790–801.

Jorgenson JP. Maya subsistence hunters in Quintana Roo, Mexico. Oryx. 1995;29:49–57.

Lechuga J. The feasibility of sport hunting as a wildlife conservation and sustainable development tool in southern Mexico. Master of Science Thesis. University of Florida. 2001.
León P, Montiel S. Wild Meat Use and traditional hunting practices in a rural Mayan community of the Yucatán Peninsula, Mexico. Hum Ecol. 2008;36:249–57.
Manzón-Che MJ. Uso de fauna silvestre en relación con las actividades productivas en dos comunidades forestales de Calakmul, Campeche, México. Honor's thesis. Instituto Tecnológico de Chetumal, Chetumal, Quintana Roo; 2010.
Ojasti J. Manejo de fauna silvestre Neotropical. Washington, DC: Smithsonian Institution/Man and Biosphere Program; 2000.
Peres CA. Synergistic effects of subsistence hunting and habitat fragmentation on Amazonian forest vertebrates. Conserv Biol. 2001;15:1490–505.
Peres CA, Barlow J, Laurance WF. Detecting anthropogenic disturbance in tropical forests. Trends Ecol Evol. 2006;21:227–9.
Pérez-Gil R, Jaramillo F, Muñiz AM, Torres MG. Importancia económica de los vertebrados silvestres de México. PG7 Consultores CONABIO, México; 1995.
Mandujano S, Rico-Gray V. Hunting, use, and knowledge of the biology of the white-tailed deer, Odocoileus virginianus (Hays), by the maya of central Yucatán, Mexico. J Ethnobiol. 1991;11:175–83.
Masson MA, Peraza CL. Animal use at the Postclassic Maya center of Mayapan. Quat Int. 2008;191:170–83.
Milner-Gulland EJ, Bennett EL, SAMWM Group. Wild meat: the bigger picture. Trends Ecol Evol. 2003;18:351–7.
Montiel S, Arias LM, Dickinson F. La cacería tradicional en el norte de Yucatán: una práctica comunitaria. Geografía Agrícola. 1999;29:42–52.
Morales C. Cacería de subsistencia en tres comunidades de la zona maya de México y Guatemala. Master Thesis, ECOSUR, Chetumal; 2000.
Nasi R, Taber A, van Vliet N. Empty forests, empty stomachs? Bushmeat and livelihoods in the Congo and Amazon Basins. Int Forest Rev. 2011;13:355–68.
Quijano-Hernández E, Calmé S. Patrones de cacería y conservación de la fauna silvestre en una comunidad Maya de Quintana Roo, México. Etnobiología. 2002;2:1–18.
Ramírez-Barajas PJ, Torrescano-Valle N, Tecpa A, Vázquez J. Importancia y uso del entorno natural en una comunidad indígena maya Petcacab, Quintana Roo, México. Tip Revista Especializada en Ciencias Químico-Biológicas. 2001;4:61–71.
Ramírez-Barajas PJ, Torrescano-Valle N, Chan-Rivas C. Diagnóstico del aprovechamiento de flora y fauna por los mayas del Ejido Petcacab y evaluación de la cacería y pesca, Felipe Carrillo Puerto, Quintana Roo. Sociedad de Productores Forestales Ejidales de Quintana Roo S.C. Informe final SNIB-CONABIO proyecto No. BJ008. México, DF; 2006.
Ramírez-Barajas PJ, Naranjo E. La cacería de subsistencia en una comunidad de la Zona Maya, Quintana Roo, México. Etnobiología. 2007;5:65–85.
Ramírez-Barajas PJ, Islebe GA, Calmé S. Impact of Hurricane Dean (2007) on game species of the Selva Maya. Biotropica. 2012a;44:402–11.
Ramírez-Barajas PJ, Islebe GA, Torrescano-Valle N. Perturbación post-huracán Dean en el hábitat y la abundancia relativa de vertebrados mayores en la Selva Maya, Quintana Roo, México. Rev Mex Biodivers. 2012b;83:1194–207.
Redford KH. The empty forest. Bioscience. 1992;42:412–22.
Reyna-Hurtado R, March I, Naranjo E, Mandujano S. Pecaríes en México. In: Valdéz R, Ortega JA, editors. Ecología y manejo de fauna silvestre en México. Colegio de Postgraduados and New Mexico State University. Texcoco, México; 2014.
Rodríguez M, Montiel S, Cervera MD, Castillo MT, Naranjo EJ. The practice and perception of batida (group hunting) in a Maya community of Yucatán, Mexico. J Ethnobiol. 2012;32:212–27.
Santos-Fita D, Naranjo E, Rangel-Salazar JL. Wildlife uses and hunting patterns in rural communities of the Yucatán Peninsula, Mexico. J Ethnobiol Ethnomed. 2012;8:31–17.

Santos-Fita D, Naranjo E, Bello-Baltazar E, Estrada-Lugo EI, Mariaca-Méndez R, Macario-Mendoza P. La milpa comedero-trampa como una estrategia de cacería tradicional maya. Estudios de cultura maya. 2013;XLII:89–118.
Segovia-Castillo A, Hernández-Betancourt S. La cacería de subsistencia en Tzucacab, Yucatán, México. Trop Subtrop Agroecosyst. 2003;2:49.
Segovia-Castillo A, Chablé-Santos J, Delfín-González H, Sosa-Escalante J, Hernández-Betancourt SF. Aprovechamiento de la fauna silvestre por comunidades mayas. In: Durán R, Méndez M, editors. Biodiversidad y Desarrollo Humano en Yucatán. CICY, PPD-FMAM, CONABIO, SEDUMA. 2010. p. 496.
Stearman AM. Cambio social, cacería y conservación en pueblos indígenas: puntos de conflicto y caminos hacia la resolución. In: Fang GT, Montenegro LO, Bodmer ER, editors. Manejo y conservación de la fauna silvestre en América Latina. Bolivia. 1999. p. 496.
Stearman AM. A pound of flesh: social change and modernization as factors in hunting sustainability among neotropical indigenous societies. In: Robinson JG, Bennett EL, editors. Hunting for sustainability in tropical forests. New York, NY: Columbia University Press; 2000. p. 582.
Terán S, Rasmussen CH. La milpa de los mayas. La agricultura de los mayas prehispánicos y actuales en el Noreste de Yucatán. Ministerio de Relaciones Exteriores de Dinamarca (DANIDA) – Centro de Investigación Científica de Yucatán, Mérida, Yucatán, México; 1994.
Thornton EK, Emery KF, Steadman DW, Speller C, Matheny R, Yang D. Earliest Mexican turkeys (Meleagris gallopavo) in the Maya region: implications for pre-Hispanic animal trade and the timing of turkey domestication. PLoS One. 2012;7, e42630.
Toledo VM. What is ethnoecology? Origins, scope and implications of a rising discipline. Etnoecológica. 1992;1:5–21.
Toledo VM, Barrera-Bassols N, García-Frapolli E, Alarcón-Chaires P. Uso múltiple y biodiversidad entre los Mayas Yucatecos (México). Interciencia. 2008;33:345–52.
Toledo VM. Prologo: Biodiversidad útil y uso múltiple en la Península de Yucatán. In: Elizondo C, López-Merlín D, editors. Las áreas comunitarias de conservación en Quintana Roo Corredor Biológico Mesoamericano México. México: Comisión Nacional para el Conocimiento y Uso de la Biodiversidad; 2009.
Weber M, Retana-Guiascón O. Usos generales de la fauna silvestre. In: Villalobos-Zapata GJ, Mendoza-Vega J, editors. La Biodiversidad en Campeche: Estudio de Estado. 2010. Comisión Nacional para el Conocimiento y Uso de la Biodiversidad (CONABIO), Gobierno del Estado de Campeche, Universidad Autónoma de Campeche, El Colegio de la Frontera Sur. 2010. p. 730.
Villa-Rojas A. Los elegidos de Dios: etnografía de los mayas de Quintana Roo. México, Instituto Nacional Indigenista. Serie Antropología Social 56; 1987.
Wilkie DS, Bennett EL, Peres CA, Cunningham AA. The empty forest revisited. Ann N Y Acad Sci. 2011;1223:120–8.
Wright SJ, Zeballos H, Dominguez I, Gallardo MM, Moreno MC, Ibánez R. Poachers alter mammal abundance, seed dispersal and seed predation in a Neotropical forest. Conserv Biol. 2000;14:227–39.

Part IV
Ecosystems and Conservation

Chapter 14
Transverse Coastal Corridor: From Freshwater Lakes to Coral Reefs Ecosystems

Héctor A. Hernández-Arana, Alejandro Vega-Zepeda, Miguel A. Ruíz-Zárate, Luisa I. Falcón-Álvarez, Haydée López-Adame, Jorge Herrera-Silveira, and Jerry Kaster

Abstract Understanding the processes and mechanisms that determine coastal connectivity may help us to propose and establish more effective strategies for maintaining the environmental goods and services from coastal ecosystems. The eastern Yucatán Peninsula, from inland to sea range from: karstic freshwater lakes, brackish lagoons, estuarine coastal lagoons to reef lagoons. These ecosystems coalesced to form the largest and complex coastal ecosystems of the Yucatán Peninsula. The spatio-temporal structure of this complex ecosystem can be used as a model system to study connectivity processes through its hydrological, bio-geochemical and ecological interactions. Based on its structure we propose a conceptual model, which we call the "Transverse Coastal Corridor" as a meaningful strategy for research, coastal development and conservation in the southeastern Yucatán Peninsula. The corridor includes the second largest freshwater karstic lake in Mexico, Laguna Bacalar; the largest estuarine-coastal lagoon, Chetumal Bay, the largest extant freshwater microbialite ecosystem and the structurally most complex coral reef ecosystem of the Mesoamerican Barrier Reef System. Physical, chemical and biological exchanges and interactions across the corridor of terrestrial,

H.A. Hernández-Arana (✉) • A. Vega-Zepeda • M.A. Ruíz-Zárate • H. López-Adame
Departamento de Sistemática y Ecología Acuática, El Colegio de la Frontera Sur unidad Chetumal, Av. Centenario Km. 5.5, CP 77014 Chetumal, Quintana Roo, Mexico
e-mail: hhernand@ecosur.mx; alvegaz@ecosur.mx; maruizzar@ecosur.mx; halopez@ecosur.edu.mx

L.I. Falcón-Álvarez
Instituto Nacional de Ecología, Universidad Nacional Autónoma de México, Cd. Universitaria, Circuito exterior s/n anexo al Jardín Botánico Exterior, México DF 70-275, Mexico
e-mail: falcon@ecologia.unam.mx

J. Herrera-Silveira
Departamento de Recursos del Mar, Centro de Investigación y de Estudios Avanzados del IPN unidad Mérida, Antigua Carretera a Progreso km 6.5, Mérida, Yucatán CP 97310, Mexico
e-mail: jherrera@mda.cinvestav.mx

J. Kaster
School of Freshwater Science, University of Wisconsin, Milwaukee, 600 East Greenfield Ave., Lampall Hall S287, P.O. Box 413, 53204 Milwaukee, WI, USA
e-mail: jlk@uwm.edu

G.A. Islebe et al. (eds.), *Biodiversity and Conservation of the Yucatán Peninsula*,
DOI 10.1007/978-3-319-06529-8_14

freshwater, marine environments drive a diverse and complex set of ecological processes that determine biodiversity and ecosystem functioning. We consider that the conceptual model of The Transverse Coastal Corridor may be more appealing to the different actors involved in the regional scale as it recognizes the particular biological and ecological characteristics of the interrelated ecosystems by seeking a balance from protection, management efforts, and sustainable use of the natural resources whilst preserving ecosystem functioning.

Keywords Coastal ecosystems • Connectivity • Transversal corridor • Coral reef • Sink holes • Mangroves • Microbialites • Hydrology • Biogeochemical processes • Ecological interactions

14.1 Coastal Ecosystem Connectivity

Coastal ecosystem connectivity is readily recognizable by environmental scientists although the quantification and determination of the mechanisms that define its structural patterns and associated connectivity processes are hard to measure. Understanding the processes and mechanisms that determine coastal connectivity may help us to propose and establish more effective strategies for maintaining the environmental goods and services from coastal ecosystems. Sometimes the connections are not evident, but there is a plethora of interrelationship between elements and processes of different ecosystems. The hydrologic connectivity is the most obvious for river, coastal and terrestrial ecosystems (Freeman et al. 2007). However, ecological connections are not always obvious and direct even inside just one specific ecosystem (e.g. Adam et al. 2011; Rogers et al. 2014; Schneider et al. 2014); and the consequences of historical natural disturbances and human influences are usually neglected (e.g. Jackson 1997; Jackson et al. 2001).

14.2 Uniqueness of Coastal Ecosystems in Southern Quintana Roo

The Yucatán Peninsula displays several features characteristic of karst landscapes such as sinkholes and freshwater lakes associated with the groundwater flow (Perry et al. 2009; Pérez-Ceballos et al. 2012). Freshwater lakes are distributed preferentially on the eastern side of the Yucatán with the largest ones in the southeast of the peninsula. These lakes share a similar geologic origin related to a geomorphic district originated by block faulting that created a series of block fault basins that extend from northern Q. Roo, south to Belize. Thus, the southern Q. Roo coastal lagoons complex is the expression of a geological fault system that runs from Holbox to Bacalar, known as the Rio Hondo fault. This fault's basis filled up due to the karstic environmental setting giving origin to a set of parallel coastal lagoons, including Chetumal Bay, running along a SW-NE axis. Some of these are freshwater lakes such is the case for Bacalar, Chile Verde, and Milagros being Bacalar lagoon the most extensive one.

The eastern Yucatán Peninsula, which is comprised by the state of Q. Roo, borders the Caribbean Sea and has outstanding coastal features such as floodplains, mangrove forests, coastal lagoons, and fringing reefs. The intricate connections of those ecosystems have resulted in unique coastal environments where transversal corridors allow the exchange of energy and matter as it occurs at northern, central and southern Q. Roo. These complex assemblages of ecosystems are unique in their own, but the one at southern Q. Roo stands out as the most extensive and intricately connected. Here, a well-developed reef ecosystem, several interconnected coastal lagoons, mangrove forests, floodplains and karstic freshwater lakes coalesced to form the largest and complex coastal ecosystems of the Yucatán Peninsula.

In a relatively small area of 459,102 ha, these ecosystems with singular characteristics, complex interconnections and with a traceable human influence form an important transverse corridor from inland through floodplains, karstic lakes estuarine-coastal lagoons and the southern Mexican Caribbean coast. This eco-complex can be conceptualized as a highly dynamic spatio-temporal structure where the degree of coupling based on hydrological connections drives the processes that sustain its ecological functioning.

14.3 Purpose of the Chapter

We propose the aforementioned transverse corridor to be used as a model system to study connectivity processes through its hydrological, biogeochemical and ecological interactions. Thus, the purpose of the chapter is to describe a series of ecosystems with singular characteristics and interconnections present in an area relatively small in the SE Yucatán Peninsula. We highlight the relevance of preserving the hydrological, biological and ecological connections because biodiversity, ecosystem functioning and the provision of ecosystems goods and services largely depends on connectivity processes. We also propose a conceptual model which we call the "Transverse Coastal Corridor" as a meaningful strategy for research, coastal development and conservation in the southeastern Yucatán Peninsula.

14.4 Description of Ecosystems Within "The Transverse Coastal Corridor"

The ecosystems represented in "The Transverse Coastal Corridor" from inland to sea range from: karstic freshwater lakes, brackish lagoons, estuarine coastal lagoons to reef lagoons. The corridor includes the second largest freshwater karstic

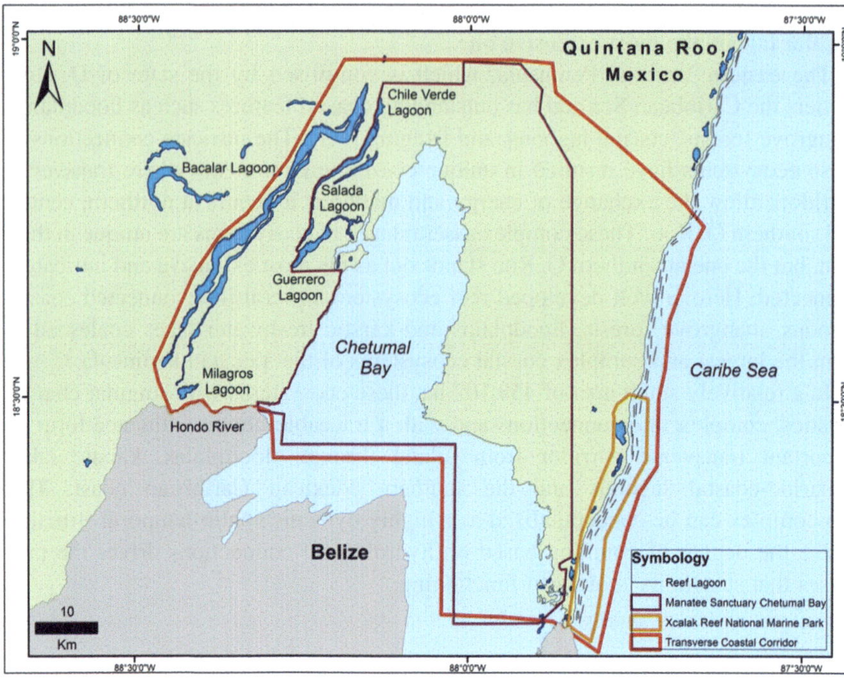

Fig. 14.1 Map depicting the location of the Transverse Coastal Corridor, showing the main water bodies and protected areas. Approximate area of 459,102 ha

lake in Mexico, Laguna Bacalar; the largest estuarine-coastal lagoon, Chetumal Bay, and the structurally most complex coral reef ecosystem of the Mesoamerican Barrier Reef System (Fig. 14.1).

14.4.1 Laguna Bacalar

Laguna Bacalar is an oligotrophic, nitrogen limited and carbonate because supersaturated system. Its ecology and biogeochemistry are starting to be understood of its natural beauty attracts tourists, which translate into commercial development. Along the southwestern shore of Laguna Bacalar five sinkholes are located, vertical karst expressions that transverse through the underground water system, with a recorded maximum depth of 90 m. The existence of strong conductivity, isotopic and radiochemical gradients shows great potential as untapped tool for quantifying the hydrologic balance critical for the understanding of the interactions between surface and groundwater (Klump et al. 2010). A remarkable feature of the biodiversity of freshwater lakes is reef building microbial communities; considered the largest recent microbialite structures known (Gischler et al. 2011). These authors recorded an extension of over 10 km of microbialites structures at the southern part

Fig. 14.2 (**a**) Microbialite formation in the Blue Cenote, sinkhole located at the western shore of Bacalar freshwater lake. Depth of the photograph 18 m (Photographer H. Bahena-Basave, ECOSUR). (**b**) Mangrove on top of stromatolite formation in southern Bacalar freshwater lake (Photographer H. Hernández-Arana, ECOSUR). (**c**) Ecological interaction of the black-striped mussel (*Mytilopsis sallei*) and stromatolites formations in southern Bacalar freshwater lake

of Laguna Bacalar. However, field observations (Hernández-Arana pers. comm.) indicate an even wider distribution of microbial structures, some of them of spectacular dimensions forming the walls of the sinkhole "The blue cenote" reaching depths of more than 20 m (Fig. 14.2). Microbialite structures are widely distributed from the freshwater lakes into the estuarine-coastal lagoon system of Chetumal Bay. Bacterial communities from microbialites are different and highly diverse compared to other karstic systems (Centeno et al. 2012). This microbial ecosystem could be the basis for the trophic network of Laguna Bacalar making it a unique ecosystem regarding energy flows.

14.4.2 Channel Systems

Laguna Bacalar communicates with Chetumal Bay, the estuarine-coastal lagoon, through two main systems of surface channels and flooding plains clearly identifiable. (1) Northern Bacalar that flows into Laguna Chile Verde and continues uninterruptedly into Laguna Guerrero until it reaches the north-western coast of Chetumal Bay. (2) Southern Bacalar that flows into Laguna Mariscal, follows into a tributary of Rio Hondo, feeds Laguna Milagros another freshwater lake, and exits to southern Chetumal Bay via the Rio Hondo. Both the northern and southern channel systems are permanently flooded and the flow intensity varies seasonally in relation to the amount of rainfall. Scientific information related to hydrology and biology of the channel systems and associated smaller karstic lakes is extremely limited. Direct field observations indicate some differences in water transparency at Laguna Chile Verde and Laguna Milagros as compared to Laguna Bacalar.

14.4.3 Underground Flow System

The information about the network of groundwater flows in the peninsula is localized in certain areas and the underground connection system is poorly understood (Pérez-Ceballos et al. 2012). It is likely that the southern Q. Roo groundwater network could be as complex as in other areas of the peninsula based on information related to the different depths of cenotes, freshwater Laguna's and brackish lagoons. The horizontal and vertical gradients and the marine/freshwater transition are related to its physical location at the junction of a hydraulic transitional seam and the Rio Hondo-Bacalar fault (Bauer-Gottwein et al. 2011) (see Fig. 2.2).

14.4.4 Chetumal Bay

Chetumal Bay, an estuarine-coastal lagoon ecosystem, is an extensive, relatively shallow and oligotrophic coastal lagoon (Espinoza-Ávalos et al. 2009). The system also presents the typical karstic expressions of sinkholes sparsely distributed within the main water body. This peculiar expressions range in depths of 12–40 m, with some seeping freshwater or marine water depending on its closeness to the sea. The bay presents a marked spatial gradient of salinity coupled with a large seasonal oscillation as a result of a varying degree of marine influence and the supply of freshwater through riverine input, non-point runoff and underground seepage. Its spatial and temporal gradients of salinity favors a wide spatial and temporal variation in biodiversity ranging from dominant marine benthic assemblages (Hernández-Arana and Ameneyro-Angeles 2011) through highly tolerant marine and freshwater benthic biota on its western shores (Fig. 14.3). Its trophic status is also spatio-temporally variable from oligotrophic to eutrophic depending on the inter-annual climatic variation and its interaction with the riverine runoff and human influence (Herrera-Silveira et al. 2009). Based on a preliminary trophic model of Chetumal Bay, this estuarine-coastal lagoon is notoriously poor in submerged macrophyte vegetation, despite the fact that it supports a considerable population of the endangered Caribbean manatee (Castelblanco-Martínez et al. 2011). Biodiversity, distribution, and biomass of submerged vegetation are also driven by the degree of freshwater-brackishwater interactions (Fig. 14.4). The trophic model suggested as well that the system might be detritus controlled and

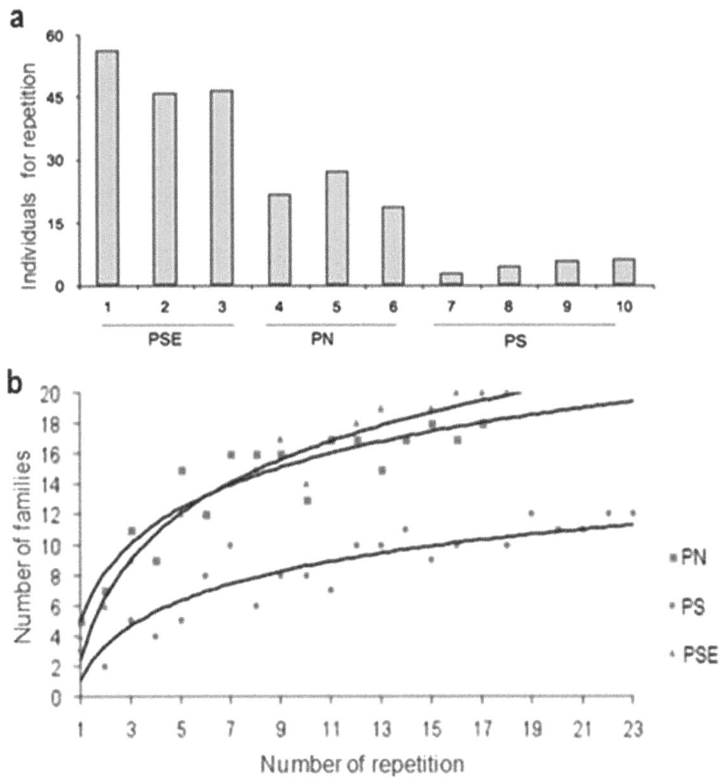

Fig. 14.3 Benthic biota distribution according to seawater influence in three portions of Chetumal Bay. (**a**) Density of benthic biota. (**b**) Diversity of families of benthic biota. The locality PSE has the highest density and diversity of benthic biota. *PSE* southeast portion with marine influence, *PS* south portion with freshwater influence, *PN* north portion with brackish condition (Unpublished data from Hernández-Arana)

highlight unknown issues related to detritus dynamics, the exchange of matter between adjacent systems, and the seagrass and mangrove production.

14.4.5 Mangrove Ecosystem

One of the most notable ecological attributes present across the entire coastal ecosystem in southern Q. Roo is the extensive mangrove ecosystem that structurally defines the primary vegetation associated with the floodplains, freshwater karstic lakes, brackish lagoons and most of the shorelines of the Caribbean sea. The mangrove system contributes to the functional processes of soil formation, organic matter production and remineralisation. Estimates for mangrove area cover that corresponds to the two municipalities in southern Q. Roo is approximately

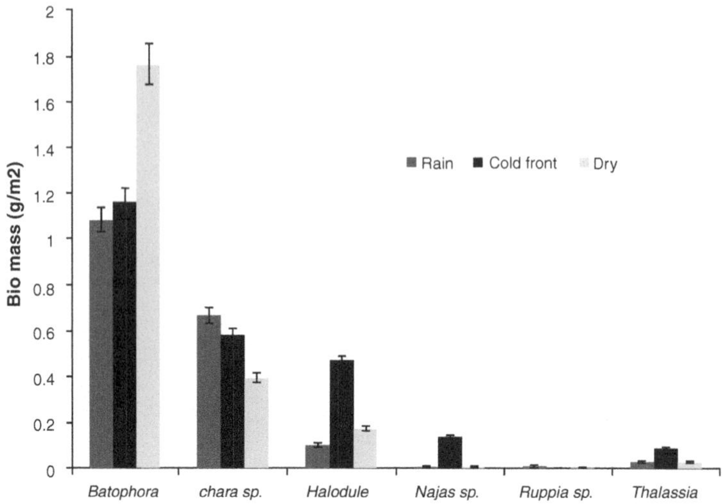

Fig. 14.4 Total dry biomass of six macrophytes species found in Chetumal Bay during three climatic seasons. *Batophora* sp. *Halodule* sp. and *Thalassia* sp. distributes from marine to brackish environments. *Chara* sp. *Najas* sp. and *Ruppia* sp. distributes from freshwater to brackish environments (Unpublished data used with permission from Castelblanco-Martinez D.N.)

59,268 ha (SEMARNAT-CONAFOR 2014), however this information relates to a variable association of mangrove with tropical forest vegetation which can achieve a variety of assemblages (Fig. 14.5). The main mangrove development in terms of forest structure in southern Q. Roo is located along the shorelines of the Caribbean Sea and Chetumal Bay. The mangrove association is composed of four species *Rhizophora mangle* (L. 1753), *Laguncularia racemosa* ((L.) C. F. Gaetn 1807), *Avicennia germinans* ((L.) L. 1764) and the false mangrove *Conocarpus erectus* (L. 1753) (Ellison et al. 2010). The mangrove association is reduced to two species (*R. mangle* and *C. erectus*) at the shorelines of karstic freshwater systems, interconnecting channels and associated floodplains. Large extensions in southern Q. Roo, an approximate area of 7500 ha, are covered by the dwarf variety of *R. mangle*, adapted to extreme conditions of prolonged flooded and dry periods, and also to shallow and poor soils. In Laguna Bacalar, extensive growths of mangroves exist on the stromatolites, i.e. using the carbonate deposits as a growth substrate (Fig. 14.2b). The data from some sites along the western coast of Chetumal Bay shows that litter production responds to hydrological characteristics (Herrera-Silveira et al. 2013), and study sites reflect the hydrological variation that can be found along the western coast of Chetumal Bay. Other crucial aspects that influence overall litter production and mangrove structure are tropical storms and hurricanes (Adame et al. 2012) that are major climatic drivers of ecosystem structure, processes, and function. For Chetumal Bay area, frequently storms and hurricanes of moderate intensity, together with low interstitial salinity and low total phosphorus due to its precipitation by supersaturated calcium carbonate

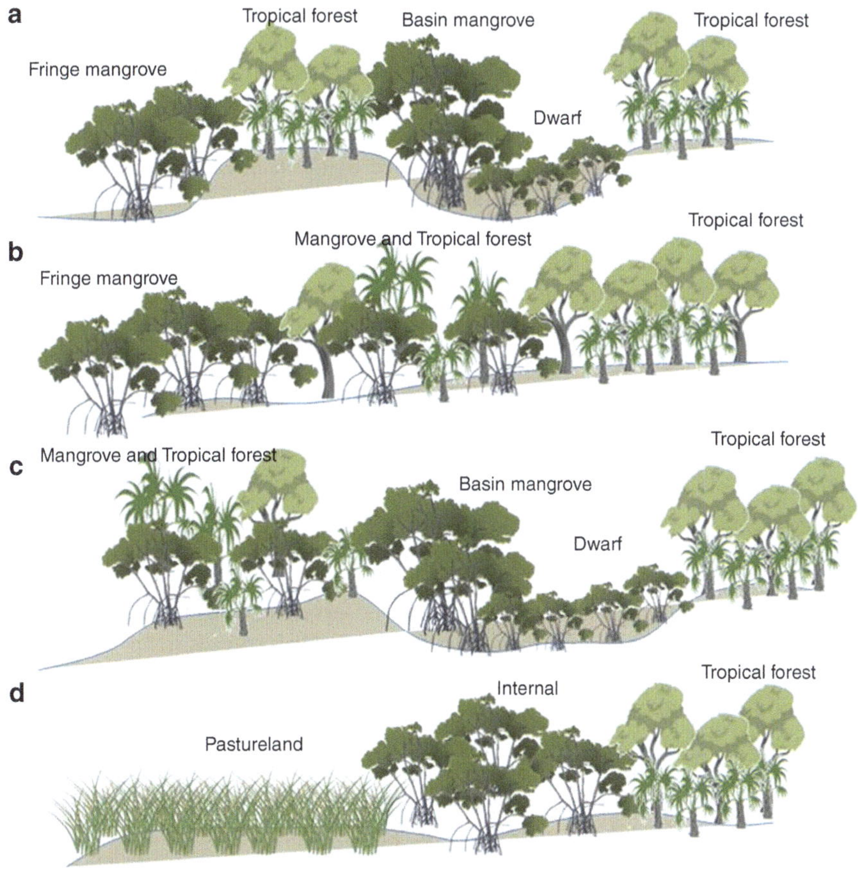

Fig. 14.5 Structural types of mangrove and tropical forest assemblages along the Transverse Coastal Corridor

environments lead to a low litterfall (1.96 ± 0.23 g/m^2 d) compared to other areas in the Yucatán Peninsula (Adame et al. 2012).

14.4.6 Coral Reefs

The coral reef system in the Mexican Caribbean represents the northern portion of the Mesoamerican Barrier Reef System, and goes from the Belize border up to Cabo Catoche in northern Quintana Roo's state's coast. The area of influence of the transverse coastal corridor on these coral reefs probably extends from the southern portion of the Sian Ka'an Biosphere Reserve to the northern extreme of the Belize reef system (Fig. 14.1). In terms of its physiographic coastal characteristics and differences in the reef structure, this reef development has been identified as the central and southern areas of the Mexican Caribbean (Ruiz-Zárate et al. 2003; Ruiz-Zárate and Arias-González 2004; Rodríguez-Zaragoza and Arias-González 2015).

These areas present a greater coastal reef development, the highest coral richness, the greatest coral cover, and the highest topographic complexity in contrast to the northern area, with stony coral dominance of *Orbicella annularis* (Ellis and Solander 1786), *O. faveolata* (Ellis and Solander 1786), *Pseudodiploria strigosa* (Dana 1846) and *Siderastrea* spp. (Vaughan and Wells 1943), *Agaricia agaricites* (Linnaeus 1758), and *A. tenuifolia* (Dana 1848) (Rodríguez-Zaragoza and Arias-González 2015). However, some particular reefs can present differences, like the Uvero-Punta Piedras section, where the coral cover is lower, macroalgae are dominant, and the coral community is mainly represented by *A. agaricites*, *S. siderea* and *Porites astreoides* (Lamark 1816) (Figueroa-Zavala et al. 2015).

The reef fish community is dominated in terms of abundance by families typical of the Caribbean reefs: Labridae, Scaridae, Pomacentridae and Acanthuridae (Nuñez-Lara et al. 2005). However, geomorphological structure, "reefscape" attributes at different scales, and depth are important variables for shaping beta/delta diversity in this areas and the entire Mexican Caribbean (Arias-González et al. 2008). The reefscape features (reef area, live coral cover, and habitat complexity), are the main factors influencing the observed biogeographic patterns of fish and coral biodiversity in this areas with coral and fish species richness strongly correlate (Arias-González et al. 2011). There are several coastal features, usually not mentioned in the literature such as water drainages from coastal lagoons and springs along the reef lagoons with unknown effects over the reef biota and coastal water quality. Evidences of a study done in northern Q. Roo on the benthic community structure and water quality analyses in two springs in a reef lagoon indicates significant differences between benthic communities next to springs and water quality in a distance gradient. Besides, possible contaminants like Cu and Fe were detected in sediments and *Siderastrea* sp. tissue in the same distance gradient (Rosado-Torres 2013). Drainages from coastal lagoons and freshwater springs in reef lagoons highlight the need for quantifying and understanding the coastal ecosystems interconnectivity processes as the basis for management and conservation of coral reef ecosystems.

14.5 Relevance of Preserving Connectivity Processes

14.5.1 Emblematic Interconnections and Ecological Interactions

The biogeochemical and hydrogeological processes within this region drive a complex biological system that sustains a range of human livelihoods that rely on the use and exploitation of natural resources. Besides, this environmental setting favors the coexistence of interconnected unique aquatic ecosystems from mangroves, microbialites to coral reefs. The horizontal and vertical gradients and the marine/freshwater transition facilitate a range of biological and ecological interactions. The intricate network of permanently and temporarily flooded channels facilitates biological connections and allows for the transportation, movement and establishment of biota from freshwater, brackishwater and marine habitats. Prominent examples include: Tarpon (*Megalops atlanticus* (Valenciennes 1847)) occurs in karstic sinkholes (cenotes) as well as records of larvae, juveniles and adults in reefs, lagoons and wetlands along the Mexican Caribbean coast (Schmitter-Soto et al. 2002), including the transverse coastal corridor. Cichlid fishes like *Cichlasoma* sp. with a distribution that ranges from the freshwater lagoons to the western shore of Chetumal Bay. The white turtle *Dermatemys mawii* (Gray 1847), the largest freshwater turtle in Mexico and listed as an endangered species is distributed in all of the freshwater lagoons, rivers and creeks that border the western shores of Chetumal Bay. The apple snail *Pomacea* sp. is distributed throughout the freshwater lakes reaching brackish environments like Laguna Guerrero. *Mytilopsis sallei* (Recluz 1894), the black striped mussel, native in the Caribbean and adapted to brackish and freshwater environments from Chetumal Bay into Laguna Bacalar and neighboring sinkholes. The degree of marine influence is one of the drivers for species distribution and establishment (Fig. 14.6). Salinity patterns along the western shore of Chetumal Bay ranges from 5 to 15 practical salinity units (psu) during the wet season and 5–20 psu during the dry season (Carrillo et al. 2009). Records of salinity in the south of Laguna Guerrero indicate that salinity during the dry season reaches 5 psu, the limit of marine influence hence the distributional limit of freshwater biota. Some other examples of biological connections are represented by highly mobile species with a wider distribution from the reef ecosystems and well into the brackish water and freshwater lakes. For example fish that use the

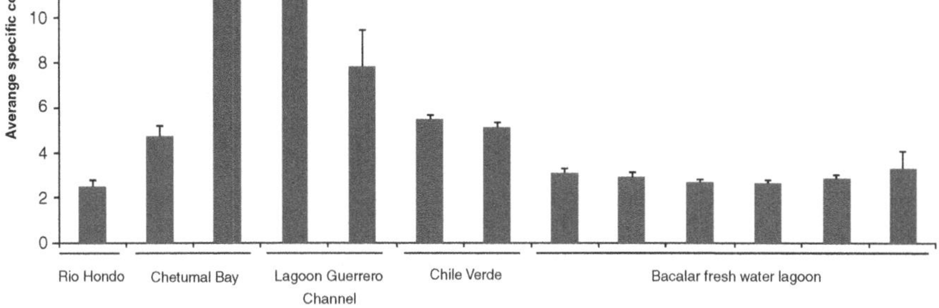

Fig. 14.6 Average specific conductivity in different areas along the Transversal Coastal Corridor

system as feeding grounds and shelter; clearly mangrove systems in the Caribbean strongly influence the community structure of fish on neighboring coral reefs and, the biomass of several commercially important species is more than doubled when adult habitat is connected to mangroves (Mumby et al. 2004); crocodiles that modify the landscape and build pathways for water and matter flows, are considered being ecosystem engineers (Naiman and Rogers 1997; Moore 2006).

14.5.2 Natural and Human Threats that Affect Interconnections and Ecological Interactions

The Yucatán Peninsula is an extensive carbonate platform formed of Quaternary (Holocene-Pleistocene) sediments. Due to the carbonate origin of the surface rocks carbonate dissolution process forms conduits and caverns trough that groundwater flows. Thus, the karstic aquifer is the main water reservoir for the peninsula and constitutes a defining input for the region's economic development. Management of water resources has long had a major influence on the Mayan culture and economic development of the Yucatán. Extensive use of groundwater allowed the Mayan culture to flourish in the northern Yucatán. Given the often severe drought conditions and its threats to agriculture the Mayan priests prayed to Chac, the water god, for assistance in water management primarily to decrease the severity of droughts. The Spaniards arrived in 1517 and augmented water availability by digging wells, which remained a common practice for more than 300 years. Many of the wells have been abandoned by now because of serious pollution problems resulting from the sewage disposal well adjacent to many supply wells. The most recent phase of water management began in 1959 when the Ministry of Hydraulic Resources was charged with the responsibility for both scientific investigations and the implementation of programs for water-supply and sewage-disposal systems for cities, villages and islands (Back and Lesser 1981). Today in many parts of the Yucatán Peninsula, and specifically in the southern Q. Roo area, the supply of freshwater has been assumed to be abundant and adequate for local needs. The fragility of this system, however, has become increasingly recognized as development of cities and tourism accelerates the demand for a sustainable supply of freshwater and the concomitant need for sewage treatment and disposal. Expanding urban and tourist coastal development, both current and projected, is increasing the demand for water. The over extraction of the aquifer, combined with the practice of sewage disposal by injection into the saline aquifer below the freshwater lens may cause local salinization of the freshwater lens via saltwater intrusion and the destabilization of the halocline density barrier between the two aquifers. As a result, the area is rated at the highest conservation priority, for freshwater supply, of the entire Caribbean ring (Huggins et al. 2007) indicating high vulnerability of the freshwater reservoir to human activities.

Quintana Roo state contributes over one third of the total national income from its tourism revenues (SEDETUR 2015). The state relies on its natural attractions (mainly sandy beaches and coral reefs) to boost its local economy based on the tourist industry. It started in the late 1980s with Cancun in the northern part of the state, followed, in the late 1990s by the promotion of the Riviera Maya. At the beginning of the twenty-first century federal and state government have made explicit its coastal development plans for Southern Q. Roo. From the lessons learned in northern Q. Roo society is now aware that coastal development comes with many negative impacts on ecosystem goods and services. The lack of understanding of the relevance of the connectivity processes has led to an overall deterioration of the coastal ecosystems in northern Q. Roo. The direct and visible effects are deforestation of tropical forest and mangrove forest, fill up of floodplains for construction, alterations of the local hydrological regime, dunes and beach erosion, overexploitation of natural resources, structural and functional changes in coral reef ecosystems (Calderón-Aguilera et al. 2012). For example, 160 km of the Yucatán Peninsula coastline of the 600 km facing the Caribbean Sea show permanent damages associated with 30 years of coastal development, including growing infrastructure for urban services and tourism resorts, increasing pollution, and an increased demand for natural resources (Murray 2007; Calderón-Aguilera et al. 2012). Affected ecosystems ranged from tropical forest, mangrove forest, floodplains, dunes and sandy beaches, coastal lagoons and coral reefs ecosystems.

There is consensus, but with a big uncertainty, that all natural ecosystems are threatened by climate change and anthropogenic activities (Hoegh-Guldberg and Bruno 2010; Kerr and Dobrowski 2013). However, threats are typically assessed and prioritized via expert opinion workshops that often leave no record of the rationale for decisions, making it difficult to update recommendations with new information (Halpern et al. 2007). To our knowledge this is the first formal attempt to identify the natural and anthropogenic threats in the transverse coastal corridor, which are based on personal observations, research reports and grey-literature that provide data of some of the ecosystems. The natural threats mainly include the impact of storms and hurricanes (Fig. 14.7) that alter and modify the hydrodynamic of the watershed and groundwater; this thread has relevant implication for the connectivity of coastal ecosystems and also in the drinking water supply for cities and rural communities. Besides, the strength of storms and hurricanes has structured coral reefs, mangroves and tropical forest ecosystems in the Yucatán Peninsula (Whigham et al. 1991; Gardner et al. 2005; Farfán et al. 2014). For example, accumulated damaged estimated via tree mortality from the impact of hurricane Dean in 2007 (category 5 on the Saffir-Simpson scale), Tropical storm Karl in 2010 and Category 1 hurricane Ernest in 2012 was 1753 dead trees in 55 ha in the Northwestern shore of Chetumal Bay (López-Adame and Hernández-Arana unpublished data). Hurricane damage on coral reefs, mainly on coral cover, depends on hurricane intensity and time elapsed since the last impact (Gardner et al. 2005). However, hurricane Gilbert had a very low impact over the Cozumel reef despite of its category 5 classification (Fenner 1991), but after hurricanes

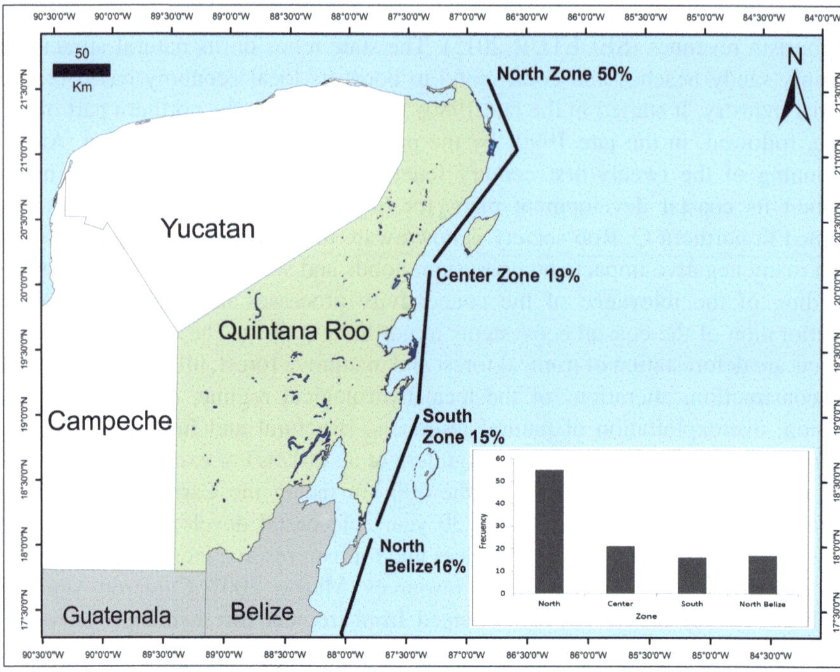

Fig. 14.7 Frequency of storms and hurricanes from 1850 to 2010 along the Mexican Caribbean and North of Belize Coast

Emily and Wilma the coral cover had a cumulative decline of 56 % in living coral cover (Álvarez-Filip et al. 2009).

Empirical evidence shows that the capacity of the system to return to its previous state is seriously threatened by the interaction of human and natural disturbances (Salazar-Vallejo 2002; Ruiz-Zárate et al. 2003; Nuñez-Lara et al. 2005). Recurrent patterns of natural disturbance along with human induced disturbance are directly affecting the capacity of coastal regions to return to previous states and resulting in new ecosystem structure trajectories not seen before, particularly for coral reef ecosystems (Aronson et al. 2002; Mumby 2006). Facing the threats of global and climate change requires new fundamental understanding of the connectivity and drivers that operate across the gradient of coastal ecosystems. According to the IPCC 2013 report, sea level for the Caribbean will increase in average 50 cm (within a range of 20–80 cm) by the end of the century (Church et al. 2013). The central and southern Q. Roo coastal areas are considered critical areas in México where sea level rise is very likely to impact, since this region lies within one of the most active hurricane corridors in the Caribbean, and storm surges, sea level rise, and saltwater intrusion/inundation are likely to have major impacts within the next several decades.

14.5.3 Research Priorities and Approaches to Understand Connectivity Processes, Ecological Interactions and Ecosystem Functioning

Physical, chemical and biological exchanges and interactions across the corridor of terrestrial, freshwater, marine environments drive a diverse and complex set of ecological processes. Scientific research efforts along the southern coastal corridor ecosystems have been valuable but limited in scope as they focused on particular ecosystems. It is necessary to produce new and essential understandings of the connectivity and the drivers that operate across this gradient of coastal ecosystems. The basic hydrology and hydrological budgets of this system have not been systematically studied. Nutrient budgets and nutrient load thresholds are completely unknown, but with projected increases in runoff and discharges, the freshwater lakes will likely reach a tipping point in ecological/biogeochemical status and function. The information about the groundwater flow network in the peninsula is localized in certain areas and the underground connection system is poorly understood (Pérez-Ceballos et al. 2012). It is likely that the southern Q. Roo groundwater network is as complex as in other areas of the peninsula based on information related to the different depths of sinkholes, freshwater Laguna's and brackish lagoons (Bauer-Gottwein et al. 2011).

Little was known of the stromatolites (*thrombolytic microbialites*) of Laguna Bacalar until Kaster (2007) brought them wider attention as being especially unique and thus coined Laguna Bacalar as Mexico's "world class lake" because of their large size and extensive distribution in the Laguna. Importantly, extant microbialite presence has been most often identified in a variety of environments (Domínguez-Escobar et al. 2011; Beltrán et al. 2012; Centeno et al. 2012) including harsh (e.g., hyper-saline, or high alkalinity) geochemical environments (e.g., Des Marais 2003; Garcia-Pichel et al. 2004; Kazmierczak and Kempe 2006), where conditions preclude the presence of metazoan grazers that disrupt microbialite development. It is theorized that environmental exclusion of these grazers is one of the factors necessary for modern microbialite formation. The geochemistry of these extreme environments is of particular interest given the potential insights into early Earth oceanic chemistry (Kazmierczak and Kempe 2006). Laguna Bacalar is one of a handful of extant limnological settings that are not immediately identifiable as either an extreme or harsh environment for microbial or metazoan life. As such, the presence of thrombolites in Laguna Bacalar presents an important opportunity to test the aforementioned theories about aquatic conditions that are conducive to the morphogenesis of microbialites. Furthermore, the broad coverage of microbialites in Laguna Bacalar will allow for hypothesis testing across similar and varying environmental conditions.

An important ecological interaction is the space/food/other competition between the stromatolites and the black-striped dreissenid mussel *Mytilopsis sallei*. A primary concern is that increased urban and coastal development to the area will increase nutrient flow into the Laguna, which will then reach the tipping point for increasing nutrients, accelerating phytoplankton growth, and spurring excessive mussel growth with the potential to blanket and kill or suppress the Laguna's stromatolites (Kaster 2012). This mussel is originally a native marine group with circum-Caribbean distribution that invaded Laguna Bacalar. The time frame of the invasion is unknown but could have happened thousands of years ago. Stromatolites grow so fast (relative to mussel colonization) that these mussels are often encapsulated and preserved within the interior of the stromatolite thus providing a possible historical record. In Laguna Bacalar, the black-striped mussel speckles the stromatolites with growth (up to 3000/m^2); however, they appear to be held in check by the Laguna's oligotrophic nutrient regime that is directly connected to the land-ocean transverse corridors transecting Laguna Bacalar (Fig. 14.2c).

Mangroves are among the most carbon-rich forests in the tropics (Donato et al. 2011), and below ground carbon (C) pools in mangrove forests are an important component of global carbon pools, storing 3–5× more C than any other tropical forested ecosystem. High rates of tree growth coupled with low-oxygenated, waterlogged soils result in massive and long term stores of C. Identification of mangroves that are more resilient to the impacts of climate change are a priority for conservation and restoration initiatives globally (Donato et al. 2011). The shear area covered by the dwarf *R. mangle* variety makes the transverse coastal corridor a potential area for large quantities of C sequestration. Similarly, coral reefs are known to contribute as sink ecosystems for C storage at a scale of millennia. The key ecosystems mangrove, coral reefs and microbialites; within this ecological corridor, may contribute substantially as C sink. Therefore, measuring and understanding the carbon flow and paths within the corridor is essential to provide the economic and environmental value of ecosystem functioning. The impact of future climate change is likewise an unknown, and it is imperative to provide more realistic scenarios of global and climatic change forcing, such as the sea level and sea surface temperature rise and its effect on coastal ecosystems. To achieve these scenarios it is necessary to gather information on ecosystems processes such as growth potential of structural and functional relevant ecosystems such as those described above. At the marine end member of the gradient of ecosystems, research on the coral reef ecosystems urgently requires a long term monitoring scheme for continuous observations of structural changes in communities as well as changes in recruitment patterns of hard coral species aiming to understand the main drivers of the present coral reefs decline in the Mexican Caribbean (Hernández-Arana et al. 2014).

14.5.4 The Conceptual Model and the Managing Challenges of this Eco-complex Transverse Coastal Corridor

The Transverse Coastal Corridor of southern Q. Roo epitomizes one of the great challenges of our time—understanding and managing socio-ecological systems.

The way in which urban and coastal development interacts with water and natural resources of the area is an economic, ecological, and environmental management challenge. Natural resource management and conservation requires a multidimensional approach and its methods are in constant debate (Gavin et al. 2015). Independently of that debate and particularly to the transverse coastal corridor the applied management and conservation strategies need to consider that connectivity process are likely to be the determinant process for ecosystem biodiversity and functioning. Thus, a scientific understanding of the connectivity processes may help us to establish a quantitative basis for understanding and managing the interactions between human activity and vulnerable environmental systems. This characteristic is challenging because there are several ecosystems involved with tight connections and with ancient evolving biotic communities (like stromatolites, mangrove and coral reefs), and different natural resources exploited for different purposes Moreover, the area is subjected to a variety of governmental jurisdictions where different levels of government actors are involved in the natural resources management and conservation.

Sustainable development of coastal zones and the conservation of dynamic habitats is nowadays a common language among resources managers and governmental agencies. However, to achieve such an elusive goal we need to measure and understand the response of coastal ecosystems to external pressures, whilst maintaining ecosystems functioning. The coastal ecosystems of the eastern Yucatán Peninsula appear to us ideal for the role they play in translating and transporting energy and matter. The spatial structure of coastal ecosystems in the Mexican Caribbean can be represented based on a generalized conceptual model of Transverse Coastal Corridors (Fig. 14.8), in which complexity and extension increase from north to south, coincidently with a coastal development gradient from highly to relatively undeveloped at the south. However, with continued tourism development, its impact begins to spill over to the southeastern Yucatán.

Although, inside the corridor there are two natural protected areas, one established by federal decree (Xcalak Reefs Marine Park) and the other by the State government (The Manatee Sanctuary in the Chetumal Bay), both representing around the 71 % of the total area of the transverse coastal corridor. One third of the corridor has no protection status (including Laguna Bacalar) and society is reluctant to implement any protection status due to the restrictive policy of resource use for natural protected areas in Mexico. Despite that Natural Protected Areas have been widely adopted as a leading tool for achieving conservation targets, its effectiveness in achieving their goals and objectives remain controversial, especially its effectivity for preventing habitat degradation (McClanahan et al. 2006; Selig and Bruno 2010; Dalton et al. 2015), particularly for Marine Protected Areas. We consider that the conceptual model of The Transverse Coastal Corridor may be more appealing to the different actors involved in the regional scale as it recognizes the particular biological and ecological characteristics of the interrelated ecosystems by seeking a balance from protection, management efforts, and sustainable use of the natural resources whilst preserving ecosystem functioning. We aim at a meaningful and useful strategy that can contribute to develop a different management instrument

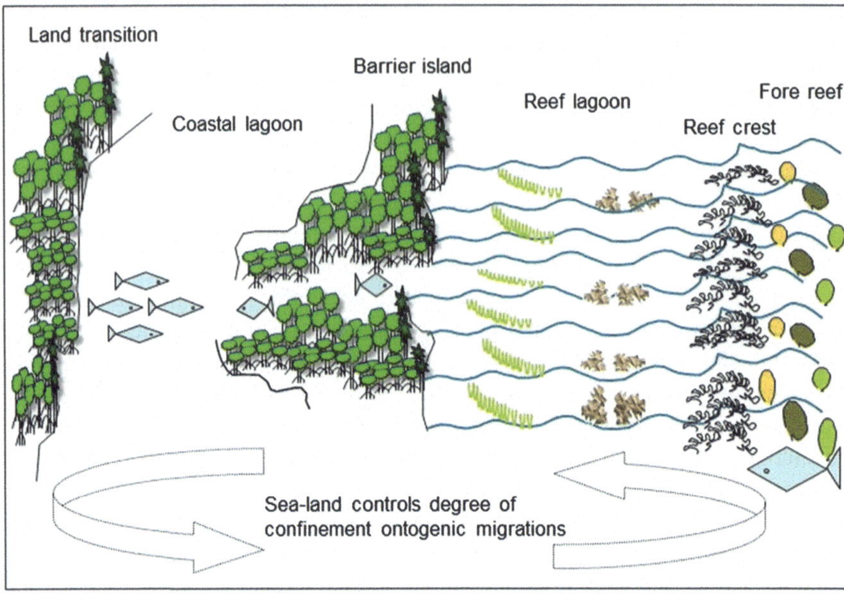

Fig. 14.8 Generalized structure of coastal ecosystems in the Mexican Caribbean, representing the model of the transverse coastal corridor

that goes beyond the present managing instruments such as Natural Protected Areas and geopolitical boundaries. Our collective goal is to further the development of a concrete, long term research program that will open an entirely new avenue for interlacing scientific researchers, conservation groups, municipal leaders, and the local community for promoting regional, national and international cooperation and advancing scientific studies in this globally unique biome.

Acknowledgements We thank to G. Alfaro-Padilla for map and figures edition, DN Castelblanco-Martinez for kindly providing data for Fig. 14.4 and to E. Ramirez-Canul for generate the Fig. 14.5.

References

Adam TC, Schmitt RJ, Holbrook SJ, Brooks AJ, Edmunds PJ, Carpenter RC, Bernardi G. Herbivory, connectivity, and ecosystem resilience: response of a coral reef to a large-scale perturbation. PLoS One. 2011;6(8), e23717. doi:10.1371/journal.pone.0023717.

Adame MF, Zaldívar-Jiménez A, Teutli C, Caamal JP, Andueza MT, López-Adame H, Cano R, Hernández-Arana HA, Torres-Lara R, Herrera-Silveira J. Drivers of mangrove litterfall within a karstic region affected by frequent hurricanes. Biotropica. 2012. doi:10.1111/btp.12000.

Álvarez-Filip L, Millet-Encalada M, Reyes-Bonilla H. Impact of hurricanes Emily and Wilma on the coral community of Cozumel Island, Mexico. Bull Mar Sci. 2009;84:295–306.

Arias-González JE, Legendre P, Rodríguez-Zaragoza FA. Scaling up beta diversity on Caribbean coral reefs. J Exp Mar Biol Ecol. 2008;366:28–36.
Arias-González JE, Núñez-Lara E, Rodríguez-Zaragoza FA, Legendre P. Reefscape proxies for the conservation of Caribbean coral reef biodiversity. Ciencias Marinas. 2011;37:87–96.
Aronson RB, Macintyre IG, Precht WF, Murdoch TJT, Wapnick CM. The expanding scale of species turnover events on coral reefs in Belize. Ecol Monogr. 2002;72:233–49. doi:10.1890/0012-9615(2002)072[0233:TESOST]2.0.CO;2.
Back W, Lesser JM. Chemical constraints of groundwater management in the Yucatán Peninsula, Mexico. J Hydrol. 1981;51:119–30.
Bauer-Gottwein P, Gondwe BRN, Charvet G, Marín LE, Rebolledo-Vieyra M, Merediz-Alonso G. Review: the Yucatán Peninsula karst aquifer, Mexico. Hydrogeol J. 2011. doi:10.1007/s10040-010-0699-5.
Beltrán Y, Centeno CM, García-Oliva F, Legendre P, Falcón LI. Nitrogenase activity and associated diversity of microbialite and mat-forming communities from different environments in Mexico. AME. 2012;67:15–24.
Calderón-Aguilera LE, Rivera-Monroy VH, Porter-Bolland L, Martínez-Yrízar A, Ladah LB, Martínez-Ramos M, Alcocer J, Santiago-Pérez AL, Hernández-Arana HA, Reyes-Gómez VM, Pérez-Salicrup D. An assessment of natural and human disturbance effects on Mexican ecosystems: current trends and research gaps. Biodivers Conserv. 2012;21:589–617.
Carrillo L, Ramírez AM, Morales JB. Características hidrometeorológicas y batimétricas. In: Espinoza-Avalos J, Islebe GA, Hernández-Arana HA, editors. El sistema ecológico de la bahía de Chetumal/Corozal: costa occidental del Mar Caribe. El Colegio de la Frontera Surl; 2009.
Castelblanco-Martínez DN, Barba-Macias E, Schmitter-Soto JJ, Hernández-Arana HA, Morales-Vela B. The trophic role of the endangered Caribbean manatee Trichechus manatus in an estuary with low abundance of seagrass. Estuar Coasts. 2011. doi:10.1007/s12237-011-9420-8.
Centeno CM, Legendre P, Beltrán Y, Alcántara-Hernández RJ, Lidstrom U, Ashby MN, Falcón LI. Patterns in microbialite genetic composition relate to environmental variables. FEMS Microbiol Ecol. 2012;82:724–35.
Church JA, Clark PU, Cazenave A, Gregory JM, Jevrejeva S, Levermann A, Merrifield MA, Milne GA, Nerem RS, Nunn PD, Payne AJ, Pfeffer WT, Stammer D, Unnikrishnan AS. Sea Level Change. In: Stocker TF, Qin D, Plattner G-K, Tignor M, Allen SK, Boschung J, Nauels A, Xia Y, Bex V, Midgley PM, editors. Climate Change 2013: The Physical Science Basis, Contribution of Working Group I to the Fifth Assessment Report of the Intergovernmental Panel on Climate Change. Cambridge: Cambridge University Press; 2013.
Dalton T, Forrester G, Pollnac R. Are Caribbean MPAs making progress toward their goals and objectives? Mar Policy. 2015;54:69–76.
Des Marais DJ. The biogeochemistry of hypersaline microbial mats illustrates the dynamics of modern microbial ecosystems and the early evolution of the biosphere. Biol Bull. 2003;204:160–7.
Domínguez-Escobar J, Beltrán Y, Bergman B, Díez B, Ininbergs K, Souza V, Falcón LI. Phylogenetic and molecular clock inferences of cyanobacteria within Rivulariaceae from distant environments. FEMS Microbiol Lett. 2011;316:90–9.
Donato DC, Kauffman JB, Murdiyarso D, Kurnianto S, Stidham M, Kanninen M. Mangroves among the most carbon-rich forests in the tropics. Nat Geosci. 2011;4:293–7.
Ellison A, Farnsworth E, Moore G. *Conocarpus erectus*. The IUCN Red List of Threatened Species; 2010.
Espinoza-Ávalos J, Islebe GA, Hernández-Arana HA, editors. El sistema ecológico de la bahía de Chetumal/Corozal: costa occidental del Mar Caribe, México; 2009. p. 252.
Farfán LM, D'Sa EJ, Liu K, Rivera-Monroy VH. Tropical cyclone impacts on coastal regions: the case of the Yucatán and the Baja California Peninsulas, Mexico. Estuar Coasts. 2014. doi:10.1007/s12237-014-9797-2.

Fenner DP. Effects of hurricane Gilbert on coral reefs, fishes and sponges at Cozumel, Mexico. Bull Mar Sci. 1991;48:719–30.
Figueroa-Zavala B, Correa-Sandoval J, Ruiz-Zárate MA, Weissenberger H, González-Solís D. Environmental and socioeconomic assessment of a poorly known coastal section in the southern Mexican Caribbean. Ocean Coast Manag. 2015;110:25–37.
Freeman MC, Pringle CM, Jackson CR. Hydrologic connectivity and the contribution of stream headwaters to ecological integrity at regional scales. J Am Water Resour Assoc. 2007;43:5–14. doi:10.1111/j.1752-1688.2007.00002.x.
Garcia-Pichel F, Al-Horani FA, Farmer JD, Ludwig R, Wade BD. Balance between microbial calcification and metazoan bioerosion in modern stromatolitic oncolites. Geobiology. 2004;2:49–57.
Gardner TA, Côté IM, Gill JA, Grant A, Watkinson AR. Hurricanes and Caribbean coral reefs: impacts, recovery patterns, and role in long-term decline. Ecology. 2005;86(1):174–84.
Gavin MC, McCarter J, Mead A, Berkes F, Stepp JR, Peterson D, Tang R. Defining biocultural approaches to conservation. Trends Ecol Evol. 2015;30:140–5.
Gischler E, Golubic S, Gibson MA, Oschmann W, Hudson JH. Microbial mats and microbialites in the freshwater Laguna Bacalar, Yucatán Peninsula, Mexico. Lect Notes Earth Sci. 2011;131:187–205.
Halpern BS, Selkoe KA, Micheli F, Kappel CV. Evaluating and ranking the vulnerability of global marine ecosystems to anthropogenic threats. Conserv Biol. 2007;21(5):1301–15.
Hernández-Arana HA, Ameneyro-Angeles B. Benthic biodiversity changes due to the opening of an artificial channel in a tropical coastal lagoon (Mexican Caribbean). J Mar Biol Assoc UK. 2011;91:969–78.
Hernández-Arana HA, López-Adame H, Vega-Zepeda A. An overview of the coral bleaching event in central and southern Mexican Caribbean in 2011. Reef Encounter. 2014;29(1):32–3.
Herrera-Silveira JA, Arreola-Lizárraga JA, Ramírez-Ramírez J. Cambios hidrológicos y de estado trófico entre los años 2000 y 2006. In: Espinoza Ávalos J, Islebe GA, Hernández Arana HA, editors. El sistema ecológico de la bahía de Chetumal/Corozal: costa occidental del Mar Caribe; 2009. p. 21–7.
Herrera-Silveira JA, Teutli-Hernández C, Zaldívar-Jiménez A, Pérez-Ceballos R, Cortés-Balán O, Osorio-Moreno I, Ramírez-Ramírez J, Caamal-Sosa J, Andueza-Briceño MT, Torres R, López-Adame H, Hernández-Arana H. Programa regional para la caracterización y el monitoreo de ecosistemas de manglar del Golfo de México y el Caribe Mexicano: Inicio de una Red Multi-Institucional. Península De Yucatán. CINVESTAV-ECOPEY, Noviembre 2013, Informe Final SNIB CONABIO. Proyecto FN009. México, D.F.; 2013.
Hoegh-Guldberg O, Bruno JF. The impact of climate change on the world's marine ecosystems. Science. 2010;328(5985):1523–8.
Huggins AE, Keel S, Kramer P, Núñez F, Schill S, Jeo R, Chatwin A, Thurlow K, McPearson M, Libby M, Tingey R, Palmer M, Seybert R. Biodiversity conservation assessment of the insular Caribbean using the Caribbean decision support system, summary report, The Nature Conservancy; 2007.
Jackson JBC. Reefs since Columbus. Coral Reefs. 1997;16(1 Supplement):S23–32.
Jackson JB, Kirby MX, Berger WH, Bjorndal KA, Botsford LW, Bourque BJ, Bradbury RH, Cooke R, Erlandson J, Estes JA, Hughes TP, Kidwell S, Lange CB, Lenihan HS, Pandolfi JM, Peterson CH, Steneck RS, Tegner MJ, Warner RR. Historical overfishing and the recent collapse of coastal ecosystems. Science. 2001;293:629–37.
Kaster JL. Stromatolites of Laguna Bacalar: economic and environmental importance. Presentation at Rancho Encantado, Bacalar, MX; 2007.
Kaster JL. Ecological consequences of microbialite encrustation by the invasive black striped mussel, *Mytilopsis sallei*. In: Proceedings of the International Stromatolite Symposium, Bacalar, Mexico. January 2012. http://www.lagunabacalarinstitute.com/Proceedings.html
Kazmierczak J, Kempe S. Genuine modern analogues of Precambrian stromatolites from caldera lakes of Niuafo'ou Island, Tonga. Naturwissenschaften. 2006;93:119–26.

Kerr JT, Dobrowski SZ. Predicting the impacts of global change on species, communities and ecosystems: it takes time. Glob Ecol Biogeogr. 2013;22:261–3.
Klump JV, Paddock RW, Anderson P, Ruberg S, Johengen T, Biddanda B. Tracking the rate of groundwater mixing in a lake huron sinkhole using Rn-222. IAGLR Annual Conference, Toronto, Canada, 24–28 May 2010.
McClanahan TR, Marnane MJ, Cinner JE, Kiene WE. A comparison of marine protected areas and alternative approaches to coral-reef management. Curr Biol. 2006;16:1408–13.
Moore JW. Animal ecosystem engineers in streams. BioScience. 2006;56:237–46.
Mumby PJ, Edwards AJ, Arias-González JE, Lindeman KC, Blackwell PG, Gall A, Gorczynska MI, Harborne AR, Pescod CL, Renken H, Wabnitz CCC, Llewellyn G. Mangroves enhance the biomass of coral reef fish communities in the Caribbean. Nature. 2004;427:533–6.
Mumby PJ. The impact of exploiting grazers (Scaridae) on the dynamics of Caribbean coral reefs. Ecol Appl. 2006;16:747–69.
Murray G. Constructing Paradise: the impacts of big tourism in the mexican coastal zone. Coast Manag. 2007;35(339e):355.
Naiman RJ, Rogers KH. Large animals and system-level characteristics in river corridors: implications for river management. BioScience. 1997;47:521–9.
Nuñez-Lara E, Arias-González EJ, Legendre P. Spatial patterns of Yucatán reef fish communities: testing models using a multi-scale survey design. J Exp Mar Biol Ecol. 2005;324:157–16.
Pérez-Ceballos R, Pacheco-Ávila J, Euán-Ávila J, Hernández-Arana H. Regionalization based on water chemistry and physicochemical traits in the "Ring of Cenotes", Yucatán, Mexico. J Cave Karst Stud. 2012;74:90–102.
Perry E, Paytan A, Pedersen B, Velazquez-Oliman G. Groundwater geochemistry of the Yucatán Peninsula, Mexico: constraints on stratigraphy and hydrogeology. J Hydrol. 2009;367:27–40.
Rogers CS, Willette DA, Miller J. Rapidly spreading seagrass invades the Caribbean with unknown ecological consequences. Front Ecol Environ. 2014;12:546–7. doi:10.1890/14.WB.016.
Rodríguez-Zaragoza FA, Arias-González JE. Coral biodiversity and bio-construction in the northern sector of the Mesoamerican reef system. Front Mar Sci. 2015;2:13. doi:10.3389/fmars.2015.00013.
Rosado-Torres AA. Efecto de las descargas de agua subterránea en la laguna arrecifal de Puerto Morelos, Quintana Roo. Master Degree Thesis, Universidad Autónoma de Yucatán; 2013.
Ruiz-Zárate MA, Hernández-Landa RC, González-Salas C, Núñez-Lara E, Arias-González JE. Condition of coral reef ecosystems in central-southern Quintana Roo, Mexico. Part 1: Stony corals and algae. In: Lang JC, editor. Status of coral reefs in the western Atlantic: results of initial surveys. Atlantic and Gulf Rapid Reef Assessment (AGRRA) program. Atoll Research Bulletin 496; 2003. p. 318–37.
Ruiz-Zárate MA, Arias-González JE. Spatial study of juvenile corals in the Northern region of the Mesoamerican Barrier Reef System (MBRS). Coral Reefs. 2004;23:584–94. doi:10.1007/s00338-0040420-z.
Salazar-Vallejo SI. Huracanes y biodiversidad costera tropical. Rev Biol Trop. 2002;50(2):415–28.
SEDETUR. http://sedetur.qroo.gob.mx/. Secretaria de turismo de Quintana Roo. Fecha de acceso: 5 de mayo del 2015.
Schmitter-Soto JJ, Arce-Ibarra AM, Vásquez-Yeomans L. Records of *Megalops atlanticus* on the Mexican Caribbean coast. Contrib Mar Sci. 2002;35:34–42.
Schneider B, Cunha ER, Marchese M, Thomaz SM. Explanatory variables associated with diversity and composition of aquatic macrophytes in a large subtropical river floodplain. Aquat Bot. 2014. doi:10.1016/j.aquabot.2014.11.003.
Selig ER, Bruno JF. A global analysis of the effectiveness of marine protected areas in preventing coral loss. PLoS One. 2010;5(2), e9278. doi:10.1371/journal.pone.0009278.

SEMARNAT (Secretaría de Medio Ambiente y Recursos Naturales), CONAFOR (Comisión Nacional Forestal). Inventario Estatal Forestal y de Suelos – Quintana Roo 2013. Secretaría de Medio Ambiente y Recursos Naturales, Comisión Nacional Forestal; 2014.

Whigham DF, Olmsted I, Cabrera-Cano E, Harmon ME. The impact of Hurricane Gilbert on trees, litterfall, and woody debris in a dry tropical forest in the Northeastern Yucatán Peninsula. Biotropica. 1991;2(4, Part A):434–41. Special Issue: Ecosystem, Plant, and Animal Responses to Hurricanes in the Caribbean.

Chapter 15
Forest Ecosystems and Conservation

Luciana Porter-Bolland, Martha Bonilla-Moheno, Eduardo Garcia-Frapolli, and Swany Morteo-Montiel

Abstract Forest ecosystems, as part of land systems, are strongly determined by governance structures that are linked at different spatio-temporal scales. The Yucatán Peninsula is an important region in which to study processes affecting forest conservation, as its geographical extent and differing local histories produce specific land use and vegetation trajectories that help us understand different socioeconomic and environmental drivers. In this chapter, we analyze the principal processes that determine and affect land systems, including conservation of forests, within the Yucatán Peninsula. In particular, we evaluate how productive, development and conservation trends have influenced forest change. To do so, we use data available from official sources. The results show differentiated regional trends. In some areas forest change shows a pattern of decrease responding to agricultural intensification and urbanization; however, other regions have showed forest conservation or even forest increase due to various specific factors. We argue that economic development should favor activities and local institutions that work towards productive landscapes that integrate forest ecosystems.

Keywords Land systems • Forest change • Forest conservation • Drivers • Local institutions • Productive landscapes

L. Porter-Bolland (✉)
Instituto de Ecología, A.C., Red de Ecología Funcional, Carretera Antigua a Coatepec 351, CP 91070 Xalapa, Veracruz, Mexico
e-mail: luciana.porter@inecol.mx

M. Bonilla-Moheno • S. Morteo-Montiel
Instituto de Ecología, A.C., Red de Ambiente y Sustentabilidad, Carretera Antigua a Coatepec 351, CP 91070 Xalapa, Veracruz, Mexico
e-mail: martha.bonilla@inecol.mx; swany.morteo@posgrado.inecol.edu.mx

E. Garcia-Frapolli
Universidad Nacional Autónoma de México, Instituto de Investigaciones en Ecosistemas y Sustentabilidad, CP 58087 Morelia, Michoacán, Mexico
e-mail: garcia.frapolli@gmail.com

G.A. Islebe et al. (eds.), *Biodiversity and Conservation of the Yucatán Peninsula*,
DOI 10.1007/978-3-319-06529-8_15

15.1 Introduction

The conservation of forest ecosystems is determined by non-linear processes that are associated both with biophysical and societal changes (Lambin and Meyfroidt 2010). These changes are strongly influenced by human-environment relations at the landscape level, defined as land systems (Turner 2010). In this sense, forests ecosystems as a part of land systems are strongly determined by governance structures that are linked at different spatio-temporal scales, where past and present social institutions (i.e., the systems of rules and decision-making procedures that give rise to social practices; Young 1999) have had a major influence in defining the structure and composition of current forests (Turner 2010). At global and regional scales, markets and policies (including international agreements) influence the development of contradictory strategies that can generate tensions at the regional level (Harvey et al. 2008). For example, the promotion of programs incentivizing the expansion of agriculture, versus programs directed towards conservation. At the local scale, formal and informal institutions, constrained by these tensions generated at regional levels, frame livelihood strategies that ultimately determine forest conservation practices. In the words of Hecht (2010: 163), "interaction of large scale and local forces, agrarian and non-agrarian livelihoods, formal and informal economies, and national and international processes, interact to give way to actions and politics that produce observed forest trends."

Evidence suggests that factors conducive to the processes of land use/land cover change vary across regions (Rudel 2008), resulting in particular land systems at the local scale. Although there are many examples of forest transition processes leading to forest regrowth in many parts of the world (Rudel et al. 2005), the global tropics are still experiencing significant forest loss (FAO and JRC 2012; Bawa and Seidler 2015). It has been documented that agriculture, including pasture expansion, has been a major proximate cause of tropical deforestation and a major driver of global environmental change (e.g., Geist and Lambin 2002; Wassenaar et al. 2007; Aide et al. 2013). In fact, in the last decades, these trends have been exacerbated though the industrialization of production, with copious increases in large-scale agriculture (including processes resulting from biofuel promotion), cattle ranching, plantations, illegal logging, mining, and infrastructure growth (Alvarez-Berríos and Aide 2015; Laurence 2015; Bawa and Seidler 2015).

Laurence (2015) argues that the increasing importance of the above-mentioned industrial drivers gives way to structural changes that imply rises of per-capita forest loss rates while de-linking the relationship between population density and deforestation. In this sense, forest transitions do not usually respond to linear processes, and therefore, land change science should focus on understanding the complexities of landscape processes (Angelsen and Kaimowitz 1999; Lambin and Meyfroidt 2010). Given the current worldwide interconnectedness of social and political institutions, it is important to conduct local and regional scale studies to understand the processes influencing forests conservation.

In this chapter we analyze the principal processes that determine and affect land systems (including conservation of forests) within the Yucatán Peninsula. For this, we use data from official sources to analyze current socioeconomic information and draw conclusions on the factors affecting forest conservation. The present chapter constitutes an account of institutional factors that have influenced current land use and land cover trajectories in different regions of the peninsula. We discuss these trends in light of local studies regarding the different processes occurring in the region reflecting the complexities of land use systems and forest conservation.

15.2 The Yucatán Peninsula

The Yucatán Peninsula is an important region in which to study processes affecting forest conservation, as its geographical extent and the differing local histories within its territory give way to specific land use/cover trajectories that help us to understand different socioeconomic and environmental drivers. The region harbors diverse vegetation types, the dominant one being dry tropical forest (Chap. 3). These vegetation types significantly contribute to regional biodiversity, and to the provision of important ecological services (Koleff et al. 2012). The still extensive forested areas in the region are the result of dynamic vegetation processes strongly influenced by an intricate history of human and natural disturbances (Chaps. 3, 7 and 8). This history can be dated back to the Classic Mayan occupation period, in which high population densities caused intensive activities that greatly transformed landscapes (Chap. 2). After the decline of the Classic Mayan era (ca. 250–850 AD), which had been characterized by extensive deforestation, forests recovered in a context of lower population densities and influence by disturbance regimes (Calderon-Aguilera et al. 2012). During the last century, human-environment interactions have significantly determined the current land systems in the area (Turner 2010). Some of the important historical factors that determined the current landscape configuration include formalized land tenure and political demarcation, the expansion of modern agriculture, urbanization, land use and conservation policies, and in general, population growth through migration and redistribution (Porter-Bolland et al. 2008; Turner et al. 2001).

15.3 Current Factors Influencing Landscape Processes

In order to analyze the factors influencing land use/cover change, we compiled current information regarding production, development, conservation, and land cover and vegetation data available from official sources.

15.3.1 Forest Change

To determine forest change at the municipal level in the Yucatán Peninsula, we used land use and vegetation information from INEGI.[1] Specifically, we used information from digital land use and vegetation maps (INEGI's Land-Use and Vegetation Charts) available in vector format and corresponding to the states of Campeche, Quintana Roo and Yucatán (2002 (Series II; INEGI 2003) and 2011 (Series V; INEGI 2013b); 1:250,000 scale). We also used the municipality layer to extract information at that scale (INEGI 2013c). All geographical information was projected as UTM Datum ITRF92 and analyzed using ArcMap 10.2.2.

To determine the land use/cover change at the municipality level, we first standardized the cover classes used in both vegetation series. We grouped all of the classes into eight categories: (i) productive land, including agriculture, active pastures for cattle rearing, and tree plantations (including introduced palms); (ii) water; (iii) grasslands (including induced grassland, hydrophilic vegetation and savannas); (iv) forest (all forest categories determined for the peninsula); (v) bare soil; (vi) coastal vegetation; (vii) hydrophilic vegetation; and (viii) urban areas. We used the ArcMap "merge" and "dissolve" tools in order to generate a map for 2002 and 2011.

For both years, we calculated the areas (ha) of each municipality that was covered by different land and vegetation categories. To determine the extent of land use within each municipality, we used the identity function and extracted information for each year into two tables. Each table compiled all the information on land cover and vegetation area for each municipality in the two study years. With this information, we calculated the area of forest change in hectares. We calculated the percent of forest change over the study period, using 2002 as the baseline forested area.

We identified municipalities that showed forest change (loss or gain) in area (ha) and percentage. For this discussion, we report forest change in area for municipalities that lost at least 10,000 ha of forest cover, and those that gained at least 1000 ha of forest cover in Yucatán and 5000 ha of forest cover in Campeche and Quintana Roo. For changes in percentage cover we considered those municipalities that changed by at least 10 % of the amount of forest shown in 2002 (gain or loss) or had no change (those that showed 1 % forest change or less).

[1] INEGI is the National Institute of Geography and Statistics (Instituto Nacional de Estadística y Geografía); www.inegi.gob.mx.

15.3.2 Socioeconomic Factors

Agricultural Intensification Data was obtained from the official governmental source for information on agriculture (SIAP[2]). For each state we compiled the area (ha) dedicated to agriculture, identifying the main crops and their changes for each municipality for Yucatán, Quintana Roo, and Campeche. Data is available at the state level from 1980 onwards, for the district level from 1999, and for municipality level from 2003.

Tourism Most information was obtained from INEGI. Data on tourist arrivals and hotel rooms were obtained from statistical and geographical yearbooks for each state,[3] and (INEGI 2013a, 2014, 2015).

Road Infrastructure A compilation of information was provided by INFOMEX[4] who reported at the state level regarding investment in road infrastructure (construction, maintenance, and conservation of differing road types). The information is derived from statistical yearbooks provided by SCT[5] and PNI.[6]

Forest Management information refers to information provided by SEMARNAT[7] through CONAFOR, and also received via INFOMEX, about permits granted for extracting timber within each state.

Conservation. information was obtained from several sources. Regarding protected areas, the information was gathered from the CONANP,[8] SEDUMA[9] and from the State System of Protected Areas from Campeche and Quintana Roo.[10]

---

[2] SIAP is the Information Service regarding Agriculture, Food and Fisheries (Servicio de Información Agroalimentaria y Pesquera); http://www.siap.gob.mx/cierre-de-la-produccion-agricola-por-estado/.

[3] INEGI Statistical and Geografical Yearbooks for each State (Anuario Estadístico del Estado de Campeche, Quintana Roo and Yucatán: Cam: 1984, 1991, 1996, 2001, 2006, 2011, 2012, 2014; QR: 1984, 1991, 1996, 2001, 2006, 2008, 2010, 2011, 2012, 2013, 2014; Yuc: 1986, 1992, 1996, 2001, 2006, 2008, 2010, 2011, 2012, 2013, 2014).

[4] INFOMEX is a governmental consulting system that has the objective of providing access to the general public regarding government information. It is part of the Instituto Federal de Acceso a la Información y Protección de Datos and responds to the Ley Federal de Transparencia y Acceso a la Información Pública Gubernamental (https://www.infomex.org.mx/).

[5] Secretary of Roads and Transports (SCT—Secretaría de Caminos y Transportes).

[6] Infrastructure National Planning (PNI—Planeación Nacional de Infraestructura).

[7] Secretary of Environment and Natural Resources (SEMARNAT Secretaría de Medio Ambiente y Recursos Naturales) of which CONAFOR—Comisión Nacional Forestal (National Commission on Forests) is a part.

[8] National Comission on Protected Areas (CONANP Comisión Nacional de Áreas Naturales Protegidas); www.conanp.gob.mx.

[9] Secretary of Urban Development and the Environment of Yucatán State (SEDUMA Secretaría de Desarrollo Urbano y Medio Ambiente del Estado de Yucatán).

[10] Sistema Estatal de Áreas Naturales Protegidas del Estado de Campeche y de Quintana Roo.

Information on the number of hectares financed by PES (Payment for Ecosystem Services) was gathered from CONAFOR (www.conafor.gob.mx).

15.4 Observed Socioecological Trends

15.4.1 Forest Change

According to our change analysis, the state with the largest extent of forested area in 2002 was Campeche (ca. 4,000,000 ha), followed by Quintana Roo (ca. 3,600,000 ha) and lastly Yucatán (ca. 2,800,000 ha). The analysis of changes in land cover at the state level indicates in general that in the 9-year period (2002–2011), forested areas decreased while productive land and urban cover classes increased (Table 15.1). This loss in forests was most noticeable in the state of Yucatán, where the area lost was equivalent to 4.06 % of the forested area in 2002. Most of this loss can be attributed to an increase in the area under the productive cover class, in which the state gained about 3.4 %. Both Quintana Roo and Campeche lost similar amounts of forest cover in that short period (ca. 75,000 and 79,000 ha, respectively) that corresponded to 1.68 % and 1.37 %, respectively, of its forest in 2002. Although Campeche's increase in productive cover area was higher than that observed in Quintana Roo (1.42 % and 1.05 %, respectively), the latter recruited more land to urban cover (0.64 %, vs. 0.13 % in Campeche).

In an earlier study evaluating land use/cover changes in the Yucatán Peninsula for the period 1985–1994, Sánchez-Aguilar and Rebollar-Domínguez (1999), using information from the Inventario Nacional Forestal, report greater forest losses for the period 1985–1994 (also 9 years), accounting for up to 3,600,000 ha, for both Campeche and Quintana Roo States; this represents more than 35 % of the total land area of both states. In addition, they indicated that the state of Yucatán did not follow a similar trend because forests had already been cut down before that period. They attribute these trends mainly to agricultural expansion and an increase in the

Table 15.1 Changes in land cover area (ha) in the period 2002–2011 according to Vegetation Series (INEGI)

	Forests	Productive land (agriculture, pasture, forest plantations)	Urban
Quintana Roo	−75,395.22	+47,411.94	+28,827.13
	(1.68 %)	(1.05 %)	(0.64 %)
Campeche	−79,123.47	+81,928.40	+7525.18
	(1.37 %)	(1.42 %)	(0.13 %)
Yucatán	−160,714.70	+134,644.86	+31,567.38
	(4.06 %)	(3.40 %)	(0.79 %)

Numbers are hectares gained (+) or lost (−) in the period. Percentages refer to area changed with respect to total land area of each state

tourist sector since the 1970s, but also to forest exploitation and to the effects of hurricanes, including their triggering effects on forest fires.

Municipalities with forest loss were aggregated in specific areas within the peninsula (Fig. 15.1). They are mostly concentrated in the northeast (largely in several municipalities of Yucatán but also in Benito Juarez, Quintana Roo) and in the center and south of the peninsula (José Maria Morelos, Bacalar and Othón P. Blanco in Quintana Roo and Tenabo and Hopelchen in Campeche), though forest loss also occurred in a municipality in the extreme southwest of the peninsula (Candelaria). In the case of the state of Yucatán indicated municipalities lost at least 10 % of forest with respect to the baseline year of the analysis (2002). Many other municipalities (mainly located in the central region of Yucatán) that did not lose large absolute amounts of forested area because of their small areas, still experienced a loss of 10 % or more of their previous (2002) forest cover (Fig. 15.2). A similar situation occurred in Campeche for the municipality of Palizada.

For municipalities that gained forest (>1000 ha for Yucatán and >5000 ha for Quintana Roo and Campeche), changes are less spatially concentrated (Fig. 15.1).

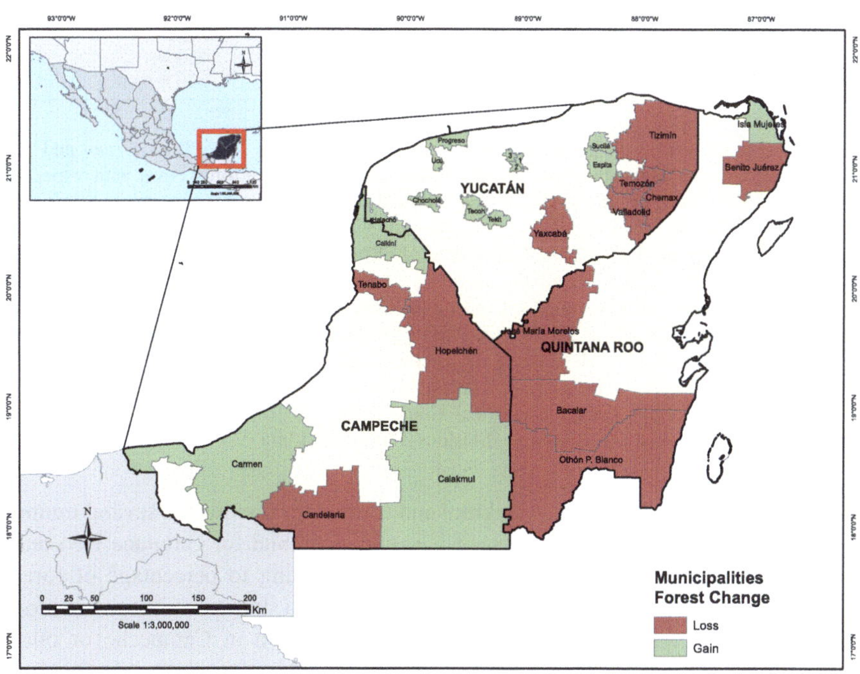

Fig. 15.1 Municipalities with forest change in the period 2002–2011. The *red* corresponds to municipalities that lost more than 10,000 ha of forests. *Green* municipalities are those that gained more than 1000 ha of forest cover in Yucatán, 5000 ha in Quintana Roo and Campeche. *Numbers* in map represent municipalities 1. Teya, 2. Tekantó, and 3. Suma. *Source:* Cartographic projection: Lambert Conformal Conic projection. Datum: ITRF92. Cartographic base: State and municipal administrative divisions, INEGI (2013a)

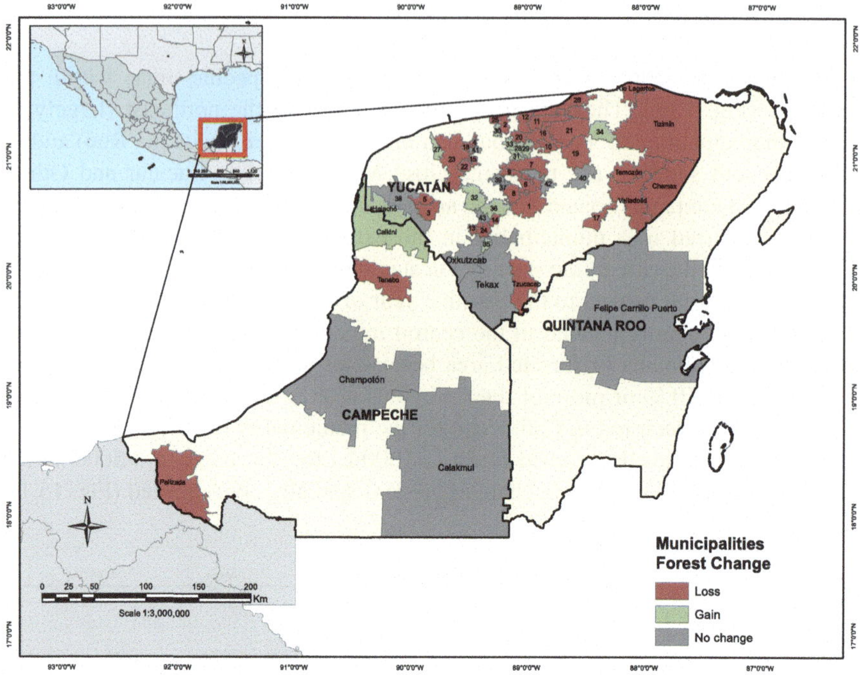

Fig. 15.2 Municipalities with relative forest change in the period 2002–2011. *Green* and *red* correspond to municipalities that lost or gained more than 10 % of their forest cover with respect to the baseline year (2002) of the analysis. *Gray* colored municipalities did not present significant forest loss or gain (less than or equal to 1 % change). *Numbers* in map represent municipalities: 1. Sotuta, 2. Sinanché, 3. Opichén, 4. Mocochá, 5. Kopomá, 6. Kantunil, 7. Izamal, 8. Huhí, 9. Hoctún, 10. Dzoncauich, 11. Dzilam González, 12. Dzidzantún, 13. Dzán, 14. Chumayel, 15. Tixpéhual, 16. Temax, 17. Tekom, 18. Conkal, 19. Cenotillo, 20. Cansahcab, 21. Buctzotz, 22. Kanasín, 23. Mérida, 24. Maní, 25. Telchac Puerto, 26. Dzilam de Bravo. 27. Ucú, 28. Teya, 29. Tepakán, 30. Telchac Pueblo, 31. Tekantó, 32. Tecoh, 33. Suma, 34. Sucilá, 35. Akil, 36. Tekit. 37. Sanahcat, 38. Maxcanú, 39. Hocabá, 40. Dzitás, 41. Yaxkukul, 42. Sudzal, 43. Mama. *Source*: Cartographic projection: Lambert Conformal Conic projection. Datum: ITRF92. Cartographic base: State and municipal administrative divisions, INEGI (2013a)

Specific cases include Calkini, Calakmul and Carmen in Campeche, several municipalities in the northwestern portion of Yucatán state, and for Quintana Roo only Isla Mujeres. However, if we consider trends according to percentage of forest cover in the baseline year (2002), Felipe Carrillo Puerto shows no change (no more than 1 % cover from year 2002), nor does Champoton in Campeche or other municipalities in the center of Yucatán.

The municipalities that were identified as having lost forest while gaining productive area were Hopelchen and Candelaria in Campeche, Jose Maria Morelos and Bacalar in Quintana Roo, and the northeastern municipalities in Yucatán (Tizimin, Valladolid and surrounding areas). The municipalities that were identified as having lost forest in favor of a gain of more than 10,000 ha of urban area are

Benito Juarez in Quintana Roo and Merida in Yucatán. In the following sections we present some of the socioeconomic information that explains part of the observed trends in forest change.

15.4.2 *Agricultural Intensification*

In the last three decades (1980–2013), the Yucatán Peninsula has shown a differentiated increase in the land area used for agriculture (Fig. 15.3). Yucatán has been the state with the largest area dedicated to this activity, as well as the one that had the greatest increase in cultivated land area (from 412,645 to 753,524 ha; an increase of more than 80 %). Since 1999, the section with the largest cropped area in Yucatán state is in the northeast region, specifically Tizimin and surrounding municipalities. This region showed high rates of deforestation, as explained above. Interestingly, the dominant agricultural use in this area has been pasture for cattle (pasture alone occupies more than 500,000 ha). Although the state of Yucatán was the principal cattle producer of the peninsula, the number of head decreased considerably over the last 30 years (ca. 300,000 head). Campeche, with an increase of at least 200,000 head, is now the principal cattle producer. In contrast, the greatest crop loss registered in the state was that of the henequen (or sisal, *Agave fourcroydes* Lem.), a crop widely used as fiber and an export product before its substitution by synthetic materials. This reduction occurred in the northeastern region of the peninsula (i.e., Merida and surrounding municipalities). The area dedicated to this crop changed from >100,000 ha in 1980 to just over 10,000 ha in 2013. According to González-Iturbe et al. (2002), henequen in the northeastern region had been planted on at least 200,000 ha in the region, but by the beginning of the twenty-first century, only 7 % of the henequen plantations were still active.

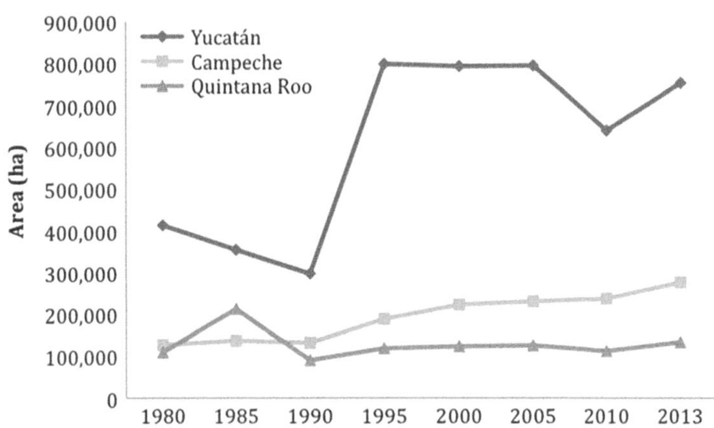

Fig. 15.3 Area cultivated in Campeche, Quintana Roo and Yucatan from 1980 to 2013. Data available from SIAP

In our change analysis this area corresponds to the area where several municipalities showed a tendency towards forest gain, or showed little change in forest cover.

During this same period (1980–2013), the area dedicated to agriculture in Campeche state has been steadily, doubling from 126,000 ha in 1980 to 276,000 ha in 2013. Most of this agricultural production can be found in the central region of the state, specifically in the municipalities of Champoton, Hopelchen and Campeche. The last two municipalities were the ones that showed the greatest increases in cultivated area. Nevertheless, only Hopelchen exhibited significant forest loss during the study period. In these areas, maize (mostly for local consumption) has recently been the dominant crop, followed by more commercial crops (e.g., sorghum and soya). These changes have paralleled an increasing Mennonite population. According to INEGI (2006) the Mennonite population accounted for up to 12 % of the population of Hopelchen by 2005 while in 2000 it represented only 8 %. In a study conducted in the southern part of Hopelchen (the forested part of the municipality known as the La Montaña area) Porter-Bolland et al. (2007) reported that the deforestation rate is increasing, and estimated that forest loss rates increased from 0.3 % in the period 1988–2000 to more than 0.7 % for 2000–2005. Part of this pattern resulted from changes in agricultural activities and an increase in cattle ranching; it was also due to the expanding Mennonite population, who in one of the communities of La Montaña bought 5000 ha of forested lands.

Quintana Roo has been the state with the least land in agriculture, maize and sugar cane being the principal crops. The major agricultural area has been concentrated in the central region known as the "Mayan zone", specifically in the municipalities of Othón P. Blanco and Felipe Carrillo Puerto. Coincidentally, over the last decade these municipalities lost 10,842 and 8450 ha, respectively, of area dedicated to agriculture, although Othón P. Blanco (and two other municipalities: Bacalar and Jose Ma. Morelos) continue to show forest losses. The northern region found near Cancun not only presented the smallest cropped area, but this area also decreased in the last decade. Overall, the state has shown a marginal increase in cropped area (from 109,037 ha in 2003 to 133,505 ha in 2013; SIAP).

15.4.3 Tourism Growth

Over the last decades, the tourist sector has been one of the main driving forces of social, economic and environmental transformations in the Yucatán Peninsula. This situation has not been unique to the region. At the national level, tourism has become the third-ranked sector in importance for employment generation and foreign exchange earnings (INEGI 2015). The expansion of the tourist industry in the Yucatán Peninsula can be seen in a diagram of the intensification of tourist visits to the area (Fig. 15.4). Currently, Mexico ranks as fifteenth worldwide in tourist visits (INEGI 2015). In 2013, it captured as many as 24 million visitors, for more than 50 % of whom the Yucatán Peninsula was their final destination. These

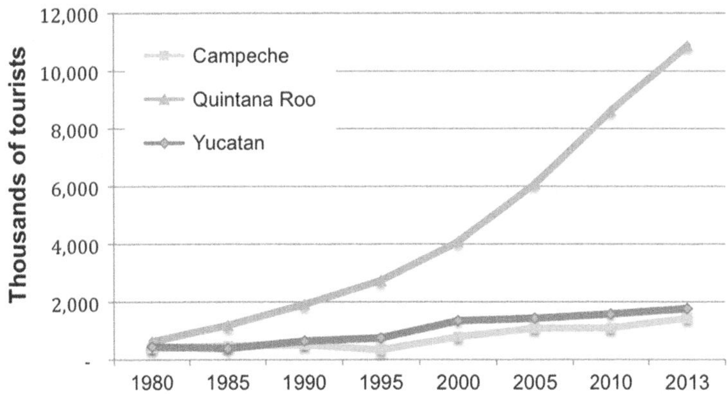

Fig. 15.4 Number of tourist arrivals to Campeche, Quintana Roo and Yucatan. *Source:* INEGI

numbers have been on the rise dramatically in recent years. From 1980 to date, the number of tourists visiting the Yucatán increased from 1.4 million to a little more than 14 million in 2013, accounting for a mean annual increment of 7 % (SECTUR 2015).

Tourism growth has come hand in hand with increases in infrastructure and service development. Infrastructure development has included the construction of roads (see below), hotels, commercial establishments, potable water services and sewage treatment plants, port infrastructure development, and other elements. These patterns are underscored by the number of hotel rooms available for tourists in the area. Between 1980 and 2013 hotel rooms grew from almost 11,000 to more than 93,000; again showing an average annual growth rate of 7 %. Port services offered for the tourist sector also developed significantly. Even though Cozumel (Quintana Roo) is still by far the most important destination for cruise ships in Mexico, ports in Mahahual (Quintana Roo) and Progreso (Yucatán) are now situated among the five most important in the country (SECTUR 2015).

Not all regions in the Yucatán have shown the same incremental tendency. Campeche and Yucatán, although they have experienced a significant increase in tourists, are far from seeing the level of transformation experienced in the state of Quintana Roo, particularly along the Cancun-Tulum tourist corridor and the Riviera Maya. For example, along this 130-km coastal fringe, more than 65,000 hotel rooms have been built in the last 30 years. The Riviera Maya alone accounts for more than twice the number of hotel rooms in Campeche and Yucatán combined (INEGI 2014).

As has been amply documented (Hunter and Green 1995; Hillery et al. 2001; Gössling and Peeters 2015), the growth of the tourist sector has significant consequences for the use of natural resources and the transformation of ecosystems. This pattern has been true in the area. Tourism has impacted forests significantly through the conversion of forests into urban areas. For example, for the construction of Cancun in the 1970s, approximately 1000 ha of forests and 400 ha of mangroves were cleared (Pérez Villegas and Carrascal 2000). During 2002–2011, the Municipality of Benito Juárez, where Cancun is located, lost more than 10,000 ha of forest. This municipality showed a significant urban gain in our change analysis. The same pattern resulted more recently from the construction of Playa del Carmen, and with the growth of the cities of Campeche, Chetumal and Merida. In fact, the urbanized area surrounding Merida includes two of the municipalities with the highest deforestation figures in Yucatán (Kanasin and Conkal). According to our analyses these municipalities lost 34 % and 26 %, respectively, of their forest during the period 2002–2011. On the other hand, tourism has impacted mangrove forests significantly as a result of the construction of hotel complexes, residential areas and port infrastructure. For example, Hirales-Cota et al. (2010) calculated that during the period of 1995–2007 the coastal fringe of Mahahual-Xcalak in Quintana Roo (nearly 50 km long) lost more than 1000 ha of mangrove forests. Despite being a protected area, this represents a mean annual deforestation rate of 0.85 %. Hirales-Cota et al. (2010) argue that the main drivers of deforestation are the construction of roads and hotels, and the delimitation of private properties.

15.4.4 *Road Development*

According to official information, annual investment in road infrastructure in the region has been rising steadily in the last few years (from 2000 and at least until the year 2011; Fig. 15.5). Investment includes the construction, improvement, and maintenance of different road types that run the gamut from federal highways to less major state roads, as well as the opening of rural farm access roads (*caminos*

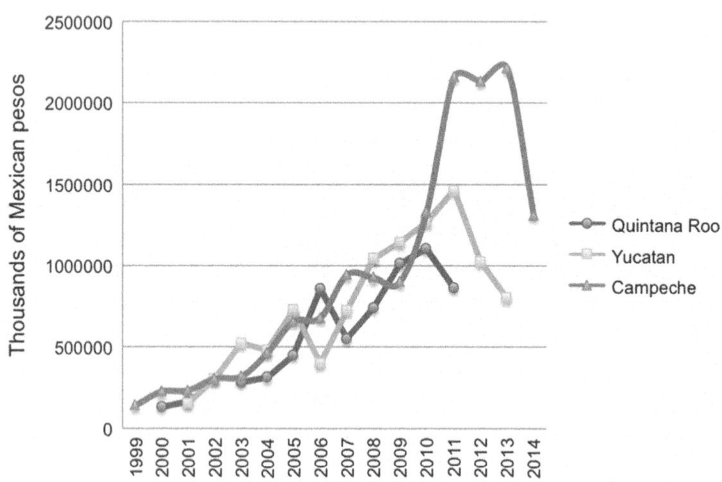

Fig. 15.5 Thousands of Mexican pesos invested in road infrastructure per state (Quintana Roo, Yucatán, Campeche). *Source:* Secretaría de Caminos y Transportes

sacacosecha). Investment also includes a portion of Temporary Employment (*empleo temporal*, which refers to money paid to local inhabitants for road maintenance—mostly clearing the vegetation bordering roads, a type of social welfare).

This effort to emphatically increase road infrastructure during recent years was a response to the goals established by the government through the Plan Puebla Panama (PPP). The PPP was established in 2001 as an accord among several countries of Central America. It is inserted into the Declaration of Tuxtla (established in 1991) that brings special attention to the Program for Free Trade (Programa de Liberación Comercial). Financed by the Inter-American Development Bank (IDB), the PPP in Mexico had the objective of bringing economic and social development to the southeast of the country. It considered different goals, including social and environmental ones (The Mesoamerican Biological Corridor[11] being a result), but primarily focused on the commercial sector including tourism, transportation, and telecommunications. Financially, it is notable that most of the total amount of money available to the PPP (more than 86 %) was designated for use in the transportation sector (SIECA 2004). While it is difficult to directly attribute forest changes at the level of the municipality to road development, road infrastructure grows hand in hand with productive and commercial sectors (including agriculture, markets and tourist expansion) and has been identified as a driver of deforestation (Chomitz and Grey 1996).

15.4.5 Forest Management

Forest management in the Yucatán Peninsula differs greatly among states. Quintana Roo and Campeche have forests valuable for their commercial potential for timber production, as well as conservation goals. In Yucatán state, forests have mostly been replaced by agricultural landscapes (Durán García and García Contreras 2010), or hold little commercial value. Nevertheless, the forestry history and current forest production of the two aforementioned states differs greatly. While most of its forested areas are under a collective land tenure system (mostly in the form of ejidos), as occurs throughout the country (Bray et al. 2003), not all these communities have engaged in forestry activities. Information about permits for forest extraction that have been approved by the secretary in charge of the forest sector (SEMARNAT through CONAFOR) shows differences in forestry activity in the last 15 years between the two states. These permits indicate potential official forestry production or the timber volumes authorized for extraction (Fig. 15.6). For

[11] The Mesoamerican Biological Corridor was an initiative administered by the government through the National Commission regarding the Knowledge and Use of Biodiversity (CONABIO—Comisión Nacional para el Conocimiento y Uso de la Biodiversidad) with the objective of financing activities related to the conservation and management of biodiversity as well as for local development, in specific areas of the southeast of Mexico. The initial phase of the program lasted from 2001 to 2008 (World Bank 2000).

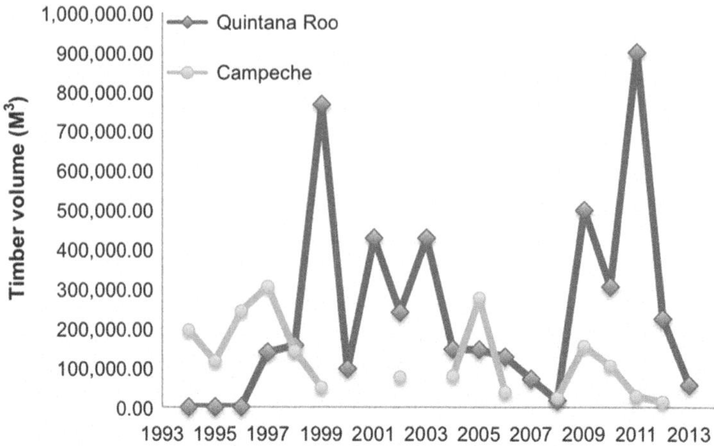

Fig. 15.6 Timber volume authorized per year to *ejidos* and private owners, reflected in permits granted by the Forest Service per year in Campeche and Quintana Roo (*Source:* SEMARNAT)

the years that we found information for both states (1997–2012), 328 permits were approved in Quintana Roo as opposed to 103 in Campeche. Correspondingly, the total volume of timber to be extracted under all permits accounted for 293,322 m^3 in Quintana Roo and 107,737 m^3 in Campeche.

The above information comes as no surprise. Quintana Roo has a long history of community forestry development rooted in the Forestry Pilot Plan project. This project started at the beginning of the 1980s and was sponsored by the Mexican government and a German cooperation agency (GTZ). The Forestry Pilot Plan worked towards the development of community forestry at a time when forestry concessions to private companies were ending and forestry law was changing in favor of the holders of land rights (Bray et al. 2005, 2006; Merino-Pérez 2004). Mexico's weak forestry policies (Merino-Pérez 2013)[12] are compounded by contemporary challenges faced by that sector, including land reforms under Article 27 of the Constitution that favor the privatization of forests lands (which at times leads to its parcelization and weakens local institutions for communal management), poor administration, and climate change, among other factors (Ellis et al. 2014). Despite these challenges, there are positive experiences of community forestry in the state, without losses in forest cover (Bray et al. 2003). Bray et al. (2004) found that land use/cover change analysis in central Quintana Roo indicate sustainable landscapes (with no net forest change), in large part as a result of forest use and management by local communities. Their study was conducted in Felipe Carrillo Puerto, a municipality that in our change analysis for 2011 still

[12] Merino-Pérez (2013) explains how since the late 1980 forest policy has tended to place less emphasis on productive goals (although much investment has focused on reforestation projects and forestry plantations) and more on trying to reduce extractive pressure on natural forests, therefore little development has occurred in the forest sector, except in a few cases.

showed no change in terms of percentage of forest cover from the baseline year of our study (2002).

Contrary to the Quintana Roo experience, the forest sector in Campeche has not had the same push towards community forestry. Campeche's forests have been economically important at the state and national levels, particularly with *chicle* production during the first half of the twentieth century. However, an impoverished forest after unregulated forestry concessions during the last half of the last century (Kú-Quej et al. 2010), and a poor forest policy that discouraged the development of forest management, has given way to deforestation and even illegal logging in some areas, such as La Montaña in Hopelchen (Porter-Bolland et al. 2007, 2008). According to Kú-Quej et al. (2010), forestry in Campeche is currently limited and focuses mostly on charcoal production, which is increasing steadily (from one ton in 1998 to more than 80,000 tons in 2006). Moreover, there is little production of forest products such as veneer and plywood, posts, hardwood logs, and little decorative timber. Ellis and Porter-Bolland (2008) compared land use/cover change processes in the La Montaña area in Campeche and the Maya zone of Quintana Roo and concluded that, although La Montaña is north of the Calakmul Biosphere Reserve and is part of the Balan Ka'ax State Reserve, it experiences increasing forest loss, as opposed to the Mayan zone of Quintana Roo. While La Montaña should be influenced by conservation policies, it is part of the Hopelchen municipality, an agricultural frontier area. Efforts in the area have not been focused towards the development of local institutions that lead to productive activities that could be compatible with forest conservation (Porter-Bolland et al. 2008). Activities such as forest management or agroforestry, incorporating the traditional ecological knowledge of the local Mayan population, could help stimulate local economies based on forest livelihoods.

15.4.6 Conservation Policies

Over the last 30 years, forest conservation in the Yucatán Peninsula happened in several ways. The first and most important has been through traditional activities practiced by the Mayan population, which have been widely documented (see Gómez-Pompa 1987; Faust 2001; Jiménez Osornio et al. 2003; Barrera-Bassols and Toledo 2005). Specifically, management strategies for productive diversification (Gómez-Pompa and Bainbridge 1995; Porter-Bolland et al. 2006), the maintenance of forest corridors (*tolche*' in Yucatec Mayan; Remmers and Ek' 1996) surrounding *milpas* and urban areas, low demographic pressures (Lutz et al. 2000), and a balance between commercial and subsistence activities (García-Frapolli et al. 2008), to mention only a few, are beneficial for conservation. As part of this management strategy, communal reserves that can be either *ejidal* or *comunitaria*, and which are designated forest areas that can be considered Indigenous Peoples' and Community Conserved Territories and Areas (ICCAT; Martin et al. 2011) play a very important conservation role. In fact, Elizondo and López-

Merlín (2009) documented the conservation of more than 50,000 forest hectares within 60 ejidal reserves in two municipalities of Quintana Roo (Felipe Carrillo Puerto and Othón P. Blanco).

The second way in which conservation in the Yucatán Peninsula takes place is through state conservation policies with the goal of biodiversity or ecosystem service preservation. This has occurred through the establishment of Protected Areas (PAs) at the federal, state and municipal levels, and more recently at the ejidal, community, or private property levels. The history of PAs in the peninsula began in the 1980s, with the decree of the Tulum National Park. After that, the PA policy took off with the gazetting of the Sian Ka'an Biosphere Reserve (1986) and the Calakmul Biosphere Reserve (1989). Figure 15.7 shows a substantial increase in the number of hectares under some type of PA, be it National Parks, Flora and Fauna Protection Areas (Areas de Protección de Flora y Fauna), Biosphere Reserves, Sanctuaries, State Reserves, Biocultural Reserves, or Voluntary Conservation Areas, amongst other types. Currently, there are nearly 50 PAs in the Yucatán Peninsula, comprising up to 5,000,000 ha, both land and sea area. However, as has been amply discussed (Porter-Bolland et al. 2012; García-Frapolli et al. 2009; Smardon and Faust 2006), PAs in and of themselves do not necessarily guarantee effective conservation.

The third road to conservation is much more recent and takes the form of official programs such as Payments for Environmental Services (PES) and the like. The goal of these conservation programs is to provide monetary compensation to landowners or those holding land rights for conducting stewardship activities that protect or enhance ecosystem services, mostly hydrology or biodiversity. These conservation schemes have been implemented in the peninsula since 2004 and have grown significantly in number (Fig. 15.8). In a decade, these programs have financed the conservation of more than 600,000 ha, although there are no estimates of their effectiveness regarding forest conservation in the longer term. A very

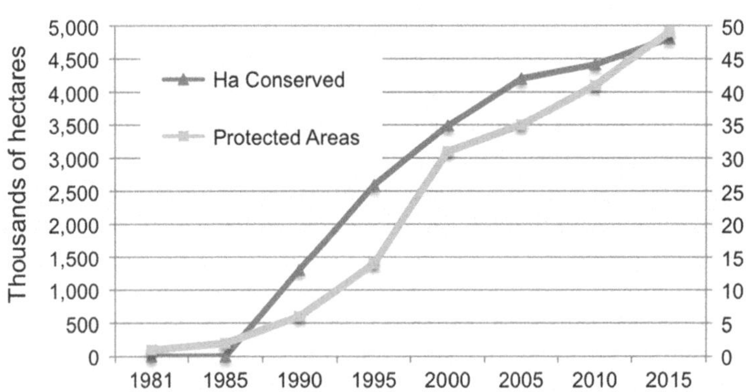

Fig. 15.7 Total area conserved (ha) by protected areas in the Yucatan Peninsula and number of protected areas (*right-hand axis*). *Source:* CONANP (2015), PRONATURA (2012), SEDUMA (2015)and Pozo et al. (2011)

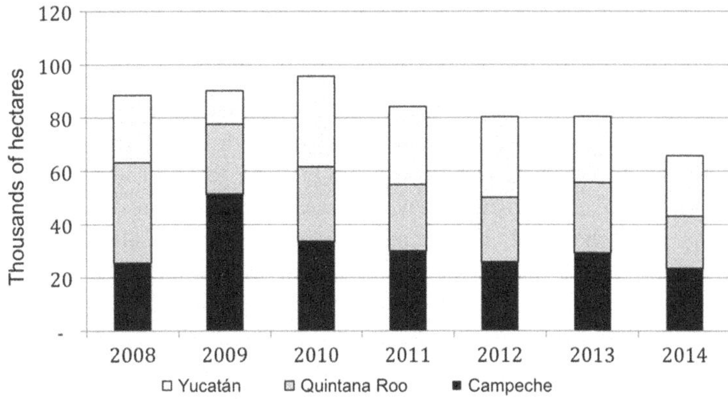

Fig. 15.8 Total area (ha) under the program of hydrological and biodiversity payments for environmental services in the Yucatan Peninsula. *Source:* CONAFOR (2015)

noteworthy aspect of these programs is that the areas under PES can also fall within a PA or an ICCAT. Therefore, the areas affected cannot necessarily be compared or added to those in PAs. Another important aspect is that properties under PES receive a payment for only 5 consecutive years, and the commitment to conserve these areas is only for that period. After the initial 5 years, if landowners make a request, which is not always the case, they can renew benefits for another 5 years under the program. Taking into consideration that the program has been in operation for almost 10 years on the peninsula, areas approved since 2004 could already be in their second round of a PES program.

In our change detection analysis the municipality of Calakmul, harboring the Calakmul Biosphere Reserve, showed no forest loss. At the broad scale of our analysis, it is difficult to see the positive effects derived from different conservation schemes. However, as we previously indicated, it is important to have different conservation initiatives, particularly ones developed through the initiative or participation of local inhabitants, for conserving forests and enhancing livelihood strategies (Tolentino et al. 2015).

15.5 Conclusions

We have presented general tendencies in forest conversion, persistence, and conservation in the Yucatán Peninsula derived from official information regarding the main socioeconomic drivers and their relation to processes occurring at the landscape level. Although simplified into general terms because of the limitations of official information and the broad scale of the analysis, these factors are indicative of the complexities and rapid pace of the changes that exert pressures on forest ecosystems. In general terms, our change detection analysis indicate that in many parts of the peninsula, forests are receding to give way to human-dominated productive landscapes. Agriculture and urbanization are expanding significantly in many areas. This growth responds in large part to the adoption of commercial forms of production (i.e., crops such as sorghum and soy) but also to factors related to infrastructure development. Tourism is a strong driver of the latter process, exerting pressure on forests, particularly in Quintana Roo but also in Campeche and Yucatán. However, tourism does not necessarily cause negative environmental impacts. As Riensche et al. (2015) argue, when tourism is aligned with other productive activities and development goals, it can benefit local communities without negatively affecting landscapes and ecosystems. The growth of road infrastructure in the last 10 years has no precedent, and will surely drive the way production and population patterns change in the years to come.

However, there are other processes that favor forest conservation in many parts of the region, and which are visible at the landscape scale. These processes include formal conservation policies, notably including those that have adopted more participatory approaches. Although PAs are not a silver bullet, they are in fact benefiting forest conservation. As Blackman et al. (2015) show, this is especially true in the Yucatán Peninsula. Landscape-level processes that favor the conservation of forests should strengthen these efforts not only by inducing local inhabitants to conserve forests, but also by strengthening productive activities that favor forest management as well as agroforestry activities inherited from traditional practices. This is what PES programs seek to accomplish, with recent changes in their operating rules that favor productive activities as well as strict conservation. In this way, economic development should reinforce activities and local institutions that favor conservation, and places where productive landscapes integrate forest ecosystems. These findings are intended to raise questions that should be addressed at more local scales, to shed light into processes that need to be strengthened in order to favor the conservation of forests.

Acknowledgment The authors acknowledge the contribution of Dr. Edward A. Ellis for providing some of the information used for the discussion.

References

Aide TM, Clark ML, Grau HR, López-Carr D, Levy M, Redo DJ, Bonilla-Moheno M, Riner G, Andrade Núñez MJ, Muñiz M. The deforestation and reforestation of Latin America and the Caribbean (2001–2010). Biotropica. 2013;45:262–71.

Alvarez-Berríos NL, Aide M. Global demand for gold is another threat for tropical forests. Environ Res Lett. 2015;10:014006.

Angelsen A, Kaimowitz D. Rethinking the causes of deforestation: lessons from economic models. World Bank Res Observ. 1999;14:73–98.

Barrera-Bassols N, Toledo VM. Ethnoecology of the Yucatec Maya: symbolism, knowledge and management of natural resources. J Lat Am Geogr. 2005;4:9–41.

Bawa KS, Seidler R. Deforestation and sustainable mixed-use landscapes: a view from the eastern Himalaya. Ann Mo Bot Gard. 2015;100:141–9.

Blackman A, Pfaff A, Robalino J. Paper park performance: Mexico's natural protected areas in the 1990s. Glob Environ Chang. 2015;31:50–61.
Bray DB, Antinori C, Torres-Rojo JM. The Mexican model of community forest management: the role of agrarian policy, forest policy and entrepreneurial organization. Forest Policy Econ. 2006;8:470–84.
Bray DB, Merino-Pérez L, Negreros-Castillo P, Segura-Warnholtz G, Torres-Rojo JM, Vester HFM. Mexico's community-managed forests as a global model for sustainable landscapes. Conserv Biol. 2003;17(3):672–7.
Bray DB, Merino-Pérez L, Negreros-Castillo P, Segura-Warnholtz G, Torres-Rojo JM, Vester HFM. Mexico's community-managed forests as a global model for sustainable landscapes. Conserv Biol. 2005;17:672–7.
Bray DB, Ellis EA, Armijo-Canto N, Beck CT. The institutional drivers of sustainable landscapes: a case study of the 'Mayan Zone' in Quintana Roo, Mexico. Land Use Policy. 2004;21:333–46.
Calderon-Aguilera LE, Rivera-Monroy VH, Porter-Bolland L, Martínez-Yrízar LA, Ladah LB, Martínez-Ramos M, Alcocer J, Santiago-Pérez AL, Hernández-Arana HA, Reyes-Gómez VM, Pérez-Salicrup DR, Díaz-Nuñez V, Sosa-Ramírez J, Herrera-Silveira J, Búrquez A. An assessment of natural and human disturbance effects on Mexican ecosystems: current trends and research gaps. Biodivers Conserv. 2012;21:589–617.
Chomitz KM, Gray DA. Roads, land use, and deforestation: a spatial model applied to Belize. World Bank Econ Rev. 1996;10:487–512.
CONAFOR. Comisión Nacional Forestal. 2015. http://www.conafor.gob.mx/portal/index.php/tramites-y-servicios/apoyos. Accessed 14 Apr 2015.
CONANP. Áreas protegidas decretadas. Comisión Nacional de Áreas Naturales Protegidas. 2015. http://www.conanp.gob.mx/que_hacemos/. Accessed 14 Apr 2015.
Durán-García R, García-Contreras G. Distribución espacial de la vegetación. In: Durán-García R, Méndez-González M, editors. Biodiversidad y desarrollo humano en Yucatán. CICY, PPD-FMAM, CONABIO, Seduma, Mérida; 2010. p. 131–135, 496 pp.
Elizondo C, López-Merlín D. Las áreas voluntarias de conservación en Quintana Roo. México: CONABIO; 2009.
Ellis EA, Kainer KA, Sierra-Huelsz JA, Negreros-Castillo P, DiGiano ML. Community-based forest management in Quintana Roo, Mexico. 2014.
Ellis EA, Porter-Bolland L. Is community-based forest management more effective than protected areas? A comparison of land use/land cover change in two neighboring study areas of the central Yucatán Peninsula, Mexico. Forest Ecol Manage. 2008;256:1971–83.
FAO, JRC. Global forest land-use change 1990–2005. In: Lindquist EJ, D'Annunzio R, Gerrand A, MacDicken K, Achard F, Beuchle R, Brink A, Eva HD, Mayaux P, San-Miguel-Ayanz J, Stibig H-J, editors. FAO Forestry Paper No. 169. Food and Agriculture Organization of the United Nations and European Commission Joint Research Centre. Rome, FAO; 2012.
Faust BB. Maya environmental successes and failures in the Yucatán Peninsula. Environ Sci Policy. 2001;4:153–69.
García-Frapolli E, Ramos-Fernández G, Galicia E, Serrano A. The complex reality of biodiversity conservation through natural protected area policy: three cases from the Yucatán Peninsula, Mexico. Land Use Policy. 2009;26:715–22.
García-Frapolli E, Toledo VM, Martinez-Alier J. Adaptations of a Yucatec Maya multiple-use ecological management strategy to ecotourism. Ecol Soc. 2008;13:31.
Geist HJ, Lambin EF. Proximate causes and underlying driving forces of tropical deforestation. BioScience. 2002;52(2):143–50.
Gómez-Pompa A. On Maya silviculture. Estudios Mexicanos. 1987;3:1–17.
Gómez-Pompa A, Bainbridge DA. Tropical forestry as if people mattered. In: Lugo A, Lowe C, editors. Tropical forests: management and ecology. New York, NY: Springer; 1995.
González-Iturbe JA, Olmsted I, Tun-Dzul F. Tropical dry forest recovery after long term Henequen (sisal, *Agave fourcroydes* Lem.) plantation in northern Yucatán, Mexico. Forest Ecol Manage. 2002;167:67–82.

Gössling S, Peeters P. Assessing tourism's global environmental impact 1900–2050. J Sustain Tour. 2015;23:639–59.
Harvey CA, Komar O, Chazdon R, Ferguson BG, Finegan B, Griffith DM, Martínez-Ramos M, Morales H, Nigh R, Soto-Pinto L, Van Breugel M, Wishnie M. Integrating agricultural landscapes with biodiversity conservation in the Mesoamerican hotspot. Conserv Biol. 2008;22:8–15.
Hecht S. The new rurality: globalization, peasants and the paradoxes of landscapes. Land Use Policy. 2010;27:161–9.
Hillery M, Nancarrow B, Griffin G, Syme G. Tourist perception of environmental impact. Ann Tour Res. 2001;28:853–67.
Hirales-Cota M, Espinoza-Avalos J, Schmook B, Ruiz-Luna A, Ramos-Reyes R. Agentes de deforestación de manglar en Mahahual-Xcalak, Quintana Roo, sureste de México. Ceinc Mar. 2010;36:147–59.
Hunter C, Green H. Tourism and the environment: a sustainable relationship? London: Routledge; 1995.
INEGI. II Conteo de Población y Vivienda 2005. Tabulados Básicos de los Estados Unidos Mexicanos Tomo I. InstitutoNacional de Estadística, Geografía e Informática, México; 2006. 648 p.
INEGI. Cartas de uso del suelo y vegetación, escala 1: 250,000 Serie III. 2003. http://www.inegi.org.mx/geo/contenidos/recnat/usosuelo/
INEGI. Anuario estadístico y geográfico por entidad federativa 2013/Instituto Nacional de Estadística y Geografía, México; 2013a.
INEGI. Cartas de uso del suelo y vegetación, escala 1: 250,000 Serie V. 2013b. http://www.inegi.org.mx/geo/contenidos/recnat/usosuelo/
INEGI. Marco geoestadístico 2013: áreas geoestadísticas estatales y áreas geoestadísticas municipales; 2013c. http://www.inegi.org.mx/geo/contenidos/geoestadistica/m_geoestadistico_2013.aspx
INEGI. Sistema de cuentas nacionales de México: cuenta satelite de turismo de México 2013. Instituto Nacional de Estadística y Geografía, Mexico; 2014.
INEGI. Sistema de cuentas nacionales de México: cuenta satelite de turismo de México 2013. Instituto Nacional de Estadística y Geografía, México; 2015.
Jiménez Osornio JJ, Caballero A, Quezada D, Bello Baltazar E. Estrategias tradicionales de apropiación de los recursos naturales. In: Colunga-García Marín P, Larque Saavedra A, editors. Naturaleza y sociedad en el área maya. Pasado, presente y futuro. México: Academia Mexicana de Ciencias and Centro de Investigaciones Científicas de Yucatán; 2003.
Koleff P, Urquiza-Haas T, Contreras B. Prioridades de conservación de los bosques tropicales en México: reflexiones sobre su estado de conservación y manejo. Ecosistemas. 2012;21:6–20.
Kú-Quej V, Mendoza-Vega J, Silva-Duarte C. Aprovechamientos forestales maderables y no maderables. In: Villalobos-Zapata GJ, Mendoza-Vega J, editors. La biodiversidad de Campeche: estudio de Estado. CONABIO, Gobierno del Estado de Campeche, Universidad Autónoma de Campeche, El Colegio de la Frontera Sur. Mexico; 2010.
Lambin EF, Meyfroidt P. Land use transitions: socio-ecological feedback versus socio-economic change. Land Use Policy. 2010;27:108–18.
Laurence WF. Emerging threats to tropical forests. Ann Mo Bot Gard. 2015;100:159–69.
Lutz W, Prieto L, Sanderson WC. Population, development, and environment on the Yucatán Peninsula: from ancient Maya to 2030. Research Report RR-00-14. Laxenburg, Austria: IIASA; 2000.
Martin GJ, Camacho-Benavides CI, Del Campo García CA, Anta-Fonseca S, Chapela MF, González Ortíz MA. Indigenous and community conserved areas in Oaxaca, Mexico. Manage Env Qual: Int J. 2011;22(2):250–66.
Merino-Pérez L. Las políticas forestales y de conservación en México y en Quintana Roo. In: Armijo N, Llorens C, editors. Uso, conservación y cambio en los bosques de Quintana Roo. Mexico: Universidad de Quintana Roo; 2004.

Merino-Pérez L. Conservation and forest communities in Mexico: experiences, visions, and rights. In: Porter-Bolland L, Ruiz-Mallén I, Camacho-Benavides C, McCandless S, editors. Community action for conservation: Mexican experiences. New York, NY: Springer; 2013.

Pérez Villegas G, Carrascal E. El desarrollo turístico en Cancún, Quintana Roo y sus consecuencias sobre la cubierta vegetal. Investigaciones Geográficas. 2000;43:145–66.

Planeación Nacional de Infraestructura 2014–2018 Subsecretaría de Infraestructura y Coordinación. http://www.presidencia.gob.mx

Porter-Bolland L, Drew AP, Vergara-Tenorio C. Analysis of a natural resources management system in the Calakmul Biosphere Reserve. Landsc Urban Plan. 2006;74:223–41.

Porter-Bolland L, Ellis EA, Gholz HL. The maya landscape at La Montaña region, Campeche, México: drivers of deforestation. Landsc Urban Plan. 2007;82:198–207.

Porter-Bolland L, Ellis EA, Guariguata MR, Ruiz-Mallén I, Negrete-Yankelevich S, Reyes-García V. Community managed forests and forest protected areas: an assessment of their conservation effectiveness across the tropics. Forest Ecol Manage. 2012;268:6–17.

Porter-Bolland L, Sánchez González MC, Ellis EA. La conformación del paisaje y el aprovechamiento de los recursos naturales por las comunidades mayas de La Montaña, Hopelchén, Campeche. Investigaciones Geográficas. 2008;66:65–80.

Pozo C, Armijo Canto N, Calmé S, editors. Riqueza Biológica de Quintana Roo. Un análisis para su conservación, Tomo I. ECOSUR, CONABIO, Gobierno del Estado de Quintana Roo, PPD. Mexico; 2011.

PRONATURA. Sistema estatal de áreas naturales protegidas del estado de Campeche. Medio Ambiente y Aprovechamiento Sustentable del Estado de Campeche. PRONATURA. Mexico; 2012.

Remmers GGA, Ek' EU. La roza-tumba-quema maya: evaluación de un sistema agroecológico tradicional bajo presiones de cambio tecnológico. Etnoecológica. 1996;3:97–101.

Riensche M, Castillo A, Flores-Díaz A, Maass M. Tourism at Costalegre, Mexico: An ecosystem services-based exploration of current challenges and alternative futures. Futures. 2015;66:70–84.

Rudel TK. Meta-analyses of case studies: a method for studying regional and global environmental change. Glob Environ Chang. 2008;18:18–25.

Rudel TK, Coomes OT, Moran E, Achard F, Angelsen A, Xu J, Lambin E. Forest transitions: towards a global understanding of land use change. Glob Environ Chang. 2005;15:23–31.

Sánchez-Aguilar R, Rebollar-Domínguez S. Deforestación en la Península de Yucatán, los retos que enfrentar. Madera Bosques. 1999;5:3–17.

SCT – Documentos del Sector 2001–2006, Documentos del Sector 2007–2012, Documentos del Sector 2013–2018. http://www.sct.gob.mx/

SECTUR. Resultados de la actividad turística. México año 2014. SECTUR. Mexico; 2015.

SEDUMA. Áreas naturales protegidas. Secretaría de Desarrollo Urbano y Medio Ambiente del Estado de Yucatán. 2015. http://www.seduma.yucatan.gob.mx/areas-naturales/index.php. Accessed 14 Apr 2015.

SIAP – Sistema de Información Agroalimentaria y Pesquera. http://www.siap.gob.mx/cierre-de-la-produccion-agricola-por-estado/

SIECA – Secretaría de Integración Económica Centroamericana. Resumen: El Plan Puebla Panama; 2004. http://www.sieca.org.gt/SIECA.htm, http://www.hacienda.go.cr/centro/datos/Articulo/El%20Plan%20Puebla-Panam%C3%A1-marzo%202004-Resumen.doc

Smardon RC, Faust BB. Introduction: international policy in the biosphere reserves of Mexico's Yucatán Peninsula. Landsc Urban Plan. 2006;74:160–92.

Tolentino O, García-Frapolli E, Porter-Bolland L, Ruiz-Mallén I, Reyes-García V, Sánchez-González MC, López Méndez ME. Triggering community conservation through the trade of carbon offsets: the case of the ejido Felipe Carrillo Puerto, Mexico. J Environ Dev. 2015. doi:10.1177/1070496514565460.

Turner II BL. Sustainability and forest transitions in the southern Yucatán: the land architecture approach. Land Use Policy. 2010;27:170–9.

Turner II BL, Cortina-Villar S, Foster D, Geoghegan J, Keys E, Klepeis P, Lawrence D, Macario Mendoza P, Manson S, Ogneva-Himmelberger Y, Plotkin AB, Perez-Salicrup D, Roy Chowdhury R, Savitsky B, Schneider L, Schmook B, Vance C. Deforestation in southern Yucatán peninsular region: an integrative approach. Forest Ecol Manage. 2001;154:353–70.

Wassenaar T, Gerber P, Verburg PH, Rosales M, Ibrahim M, Steinfeld H. Projecting land use changes in the Neotropics: the geography of pasture expansion into forest. Global Environ Chang. 2007;17:86–104.

World Bank. Project appraisal document on: a proposed grant from the global environment facility trust fund in the amount of SDR 11.5 million to Nacional Financiera, S.N.C. for Mexico - Mesoamerican Biological Corridor Project. Report No: 21136-ME; 2000. http://www.biodiversidad.gob.mx/corredor/cbmm/DOC/pad_english.pdf

Young OR. Governance in World Affairs. New York, NY: Cornell University Press; 1999.

Index

G.A. Islebe et al. (eds.), *Biodiversity and Conservation of the Yucatán Peninsula*,
DOI 10.1007/978-3-319-06529-8

The manufacturer's authorised representative in the EU is Springer Nature Customer Service Centre GmbH, Europaplatz 3, 69115 Heidelberg, Germany. If you have any concerns regarding our products, please contact ProductSafety@springernature.com

Printed and bound by CPI Group (UK) Ltd, Croydon, CR0 4YY

25/09/2026

02222851-0004